International Conference on Exotic Atoms and Related Topics (EXA 2017)

EPJ Web of Conferences Volume 181 (2018)

Vienna, Austria
11-15 September 2017

Editor:

P. Buhler

ISBN: 978-1-5108-6453-5

Printed from e-media with permission by:

Curran Associates, Inc.
57 Morehouse Lane
Red Hook, NY 12571

Some format issues inherent in the e-media version may also appear in this print version.

Printed by Curran Associates, Inc. (2018)

For additional information, please contact EDP Sciences – Web of Conferences
at the address below.

EDP Sciences – Web of Conferences
17, Avenue du Hoggar
Parc d'Activité de Courtabœuf
BP 112
F-91944 Les Ulis Cedex A
France

Phone: +33 (0) 1 69 18 75 75
Fax: +33 (0) 1 69 28 84 91

contact-edps@webofconferences.org

Additional copies of this publication are available from:

Curran Associates, Inc.
57 Morehouse Lane
Red Hook, NY 12571 USA
Phone: 845-758-0400
Fax: 845-758-2633
Email: curran@proceedings.com
Web: www.proceedings.com

TABLE OF CONTENTS

edp sciences

EPJ Web of Conferences

Statement of Peer review

In submitting conference proceedings to EPJ Web of Conferences, I certify to the Publisher that I adhere to the **Policy on Publishing Integrity** of the journal in order to safeguard good scientific practice in publishing.

1. All articles have been subjected to peer review administered by the proceedings editors.
2. Reviews have been conducted by expert referees, who have been requested to provide unbiased and constructive comments aimed, whenever possible, at improving the work.
3. Proceedings editors have taken all reasonable steps to ensure the quality of the materials they publish and their decision to accept or reject a paper for publication has been based only on the merits of the work and the relevance to the journal.

Title, date and place of the conference

International Conference on Exotic Atoms and Related Topics - EXA2017
11.-15. September 2017
Vienna, Austria

Proceedings editor(s):

Dr. Paul Bühler

Date and editor's signature 17.5.18

17, avenue du Hoggar - PA de Courtabœuf – BP 112 - 91944 Les Ulis Cedex A (France)
Tél. : 33 (0)1 69 18 75 75 - Fax : 33(0)1 69 07 45 17 – www.edpsciences.org

EPJ Web of Conferences **181**, 01001 (2018) https://doi.org/10.1051/epjconf/201818101001
EXA2017

Precision laser spectroscopy experiments on antiprotonic helium

Masaki Hori[1,*]

[1]Max-Planck-Institut für Quantenoptik, Hans-Kopfermann-Strasse 1, D-85748 Garching, Germany

Abstract. At CERN's Antiproton Decelerator (AD) facility, the Atomic Spectroscopy and Collisions Using Slow Antiprotons (ASACUSA) collaboration is carrying out precise laser spectroscopy experiments on antiprotonic helium ($\overline{p}He^+ \equiv \overline{p} + He^{2+} + e^-$) atoms. By employing buffer-gas cooling techniques in a cryogenic gas target, samples of atoms were cooled to temperature T = 1.5–1.7 K, thereby reducing the Doppler width in the single-photon resonance lines. By comparing the results with three-body quantum electrodynamics calculations, the antiproton-to-electron mass ratio was determined as $M_{\overline{p}}/m_e$ = 1836.1526734(15). This agreed with the known proton-to-electron mass ratio with a precision of 8×10^{-10}. Further improvements in the experimental precision are currently being attempted. The high-quality antiproton beam provided by the future Extra Low Energy Antiproton Ring (ELENA) facility should further increase the experimental precision.

1 Introduction

The metastable antiprotonic helium ($\overline{p}He^+ \equiv \overline{p} + He^{2+} + e^-$) atom is an exotic long-lived system made of a helium nucleus, an electron in the ground state, and an antiproton occupying a Rydberg state of principal and orbital angular momentum quantum numbers $n \sim \ell - 1 \sim 38$ [1–3]. The unusual longevity of the atom allows us to measure its transition frequencies by laser spectroscopy. By comparing the values with the results of three-body quantum electrodynamics (QED) calculations, the antiproton-to-electron mass ratio $M_{\overline{p}}/m_e$ can in principle be determined with a relative precision of $\sim 10^{-11}$ [4–6]. This corresponds to the best determinations of the proton-to-electron mass ratio M_p/m_e obtained from Penning trap experiments [7–11], or laser spectroscopy of HD^+ molecular ions [12–14]. The $\overline{p}He^+$ experiments also provide a consistency test of CPT symmetry [15], which may be complementary to the spectroscopy experiments on antihydrogen atoms [16–18]. It provides experimental constraints on any exotic fifth force that may exist at the $\sim$ 1 Å length scale [19–22].

The atoms can be readily synthesized via the reaction, $\overline{p} + He \rightarrow \overline{p}He^+ + e^-$, by allowing an antiproton beam [23] to come to rest in a helium gas target [24–26]. The transition frequencies of $\overline{p}He^+$ spanning the infrared to ultraviolet range have been calculated [4–6] to a relative precision of $\sim 10^{-10}$ by evaluating the quantum electrodynamics (QED) corrections up to order $m_e\alpha^7$ in atomic units. Here m_e and α respectively denote the electron mass and the fine structure constant. These calculations used the International Council for

*e-mail: Masaki.Hori@mpq.mpg.de

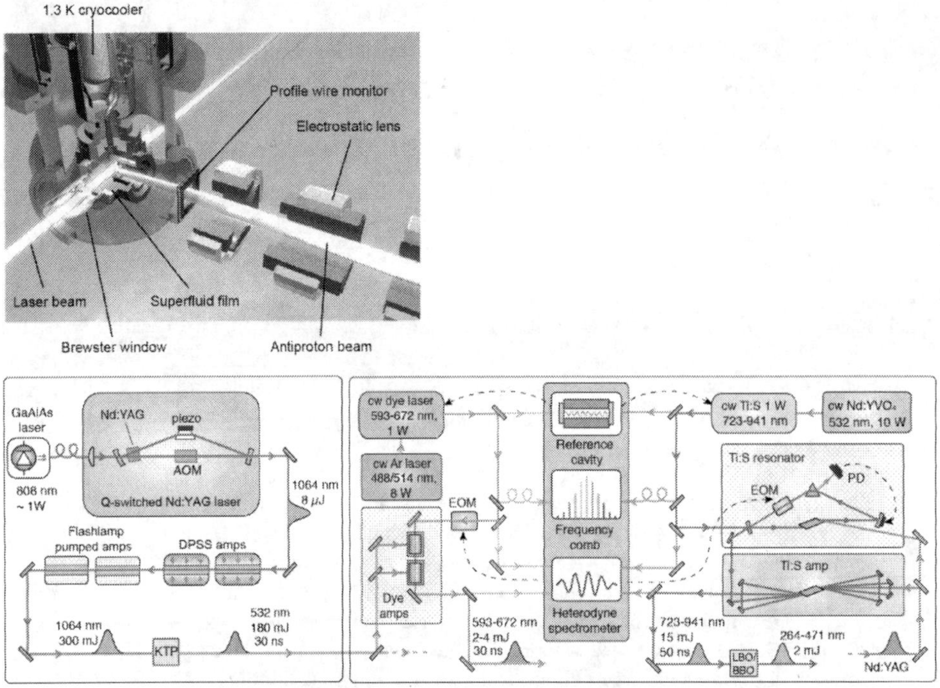

Figure 1. Experimental layout [3] used to synthesize $\overline{p}He^{+}$ and cool them to temperature T = 1.5–1.7 K (top). Laser system used for single-photon spectroscopy (bottom).

Science Committee on Data for Science and Technology (CODATA) 2010 recommended values of the fundamental constants [27], including the fine structure constant α, the ^{3}He- and ^{4}He-to-electron mass ratios, the Bohr radius, and the Rydberg constant. The corrections to the transition frequencies that arise from the finite charge radii of the helium nucleus (4 to 7 MHz) and of the antiproton [28] (< 1 MHz) are small because the spatial overlap between the Rydberg antiproton orbital and the nucleus is relatively small, and because the antiproton is polarized away from the $1s$ electron in the atom.

2 Buffer gas cooling of $\overline{p}He^{+}$ atoms

The thermal motions of $\overline{p}He^{+}$ in the experimental target at temperature T broadens the width of the measured single-photon laser resonances by a factor, $\nu\sqrt{8k_BT\log 2/Mc^2}$. Here ν denotes the transition frequency, k_B the Boltzmann constant, M the atom's mass, and c the speed of light. This loss in the spectral resolution limited [29, 30] the precision of determining the resonance centroid to around 10^{-7}–10^{-8}. One way to reach a precision beyond this Doppler limit was provided by two-photon spectroscopy [2, 31], in which the $\overline{p}He^{+}$ was irradiated by two counterpropagating ultraviolet laser beams.

An alternative method involved cooling some 2×10^{9} $\overline{p}He^{+}$ atoms to a temperature T = 1.5–1.7 K, by allowing the $\overline{p}He^{+}$ to undergo elastic collisions with cryogenic helium gas [3]. This cooling behavior is in contrast to some other kinds of hadronic exotic atoms, such as pionic hydrogen [32], which was found to be heated by collisions with H_2 molecules that deexcite the atom. The density of the buffer gas ($T\sim1.5$ K and P = 40–170 Pa) used in the experiment was carefully adjusted so that the $\overline{p}He^{+}$ atoms, once formed, rapidly underwent a

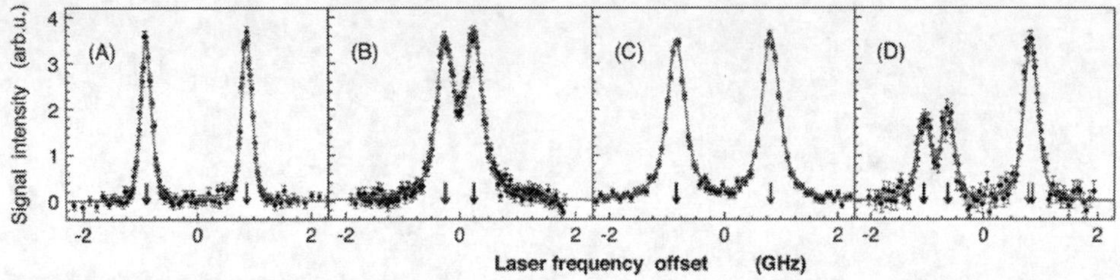

Figure 2. The resonance profiles of the single-photon transitions (A) $(n, \ell) = (37, 35) \to (38, 34)$, (B) $(n, \ell) = (39, 35) \to (38, 34)$, and (C) $(n, \ell) = (38, 35) \to (39, 34)$ of buffer-gas cooled $\overline{p}^4He^+$. (D) The resonance profile $(36, 34) \to (37, 33)$ of $\overline{p}^3He^+$. The x-abscissa indicates the offset of the laser frequency relative to the resonance centroid. Solid curves indicate the best fit of an ab initio model based on the optical Bloch equation. From Ref. [3].

few hundred or more cooling collisions. It was then interrogated by the resonant laser beam. The $1s$ electron protected a significant fraction of the $\overline{p}He^+$ from annihilation during this cooling.

The experiment (Fig. 1) was carried out by utilizing the pulsed beam of AD that contained between 2×10^7 and 3×10^7 antiprotons with a kinetic energy $E = 5.3$ MeV and repetition rate $f = 0.01$ Hz. Some of the antiprotons were slowed down to $E = 75$ keV by allowing them to traverse a 3 m long radiofrequency quadrupole decelerator [29, 33]. This setup was also recently used to attempt to measure the annihilation cross sections of antiprotons that traversed some thin target foils [34–38]. The emerging 75-keV antiprotons were transported by a beamline and allowed to enter the cryogenic helium gas target. The target was in thermal contact with an open-cycle Joule-Thomson cryocooler at temperature $T = 1.3$ K. The $\overline{p}He^+$ were irradiated by $\Delta t = 40$ to 100 ns long laser pulses [39] with peak powers $P = 0.5$ to 10 kW and wavelengths $\lambda = 264$ to 841 nm, which were generated by Ti:Sapphire and dye laser systems.

The spectra in Figs. 2 (A)–(C) show the resonance profiles of the transitions $(n, \ell) = (37, 35) \to (38, 34)$, $(39, 35) \to (38, 34)$, and $(38, 35) \to (39, 34)$ of $\overline{p}^4He^+$. They were obtained by plotting the intensities of the antiproton annihilation signals which were induced at multiple laser frequencies between -1 and 1 GHz around the resonance centroid. The arrows indicate the positions of the four hyperfine sublines that arise from the spin-spin interaction between the antiproton and electron in $\overline{p}^4He^+$. The single photon resolution seen here exceeds those of sub-Doppler two-photon spectroscopy experiments using higher-temperature atoms [2] described above; this is due to the low $T = 1.5$–1.7 K temperature of the atoms. Fig. 2 (D) shows the profile of the $\overline{p}^3He^+$ resonance $(n, \ell) = (36, 34) \to (37, 33)$. The three-peak structure arises from the eight unequally spaced hyperfine sublines caused by the interactions between the ^3He nuclear, electron, and antiproton spins. The spin-independent transition frequencies were determined by fitting the profiles with a theoretical line shape (indicated by blue lines). The line shapes were obtained by solving the optical Bloch equations [31] that model the single-photon transition. The experimental uncertainties in the determinations of the $\overline{p}He^+$ frequencies include contributions from the statistical uncertainty (± 1 MHz) that arises from the limited number of measured atoms, and the systematic uncertainty of 0.4–3 MHz caused by the fitting function. The dye and Ti:Sapphire lasers used to excite the atoms contain a spurious frequency modulation. These were measured with a precision of 0.4–1.0 MHz. The ac Stark effects induced by the lasers, and magnetic Zeeman shifts due to residual magnetic fields, caused systematic effects of < 0.1 MHz and < 0.2 MHz, respectively.

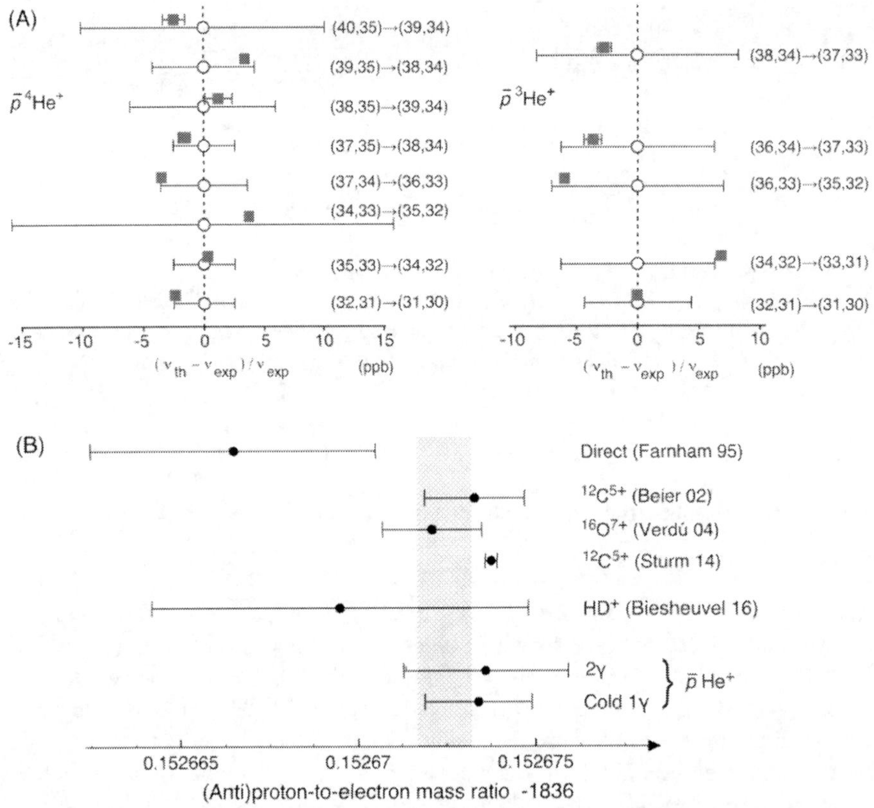

Figure 3. (A) Comparison of experimental and calculated transition frequencies, showing the fractional differences between the experimental (open circles) and theoretical (squares) values of 13 transition frequencies of cooled $\overline{p}^4He^+$ and $\overline{p}^3He^+$ atoms. (B) Proton-to-electron mass ratios measured in Penning traps and laser spectroscopy of HD^+ molecular ions, compared with the antiproton-to-electron mass ratio determined by spectroscopy of $\overline{p}He^+$. The shaded area represents the CODATA 2010 recommended value. From Ref. [3].

Eight transition frequencies of $\overline{p}^4He^+$ atoms and five frequencies of $\overline{p}^3He^+$ were measured with relative uncertainties between 2.5×10^{-9} and 15×10^{-9}. The frequencies ν_{exp} (Fig. 3A, open circles with error bars) agree with theoretical ν_{th} values (filled squares). Due to the cooling techniques described here, this agreement is a factor of 1.4 to 10 times better than previous single-photon experiments [30] of $\overline{p}He^+$. The uncertainties for most of the theoretical frequencies ν_{th} arise from QED contributions of orders higher than $m_e\alpha^7$ which have not been calculated yet.

When the antiproton-to-electron mass ratio $M_{\overline{p}}/m_e$ used in the calculations was changed by 1×10^{-9}, the theoretical $\overline{p}He^+$ frequencies ν_{th} changed by 2.6×10^{-9} to 2.7×10^{-9}. The mass ratio was determined as, $M_{\overline{p}}/m_e = 1836.1526734(15)$, by minimizing the difference between the frequencies ν_{exp} and ν_{th}. The one-standard deviation uncertainty in the parenthesis includes the three contributions 9×10^{-7}, 11×10^{-7}, and 3×10^{-7} of the experimental statistical and systematic uncertainties, and the theoretical uncertainty, respectively.

The atomic mass of the electron was recently determined [10] with a relative precision of 3×10^{-11} by confining a $^{12}C^{5+}$ ion in a Penning trap. The cyclotron frequency of its motion in a magnetic field and the precession frequency of the electron spin was then measured, and the results compared with the latest QED calculations of its g-factor [40]. From this and the proton mass which was recently measured in a separate measurement involving a Penning trap [11] from the same collaboration, the proton-to-electron mass ratio was determined as, $M_p/m_e = 1836.152673346(81)$. In Fig. 3(B), the latest M_p/m_e mass ratios are shown together with the previous experimental values that were determined by comparing the cyclotron frequencies of protons and electrons in a Penning trap [7], laser spectroscopy of cold HD^+ molecular ions [12], and the CODATA 2010 recommended value [27]. These recent high-precision values are in good agreement with the $M_{\overline{p}}/m_e$ ratio determined from $\overline{p}He^+$. The TRAP and BASE experiments of CERN have compared the cyclotron frequencies of antiprotons and H^- ion pairs confined in a Penning trap [41–43]; a limit of 5×10^{-10} was set [3, 44] on any deviation between the antiproton and proton masses and charges by combining the results with the $\overline{p}He^+$ spectroscopic data.

The $\overline{p}^4He^+$ transition $(n, \ell) = (40, 36) \rightarrow (41, 35)$ was also studied by laser spectroscopy [45]. A stimulated first-order Raman scattering process in a H_2 gas cell was utilized to generate the 7 ns long laser pulses of wavelength $\lambda = 1154.9$ nm needed to excite this transition. The measurements revealed that most of the metastable populations are concentrated into states of principal quantum number $n \geq 40$, whereas the states $n > 41$ contain very few antiprotons. A laser-microwave-laser triple resonance method was used to study the hyperfine structure of the $(n, \ell) = (36, 34)$ state of $\overline{p}^3He^+$ [46].

3 Future perspectives

Currently the precision on the calculations of $\overline{p}He^+$ energies are more than an order of magnitude higher than the experimental uncertainties. Further improvements in the theoretical precision are expected in the next 5 years. The Extra Low Energy Antiproton Ring (ELENA) facility due to begin operation in 2021 will provide a high-quality, cooled antiproton beam of energy $E = 100$ keV. Samples of $\overline{p}He^+$ would then be formed in a smaller volume, so that lasers of lower power and higher spectral resolution could be used in the experiments. The high stability of the antiproton beam should ensure an improved signal-to-noise ratio on the $\overline{p}He^+$ spectral lines. This may allow us to resolve weak atomic transitions between metastable states of smaller natural width, which can be measured with a higher precision. Intensive efforts are currently underway to achieve this goal.

Metastable pionic helium (πHe^+) is a three-body atom [47–49] consisting of a helium nucleus, an electron in the 1s ground state, and a negative-charged pion in a Rydberg state of quantum numbers $n \sim \ell - 1 \sim 16$. The spectral lines of these atoms have never been directly observed, and so the existence of πHe^+ is so far hypothetical. Laser spectroscopy of πHe^+ is currently being attempted at the 590 MeV ring cyclotron facility of the Paul Scherrer Institute. By comparing the experimental frequencies with those derived from QED calculations, the π^- mass can in principle be determined with a fractional precision of 10^{-8} to 10^{-6}.

References

[1] R.S. Hayano, M. Hori, D. Horváth, E. Widmann, Rep. Prog. Phys. **70**, 1995 (2007)
[2] M. Hori *et al.*, Nature **475**, 484 (2011)
[3] M. Hori *et al.*, Science **371**, 610 (2016)
[4] V.I. Korobov, L. Hilico, J.-P. Karr, Phys. Rev. Lett. **112**, 103003 (2014)
[5] V.I. Korobov, L. Hilico, J.-P. Karr, Phys. Rev. A **89**, 032511 (2014)
[6] M-H. Hu *et al.*, Chem. Phys. Lett. **654**, 114 (2016)

[7] D.L. Farnham, R.S. Van Dyck Jr., P.B. Schwinberg, Phys. Rev. Lett. **75**, 3598 (1995)
[8] T. Beier *et al.*, Phys. Rev. Lett. **88**, 011603 (2002)
[9] J. Verdú *et al.*, Phys. Rev. Lett. **92**, 093002 (2004)
[10] S. Sturm *et al.*, Nature, **506**, 467-470 (2014)
[11] F. Heiße *et al.*, Phys. Rev. Lett. **119**, 033001 (2017)
[12] J. Biesheuvel *et al.*, Nat. Commun. **7**, 10385 (2016)
[13] J.C.J. Koelemeij, B. Roth, A. Wicht, I. Ernsting, S. Schiller, Phys. Rev. Lett. **98**, 173002 (2007)
[14] S. Alighanbari, M.G. Hansen, V. Korobov, S. Schiller, arXiv:1802.03208 (2018)
[15] V.A. Kostelecký, A.J. Vargas, Phys. Rev. D **92**, 056002 (2015)
[16] M. Ahmadi *et al.*, Nature **541**, 506 (2017)
[17] M. Ahmadi *et al.*, Nature **548**, 66 (2017)
[18] G. Gabrielse *et al.*, Phys. Rev. Lett. **108**, 113002 (2012)
[19] E.J. Salumbides, W. Ubachs, V.I. Korobov, J. Mol. Spect. **300**, 65 (2014)
[20] J. Murata, S. Tanaka, Class. Quantum Grav. **32**, 033001 (2015)
[21] P. Brax, S. Fichet, G. Pignol, arXiv:1710.00850 (2017)
[22] F. Ficek *et al.*, arXiv:1801.00491 (2018)
[23] M. Hori, J. Walz, Prog. Part. Nucl. Phys. **72**, 206 (2013)
[24] K. Sakimoto, Phys. Rev. A **91**, 042502 (2015)
[25] V.I. Korobov, Z.-X. Zhong, Q.-L. Tian, Phys. Rev. A **92**, 052517 (2015)
[26] A. Adamczak, D. Bakalov, Phys. Rev. A **90**, 054501 (2014)
[27] P.J. Mohr, B.N. Taylor, D.B. Newell, Rev. Mod. Phys. **84**, 1527 (2012)
[28] R. Pohl *et al.*, Nature **466**, 213 (2010)
[29] M. Hori *et al.*, Phys. Rev. Lett. **91**, 123401 (2003)
[30] M. Hori *et al.*, Phys. Rev. Lett. **96**, 243401 (2006)
[31] M. Hori *et al.*, Phys. Rev. A **81**, 062508 (2010)
[32] A. Badertscher *et al.*, Phys. Lett. B **392**, 278 (1997)
[33] A.M. Lombardi *et al.*, 2001 First operating experience with the CERN decelerating RFQ for antiprotons. *Proc. Part. Acc. Conf.* Chicago, USA
[34] H. Aghai-Khozani *et al.*, Eur. Phys. J. Plus **127**, 125 (2012)
[35] M. Corradini *et al.*, Nucl. Instrum. Methods A **711**, 12 (2013)
[36] A. Sótér *et al.*, Rev. Sci. Instrum. **85**, 023302 (2014)
[37] K. Todoroki *et al.*, Nucl. Instrum. Methods A **835**, 110 (2016)
[38] H. Aghai-Khozani *et al.*, Nucl. Phys. A **970**, 366 (2018)
[39] M. Hori, A. Dax, Opt. Lett. **34**, 1273 (2009)
[40] K. Pachucki, A. Czarnecki, U.D. Jentschura, V.A. Yerokhin, Phys. Rev. A **72**, 022108 (2005)
[41] G. Gabrielse *et al.*, Phys. Rev. Lett. **82**, 3198 (1999)
[42] J.K. Thompson, S. Rainville, D.E. Pritchard, Nature **430**, 58 (2004)
[43] S. Ulmer *et al.*, Nature **524**, 196 (2015)
[44] R.J. Hughes, B.I. Deutch, Phys. Rev. Lett. **69**, 578 (1992)
[45] T. Kobayashi *et al.*, J. Phys. B: At. Mol. Opt. Phys. **46**, 213 (2013)
[46] S. Friedreich *et al.*, J. Phys. B: At. Mol. Opt. Phys. **46**, 125003 (2013)
[47] M. Hori, A. Sótér, V.I. Korobov, Phys. Rev. A **89**, 042515 (2014)
[48] V.I. Korobov, A.K. Bakbaev, D.T. Aznabayev, S.A. Zhaugasheva, J. Phys. B: At. Mol. Opt. Phys. **48**, 245006 (2015)

[49] B. Obreshkov, D. Bakalov, Phys. Rev. A **93**, 062505 (2016)

EPJ Web of Conferences **181**, 01002 (2018) https://doi.org/10.1051/epjconf/201818101002
EXA2017

Proposal for enhancement of antihydrogen ion production in the GBAR experiment

D. A. Cooke[1], A. Husson[2], D. Lunney[2], and P. Crivelli[1,*]

[1]ETH Zurich, Institute for Particle Physics, 8093 Zurich, Switzerland
[2]CSNSM-IN2P3/CNRS, Université de Paris Sud, 91405 Orsay, France

Abstract. The production of antihydrogen ions ($\bar{H}^+$) in the GBAR experiment will occur via a two step charge exchange process. In a first reaction, the antiprotons ($\bar{p}$) from the ELENA ring will capture a positron from a positronium (Ps) target producing antihydrogen ($\bar{H}$) atoms. Those interacting in the same Ps target will produce in a second step $\bar{H}^+$. This results in a dependence for the $\bar{H}^+$ production rate which is roughly proportional to the Ps density squared. We present a scheme to increase the anti-ion production rate in the GBAR experiment by tailoring the antiproton to the positron pulse in order to maximise the temporal overlap of Ps and $\bar{p}$. Detailed simulations show that an order of magnitude could be gained by bunching the antiprotons from ELENA. In order to avoid losses in their capture in the Paul trap due to the energy spread introduced by the bunching, debunching with a symmetrical inverted pulse can be applied to the $\bar{H}^+$ ions.

1 Introduction

Different theoretical explanations have been put forward and experimental efforts are ongoing to try to address the origin of the Baryon Asymmetry observed in our Universe. These include activities focusing on the gravitational behaviour of antimatter. No compelling theoretical argument seems to support that a difference between the gravitational behaviour of matter and antimatter should be expected [1], although some attempts have been made to show the contrary [2–5] and it would be allowed in the context of the Standard Model Extension (SME) [6, 7]. Moreover, observations and experiments have been interpreted as evidence against the existence of antigravity type forces [8–11]. However, those could be argued to be model dependent and, therefore, a simple free fall measurement with a straightforward interpretation is preferable. A first attempt in this direction was made recently by the ALPHA collaboration [12] that bound the ratio of gravitational mass to inertial mass of antihydrogen between −65 and 110.

The first proposal to measure directly the gravitational force acting on antiparticles using positrons was done 50 years ago by Witteborn and Fairbank [13, 14]. Such a measurement was later proposed at CERN with antiprotons [15]. However, those experiments have not yet been realised since the use of charged particles to measure the effect of gravity requires the elimination of stray electromagnetic fields to a level which is extremely challenging. The solution to this problem is to use neutral systems composed entirely of antimatter such as antihydrogen [16–18] or partly such as muonium and positronium [19–21].

*e-mail: crivelli@phys.ethz.ch

2 GBAR experiment

The Gravitational Behaviour of Antimatter at Rest (GBAR) is an experiment in preparation at CERN [22]. Its goal is to measure the gravitational acceleration $\bar{g}$ imparted to freely falling $\bar{\text{H}}$ atoms, in order to perform a direct experimental test of the Weak Equivalence Principle with antimatter. The objective is to reach a relative precision on $\bar{g}$ of 1% in a first stage, with the perspective to reach a much higher precision (10^{-4}–10^{-5}) using quantum gravitational states [23] in a second stage.

The principle of the experiment is described in detail in [17, 22]. Here we review briefly the proposed experimental technique based on the original idea of T. Hänsch and J. Walz [24] to first produce $\bar{\text{H}}^+$ ions from interactions of antiprotons ($\bar{\text{p}}$) with ortho-positronium (Ps) as:

$$\text{Ps} + \bar{\text{p}} \rightarrow \bar{\text{H}} + \text{e}^- \tag{1}$$

$$\text{Ps} + \bar{\text{H}} \rightarrow \bar{\text{H}}^+ + \text{e}^- \tag{2}$$

Reaction 1 has been demonstrated for normal hydrogen by Merrison et al. [25]. The production of $\bar{\text{H}}$ in Rydberg states has already been proven by the ATRAP collaboration via a two step charge exchange, i.e. formation of Rydberg positronium with positrons on Cs atoms and subsequent formation of $\bar{\text{H}}^*$ [26, 27]. The cross sections were calculated by different authors [28] and those are in good agreement with the available experimental data. The cross sections of reaction 2 were computed [29, 30] but have not been measured yet.

The $\bar{\text{H}}^+$ ions produced in GBAR will then be cooled down in two steps: first in a Paul trap and finally in a linear trap to a few tens of μK through sympathetic cooling with cold beryllium ions [31]. A laser pulse will be applied to detach the excess positron so that the free-fall measurement time for the neutral atom can start. The subsequent $\bar{\text{H}}$ annihilation on a plate placed at about 10 cm from the trap is detected and provides the stop signal used to measure the free fall time and thereby extract $\bar{g}$. The choice of producing $\bar{\text{H}}^+$ ions to get ultracold antihydrogen atom is the specificity of the GBAR experiment.

The $\bar{\text{H}}^+$ production yield can be estimated with:

$$N_{\bar{\text{H}}^+} \approx N_{\bar{\text{p}}} \left(\sigma_{\bar{\text{H}}} n_{\text{Ps}} L\right) \left(\sigma_{\bar{\text{H}}^+} n_{\text{Ps}} L\right) \tag{3}$$

where $\sigma_{\bar{\text{H}}}$, $\sigma_{\bar{\text{H}}^+}$ are the cross sections of reactions 1 and 2, n_{Ps} is a characteristic Ps density, and L a characteristic length of the interaction volume. Ps will be formed by dumping a bunch of few 10^9 positrons (with pulse width $\approx$ 10 ns) onto a porous silica target. Located in a 1 × 1 × 20 mm tube, Ps emitted back into vacuum is confined by reflection from the walls to a small volume which the $\bar{\text{p}}$ beam pass through. Hence, n_{Ps} is a time- and space-varying quantity, so that even in the limit of low interaction probability, Eq. 3 is approximate. The evolution of the Ps density inside the tube was simulated using GEANT4 [32], and the resultant formation probability of $\bar{\text{H}}$ by passage of an $\bar{\text{p}}$ bunch through the evolving cloud calculated, as a function of offset time between Ps and $\bar{\text{p}}$ pulses. The antiprotons from ELENA [33] are coming in bunches of σ = 75 ns and therefore the temporal overlap is clearly not maximised. As the number of Ps atoms available decays exponentially, compression of the $\bar{\text{p}}$ bunch to the shortest possible width in order to overlap with the maximum Ps density has a dramatic effect on $\bar{\text{H}}$ production.

Here we propose to first bunch the antiprotons in order to increase the $\bar{\text{H}}^+$ formation probability, followed by time debunching (bunching in energy) the $\bar{\text{H}}^+$ to keep the energy spread below 100 eV in order not to affect the capturing efficiency in the Paul trap.

3 Simulation

The simulation, performed using SIMION 8.0, comprises a five metre beam pipe with decelerator, buncher, and debuncher stages, with additional ion focussing elements. A pulse of antiprotons was created with a Gaussian distribution of kinetic energy with mean 100 keV and σ_{KE} = 0.07 keV, beamspot size of σ_r = 0.5 mm and opening angle of 1.5 mrad, and time distribution of σ_t = 75 ns, corresponding to typical expected bunch parameters from the ELENA ring. The simulation geometry is depicted in Figure 1.

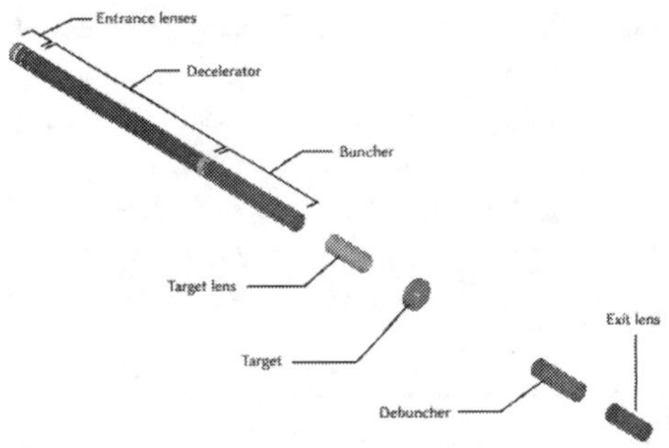

Figure 1. Schematic of apparatus layout (complete length is 5 m).

As proposed by D. Lunney for GBAR, deceleration from 100 keV to ~ 6 keV is achieved using a voltage step from -94 kV to 0 V applied to a 1 m long cylindrical electrode (diameter 80 mm) while the antiproton buncher is entirely inside it. This is modelled as an exponential decay of the voltage on the tube, with a time constant of 100 ns—allowing an antiproton drift time of approximately 10 times this time constant. Entrance lenses are required to focus the bunch through the decelerator in order to avoid significant beam divergence. Following the decelerator there is a time bunching electrode where a linear ramp potential is applied as the antiproton bunch enters, resulting in larger deceleration for early arriving particle in comparison to late arriving ones. Although an ideal buncher for monoenergetic particles with only a single electrode uses a potential varying as t^{-2} [34], a linear ramp is employed here as the energy spread on exit from the decelerator is large. As a result, no significant improvement in bunching performance can be achieved using a t^{-2} ramp, and a high-voltage linear ramp is technically easier to achieve. A single cylindrical lens is used to focus the bunch through the target, with the target itself represented by a 2 mm diameter, 20 mm long hole.[1]

After the target, the simulated particle type is transformed into $\bar{\text{H}}^+$ with the momentum directed in the same way as the incident $\bar{\text{p}}$ beam, and the bunch passes through an energy bunching (or time debunching) stage. This has an identical voltage ramp to the bunching stage, offset to later times (and with the opposite sign), and works on a similar principle. After time bunching, the $\bar{\text{p}}$ pulse has a velocity distribution correlated with the time distribution, so that after the time focus, earlier arriving particles have higher velocities than later arriving ones. This means that another time varying potential can undo the time bunching effect, restoring the original energy distribution (in an ideal case). Following the debuncher, a final exit lens is used to focus the beam at the end distance.

The time distribution at the target is shown in Figure 2(a), together with a Gaussian fit to the data, revealing compression in time by a factor of approximately 7.5. Improvement beyond this level is rendered difficult owing to the energy spread of the initial beam. The voltage ramping requirements are a rate of approximately 10 kV μs^{-1} over a voltage range of nearly 4 kV, which can be achieved using high voltage solid state switches. Focussing at the

[1] In order to reduce the memory requirements of SIMION, axial symmetry was preserved throughout.

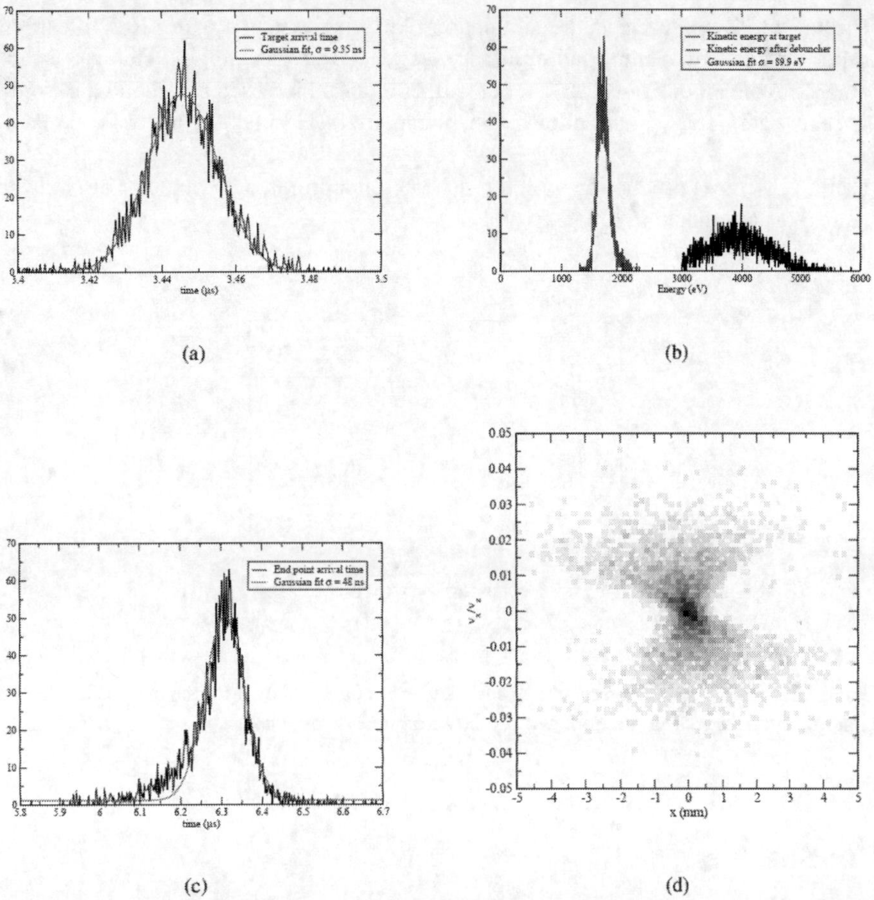

Figure 2. Various beam properties at different points in the simulation: (a) time distribution of antiproton bunch passing through the target aperture; (b) energy distribution of $\bar{p}$ bunch at target and $\bar{H}^+$ at simulation end point (equivalent to entrance of quadrupole trap); (c) time distribution of antiproton bunch at simulation end point; (d) phase space of $\left(x, \frac{v_x}{v_z}\right)$ at simulation endpoint.

target allows approximately 35% of the bunch to pass through a 1 × 1 mm hole in the centre of this (corresponding to a more realistic target geometry).

Figure 2(b) shows the kinetic energy distribution of the $\bar{p}$ bunch at the target position. This is naturally broadened by the time bunching process, which also introduces a shift to lower energies. In principle, the decelerator and buncher parameters can be optimized for arbitrary target energies, so a typical mean value in the range 0–10 keV is presented here. Also shown on Figure 2(b) is the kinetic energy distribution at the simulation end point, after the debunching stage. This is almost completely restored to the original energy spread at the input to the decelerator stage, which is the theoretical limit for this process (assuming no particle loss, and that the original energy distribution was uncorrelated with the original timing distribution). Figure 2(c) shows the time distribution for particles reaching the end of the simulation (having passed through the target aperture. The width of $\sigma = 48$ ns is consistent with the requirements for high efficiency capture in a subsequent Paul trap.

The $\left(x, \frac{v_x}{v_z}\right)$ phase space at the simulation exit is shown in Figure 2(d), illustrating a beamspot with $\sigma \approx 0.5$ mm, and emittance $\epsilon_{rms,x} = 18.5$ mm mrad. This is a beamspot comparable to the input from ELENA, but with emittance increased by around a factor of 20. These beam parameters should allow efficient capture of $\bar{H}^+$ in a centimetre-scale Paul trap [22].

The increase in $\bar{H}$ production with this scheme of approximately a factor of is illustrated in Figure 3 (as calculated in [32]).

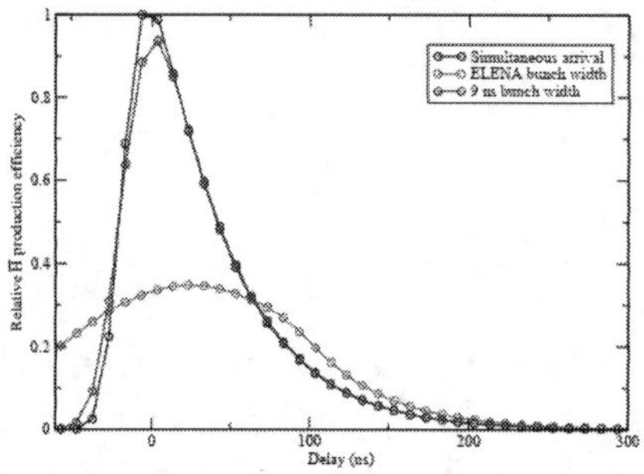

Figure 3. Effect of $\bar{p}$ bunching on relative $\bar{H}$ production efficiency, shown for the unbunched case, using the bunch width from this simulation, and for the zero bunch width case.

4 Conclusions

Simulation of $\bar{p}$ time-bunching followed by energy bunching has been performed in order to simultaneously maximize at-target density while minimizing subsequent energy spread which is very important in order to efficiently capture those ions in a Paul trap as planned in the context of the GBAR experiment. This was achieved without significantly reducing the transmitted fraction of antiprotons through the target compared to the unbunched case. An increase in $\bar{H}^+$ formation by approximately an order of magnitude could be expected by the implementation of the proposed scheme. Further improvements might be possible by using an antiproton trap which would allow for a shorter bunch owing to the smaller initial emittance assuming that the same can be achieved with the positron bunch.

5 Acknowledgements

This work is supported by the SNFS under the grant number 200021_166286.

6 References

References

[1] Nieto M M and Goldman T, Phys. Rep. **205**, 221 (1991)
[2] Scherk J, Phys. Lett. B **88**, 265 (1979)

[3] Chardin G and Rax J M, Phys. Lett. B **282**, 256 (1992)
[4] Chardin G, Nucl. Phys. A **558**, 477 (1993)
[5] Chardin G, Hyperfine Interactions **109**, 83 (1997)
[6] Kostelecky V A and Tasson J D, Phys. Rev. D **83**, 016013 (2011)
[7] Kostelecky V A and Russell N, arxiv:0801.0287 (2018)
[8] Good M L, Phys. Rev. **121**, 311 (1961)
[9] Pakvasa K, Simmons W A and Weiler T J , Phys. Rev. D **39**, 1761 (1989)
[10] Apostolakis A et al. (CPLEAR Collaboration), Phys. Lett. B **452**, 425 (1999)
[11] Gabrielse G, Khabbaz A, Hall D S, Heimann C, Kalinowsky H and Jhe W, Phys. Rev. Lett. **82**, 3198 (1999)
[12] ALPHA Collaboration and Charman A E, Nature Comms. **4**, 1785 (2013)
[13] Witteborn F C and Fairbank W M, Phys. Rev. Lett. **19**, 1049 (1967)
[14] Witteborn F C and Fairbank W M, Nature **220**, 436 (1968)
[15] Holzscheiter M H, Int. J. Mod. Phys: Conf. Ser. **30**, 1460260 (2014)
[16] Doser, M et al. (AEGIS Collaboration), Class. Quant. Grav. **29**, 184009 (2012)
[17] Pérez, P et al. (GBAR Collaboration), Hyperfine Interactions **233**, 21 (2015)
[18] Hamliton P, Zhmoginov A, Robicheaux F, Fajans J, Wurtele J and Mueller H, Phys. Rev. Lett. **112**, 121102 (2016)
[19] Kirch K and Siang Khaw K, Int J. Mod. Phys.: Conf. Ser. **30**, 1460258 (2014)
[20] Cassidy D B and Hogan S D, Int. J. Mod. Phys.: Conf. Ser. **30**, 1460259 (2014)
[21] Crivelli P, Nesvizhevsky V V and Voronin A Y, Adv. High Energy Phys. **VOL?** 173572 (2015)
[22] Pérez, P et al. (GBAR Collaboration), CERN-SPSC-2011-29 SPSC–P–342 (2011)
[23] Dufour G, Debu P, Lambrecht A, Nesvizhavsky V V and Voronin A Y, Eur. Phys. J. **C74**, 2731 (2014)
[24] Walz J and Haensch T, Gen. Rel. Grav. **36**, 561 (2004)
[25] Merrison J P, Bluhme H, Chevallier J, Deutch B I, Hvelplund P, Jørgensen L V, Knudsen H, Poulsen M R and Charlton M, Phys. Rev. Lett. **78**, 2728 (1997)
[26] Storry, C H et al. (ATRAP Collaboration), Phys. Rev. Lett. **93**, 263401 (2004)
[27] McConnell R, Gabrielse G, Kolthammer W S, Richerme P, Muellers A, Walz J, Grzonka D, Zielinski M, Fitzakerley D, George M C, Hessels E A, Storry C H and Weel M, J. Phys. B **49**, 064002 (2016)
[28] Rawlins C M, Kadyrov A S, Stelbovics A T, Bray I and Charlton M, Phys. Rev. A **93**, 012709 (2016)
[29] Walters H J R and Starrett C, Physica Status Solidi C **4**, 3429 (2007)
[30] Comini P and Hervieux P A, New J. Phys. **15**, 095022 (2013)
[31] Hilico L, Karr J P, Douillet A, Indelicato P, Wolf S and Schmidt Kaler F, Int. J. Mod. Phys.: Conf. Ser. **30**, 1460269 (2014)
[32] Crivelli P, Cooke D and Heiss M W, Phys. Rev. D **94**, 052008 (2016)
[33] S Maury, W Oelert, W Bartmann et al , Hyperfine Interact . **105**, 229 (2014)
[34] Mills A P, Appl. Phys. **22**, 273 (1980)

Recent Developments from ASACUSA on Antihydrogen Detection

B. Kolbinger[1,★], *C.* Amsler[1], *H.* Breuker[2], *M.* Diermaier[1], *P.* Dupré[2], *M.* Fleck[1], *A.* Gligorova[1], *H.* Higaki[3], *Y.* Kanai[4], *T.* Kobayashi[5], *M.* Leali[6], *V.* Mäckel[1], *C.* Malbrunot[7,1], *V.* Mascagna[6], *O.* Massiczek[1], *Y.* Matsuda[5], *D.j.* Murtagh[1], *Y.* Nagata[8], *C.* Sauerzopf[1], *M.C.* Simon[1], *M.* Tajima[4], *S.* Ulmer[2], *N.* Kuroda[5], *L.* Venturelli[6], *E.* Widmann[1], *Y.* Yamazaki[2], and *J.* Zmeskal[1]

[1]*Stefan Meyer Institute for Subatomic Physics, Austrian Academy of Sciences, Boltzmanngasse 3, 1090 Vienna, Austria*
[2]*Ulmer Fundamental Symmetry Laboratory, RIKEN, Saitama 351-0198, Japan*
[3]*Graduate School of Advanced Science of Matter, Hiroshima University, Kagamiyama, Higashi-Hiroshima, 739-8530 Hiroshima, Japan*
[4]*Nishina Center for Accelerator-Based Science, RIKEN Saitama 351-0198, Japan*
[5]*Institute of Physics, University of Tokyo, 153-8902, Japan*
[6]*Dipartimento di Ingegneria dell'Informazione, Università degli Studi di Brescia, I-25133 Brescia, Italy and Istituto Nazionale di Fisica Nucleare, Sez. di Pavia, I-27100 Pavia, Italy*
[7]*CERN 1211, Geneva 23, Switzerland*
[8]*Department of Physics, Tokyo University of Science, Shinjuku, Tokyo 162-8601, Japan*

Abstract. The ASACUSA Collaboration at CERNs Antiproton Decelerator aims to measure the ground state hyperfine splitting of antihydrogen with high precision to test the fundamental symmetry of CPT (combination of charge conjugation, parity transformation, and time reversal). For this purpose an antihydrogen detector has been developed. Its task is to count the arriving antihydrogen atoms and therefore distinguish background events (mainly cosmics) from antiproton annihilations originating from antihydrogen atoms which are produced only in small amounts. A central BGO crystal disk with position sensitive read-out detects the annihilation and a surrounding two-layered hodoscope is used for tracking charged secondaries. The hodoscope has been recently upgraded to allow precise vertex reconstruction. A machine learning analysis based on measured antiproton annihilations and cosmic rays has been developed to identify antihydrogen events.

1 Introduction

According to current cosmological theory, matter and antimatter have been created in equal parts after the Big Bang. However, the observed matter-antimatter asymmetry in the universe today cannot be explained quantitatively by the Standard Model of particle physics.

Potential clues could be found by the precise investigation of the spectrum of the antihydrogen atom ($\overline{\mathrm{H}}$) and comparison with its matter counterpart which is known to high precision [1–3]. In this

★e-mail: bernadette.kolbinger@oeaw.ac.at

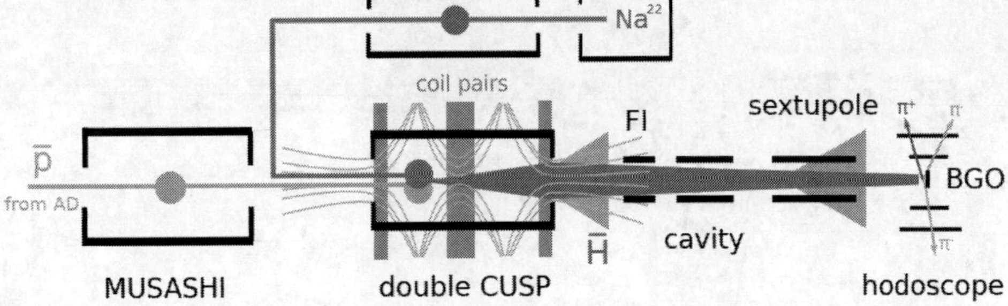

Figure 1. Sketch of the ASACUSA-CUSP setup. Antiprotons are drawn in blue, positrons in red. The produced antihydrogen atoms and the polarising effect of the double CUSP field and the sextupole magnet are indicated in purple (dark purple for low field seekers and light purple for high field seekers).

context, the ASACUSA collaboration at the Antiproton Decelerator (AD) of CERN aims to test the CPT symmetry by measuring the ground state hyperfine splitting of antihydrogen. The CPT theorem states that any local, Lorentz invariant quantum field theory should conserve CPT. From that theorem ensues that atoms and antiatoms should have the same atomic spectra.

ASACUSA is performing a Rabi-like experiment to measure the hyperfine transitions of $\overline{\mathrm{H}}$ in a beam [4]. A sketch of the beamline is displayed in Figure 1. For the production of $\overline{\mathrm{H}}$, antiprotons ($\overline{\mathrm{p}}$) extracted from the AD are further decelerated by a radio frequency quadrupole before being stored in a Penning-Malmberg type catching trap, named MUSASHI [5]. The positrons from a ^{22}Na source are accumulated in another Penning trap, before being transported to the mixing trap [6], the so-called double CUSP trap [7]. The latter is composed of multi-ring electrodes providing a static electric field, and two pairs of anti-Helmholtz coils. In the presence of a magnetic field, the ground state of (anti)hydrogen exhibits four hyperfine states which can be classified into high- and low-field seekers, according to their behaviour in an inhomogeneous magnetic field. The $\overline{\mathrm{H}}$ atoms formed in the mixing process leave the double CUSP trap as a polarised beam due to the magnetic field gradient of the trap which focuses low-field seekers and defocuses atoms in high field seeking states.

The $\overline{\mathrm{H}}$ atoms enter the spectroscopy beam line located directly downstream of the mixing trap. It consists of a microwave cavity for inducing transitions between high and low field seeking states, and a superconducting sextupole magnet for state selection [4, 8]. The spectroscopy apparatus has been tested with hydrogen [9]. The antiatoms are detected by a crystal calorimeter at the end of the beamline, surrounded by a segmented hodoscope made of plastic scintillators. As the amount of $\overline{\mathrm{H}}$ produced with the current techniques is low [10], a high detection efficiency for annihilations is needed. The antihydrogen detector and its recent upgrade will be discussed in the next section.

Since it is essential to know the amount of antihydrogen atoms exiting the production trap in the ground state, a field ioniser is placed directly after the mixing trap. Atoms above a certain principal quantum number n_{min} are ionised by the applied voltage (n_{min} = 14 at the highest field ioniser configuration used: $|\vec{E}| \approx 10$ kV/cm). By recording the rate of antihydrogen events at the detector for different voltage settings, the quantum state distribution of the produced atoms has been measured for the first time [11]. The data-driven machine learning strategy used to achieve the results will be outlined briefly below.

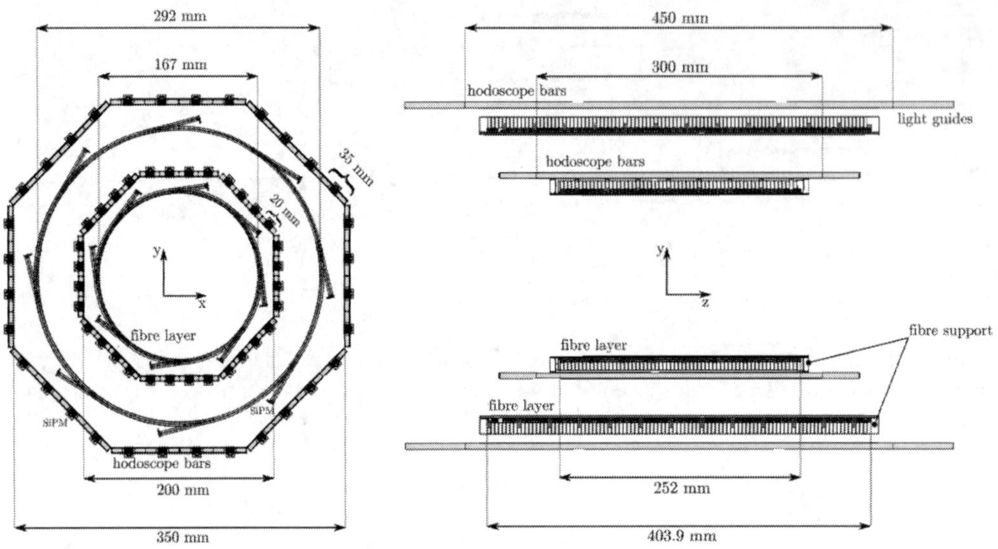

Figure 2. Drawing of the antihydrogen tracking detector. Top: cross section of the tracking detector perpendicular to the beam axis. The hodoscope bar layers and the two scintillating fibre layers are shown. Bottom: tracking detector in the y-z plane. Scintillators are coloured in light-blue, SiPMs in black.

2 The antihydrogen detector

The detector is composed of a central scintillating crystal target where the $\overline{\mathrm{H}}$ atoms annihilate, and a surrounding hodoscope for tracking the charged annihilation products (mainly pions).

The annihilating $\overline{\mathrm{H}}$ is detected with a 90 mm diameter bismuth germanate (BGO) disc with a thickness of 5 mm [12] coated with carbon on its upstream face. The crystal is placed in a vacuum vessel at UHV pressure. The BGO is read out by four Hamamatsu H8500C multianode photomultiplier tubes (MAPMT) with 8×8 channels, each connected to a Clearpulse CP80190 read-out unit. The MAPMT are separated from the BGO by a UHV viewport. The central detector is surrounded by the hodoscope detector, supported by a 1 mm thick stainless steel pipe.

A drawing of the two-layered, segmented hodoscope detector for tracking the charged annihilation products [13] is shown in Figure 2. The fibre upgrade discussed in section 4 is also shown. The hodoscope covers ≈80% of the solid angle for annihilation products emitted from the centre of the BGO. Both layers are made of eight removable panels containing four 5 mm thick scintillating bars each. The outer layer has a diameter of 350 mm, the outer bars are 450 mm long and 35 mm wide. The diameter of the inner layer is 200 mm, its bars are 300 mm long and have a width of 20 mm. Light guides are attached on both sides of the bars to match the detecting area of the KETEK 3350TS silicon photomultipliers (SiPMs). Two SiPMs operated in series are glued to each light guide to increase the light collection. The SiPMs are read out using front-end electronics [14] and CAEN V1742 waveform digitisers.

The double sided readout of the scintillating bars reduces the dark noise by requiring of a coincidence between up and downstream SiPM. Also, it allows precise time measurement of the passing particles. Of particular interest is the resolution of the time difference between bars, which is a measure of the time of flight (ToF). The characterisation of the hodoscope detector was carried out with cosmic rays. The time of flight resolution between two opposing bars has been determined to be 552

± 5 ps (FWHM) for the outer layer and 497 ± 3 ps (FWHM) for the inner one [13]. Hence, the ToF is precise enough to discriminate between particles traversing the hodoscope from outside i.e. background and particles such as pions originating from the annihilation of antiprotons on the BGO.

Furthermore, from the time difference between signals recorded by the upstream and downstream SiPMs on the same bar, the hit position of the particle on the bar in beam direction can be obtained. The position resolution for outer bars was determined to be 73 ± 3 mm (FWHM) and 59 ± 4 mm (FWHM) for the inner hodoscope bars. Considering the length of the bars, the resolution is not good enough for precise vertex reconstruction and it is not possible to distinguish upstream annihilations in the direct vicinity of the BGO disc, resulting from high field seeking antihydrogen states or during antiproton extractions for annihilations studies. In section 4 the detector upgrade using scintillating fibres is discussed which enables tracking in three dimensions and precise vertex reconstruction to overcome these shortcomings. The upgrade will also improve the performance of the analysis described in the next section.

3 Analysis strategy

The analysis uses a supervised machine learning method with boosted gradient decision trees [15]. Measured data is used for training and validation of the algorithm. The signal data sample consists of measured antiproton annihilation events of antiprotons extracted from the MUSASHI to the antihydrogen detector [5]. This is possible since the event signature of antihydrogen and antiproton annihilation in the ASACUSA detector is the same. The background, which is dominated by cosmic particles, has been recorded during beam-off periods, with no antiproton delivery to ASACUSA. Therefore, the analysis will help to discriminate against cosmic background only. About 4000 antiproton and 30000 cosmic events are available for the machine learning procedure. 2/3 of the samples are used for training and 1/3 for testing. In order to overcome class imbalance and therefore avoid a biased algorithm, oversampling is used on the antiproton events of the training sample. An ensemble of algorithms each with a randomly selected training and validation set has been built. Detection efficiency and background rejection have been determined using the class predictions of the ensemble for the validation sets.

The achieved fraction of cosmics falsely identified as signal is under 0.25% and the detection efficiency for antiproton events is close to 80%. The algorithm can then be used to identify antihydrogen events during mixing runs. The first measurement of the quantum state distribution of antihydrogen atoms at the exit of the double cusp down to $n_{min} = 14$ as resulting from this analysis was recently published [11].

In total, 43 runs have been performed with the highest field configuration of the field ioniser. With this setting, atoms are ionised down to principal quantum number $n = 14$ and they are likely to decay to the ground state before reaching the cavity. The above described analysis identified about 0.16 events per run. In order to evaluate the significance, the p-value of the background-only hypothesis was computed using Poisson statistics. The p-value was then converted into units of σ (observed significance) via the unit gaussian. The found significance of the number of events with $n < 14$ is 4.5σ.

4 Hodoscope upgrade

The tracking detector has been upgraded in order to improve its position resolution in beam direction. The upgrade consists of two additional layers of scintillating fibres perpendicular to the existing hodoscope bars, see Figure 2. Each channel is made up of four 2×2 mm^2 square fibres. They are

Figure 3. Left: assembly of the fibres on the carbon tube. Right: close up of a 4 mm×4 mm fibre bunch.

bundled together and glued to one SiPM (KETEK PM3350-EB) i.e. the read-out of the fibres is single sided. A close-up of one of the channels is shown in Figure 3 (right). The fibre bundles are wound once around the supporting carbon tube, covering one circumference with a small additional overlap to avoid loss of solid angle. They are kept in place by supporting plastic rings (indicated in Figure 2 and 3 (left)). The picture in Figure 3 (left) has been taken during the assembly of the detector.

The cladding of the fibres prevents cross talk which has been measured in the lab and was determined to be less than 1%. The inner layer houses 63 channels and has a diameter of 167 mm, whereas the outer one has 100 channels and a diameter of 292 mm. The front-end electronics powering and reading out the SIPMs is identical to that used on the scintillating bars, providing analogue and digital signal output. Two TDCs (CAEN V1190B with 64 channels, CAEN V1190A with 128 channels) are used to record the leading edge timestamp and the time over threshold of the signals.

The upgraded detector has been integrated into the experimental setup in 2017 and has been successfully used for antiproton annihilation studies. For these measurements, the BGO central detector setup has been replaced by a 2×2 array of Timepix3 [16] read-out chips bump-bonded to a 500 μm thick silicon sensor. It is capable of measuring pions and heavy nuclear fragments from the annihilation on a thin target foil in front of it.

The tracking and vertex reconstruction analysis is on-going. Figure 4 shows a cosmic event recorded with the upgraded tracking detector, where bars and fibres with a hit are indicated in colour.

On the right panel, the hodoscope bar and corresponding fibre hits are shown. The bar hodoscope hits are marked by crosses and are determined from the time difference of and up- and downstream SiPMs, the width of the crosses are the 2σ errors. The squares show fibre channels with a hit highlighting both, the consistency of the two sub-detectors and the increased position resolution provided by the new fibre upgrade.

5 Summary

We have presented the ASACUSA antihydrogen detector which monitors the rate of arriving antihydrogen atoms at the end of the beam line. A data-driven analysis employing machine learning has been developed which was used to identify antihydrogen candidates during mixing runs for the measurement of the quantum state distribution. The rate of antihydrogen atoms is currently still too low to perform spectroscopy and reach the precision goal of $\approx$1 ppm. Therefore, the current focus lies on increasing the production rate and determining characteristics of the antiatoms.

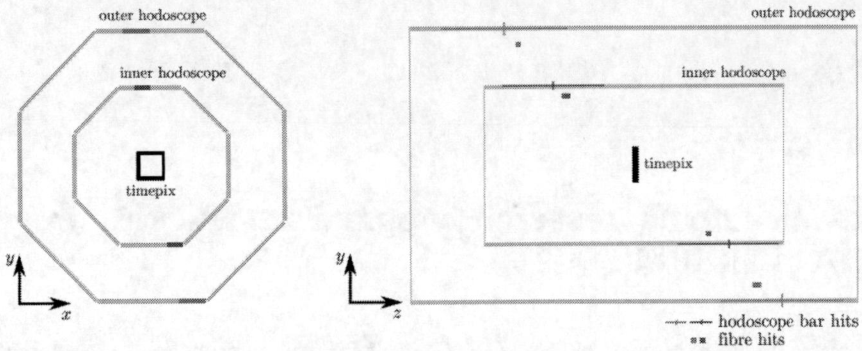

Figure 4. Cosmic event recorded during data taking in 2017. y-x plane (left) and y-z (right) plane of the detector are shown, where z is the direction of the beam. Bars and fibres with hits are indicated by crosses and squares respectively (blue for hits in the inner layers and red for the outer layers).

With the recent upgrade using scintillating fibres, the detector becomes capable of precise track and vertex reconstruction and therefore enabling the possibility to discriminate against additional background contributions such as annihilations close to the detector.

References

[1] Karschenboim S.G., Van. J. Phys. **78**, 639-678 (2000)
[2] Parthey, C.G. et al., Phys. Rev. Lett. **107**, 203001 (2011)
[3] Essen, L. et al., Nature **229**, 110-111 (1971)
[4] Widmann, E. et al., Hyperfine Interactions **215**, 1-8 (2013)
[5] Kuroda, N. et al., Phys. Rev. ST. Accel. Beams **15**, 024702 (2012)
[6] Mohri, A. et al., Europhys. Lett. **63**, 207–213 (2003)
[7] Nagata, Y. et al., Journal of Physics: Conference Series **635**, 022062 (2015)
[8] Malbrunot, C. et al., Hyperfine Interactions **228**, 61-66 (2014)
[9] Diermaier, M. et al., Nature Communications **8**, 15749 (2017)
[10] Kuroda, N. et al., Nature Communications **5**, 3089 (2014)
[11] Malbrunot, C. et al., Phil. Trans. R. Soc. A **376**, 20170273 (2018)
[12] Nagata, Y. et al., Nuclear Instrum. Methods A **840**, 153-59 (2016)
[13] Sauerzopf, C. et al., Nuclear Instrum. Methods A **845**, 579-582 (2016)
[14] Sauerzopf, C. et al., Nuclear Instrum. Methods A **819**, 163-166 (2016)
[15] Chen, T. et al., Xgboost: A scalable tree boosting system, ACM, proceedings of KDD'16 (2016)
[16] Poikela, T. et al., Journal of Instrumentation **9** C05013 (2014)

The kaonic atoms research program at DAΦNE: from SIDDHARTA to SIDDHARTA-2

A. Scordo[1,*], A. Amirkhani[2], M. Bazzi[1], G. Bellotti[2], C. Berucci[1,3], D. Bosnar[4], A.M. Bragadireanu[5], M. Cargnelli[3], C. Curceanu[1], A. Dawood Butt[2], R. Del Grande[1], L. Fabbietti[6], C. Fiorini[2], F. Ghio[1], C. Guaraldo[1], R.S. Hayano[7], M. Iliescu[1], M. Iwasaki[7], P. Levi Sandri[1], J. Marton[1,3], M. Miliucci[1], P. Moskal[8], D. Pietreanu[1,5], K. Piscicchia[1,9], H. Shi[1], M. Silarski[8], D. Sirghi[1,5], F. Sirghi[1,5], M. Skurzok[8], A. Spallone[1], H. Tatsuno[10], O. Vazquez Doce[1,6], E. Widmann[3], and J. Zmeskal[1,3]

[1] *INFN, Laboratori Nazionali di Frascati, Frascati (Roma), Italy*
[2] *Politecnico di Milano, Dipartimento di Elettronica, Informazione e Bioingegneria and INFN Sezione di Milano, Milano, Italy*
[3] *Stefan-Meyer-Institut für Subatomare Physik, Vienna, Austria*
[4] *Physics Department, University of Zagreb, Zagreb, Croatia*
[5] *Horia Hulubei National Institute of Physics and Nuclear Engineering (IFIN-HH), Magurele, Romania*
[6] *Excellence Cluster Universe, Technische Universität München, Garching, Germany*
[7] *University of Tokyo, Tokyo, Japan*
[8] *The M. Smoluchowski Institute of Physics, Jagiellonian University, Kraków, Poland*
[9] *Museo Storico della Fisica e Centro Studi e Ricerche "Enrico Fermi", Roma, Italy*
[10] *Lund Univeristy, Lund, Sweden*

Abstract. The interaction of antikaons with nucleons and nuclei in the low-energy regime represents an active research field in hadron physics with still many important open questions. The investigation of light kaonic atoms, in which one electron is replaced by a negatively charged kaon, is a unique tool to provide precise information on this interaction; the energy shift and the broadening of the low-lying states of such atoms, induced by the kaon-nucleus hadronic interaction, can be determined with high precision from the atomic X-ray spectroscopy, and this experimental method provides unique information to understand the low energy kaon-nucleus interaction at the production threshold. The lightest atomic systems, like the kaonic hydrogen and the kaonic deuterium deliver, in a model-independent way, the isospin-dependent kaon-nucleon scattering lengths. The most precise kaonic hydrogen measurement to-date, together with an exploratory measurement of kaonic deuterium, were carried out in 2009 by the SIDDHARTA collaboration at the DAΦNE electron-positron collider of LNF-INFN, combining the excellent quality kaon beam delivered by the collider with new experimental techniques, as fast and very precise X-ray detectors, like the Silicon Drift Detectors. The SIDDHARTA results triggered new theoretical work, which achieved major progress in the understanding of the low-energy strong interaction with strangeness reflected by the antikaon-nucleon scattering lengths calculated with the antikaon-proton amplitudes constrained by the SIDDHARTA data. The most important open question is the experimental determination of the hadronic energy shift and width of kaonic deuterium; presently, a major upgrade of

*e-mail: alessandro.scordo@lnf.infn.it

the setup, SIDDHARTA-2, is being realized to reach this goal. In this paper, the results obtained in 2009 and the proposed SIDDHARTA-2 upgrades are presented.

1 Introduction

The strong interaction, described by the QCD in the framework of the Standard Model, is still hiding many mysteries, especially in the low-energy limit, the so called non-perturbative regime. Particularly interesting is the strong interaction involving the strange quark which, belonging to the light quarks but having a mass of about $100\,MeV/c^2$, much heavier than the few MeV/c^2 up and down quark masses, plays a peculiar role. Since kaons and antikaons are the lowest mass particles containing strange quarks, for decades their interaction with nucleons and nuclei in the low-energy regime has been subject of intensive studies in experiment and theory (for reviews see [1, 2]). Effective field theories contain appropriate degrees of freedom to describe physical phenomena occurring at the nucleon-meson scale and chiral perturbation theory was extremely successful in describing systems like pionic atoms; however it is not directly applicable for the kaonic systems, where non-perturbative coupled-channel techniques can be used ([3]). These theories are still waiting to be verified by experimental data. There are two experimental approaches to probe the kaon-nucleus strong interaction, both exploited at LNF-INFN. One is by studying the scattering and the reaction channels between the kaon and the nucleus, to directly search for bound states and extract the potential of the strong interaction; this is the experimental method followed by the AMADEUS collaboration ([4–6]). The other method consists in the precision X-ray measurement of the shift and the broadening of the energy levels of kaonic atoms caused by the kaon-nucleus strong interaction. This latter one, exploited by the SIDDHARTA and SIDDHARTA-2 collaborations ([7–13]), is of significant importance, since it is the only method able to provide direct information on the kaon-nucleus system at threshold.

1.1 Kaonic atoms

Kaonic atoms are formed when a negatively charged kaon is stopped in a target and replaces one of the electrons in an atom; due to the much higher K^- mass with respect to the e^- one, this newly formed exotic atom radius is smaller than the one of the usual atom and the K^- is stopped at a distance from the nucleus corresponding to the $n \simeq 25$ excited state. The subsequent cascade of the antikaon to the ground level occurs via several different processes; in particular, the last transitions on the 1s level are radiative and photons are emitted in the X-ray region. For light atoms, especially for the hydrogen and the deuterium, a detectable energy shift from the electromagnetic value of the ground state is expected, as well as a broadening of the ground state level, caused by nuclear absorption (see fig.1).

By measuring these observables, kaonic atoms offer the unique possibility to determine the s-wave antikaon-nucleon scattering lengths at vanishing energy, alternatively to scattering experiments were an extrapolation to zero energy is necessary. With the advance of the experimental techniques, both in the accelerator and detector sectors, we are presently able to perform very high precision measurements, resulting in a deeper and more complete understanding of the many open questions in the QCD. In particular, with the advent of clean kaon beams, as the one provided by the DAΦNE collider, and of performant, fast, X-ray detectors as the Silicon Drift Detectors, the kaonic atoms studies entered the precision era.

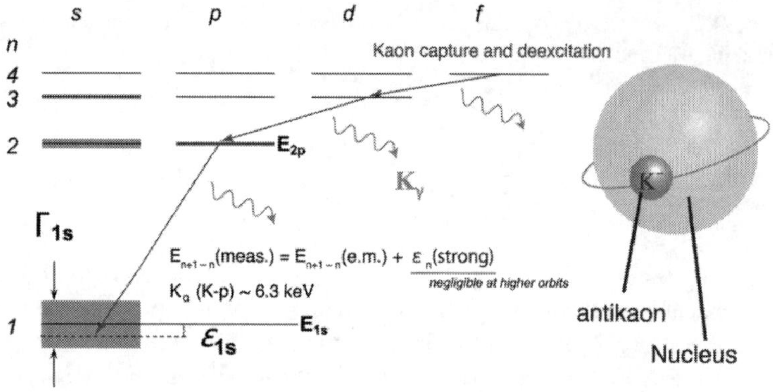

Figure 1. After the atomic capture of the kaon a kaonic atom is formed in a highly excited state and a few kaons will cascade to ground state. 1s level is shifted and broadened by the strong interaction ([14])

2 SIDDHARTA experiment

In 2009, with the SIDDHARTA experiment at DAΦNE, the strong interaction induced shift of the ground state of kaonic hydrogen atoms and the absorption width were measured with the highest accuracy up to now [8, 9]. Measurements of the 2p shift and width of kaonic helium (^4He as well as ^3He) [10–12] completed the SIDDHARTA program. Conscient about the importance of the Kd measurement an attempt was also done in this direction but, due to the very low yield of the transition, only an upper limit could be exctracted and no values for the 1s level shift and width were delivered [8].

2.1 Experimental setup

The SIDDHARTA setup consisted of two main components, the light-weight cryogenic target cell and a specially developed large area, high resolution X-ray detector system made of Silicon Drift Detectors (SDDs). The experiment made use of the K^+K^- pairs coming from the Φ decays with a 49% branching ratio. The kaons leaving the interaction point through the SIDDHARTA beam pipe were degraded in energy and entered the cryogenic gaseous hydrogen (helium) target placed above the beam pipe, forming a kaonic atom and emitting X-rays during the $2p \to 1s$ (hydrogen) and $3d \to 2p$ (helium) transitions (see fig. 2).

The K^+K^- pairs were emitted in back-to-back configuration and identified by the kaon detectors, made of two plastic scintillators placed above and below the interaction point as illustrated in fig. 2. The kaons were distinguished from the minimum ionizing particles using the time of flight information at the kaon detectors, and the coincidence of the two scintillators defined the kaon trigger, which marked the timing of the incident kaons. A fraction of the negatively charged kaons activating the kaon trigger were successfully stopped inside the volume of the gaseous target placed about 20 cm above the interaction point to form kaonic atoms. A working pressure of 0.3 MPa was achieved, which led to a hydrogen gas density of 2% of liquid hydrogen density, at a working temperature of 25 K. Special Silicon Drift Detectors were developed with excellent energy resolution ($FWHM \simeq 150\,eV$ @ 6 keV) and timing capability in the order of μs [15] when operated at 140 K temperature. Using the X-ray signal from the SDDs in coincidence with the K^+K^- pair, the continuous machine background, as well as unwanted fluorescence X-rays, could be efficiently suppressed. The SDDs had a total active

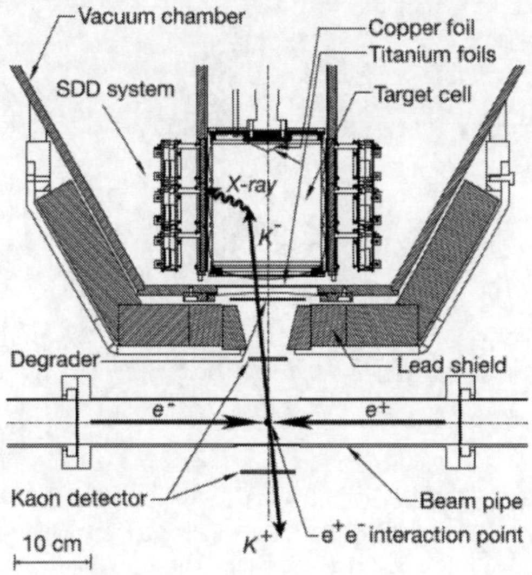

Figure 2. A schematic cutaway view of the SIDDHARTA setup at the DAΦNE interaction point. The charged kaon pairs are identified with two plastic scintillators, and the K^- induced x-rays detected by the SDDs are identified from the time correlation to the kaon pair events.

area of $144\,cm^2$, covering about 10% of the solid angle around the target cell. The energy calibrations of the SDDs were done every few hours, using the X-ray tube activated K_α lines of Ti ($4.5\,keV$), Mn ($5.5\,keV$) and Cu ($8.0\,keV$) to determine the scale of the energy spectra. The energy resolution at $6\,keV$ was stable at about $150\,eV$ (FWHM) throughout the measurement. More details about the configuration and the performance of the detectors can be found in [8–12]. Data were accumulated with gaseous targets of hydrogen ($1.3\,g/l$), deuterium ($2.50\,g/l$), helium-3 ($0.96\,g/l$), and helium-4 ($1.65g/l$ and $2.15\,g/l$).

2.2 Kaonic Helium measurements

For what concerns the kaonic helium, its transition to 1s level couldn't be observed due to the very low yield and to a transition energy out of the SDD dynamic range. Before the SIDDHARTA experiment, there existed only four measurements and the situation was rather ambiguous: three measurements [16–18], performed more than 20 years ago, giving, within few σs, results which are more than an order of magnitude higher with respect to the theoretical predictions and a more recent one, performed at KEK [19], which was, instead, compatible with the theoretical predictions ([20, 21]), but incompatible with the previous experiments. A conclusive precise measurement on K^4He was then needed in order to solve the "puzzle", together with the first measurement of K^3He, fundamental to obtain valuable information on the K^-p and the K^-n interactions. The kaonic helium spectra obtained by the SIDDHARTA experiment are shown in fig. 3 [10–12].

In the left picture, the peak seen at 6.2 keV is identified as the K^3He L_α line (the $3d \rightarrow 2p$ transition). In the right one, the peak seen at 6.4 keV is identified as the K^4He L_α line. In addition to these lines, other smaller peaks are clearly visible which are the kaonic atom X-ray lines produced by kaons stopping in the target window made of Kapton, and the Ti and Mn lines. The strong-interaction

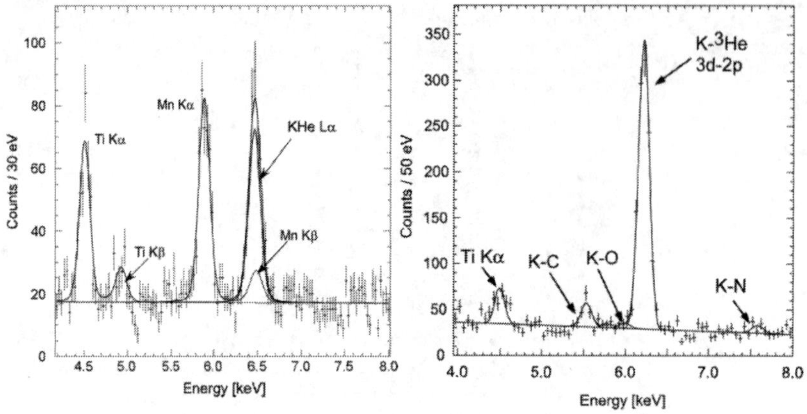

Figure 3. Fitted spectra of the kaonic ^3He (right) and ^4He (left) X-rays. The $3d \rightarrow 2p$ transitions are seen around 6 keV. Together with these peaks, others smaller are seen, which are the kaonic atom X-ray lines produced by kaons stopping in the target window made of Kapton, and the Ti and Mn lines.

shifts of the kaonic helium 2p states were obtained from the difference between the experimentally determined values and the QED calculated ones [22, 23]. The results are:

$$\varepsilon_{2p}(K^3He) = -2 \pm 2\,(stat) \pm 4\,(syst)\,eV$$

$$\varepsilon_{2p}(K^4He) = 5 \pm 5\,(stat) \pm 4\,(syst)\,eV$$

Thus a shift compatible with 0 eV of the 2p level from experiment is established, which is in agreement with the theoretical estimations and, within the errors, with the results reported by the E570 [19] collaboration, while is not in agreement with the previous measurements ([16–18]). This is probably due to the fact that, in those experiments, the helium target was a liquid one in which, as indeed happened also in the KEK experiment ([19]), a big Compton effect is present and has to be taken into account in the analysis procedure.

2.3 Kaonic Hydrogen and Kaonic Deuterium measurement

The lightest kaonic atom is the K^-p atom in which the principal electromagnetic interaction is accompanied by the strong interaction of the kaon with the proton which is measurable by X-ray spectroscopy of the radiative transitions from the np states (2p, 3p, ...) to the 1s ground state (K transitions). The K^-p scattering length a_{K^-p} can be obtain from the equation

$$\varepsilon_{1s} + \frac{i}{2}\Gamma_{1s} = 2\alpha^3\mu_c^2 a_{K^-p}(1 - 2\alpha\mu_c(ln\alpha - 1)a_{K^-p})$$

being ε_{1s} and Γ_{1s} the shift and width of the transition to the 1s level, (for more details on the various terms see [24]) and it is related to the isospin dependent scattering lengths by

$$a_{K^-}p = \frac{1}{2}(a_0 + a_1)$$

Historically there were several measurements of the strong-interaction shift ε_{1s} and width Γ_{1s} of kaonic hydrogen ([25–29]). In the 1970s and the 1980s three groups ([25–27]) reported a measured attractive shift (positive ε_{1s}), while the information extracted from the analyses of the low energy KN data ([30–32]) shows a repulsive shift (negative ε_{1s}). This contradiction has been known as the "kaonic hydrogen puzzle". In 1997, the first distinct peaks of the kaonic-hydrogen X-rays were observed by the KEK-PS E228 group [28] with a significant improvement in the signal-to-background ratio by the use of a gaseous hydrogen target, where previous experiments had employed liquid hydrogen. It was crucial to use a low-density target, namely a gaseous target, because the X-ray yields quickly decrease towards higher density due to the Stark mixing effect. The observed repulsive shift was consistent in sign with the analysis of the low energy KN scattering data, resolving the long-standing discrepancy. More recent values reported by the DEAR group in 2005 [29], with substantially reduced errors, firmly established the repulsive shift obtained in the previous E228 experiment. The latest kaonic hydrogen and deuterium X-ray energy spectra, obtained by the SIDDHARTA experiment, are shown in fig. 4 [8].

The K-series X-rays of kaonic hydrogen were clearly observed while those for kaonic deuterium were not visible. This appears to be consistent with the theoretical expectation of lower X-ray yield and greater transition width for deuterium ([33]) than for kaonic hydrogen. However, the kaonic deuterium spectrum can be used to characterize the background. The vertical dot-dashed line in fig.4 indicates the X-ray energy of kaonic hydrogen K_α calculated using only the electromagnetic interaction (EM). Comparing the kaonic hydrogen K_α measured peak and the EM value, a repulsive shift of the kaonic hydrogen 1s energy level is easily seen. Many other lines from kaonic atom X-rays were detected in both spectra as indicated with arrows in the figure. These kaonic atom lines result from high n X-ray transitions of kaons stopped in the target cell wall made of kapton ($C_{22}H_{10}O_5N_2$) and its support frames made of aluminium. There are also characteristic X-rays from titanium and copper foils installed for X-ray energy calibration. A global simultaneous fit of the hydrogen and deuterium spectra has been performed, whose results are shown in fig. 4 (b) and (c). The kaonic hydrogen lines were represented by Lorentz functions convoluted with the detector response function, where the Lorentz width corresponds to the strong interaction broadening of the 1s state. The region of interest of Kd X-rays is illustrated in fig. 4 (c). The 1s level shift ε_{1s} and width Γ_{1s} of kaonic hydrogen were determined to be [8]:

$$\varepsilon_{1s} = -283 \pm 36\,(stat) \pm 6\,(syst)\,eV$$

$$\Gamma_{1s} = 541 \pm 89\,(stat) \pm 22\,(syst)\,eV$$

The quoted systematic error is a quadratic summation of the contributions from the systematic errors on the SDD gain shift, the SDD response function, the ADC linearity, the low energy tail of the kaonic hydrogen higher transitions, the energy resolution, and the procedural dependence shown by independent analysis. This measurement represents the best precision available untill today, and it has been used to set constraints on the calculated real and imaginary part of the $K^- p$ amplitude ([34]).

3 SIDDHARTA-2 experiment

The case of kaonic deuterium is more challenging than kaonic hydrogen mainly due to the larger widths of the K lines and the lower X-ray yield expected. Experimentally the case of kaonic deuterium is still open; the SIDDHARTA experiment measured the X-ray spectrum with a pure deuterium target but, due to the limited statistics and high background, the determination of ε_{1s} and Γ_{1s} was impossible. An upper limit for the X-ray yield of the K lines could be extracted from the data: total yield < 0.0143, K_α yield < 0.0039 [13].

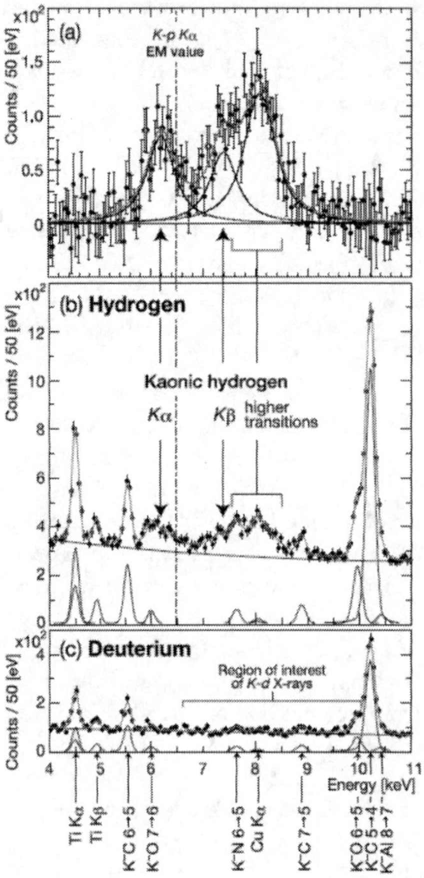

Figure 4. A global simultaneous fit result of the X-ray energy spectra of hydrogen and deuterium data. (a) Residuals of the measured kaonic-hydrogen X-ray spectrum after subtraction of the fitted background, clearly displaying the kaonic-hydrogen K -series transitions. The fit components of the K-p transitions are also shown, where the sum of the function is drawn for the higher transitions (greater than K_β). (b), (c) Measured energy spectra with the fit lines. Fit components of the background X-ray lines and a continuous background are also shown. The dot-dashed vertical line indicates the EM value of the kaonic-hydrogen K_α energy. (Note that the fluorescence K_α line consists of $K_{\alpha 1}$ and $K_{\alpha 2}$ lines, both of which are shown.)

Since the kaonic deuterium X-ray measurement represents the most important experimental information missing in the field of the low-energy antikaon-nucleon interactions today, a new experiment (SIDDHARTA-2) is planned, based on a much improved apparatus.

3.1 Experimental setup upgrade

Thanks to the knowledge acquired in 2009 with the SIDDHARTA experiment, a new version of the experimental apparatus, aiming to increase the signal over background ratio (S/B) by a factor about 20 allowing the kaonic deuterium measurement, has been developed by the SIDDHARTA-2 collaboration. The major upgrades introduced to fulfill the kaonic deuterium measurement goal, are shown in fig. 5 and listed below ([36]):

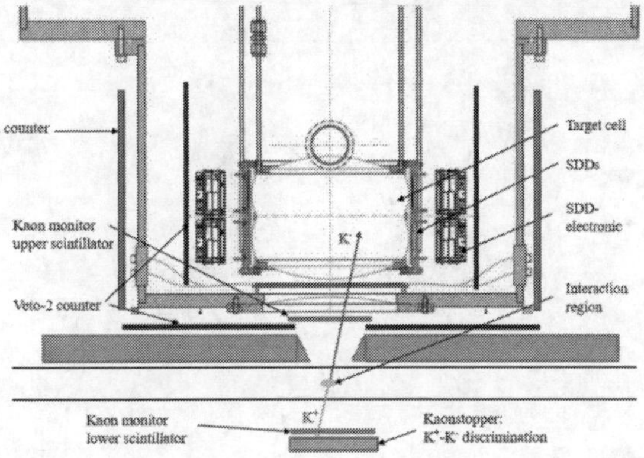

Figure 5. The SIDDHARTA-2 layout. The elements of the apparatus are indicated in the figure.

- Larger area and faster SDD detector array; the solution to improve the SDD time resolution consists in the reduction of the single element size (from $10x10$ to $8x8\,mm^2$) and the replacement of the integrated J-FET (thermally limited), by a newly-developed amplifier on the ceramics, able to operate at very low temperatures (below 50 K). The shorter path and the higher carrier mobility consent a faster charge packet drift to the anode and therefore, a reduced time window (350 ns instead of 900 ns) for each trigger, with consequent a suppression of the asynchronous background. The new SDD detectors are produced by Fondazione Bruno Kessler (FBK) in Trento, Italy.
- A veto detector to measure the prompt time of the secondaries from K^- absorption on nuclei. The system consists in scintillators surrounding the vacuum chamber, read at both ends by PMs coupled to mirrors and light-guides (to cope with the narrow space between the setup and the shielding against beam background) [35].
- A new cryogenic target in reinforced kapton (13 cm diameter, 7 cm height), operating few hundred mK above the liquid point (25 K) at a pressure of 4 bar (5% LHD), for more efficient kaon stopping.
- A veto system, consisting in scintillators read by SiPMs, placed behind each SDD array, to reject the hadronic background coming from border hits of Minimum Ionizing Particles (MIPs), depositing energy in the X-ray range.
- A kaon trigger with geometric acceptance optimized to match the kaon gas stopping distribution.
- An improved X-ray calibration system, providing low-rate in-situ calibration, as well as high rate calibration between physics runs, to compensate very small fluctuations in each single SDD response.
- Mechanical and cryogenic improvements of the vacuum chamber, necessary to add more cooling power to the SDDs and to the cryogenic target.

3.2 Kaonic deuterium expected measurement

All the above described items were optimized by GEANT4 simulation, considered reliable after having reproduced, in the same framework, the SIDDHARTA results, both in terms of signal and background, to 7% accuracy. Using theoretical values as inputs to the Monte Carlo simulation, the expected spectrum of the $K^{-}d$ transitions, for an acquired luminosity of 800 pb^{-1} and assuming a yield of 0.1%, is shown in fig. 6.

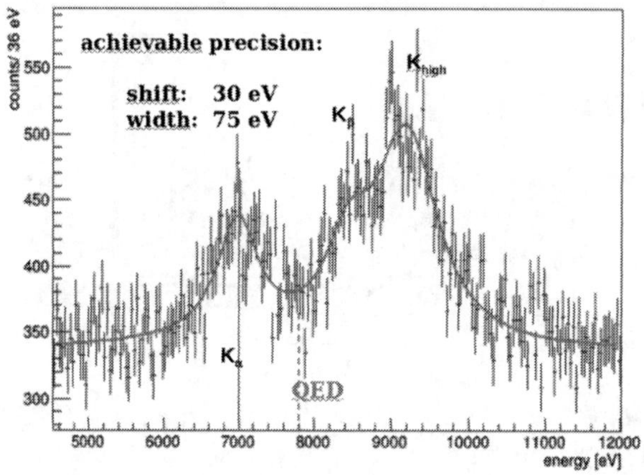

Figure 6. The simulated $K^{-}d$ spectrum for SIDDHARTA-2 for 800 pb^{-1} integrated luminosity (the K_α line is at 7 keV, while from 8 to 10 keV there is the K-complex.

The fit of the simulated spectrum shows how the $K^{-}d$ $2p \rightarrow 1s$ transition, influenced by the strong interaction, can be determined with a precision of 30 eV for the shift and 70 eV for the width.

4 Conclusions

The SIDDHARTA experiment at the DAΦNE electron-positron collider measured the strong interaction induced shift and width of kaonic hydrogen and helium transitions with unprecedented accuracy. Important implications for the theory of the strong interaction with strangeness in the low energy regime were provided by the SIDDHARTA constraints. In addition, SIDDHARTA measured for the first time ever the kaonic helium 3 transitions to the 2p level and the first, similar, measurement, of kaonic helium 4 with a gaseous target. Still open is the challenging case of kaonic deuterium which will be attacked by the follow-up experiment SIDDHARTA2, aiming at the determination of the hadronic shift and width to enable the extraction of the antikaon-nucleon isospin dependent scattering lengths a_0 and a_1. The SIDDHARTA-2 collaboration is preparing a series of modifications and upgrades of the apparatus and experimental configurations, aiming to measure the kaonic deuterium x-rays with a precision compatible with that achieved in the hydrogen target measurement starting with 2018, when the apparatus will be installed on the DAΦNE collider.

Acknowledgments

We thank C. Capoccia, G. Corradi, B. Dulach, and D. Tagnani from LNF-INFN; and H. Schneider, L. Stohwasser, and D. Stükler from Stefan-Meyer-Institut, for their fundamental contribution in designing and building the SIDDHARTA setup. We thank as well the DAΦNE staff for the excellent working conditions and permanent support. Part of this work was supported by the Austrian Science Fund (FWF) (P24756-N20); Austrian Federal Ministry of Science and Research BMBWK 650962/0001

VI/2/2009; Croatian Science Foundation under Project No. 1680; the Grant-in-Aid for Specially Promoted Research (20002003), MEXT, Japan; Ministero degli Affari Esteri e della Cooperazione Internazionale, Direzione Generale per la Promozione del Sistema Paese (MAECI), Strange Matter project; the Polish National Science Center (NCN) through grant No. UMO-2016/21/D/ST2/01155.

References

[1] E. Friedman, A. Gal, Phys. Rep. **452**, 89 (2007)
[2] C.J. Batty, E. Friedman, A. Gal, Phys. Rep. **287**, 385 (1997).
[3] J. Caro-Ramon, et al., Nucl. Phys. A **672**, 249-269 (2000).
[4] The AMADEUS collaboration, LNF preprint, LNF/9607/24(IR). *http://www.lnf.infn.it/esperimenti/siddharta/LOI_AMADEUS_March2006.pdf.*
[5] K. Piscicchia et al., PoS Bormio2013 **034** (2013).
[6] O. Vazquez Doce et al. Phys. Lett. B **758** 134-139 (2016).
[7] The SIDDHARTA-2 collaboration, The SIDDHARTA-2 proposal. *https://www.lnf.infn.it/committee/private/documenti/SIDDHARTA2-proposal_FINAL.pdf.*
[8] M. Bazzi et al., Phys. Lett. B **704**, 113 (2011).
[9] M. Bazzi et al., Nucl. Phys. A **881**, 88 (2012).
[10] M. Bazzi et al., Phys. Lett. B **681**, 310 (2009).
[11] M. Bazzi et al., Phys. Lett. B **714**, 40 (2012).
[12] M. Bazzi et al., Phys. Lett. B **697**, 199 (2011).
[13] M. Bazzi et al., Nucl. Phys. A **907**, 69 (2013).
[14] K. Phelan et al., Nucl. Inst. and Meth. in Phys. Res. A, **845**, 533-536 (2017).
[15] M. Bazzi et al., Nucl. Inst. Meth. in Phys. Res. A **628**, 264 (2011).
[16] C.E. Wiegand and R.H. Pehl, Phys. Rev. Lett **27**, 1410 (1971).
[17] C.J. Batty et al., Nucl. Phys. A **326**, 455 (1979).
[18] S. Baird et al., Nucl. Phys. A **392**, 297 (1983).
[19] S. Okada et al., Phys. Lett. B **653**, 387 (2007).
[20] C.J. Batty, Nucl. Phys. A **508**, 89c (1990).
[21] R. Seki, Phys. Rev. C **5**, 1196 (1972).
[22] V. Lucherini, et al., Nucl. Instr. Meth. A **496** (2003) 315.
[23] S.G. Karshenboim, et al., Can. J. Phys. **84** (2006) 107.
[24] T. Hyodo and D. Jido, Prog. Part. Nucl. Phys. **67**, 55 (2012).
[25] J. D. Davies, et al., Phys. Lett. B **83**, 55 (1979).
[26] M. Izycki, et al., Z. Phys. A **297**, 11 (1980).
[27] P. M. Bird, et al., Nucl. Phys. A **404**, 482 (1983).
[28] M. Iwasaki, et al. Phys. Rev. Lett. **78**, 3067 (1997).
[29] G. Beer, et al. (DEAR Collaboration), Phys. Rev. Lett. **94**, 212302 (2005).
[30] W.E. Humphrey and R.R. Rose, Phys. Rev. **127**, 1305 (1962).
[31] J.K. Kim, Phys. Rev. Lett. **19**, 1074 (1967).
[32] A.D. Martin, Nucl. Phys. B **179**, 33 (1981).
[33] T. Koike, T. Harada, Y. Akaishi, Phys. Rev. C **53**, 79 (1996).
[34] Y. Ikeda, T. Hyodo, and W. Weise, Phys. Lett. B **706**, 63 (2011).
[35] M Bazzi et al. Journ. of. Instr. **8**, T11003 (2013).
[36] M. Iliescu et al., Journal of Physics: Conference Series **770**, 012034 (2016).

Low energy interaction studies of negative kaons in light nuclear targets by AMADEUS

K. Piscicchia[1,2,*], M. Bazzi[2], G. Belloti[3], A. M. Bragadireanu[4], D. Bosnar[5], A. D. Butt[3,6], M. Cargnelli[7], C. Curceanu[2], R. Del Grande[2,8]L. De Paolis[2,8], L. Fabbietti[9,10], C. Fiorini[3,6], F. Ghio[11,12], C. Guaraldo[2], R. S Hayano[13], M. Iliescu[2], M. Iwasaki[14], P. Levi Sandri[2], J. Marton[7], M. Miliucci[2,8], P. Moskal[15], S. Okada[14], D. Pietreanu[4], A. Scordo[2], M. Silarski[15], D. L. Sirghi[2,4], F. Sirghi[2,4], M. Skurzok[15], A. Spallone[2], O. Vazquez Doce[9,10], E. Widmann[7], and S. Wycech[16]J. Zmeskal[7]

[1]CENTRO FERMI - Museo Storico della Fisica e Centro Studi e Ricerche 'Enrico Fermi', Rome, Italy
[2]INFN Laboratori Nazionali di Frascati, Frascati, Rome, Italy
[3]Politecnico di Milano, Dip. di Elettronica, Informazione e Bioingegneria, Milano, Italy
[4]Horia Hulubei National Institute of Physics and Nuclear Engineering, Magurele, Romania
[5]University of Zagreb, Zagreb, Croatia
[6]INFN Sezione di Milano, Milano, Italy
[7]Stefan-Meyer-Institut für Subatomare Physik, Wien, Austria
[8]Università degli Studi di Roma Tor Vergata, Rome, Italy
[9]Excellence Cluster 'Origin and Structure of the Universe', Garching, Germany
[10]Physik Department E12, Technische Universität München, Garching, Germany
[11]INFN Sezione di Roma I, Rome, Italy
[12]Istituto Superiore di Sanità, Rome, Italy
[13]The University of Tokyo, Tokyo, Japan
[14]RIKEN, The Institute of Physics and Chemical Research, Saitama, Japan
[15]Institute of Physics, Jagiellonian University, Kracow, Poland
[16]National Centre for Nuclear Research, Warsaw, Poland

Abstract. The AMADEUS collaboration is performing experimental investigations in the sector of the low energy strangeness hadron physics by taking advantage of the DAΦNE collider, which provides a unique source of monochromatic low-momentum kaons and exploiting the KLOE detector as an active target. The K^- single and multi-nuclear absorption on H, ^{4}He, ^{9}Be and ^{12}C, both at-rest and in-flight, are studied with the aim to determine the nature of the controversial Λ(1405), the non-resonant hyperon pion formation amplitude below the $\bar{K}N$ threshold, the yield and cross sections of K^- multi-nucleon absorptions intimately connected to the antikaon multi-nucleon clusters properties and the K^- scattering cross sections on light nuclear targets.

These studies have a strong impact on the non-perturbative strangeness QCD with implications ranging from nuclear physics to the equation of state of neutron stars.

*e-mail: kristian.piscicchia@lnf.infn.it

1 Introduction

The AMADEUS (Anti-kaonic Matter At DAΦNE: An Experiment with Unraveling Spectroscopy) [1] experiment is performing a systematic investigation of the hyperon resonances production, and hyperon pion non-resonant formation, in low energy K^- hadronic interaction with light nuclear targets (e.g. H, ^{4}He, ^{9}Be and ^{12}C). The K^- multi-nucleon absorption and bound states formation is also explored, in order to provide experimental constraints on the non-perturbative QCD in the strangeness sector. AMADEUS exploits the low momentum (about 127 MeV/c), almost monochromatic, charged kaons provided by the decay of ϕ mesons at-rest at the DAΦNE factory [2]. Two data sets are under study: one was acquired by the KLOE [3] collaboration during the 2004/2005 data taking campaign, the second was collected by KLOE in collaboration with AMADEUS in the period November-December 2012. The latter data taking was performed by implementing in the central region of the KLOE setup a dedicated pure carbon (graphite) solid target.

Chiral perturbation theory can not be applied in the low-energy QCD strangeness sector as a consequence of the existence of the broad $\Lambda(1405)$ state just few MeV below the $\bar{K}N$ threshold. The $\Lambda(1405)$ is a $J^P = 1/2^-$ isospin $I = 0$ strange baryon resonance, assigned to the lowest $L = 1$ supermultiplet of the three-quark system, which decays into $(\Sigma\pi)^0$ through the strong interaction. The nature of this state is being a puzzle since decades both from the experimental and the theoretical point of view. The three quark picture (uds) fails to reproduce the observed properties of this state. A review of the theoretical works, and references to the experimental literature can be found in [5].

According to the chiral unitary predictions [6–9] a high mass pole, coupled to the $\bar{K}N$ production channel and located around 1420 MeV, might contribute to the measured $\Lambda(1405)$ shape. Since the accessible invariant mass, in K^-p absorption processes, is diminished by the binding energy of the absorbing proton in the hosting nucleus, our strategy is to unveil the presence of the high mass pole by exploiting K^- captures in-flight [10, 11]. In this case the kinetic energy of the kaon moves the energy threshold just below the $\bar{K}N$ threshold, thus disclosing the phase space region in which the high mass pole is expected to emerge.

The $\Sigma^0\pi^0$ decay channel, which is free from the $I = 1$ contribution and from the isospin interference term, represents the cleanest signature of the $\Lambda(1405)$ resonance, because is not affected by the neighbour $\Sigma(1385)$ resonance.

The contamination of the $(\Sigma\pi)^0$ spectra due to the the non-resonant production below threshold is also very important to be accounted for. A key issue, which is addressed in the analyses described below, is the investigation of the non-resonant hyperon-pion transition amplitude below threshold.

The position of the $\Lambda(1405)$ clearly reflects the strength of the $\bar{K}N$ interaction, thus influencing the possible formation of $\bar{K}$ multi-nucleon bound states. For the di-baryonic kaonic bound state ppK^- theoretical predictions deliver a wide range of binding energies and widths [12], while the experimental results are contradictory [13–20]. Moreover, the extraction of ppK^- signal in K^- absorption experiments is strongly affected by the yield and the shape of the competing K^- double-nucleon absorption (2NA) process.

In Section 2 the features of the DAΦNE accelerator and the KLOE detector are summarized. In Section 3 the two collected data samples which were analysed are described. The particle identification procedure is summarised in Section 4. Sections 5 and 6 are dedicated to the obtained results and ongoing analyses, regarding K^- multi-nucleon absorption processes, the search for a ppK^- signal and the investigation of resonant and non-resonant Hyperon-pion production in light nuclei. The paper ends with conclusions and perspectives.

2 The KLOE detector at DAΦNE

DAΦNE (Double Anular Φ-factory for Nice Experiments) is a double ring e^+e^- collider, designed to work at the center of mass energy of the ϕ particle; the ϕ meson decay produces charged kaons with low momentum ($\simeq$ 127 MeV/c) which allows to either stop them, or to explore the products of their low-energy nuclear absorptions.

The KLOE detector is centered around the interaction region of DAΦNE and is characterised by an acceptance of $\simeq 98\%$; it consists of a large cylindrical Drift Chamber (DC) [21] and a fine sampling lead-scintillating fibers calorimeter [22], all immersed in the axially symmetrical magnetic field with intensity of 0.52 T, provided by a superconducting solenoid. The chamber is characterized by excellent position and momentum resolutions. Tracks are reconstructed with a resolution in the transverse $R-\phi$ plane $\sigma_{R\phi} \sim 200\,\mu$m and a resolution along the z-axis $\sigma_z \sim 2$ mm. The transverse momentum resolution for low momentum tracks ((50 $< p <$ 300)MeV/c) is $\frac{\sigma_{p_T}}{p_T} \sim 0.4\%$. The calorimeter is composed of a cylindrical barrel and two endcaps, providing a solid angle coverage of 98%. The volume ratio (lead/fibers/glue=42:48:10) is optimized for a high light yield and a high efficiency for photons in the range (20-300) MeV/c. The photon detection efficiency is 99% for energies larger than 80 MeV and it falls to 80% at 20 MeV due to the cutoff introduced by the ADC and TDC thresholds. The position of the clusters along the fibers can be obtained with a resolution $\sigma_\parallel \sim 1.4\,\mathrm{cm}/\sqrt{E(\mathrm{GeV})}$. The resolution in the orthogonal direction is $\sigma_\perp \sim 1.3$ cm. The energy and time resolutions for photon clusters are given by $\frac{\sigma_E}{E_\gamma} = \frac{0.057}{\sqrt{E_\gamma(\mathrm{GeV})}}$ and $\sigma_t = \frac{57\,\mathrm{ps}}{\sqrt{E_\gamma(\mathrm{GeV})}} \oplus 100$ ps.

The DC entrance wall composition is 750 μm of carbon fiber and 150 μm of aluminum foil. Dedicated GEANT Monte Carlo simulations of the KLOE apparatus show that out of the total number of kaons interacting in the DC entrance wall, about 81% are absorbed in the carbon fiber component and the remaining 19% in the aluminum foil. The KLOE DC is filled with a mixture of helium and isobutane (90% in volume ^{4}He and 10% in volume C_4H_{10}).

3 Data samples

As a first step of the AMADEUS program the data collected by the KLOE collaboration during the 2004/2005 data taking, for which the dE/dx information of the reconstructed tracks is available (dE/dx represents the truncated mean of the ADC collected counts due to the ionization in the DC gas), corresponding to ~ 1.74 fb^{-1}, were analysed. The KLOE detector was used as an active target, the hadronic interaction of negative kaons with the materials of the apparatus being investigated. The topology of these data is shown in figure 1, representing the radial position (ρ_Λ) of the $\Lambda(1116)$ decay vertex (see Section 4). Four components are recognizable, from inside to outside we observe K^- absorptions in the DAΦNE beryllium sphere ($\sim$ 5 cm), the DAΦNE aluminated beryllium pipe ($\sim$ 10 cm), the KLOE DC entrance wall (aluminated carbon fiber $\sim$ 25 cm) and the long tail originating from K^- interactions in the gas filling the KLOE DC (25-200 cm). Extremely rich experimental information is contained in this sample, with K^- hadronic interactions, both at-rest and in-flight [10], in a variety of light nuclear targets (H, ^{4}He, ^{9}Be and ^{12}C), which, of course, turned into challenging analysis and interpretation efforts.

In order to increase the statistics and as an essential interpretation tool, a high purity carbon target (graphite) was realized in summer 2012 and installed inside the KLOE DC, between the beam pipe and the DC entrance wall.

The geometry of the target was optimized by means of a GEANT3 MC simulation. The final configuration was a three-sector half-cylinder (ρ = 0.95g/cm^3) supported by an alu-

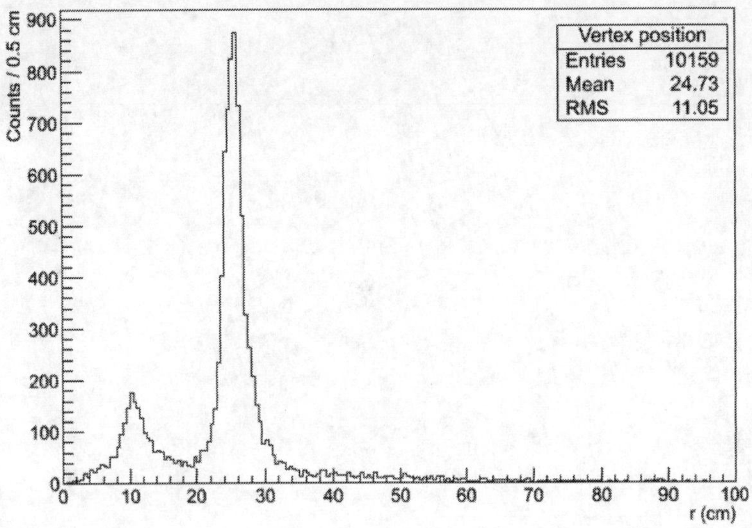

Figure 1. *Radial position distribution* ρ_Λ, *of the* Λ *decay vertex, for 2004-2005 KLOE collected data.*

minium frame, with a length of 600 mm and a mean radius of 24,7 mm. The thickness varied from 4 to 6 mm, in order to optimize the percentage of stopped kaons, taking into account the ϕ boost, derived from the crossing angle of the e^+ e^- beams. The half-cylindrical configuration was chosen in order to take advantage of the K^+ tagging in the opposite direction. The design of the realized target is shown in figure 2. We took data from 6 November to 14 December 2012, for a total integrated luminosity of ~90 pb^{-1}. Up to now we analysed a sample of 37 pb^{-1} reconstructed data. The topology of the reconstructed Λ decay vertices radial position is shown in figure 3, and as expected, the majority of kaons are stopped in the target.

4 Particle Identification

The investigation of the negatively charged kaons interactions in nuclear matter is performed through the reconstruction of hyperon-pion and hyperon-nucleon/nucleus correlated pairs productions, following the K^- absorptions in H, ^{4}He, ^{9}Be and ^{12}C. The $\Lambda(1116)$ identification proceeds through the reconstruction of the $\Lambda \to p + \pi^-$ (BR = 63.9 ±0.5%) decay vertex. A spatial resolution below 1 mm is achieved for vertices found inside the DC volume (evaluated with Monte Carlo simulations). The obtained $M_{p\pi^-}$ invariant mass mean value is 1115.753±0.002 MeV/c^2 (only statistical error is given, the systematics being under investigation), with a resolution of σ=0.5 MeV/c^2. The particle identification takes advantage of both dE/dx information from the DC wires and the measurement of the energy released in the Calorimeter, as described in [1]. Σ^0 particles are identified through their decay into $\Lambda\gamma$ as reported in [10]. The K^- absorption vertex position, obtained exploiting the correlated production of the hyperon together with an extra-particle (pion, proton etc.) is then used to select the target. As an example, the obtained resolution on the radial coordinate ($\rho_{\Lambda p}$) for the Λp vertex is 1.2 mm. Cuts on the absorption vertex radial position were optimised, based on MC simulations and a study of the Λ decay path, to select the targets with minimal contamination

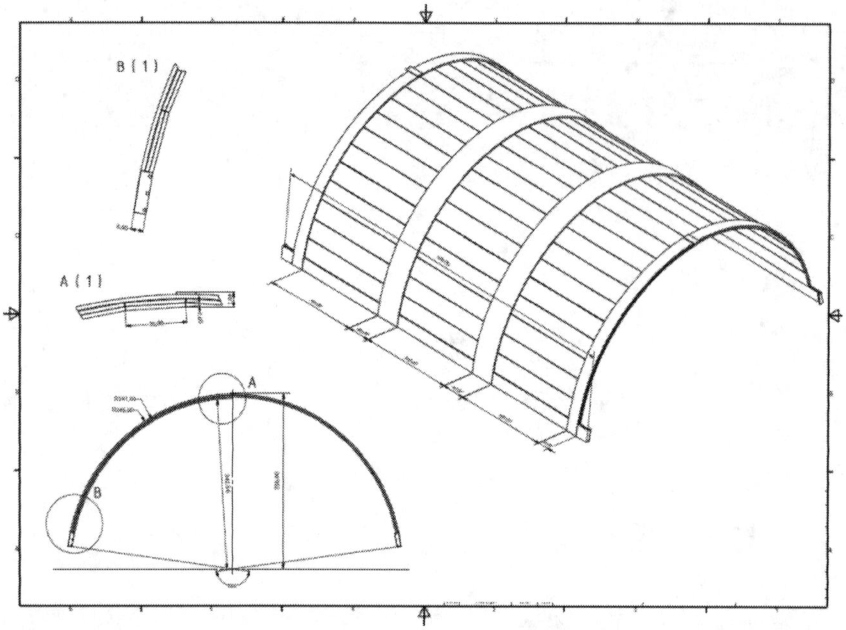

Figure 2. *Design of the graphite target inserted between the beam pipe and the KLOE DC entrance wall.*

from other components. More details on the particle identification procedure can be found in [23]

5 K^- absorption on two nucleons in the Σ^0p final state

In [23], a high purity sample of Σ^0p events, from K^- captures in the ^{12}C, was reconstructed. Σ^0p is (together with Λp) an expected decay channel of the ppK^- cluster, with the advantage to be free from the Σ N $\to$ Λ N' conversion processes. The conversions strongly affect the uncorrelated Λp production, thus distorting the observed spectra.

A simultaneous fit of the: Σ^0p invariant mass, the relative angle of the Σ^0 and proton in the laboratory system $\cos(\theta_{\Sigma^0 p})$, the Σ^0 and the proton momenta was performed to the following simulated processes:

- $K^-A \to \Sigma^0 - (\pi)p_{spec}(A')$ (1NA),
- $K^-pp \to \Sigma^0p$ (2NA),
- $K^-ppn \to \Sigma^0pn$ (3NA),
- $K^-ppnn \to \Sigma^0pnn$ (4NA).

The possible fragmentations of the residual nucleus, as well as Final State Interactions (FSI) of the Σ^0 and p emerging from a K^-pp capture, were taken into account.

The yield of the 2NA, when the produced Σ^0 and p particles are free form any FSI process, was measured for the first time, with good precision. The obtained results are summarized in Table 1. A second fit was performed including a ppK^- cluster component, decaying into Σ^0p. A systematic scan of possible binding energies and widths, varying within 15-75 MeV and 30-70 MeV respectively, was performed. The best fit resulted in a binding energy of 45 MeV and a width of 30 MeV. The resulting yield, normalised to the number of stopped K^-,

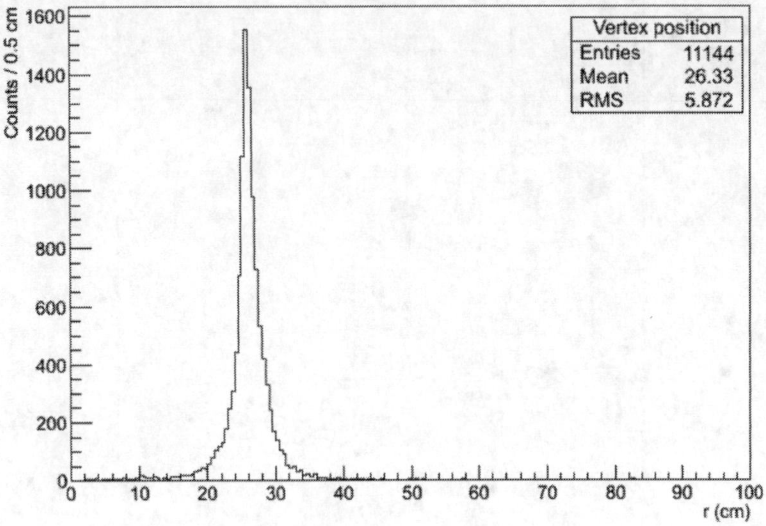

Figure 3. *Radial position distribution of ρ_Λ for carbon target events.*

is $ppK^-/K^-_{stop} = (0.044 \pm 0.009stat^{+0.004}_{-0.005}syst)\times10^{-2}$. The significance of the bound state with respect to a statistical fluctuation was checked by means of an F-test and was found to be significant at the level of 1σ only. Although the measured spectra are compatible with the hypothesis of a ppK^- contribution, the significance of the result is not sufficient to claim the measurement of the state. We refer to [23] for the details of the analysis.

Process	yield / $K^-_{stop}\times10^{-2}$	$\sigma_{stat}\times10^{-2}$	$\sigma_{syst}\times10^{-2}$
2NA-QF	0.127	±0.019	$^{+0.004}_{-0.008}$
2NA-FSI	0.272	±0.028	$^{+0.022}_{-0.023}$
Tot 2NA	0.399	±0.033	$^{+0.023}_{-0.032}$
3NA	0.274	±0.069	$^{+0.044}_{-0.021}$
Tot 3 body	0.546	±0.074	$^{+0.048}_{-0.033}$
4NA + bkg.	0.773	±0.053	$^{+0.025}_{-0.076}$

Table 1. Production probability of the Σ^0p final state for different intermediate processes normalised to the number of stopped K^- in the DC wall. The statistical and systematic errors are shown as well [23].

6 $Y\pi$ resonant and non resonant production and the shape of the $\Lambda(1405)$

When extracting the $\Lambda(1405)$ shape from K^- induced reactions in light nuclear targets (see for example [26]) the hyperon-pion spectroscopy is influenced by the energy threshold, imposed by the last nucleon binding energy. The $m_{\Sigma\pi}$ invariant mass threshold is about 1412 MeV and 1416 MeV, for K^- capture at-rest in ^4He and ^{12}C respectively, thus the K^- absorption at-rest is not sensitive to the high mass pole predicted by chiral unitary models. The $\bar{K}N$ sub-threshold region is accessible by exploiting K^-N absorptions in-flight. For a mean kaon

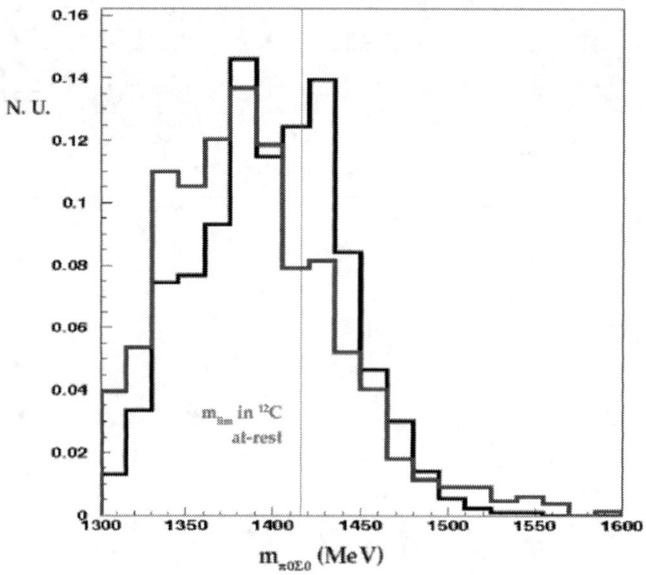

Figure 4. *(Colour online.)* $m_{\Sigma^0\pi^0}$ *invariant mass distribution from* K^- *captures in the KLOE DC wall (black curve) and pure carbon graphite target (blue curve)*

momentum of 100 MeV/c the $m_{\Sigma\pi}$ threshold is shifted upwards by about 10 MeV. A second bias is represented by the non-resonant $K^-N \to Y\pi$ formation, which gives rise to highly correlated hyperon-pion pairs production. The corresponding $m_{Y\pi}$ invariant masses spectra is narrow (of the order of 10 MeV) and peaked below the $\bar{K}$-boundN threshold. The $\Lambda\pi$ and $\Sigma\pi$ non resonant transition amplitudes, for K^- capture in light nuclear targets, were never measured. The Λ and π^- kinematic distributions for K^- captures in ^{4}He, both at-rest and in-flight, were calculated in [27]. The momentum probability distribution functions of the emerging hyperon pion pairs, following K^-n absorptions, are expressed in terms of the K^-n transition amplitudes: the isospin $I = 1$ S-wave non-resonant amplitude ($|f^{nr}|$) and the resonant $I = 1$ P-wave amplitude, dominated by the $\Sigma^-(1385)$. Since the resonant amplitude is well known from direct experiments, the measured total momentum distributions can be used to extract the non-resonant $|f^{nr}|$ amplitude module below the $\bar{K}N$ threshold. The goal of the ongoing analyses is to measure the contributions and the shapes of the non resonant $\Lambda\pi$ and $\Sigma\pi$ productions. The knowledge of the $(\Sigma\pi)^0$ isospin $I = 0$ non-resonant transition amplitude will allow to disentangle the resonant $\Lambda(1405)$ shape.

In figure 4 (black distribution) the preliminary reconstructed $\Sigma^0\pi^0$ invariant mass spectrum from K^- captures in the KLOE DC wall is shown [28]. The $m_{\Sigma0\pi0}$ spectrum can be compared with the corresponding distribution from K^- absorptions in a pure carbon target (fig. 4 blue distribution), sample which consists of K^- captures at-rest. In fig. 4 a red line indicates the energy threshold corresponding to K^- absorption in ^{12}C at-rest. In order to compare the two distributions plotted in fig. 4 the two curves are normalised to unity. A rich sample of in-flight K^{-12}C captures can be easily identified. A spectroscopic study of the kinematic region ranging between the at-rest energy threshold, and the $\bar{K}N$ threshold, opened by the low momentum in-flight capture process, is presently ongoing in order to clarify the nature of the high mass $\Lambda(1405)$ pole.

7 Conclusions and perspectives

In this work the low energy negatively charged kaons absorption processes on light nuclear targets is investigated with the aim to extract the resonant and non-resonant hyperon-pion production amplitudes. The characterization of the non-resonant $\Sigma^0\pi^0$ production from K^- captures in-flight could enable to evidence and to measure the characteristics of the high mass pole of the $\Lambda(1405)$. The measurement of the multi-nucleon absorption yields following K^- captures in ^{12}C target and the search for a K^-pp cluster are also presented.

Presently a feasibility study [31, 32] is ongoing for the realization of a dedicated AMADEUS experimental setup, in order to deepen and extend the low energy anti-kaon nuclei interaction studies and obtain fundamental input for the study of QCD with strangeness.

Acknowledgement

We acknowledge the KLOE/KLOE-2 Collaboration for their support and for having provided us the data and the tools to perform the analysis presented in this paper. We thank as well the DAΦNE staff for the excellent working conditions and permanent support. We acknowledge the CENTRO FERMI - Museo Storico della Fisica e Centro Studi e Ricerche 'Enrico Fermi', for the project PAMQ. Part of this work was supported by the Austrian Science Fund (FWF): [P24756-N20]; Austrian Federal Ministry of Science and Research BMBWK 650962/0001 VI/2/2009; the Croatian Science Foundation, under project 1680; Minstero degli Affari Esteri e della Cooperazione Internazionale, Direzione Generale per la Promozione del Sistema Paese (MAECI), Strange Matter project; Polish National Science Center through grant No. UMO-2016/21/D/ST2/01155.

References

[1] C. Curceanu, K. Piscicchia et al., Acta Phys. Polon. B46 1, (2015) 203
[2] R. Baldini et al., Proposal for a Phi-Factory, report LNF-90/031(R) (1990)
[3] F. Bossi, E. De Lucia, J. Lee-Franzini, S. Miscetti, M. Palutan and KLOE coll., Riv. Nuovo Cim. 31 (2008)
[4] C. Patrignani et al. (Particle Data Group), Chin. Phys. C, 40, (2016) 100001
[5] T. Hyodo, D. Jido, Prog. Part. Nucl. Phys. 67 (2012) 55
[6] J. A. Oller and U. G. Meissner, Phys. Lett. B500 (2001) 263
[7] T. Hyodo et al., Phys. Rev. C68 (2003) 018201
[8] D. Jido et al., Nucl. Phys. A725 (2003) 181
[9] Y. Ikeda, T. Hyodo, W. Weise, Nucl. Phys. A881 (2012) 98
[10] K. Piscicchia et al., PoS Bormio2013 (2013) 034
[11] A. Scordo et al., PoS Bormio2014 (2014) 039
[12] T. Yamazaki, et al., Phys. Rev. C 76, (2007) 045201
N.V. Shevchenko, et al., Phys. Rev. Lett. 98, (2007) 082301
P. Bicudo, Phys. Rev. D 76, (2007) 031502
A. Doté, et al., Phys. Rev. C 79, (2009) 014003
S. Wycech, et al., Phys. Rev. C 79, (2009) 014001
Y. Ikeda, et al., Phys. Rev. C 79, (2009) 035201
N. Barnea, et al., Phys. Lett. B 712, (2012) 132
E. Oset, et al., Nucl. Phys. A 881, (2012) 127
M. Bayar, E. Oset, Nucl. Phys A 914, (2013) 349

[13] G. Agakishiev, HADES Coll., Phys. Lett. B 742, (2015) 242

[14] M. Agnello, et al. (FINUDA Coll.), Phys. Rev. Lett. 94, (2005) 212303
[15] T. Yamazaki, et al., Phys. Rev. Lett. 104, (2010) 132502
[16] Y. Ichikawa, et al., Prog. Theor. Exp. Phys., (2015) 021D01
[17] A. O. Tokiyasu, et al., Phys. Lett. B 728, (2014) 616
[18] L. Fabbietti, et al., Nucl. Phys. A 914, (2013) 60
[19] T. Hashimoto et al., Prog. Theor. Exp. Phys., (2015) 061D01
[20] M. Iwasaki, EPJ Web of Conferences 130, (2016) 01023
[21] M. Adinolfi et al., [KLOE Collaboration], Nucl. Inst. Meth. A 488, (2002) 51
[22] M. Adinolfi et al. [KLOE Collaboration], Nucl. Inst. Meth. A 482, (2002) 368
[23] O. Vazquez Doce et al., Phys. Lett. B 758, 134-139 (2016)
[24] R. Roosen, J. H. Wickens, Il Nuovo Cim. A Series 11 66 (1981), 101
[25] FINUDA collaboration, Phys. Lett. B, 669 (2008) 229
[26] J. Esmaili et al., Phys. Lett. B 686 (2010) 23
[27] K. Piscicchia, S. Wycech, C. Curceanu, Nucl. Phys. A 954, (2016) 75
[28] K. Piscicchia, PhD theis (2013), Lambda(1405) measurement through the decay to $\Sigma^0\pi^0$, resulting from K^- meson absorption on ^{4}He and ^{12}C, with the KLOE detector, http://www.infn.it/thesis/thesis_dettaglio.php?tid=7097
[29] P. B. Demorest, T. Pennucci, S. M. Ransom, M. S. E. Roberts & J. W. T. Hessels, Nature (London), (2011) 467
[30] D. J. Nice, E. M. Splaver, I. H. Stairs, O. Loehmer, and A. Jessner, Astrophys. J. 634, (2005) 1242
[31] M. Bazzi, et al., Nuclear Instruments and Methods in Physics Research A 671, (2012) 125
[32] M. Bazzi et al., Journal of Instrumentation 8(05) January 2013

EPJ Web of Conferences **181**, 01006 (2018) https://doi.org/10.1051/epjconf/201818101006
EXA2017

Kaonic deuterium and low-energy antikaon-nucleon interaction

Wataru Horiuchi[1,⋆], *Tetsuo* Hyodo[2,⋆⋆], and *Wolfram* Weise[3,⋆⋆⋆]

[1] *Department of Physics, Hokkaido University, Sapporo 060-0810, Japan*
[2] *Yukawa Institute for Theoretical Physics, Kyoto University, Kyoto 606-8502, Japan*
[3] *Physics Department, Technical University of Munich, 85748 Garching, Germany*

Abstract. A new evaluation of the 1s level shift and width of kaonic deuterium is presented based on an accurate $\bar{K}NN$ three-body calculation, using as input a realistic antikaon-nucleon interaction constrained by the SIDDHARTA kaonic hydrogen data. The three-body Schrödinger equation is solved with a superposition of a large number of correlated Gaussian basis functions extending over distance scales up to several hundred fm. The resulting energy shift and width of the kaonic deuterium 1s level are $\Delta E \simeq 0.67$ keV and $\Gamma \simeq 1.02$ keV, with estimated uncertainties at the 10 % level.

1 Introduction

Kaonic hydrogen and kaonic deuterium are the prototype K^--atomic systems to be studied in the quest for constraints on the low-energy antikaon-nucleon interaction near $\bar{K}N$ threshold. Sufficiently accurate data of the strong-interaction energy shift ΔE and width Γ of the 1s level in these kaonic atoms should provide, through their theoretical analysis, key information in order to fix basic $\bar{K}N$ scattering length parameters in both isospin $I = 0, 1$ channels. An important step in that direction was taken previously by the SIDDHARTA experiment [1] which determined $\Delta E = 283 \pm 36 \pm 6$ eV and $\Gamma = 541 \pm 89 \pm 22$ eV for kaonic hydrogen. Corresponding data for kaonic deuterium do not yet exist. However, new K^-d measurements are in preparation: SIDDHARTA-2 at LNF [2] and E57 at J-PARC [3].

These developments call for advanced calculations of the coupled $K^-pn \leftrightarrow \bar{K}^0nn$ system. The K^-d atomic binding (see the schematic spectrum of Fig. 1) is produced by the attractive K^-p Coulomb interaction. The pertinent atomic length scale is determined by the kaonic deuterium Bohr radius, $a_B(K^-d) \sim 70$ fm. The strong-interaction energy shift and width, on the other hand, result from effects at distance scales of order 1 fm between the nucleons and the antikaon. Clearly, high demands on computational accuracy must be met at *all* length scales in order to deal with this challenging three-body problem involving both long-range Coulomb and short-range strong interactions.

Previous theoretical studies of kaonic deuterium basically fall into two categories. Treatments of the K^-d system as an approximate two-body problem focused on the relationship between kaonic deuterium observables and $\bar{K}N$ two-body scattering lengths [4–6]. Three-body Faddeev calculations

⋆e-mail: whoriuchi@nucl.sci.hokudai.ac.jp
⋆⋆e-mail: hyodo@yukawa.kyoto-u.ac.jp
⋆⋆⋆e-mail: weise@tum.de

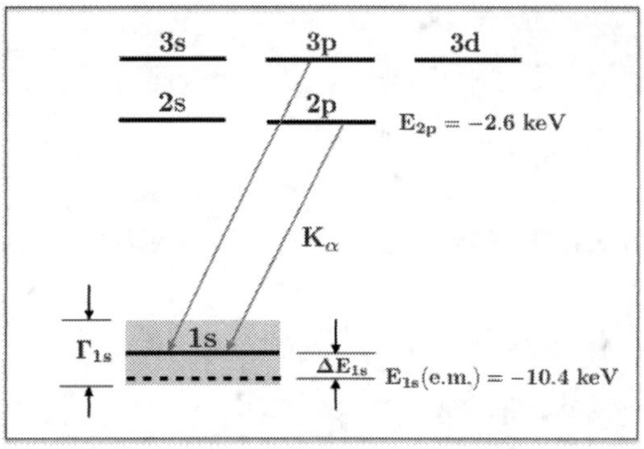

Figure 1. Schematic spectrum of kaonic deuterium showing the atomic levels and the K_α transition ($2p \to 1s$) together with the strong-interaction shift and width of the 1s level.

of kaonic deuterium have their own history - see for example Ref. [7]. Recent advanced Faddeev computations [8, 9] use separable potentials constrained by the SIDDHARTA kaonic hydrogen data and have evaluated the $1s$ K^-d atomic state assuming isospin symmetry for the $\bar{K}$ and nucleon doublets.

The present report gives a brief summary of recent work published in Ref. [10]. A novel calculation of kaonic deuterium has been performed solving the full three-body $K^-pn \leftrightarrow \bar{K}^0nn$ coupled-channels Schrödinger equation. A modern $\bar{K}N$ interaction is used as input, namely the Kyoto $\bar{K}N$ potential [11] based on chiral SU(3) effective field theory [12, 13]. A particular aim of that investigation, apart from achieving highest possible computational precision, has been to set a more stringent constraint for the isospin $I = 1$ component of the low-energy $\bar{K}N$ interaction. We recall that kaonic deuterium is expected to be a more sensitive probe than kaonic hydrogen for this $I = 1$ channel, given that the ratio of $I = 1$ to $I = 0$ components is 3:1 in the K^-d system.

2 Antikaon-NN three-body problem

In the present context the theoretical description of kaonic deuterium as a three-body system is defined by the following coupled-channels Schrödinger equation:

$$\begin{pmatrix} \hat{H}_{K^-pn} & \hat{V}^{\bar{K}N}_{12} + \hat{V}^{\bar{K}N}_{13} \\ \hat{V}^{\bar{K}N}_{12} + \hat{V}^{\bar{K}N}_{13} & \hat{H}_{\bar{K}^0nn} \end{pmatrix} \begin{pmatrix} |K^-pn\rangle \\ |\bar{K}^0nn\rangle \end{pmatrix} = E \begin{pmatrix} |K^-pn\rangle \\ |\bar{K}^0nn\rangle \end{pmatrix} , \tag{1}$$

with the Hamiltonians of the K^-pn and $\bar{K}^0nn$ channels specified as[1]

$$\hat{H}_{K^-pn} = \sum_{i=1}^{3} \hat{T}_i - \hat{T}_{\rm cm} + \hat{V}^{NN}_{23} + \sum_{i=2}^{3} \left(\hat{V}^{\bar{K}N}_{1i} + \hat{V}^{\rm EM}_{1i} \right) , \tag{2}$$

$$\hat{H}_{\bar{K}^0nn} = \sum_{i=1}^{3} \hat{T}_i - \hat{T}_{\rm cm} + \hat{V}^{NN}_{23} + \sum_{i=2}^{3} \hat{V}^{\bar{K}N}_{1i} + \Delta M . \tag{3}$$

[1] Particle indices are introduced as indicated in Fig.2(a): $i = 1$ refers to K^- or $\bar{K}^0$ and $i = 2, 3$ denotes the two nucleons.

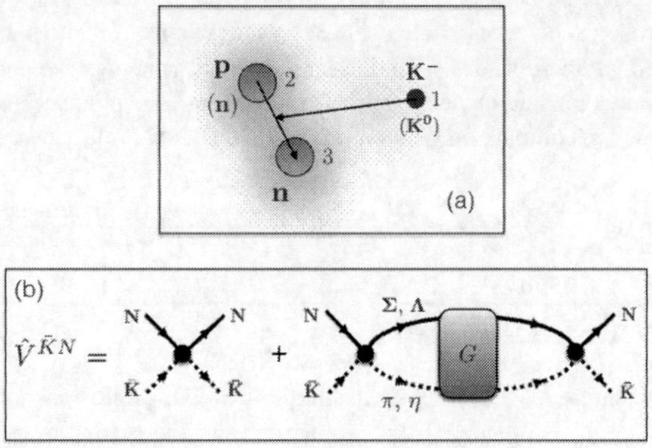

Figure 2. (a) Degrees of freedom and notations used in the description of the $K^- pn \leftrightarrow \bar{K}^0 nn$ three-body coupled-channels problem; (b) Construction of the complex effective $\bar{K}N$ potential with inclusion of $\bar{K}N$ couplings to π- and η-hyperon (Λ, Σ) continuum states.

Here $\hat{T}_i$ are the kinetic energies of the antikaon and the two nucleons. The kinetic energy $\hat{T}_{\rm cm}$ associated with the three-body center-of-mass is properly subtracted as usual. The interaction terms include the nucleon-nucleon potential, $\hat{V}^{NN}$, the electromagnetic interaction $\hat{V}^{EM}$ (i.e. the $K^- p$ Coulomb potential) and the effective antikaon-nucleon potential $\hat{V}^{\bar{K}N}$ that will now be further explained. Also included is the mass difference, ΔM, between the $K^- pn$ and $\bar{K}^0 nn$ channels. We use physical masses for nucleons and antikaons: $M_p = 938.272$ MeV, $M_n = 939.565$ MeV, $M_{K^-} = 493.677$ MeV and $M_{\bar{K}^0} = 497.648$ MeV, so as to take into account isospin-breaking effects.

2.1 Antikaon-nucleon effective potential

The present work employs the complex and energy dependent Kyoto $\bar{K}N$ potential [11], $\hat{V}^{\bar{K}N}$, as the basic antikaon-nucleon effective interaction. It incorporates important dynamics such as the coupling of the $K^- p$ and $\bar{K}^0 n$ channels to $\pi\Sigma$ and $\pi\Lambda$ continuum states as illustrated in Fig. 2(b). Effects of the decay processes into $\pi\Sigma$ and $\pi\Lambda$ are encoded in the imaginary part of $\hat{V}^{\bar{K}N}$. The r-space representation of the potential is parametrised in Gaussian form, $\hat{V}^{\bar{K}N}(r, E) \sim e^{-r^2/b^2} U(E)$, with $b \simeq 0.4$ fm. The strength of the interaction depends on the energy $E \equiv E_{\bar{K}N}$, treated self-consistently in the two-body $\bar{K}N$ Schrödinger equation. This potential is based on chiral SU(3) effective field theory [12] and constrained by empirical $\bar{K}N$ data. Thanks to systematic improvements with inclusion of higher order terms, $\hat{V}^{\bar{K}N}$ reproduces very well all available low-energy $\bar{K}N$ data ($K^- p$ total and reaction cross sections, branching ratios at $K^- p$ threshold and the SIDDHARTA data) [12] with an accuracy of $\chi^2/\text{d.o.f.} \simeq 1$.

The Kyoto $\bar{K}N$ potential in its original form was written in the isospin basis from which the particle basis potential is constructed as

$$\hat{V}^{\bar{K}N}_{ij} = \frac{1}{2}\left[\hat{V}^{\bar{K}N(I=0)}_{ij} + \hat{V}^{\bar{K}N(I=1)}_{ij}\right] - \frac{1}{2}\left[\hat{V}^{\bar{K}N(I=0)}_{ij} - \hat{V}^{\bar{K}N(I=1)}_{ij}\right] P^{ij}_{\tau} \; . \tag{4}$$

The Heisenberg operator $P^{ij}_{\tau} = (1 + \tau_i \cdot \tau_j)/2$ exchanges the i- and j-th particles in the isospin wave function. In the particle basis used throughout this work, the exchange operator acts as $P^{ij}_{\tau}|K^- n\rangle =$

Table 1. Test of the effective $\bar{K}N$ potential $\hat{V}^{\bar{K}N}$ (the Kyoto potential [11]) for the strong-interaction energy shift and width of kaonic hydrogen. The coupled-channels Schrödinger equation (1) including the K^-p Coulomb interaction has been solved with a self-consistent determination of the energy-dependence of $\hat{V}^{\bar{K}N}(E)$. Values calculated with physical antikaon and nucleon masses (1st row) and isospin-averaged masses (2nd row) are shown in comparison with empirical (SIDDHARTA [1]) results.

Masses	ΔE [eV]	Γ [eV]
physical	283	607
isospin	163	574
Exp. [1]	$283 \pm 36 \pm 6$	$541 \pm 89 \pm 22$

Table 2. Scattering lengths in the $\bar{K}N$ channels from solutions of the coupled-channels Schrödinger equation with the Kyoto effective potential $\hat{V}^{\bar{K}N}$, using physical and isospin-averaged masses.

Masses	a_{K^-p} [fm]	$a_{K^-p-\bar{K}^0n}$ [fm]	$a_{\bar{K}^0n}$ [fm]	a_{K^-n} [fm]
physical	$-0.66 + i0.89$	$-0.85 + i0.26$	$-0.40 + i1.03$	$0.58 + i0.78$
isospin	$-0.40 + i0.81$	$-0.99 + i0.04$	$-0.40 + i0.81$	$0.58 + i0.77$

$|K^-n\rangle$, $P_\tau^{ij}|K^-p\rangle = -|\bar{K}^0n\rangle$ and $P_\tau^{ij}|\bar{K}^0n\rangle = -|K^-p\rangle$. The charge exchange channel-coupling between K^-p and $\bar{K}^0n$ occurs through the isospin dependence of the $\bar{K}N$ interaction.

Before proceeding to the calculation of kaonic deuterium, the input Kyoto effective potential has to be tested in comparison with kaonic hydrogen data and $\bar{K}N$ scattering lengths. We recall that the fitting to the kaonic hydrogen results in Refs. [12] was performed making use of the improved Deser formula [14], whereas in Ref. [10] the $\bar{K}N$ coupled-channels Schrödinger equation has been solved to evaluate the K^- hydrogen level shift and width. The threshold energy difference between the K^-p and $\bar{K}^0n$ channels of about 5 MeV must properly be taken into account in the level shift calculation. It is therefore instructive to compare calculations using either physical masses or isospin averaged masses of the $(K^-, \bar{K}^0)$ and (p, n) doublets.

Results are displayed in Tables 1 and 2. Isospin-breaking mass effects are evidently important in the $\bar{K}N$ threshold physics that generates the 1s energy shift and width of kaonic hydrogen. In particular, the differences between scattering lengths obtained with physical versus isospin-averaged masses translate into a corresponding large difference in ΔE_{1s}. The computation using isospin-averaged masses cannot account for the observed energy shift.

Concerning the energy dependence of $\hat{V}^{\bar{K}N}(E)$, for kaonic hydrogen the results obtained by setting $E \equiv E_{\bar{K}N} = 0$ at threshold as input turn out to be equal to those using the self-consistent value of E. For kaonic deuterium, determining this energy of the two-body $\bar{K}N$ subsystem within the bound three-body system is a more involved issue that requires further discussion.

2.2 Solving the three-body Schrödinger equation

The coupled-channels Schrödinger equation (1) is solved using a variational approach with a large set of basis functions of the generic form

$$\Phi = \mathcal{A}[\psi^{(\text{space})} \otimes \psi^{(\text{spin})} \otimes \psi^{(\text{isospin})}] \ . \tag{5}$$

Correlated Gaussian (CG) basis functions [15, 16] are used to expand the radial part of Φ. This approach is sufficiently flexible to describe both short- and long-range properties of the wave function accurately, a necessary condition when dealing with systems such as kaonic deuterium in which very different distance scales must be handled simultaneously. The so-called global vector representation [17] is applied to describe rotational degrees of freedom, such that the form of the variational wave function does not change under linear transformations of the three-body coordinates. This permits to optimize choices of the Jacobi coordinates for the $K^- pn$ and $\bar{K}^0 nn$ channels with their different masses. Details can be found in Ref. [10].

As repeatedly emphasized, short-range strong interactions together with the long-range Coulomb force have to be treated with high numerical precision. In order to extract the detailed effects of the $\bar{K}N$ interaction from the spectrum of kaonic deuterium, its binding energy needs to be calculated with an accuracy of a few eV, a considerable computational challenge. The required precision is achieved by expanding the variational wave function (5) in a very large set of basis functions, $\{\Phi_\alpha\}$. Then the generalized eigenvalue problem

$$\sum_{\beta=1}^{N} (H_{\alpha\beta} - E B_{\alpha\beta}) C_\beta = 0, \tag{6}$$

is solved to determine the coefficients C_α and the energy E, with the Hamiltonian matrix $H_{\alpha\beta} = \langle\Phi_\alpha|H|\Phi_\beta\rangle$ and the overlap matrix $B_{\alpha\beta} = \langle\Phi_\alpha|\Phi_\beta\rangle$. Here N is the number of basis functions. To achieve energy convergence for the kaonic atom, it turns out that basis functions reaching over distance scales from a tenth to several hundreds of fm need to be included. In practice [10], convergence at the required precision is reached with $N \sim 3000$. The primary basis sizes used extend up to $N \sim 4000$.

In the actual kaonic deuterium calculations, with the energy-dependent Kyoto potential [11] as input, fixing the two-body energy $E_{\bar{K}N}$ is a non-trivial issue. While the choice $E_{\bar{K}N} = 0$ at the $\bar{K}N$ threshold is justified for kaonic hydrogen, the energy of the $\bar{K}N$ two-body subsystem within the $K^- d$ three-body system is not a well-defined concept. Different prescriptions [18–20] are available to take into account the motion of the bound nucleons while they interact with the antikaon. In our work [10] we refer to Refs. [18, 20] and start by setting $E = E_{\bar{K}N} = 0$ in the two-body potential $\hat{V}^{\bar{K}N}(E)$ also for kaonic deuterium, the choice we take as default input in the three-body calculations. It is argued in Ref. [10] that leading binding corrections to this minimal choice imply a small downward shift in the energy variable, $E \simeq -B/2$, with the deuteron binding energy $B = 2.2$ MeV, in contrast to the much larger shift suggested in Ref. [19].

3 Results for kaonic deuterium

Table 3 lists the energies of the 1s, 2p and 2s states of kaonic deuterium computed by solving the three-body equations (1). We recall that the reference energies of the $K^- d$ pure Coulomb two-body system with deuteron point charge are:

$$E_{1s} = -10.406 \text{ keV} \ , \quad E_{2p} = E_{2s} = -2.602 \text{ keV} \ . \tag{7}$$

In the three-body calculation with Coulomb interaction only, these energies are slightly shifted by the finite extension of the deuteron charge density. With inclusion of the strongly interacting $K^- pn$ and $\bar{K}^0 nn$ coupled channels, i.e. by turning on the effective $\bar{K}N$ potential $\hat{V}^{\bar{K}N}(E = 0)$, the $1s$ level experiences a complex energy shift while the $2p$ level remains unchanged. As expected the strong-interaction shift of the $1s$ energy can be directly determined by measuring the $2p \to 1s$ transition (the K_α line). In summary, our predicted kaonic deuterium $1s$ level shift is (up to corrections to be discussed below):

$$\Delta E - i\frac{\Gamma}{2} = (670 - i\,508) \text{ eV} \ , \tag{8}$$

using the Kyoto $\bar{K}N$ potential. These values are roughly consistent with those found in a recent Faddeev calculation [9], although the basic interactions used there are different from ours.

Table 3. Calculated energy spectrum of kaonic deuterium. Results of the three-body calculation are shown with Coulomb interaction only (upper row) and with inclusion of the strong $\bar{K}N$ interaction (lower row). Energies of the $1s$, $2p$ and $2s$ states are given relative to the K^-d threshold, $m_{K^-} + M_d$, with the deuteron mass M_d = 1875.613 MeV.

	E_{1s} [keV]	E_{2p} [keV]	E_{2s} [keV]
Coulomb	−10.398	−2.602	−2.600
Coulomb+$\bar{K}N$	-9.736−i0.508	-2.602−i0.000	-2.517−i0.067

The results just outlined have been obtained using physical masses for the antikaons and for proton and neutron. Notably, a calculation with isospin-averaged masses yields a $1s$ energy shift and width that differ by only a few eV from the previously mentioned values. Unlike the situation in kaonic hydrogen, effects of isospin breaking turn out to be small in kaonic deuterium. This feature is further discussed and explained in Ref. [10].

Up to this point the determination of the width Γ incorporates the decay channels $\bar{K}N \to \pi Y$, where Y stands for Λ and Σ hyperons. The question arises about possible additional contributions to the width from antikaon absorption on two nucleons, with the coupled K^-pn and $\bar{K}^0nn$ channels decaying into $\Lambda n + \Sigma^0 n + \Sigma^- p$. Early measurements at Brookhaven with K^- stopped on liquid deuterium in the BNL bubble chamber [21] demonstrated that these processes are strongly suppressed as compared to the leading single-nucleon channels, $\bar{K}N \to \pi Y$. The ratio of two-nucleon absorption reactions to the single-nucleon processes was found to be as small as $(1.2 \pm 0.1)\%$ [21]. Taking this value for orientation, the kaonic deuterium $1s$ width would increase through two-nucleon absorption by only about 10 eV, a correction that can be safely neglected within an estimated uncertainty range of approximately 10 % assigned to the calculated width of about a keV. The smallness of the two-body absorptive width can be understood as follows. Kinematical conditions for the $\bar{K}NN \to YN$ process require a large momentum transfer of order 1 GeV/c to be provided by the initial deuteron wave function at short distances. The probability is low for this to take place in such a weakly bound, dilute system.

A further source of possible uncertainties is related to the energy dependence of the $\bar{K}N$ potential, $\hat{V}^{\bar{K}N}(E_{\bar{K}N})$. As mentioned, $E_{\bar{K}N} = 0$ at threshold was set in the present calculations. The binding of the nucleons in the deuteron may cause a shift of $E_{\bar{K}N}$ towards the subthreshold region. Our estimate, derived and discussed in the Appendix of Ref. [10], suggests a small average shift, $E_{\bar{K}N} = -B_d/2 \sim -1.1$ MeV, involving the deuteron binding energy B_d. The changes thus induced by correcting the energy dependence in $\hat{V}^{\bar{K}N}(E_{\bar{K}N})$ for deuteron binding tend to increase the width of kaonic deuterium by about 10%, while the corresponding energy shift changes by only about 5%. Nonetheless, a more detailed treatment of such kinematical corrections within the $\bar{K}NN$ three-body problem is of some importance, given the strong subthreshold energy dependence of the K^-p scattering amplitude involving the $\Lambda(1405)$.

4 Test of improved Deser formulae for kaonic deuterium

The improved Deser formula [4, 14], derived from non-relativistic effective field theory (EFT), is frequently used in the investigation of strong-interaction effects in hadronic atoms. This relation connects the $1s$ level shift ΔE and width Γ of a kaonic atom with the K^--nucleus scattering length, a, as follows:

$$\Delta E - \frac{i\Gamma}{2} = -2\mu^2\alpha^3 a\,[1 - 2\mu\alpha(\ln\alpha - a],\qquad (9)$$

Table 4. Strong-interaction energy shift and width of the kaonic hydrogen $1s$ state, obtained by solving the two-body Schrödinger equation with the Kyoto $\bar{K}N$ potential, and by using the improved Deser formula and its resummed version.

Kaonic hydrogen	ΔE [eV]	Γ [eV]
Full Schrödinger equation	283	607
Improved Deser formula (9)	293	596
Resummed formula (10)	284	605

Table 5. Strong-interaction energy shift and width of the kaonic deuterium $1s$ state, obtained by solving the three-body Schrödinger equation with the Kyoto $\bar{K}N$ potential, and by using the improved Deser formula and its resummed version.

Kaonic deuterium	ΔE [eV]	Γ [eV]
Full Schrödinger equation	670	1016
Improved Deser formula (9)	910	989
Resummed formula (10)	818	1188

where μ is the kaon-nucleus reduced mass and α is the fine structure constant. The logarithmically enhanced correction term can be resummed to all orders [22], providing a "double-improved" Deser formula:

$$\Delta E - \frac{i\Gamma}{2} = -\frac{2\mu^2\alpha^3 a}{1 + 2\mu\alpha(\ln\alpha - 1)\,a} . \qquad (10)$$

A first check for kaonic hydrogen (see Table 4) using the calculated K^-p scattering length of Table 2 points out that the modified Deser formula (9) works well in this case, with a deviation of just about 10 eV from the full solution of the $\bar{K}N$ Schrödinger equation. The resummed version (10) improves the comparison further, down to a difference of only 2 eV.

Next we compare the results of the full three-body calculation for kaonic deuterium with the estimates derived from Eqs. (9) and (10). This requires as input the K^-d scattering length, $a_{K\text{-}d}$. In the fixed center approximation (FCA) for the nucleons, $a_{K\text{-}d}$ derived from a multiple scattering series is given as [4, 23]

$$a_{K\text{-}d} = \frac{\mu_{K\text{-}d}}{m_{K^-}} \int d^3r\, \rho_d(r)\, \tilde{a}_{K\text{-}d}(r) , \qquad (11)$$

$$\tilde{a}_{K\text{-}d}(r) = \frac{\tilde{a}_p + \tilde{a}_n + (2\tilde{a}_p\tilde{a}_n - \tilde{a}_{\text{ex}}^2)/r - 2\tilde{a}_{\text{ex}}^2\tilde{a}_n/r^2}{1 - \tilde{a}_p\tilde{a}_n/r^2 + \tilde{a}_{\text{ex}}^2\tilde{a}_n/r^3} , \qquad (12)$$

with the K^--deuteron reduced mass $\mu_{K\text{-}d}$, and $\rho_d(r)$ is the nucleon density distribution in the deuteron. The lab-frame scattering lengths are denoted $\tilde{a}_p \equiv \tilde{a}_{K^-p}$, $\tilde{a}_n \equiv \tilde{a}_{K^-n}$, and $\tilde{a}_{\text{ex}}^2 \equiv \tilde{a}_{K^-p\text{-}\bar{K}^0n}^2/(1 + \tilde{a}_{\bar{K}^0n}/r)$ refers to $K^-p \leftrightarrow \bar{K}^0n$ charge exchange within the multiple scattering chain. The lab frame quantities are related to the standard scattering lengths in the c.m. frame by $\tilde{a}_{\bar{K}N} \equiv \frac{m_K}{\mu_{\bar{K}N}} a_{\bar{K}N}$ with the $\bar{K}N$ reduced mass $\mu_{\bar{K}N}$. With the set of scattering lengths calculated using the Kyoto $\bar{K}N$ potential and listed in Table 2, the K^-d scattering length derived from Eqs. (11) and (12) is then:

$$a_{K\text{-}d} = (-1.42 + i\,1.60)\,\text{fm} . \qquad (13)$$

Using this value of $a_{K\text{-}d}$ we apply the improved Deser formulae to kaonic deuterium. The results are summarized in Table 5 together with those from the full three-body calculation. Unlike the kaonic hydrogen case, it is evident that estimates based on both versions, Eqs. (9) and (10), of the Deser formulae are not reliable for kaonic deuterium. In particular, the deviations in the energy shift ΔE from the full three-body calculation are significantly larger than 100 eV. Of course the K^-d scattering length (13) is estimated in the FCA limit, hence it is expected to differ from the "exact" a_{K^-d}. The importance of nucleon recoil corrections, naturally included in the full three-body calculation but neglected in FCA, is discussed in Refs. [22, 24].

5 Summary and conclusions

- Accurate three-body variational calculations have been performed for the spectrum of kaonic deuterium and the evaluation of the $1s$ level shift and width. The $\bar{K}NN$ three-body wave function is expressed as a superposition of a large set of correlated Gaussian basis functions. The computational challenge, namely a high-precision treatment of both short-range strong interactions and long-range Coulomb force, requires a very large model space covering all distance scales from 0.1 fm to several hundreds of fm.
- The $\bar{K}N$ strong interaction is treated in terms of a complex effective potential that accurately reproduces previous results of coupled-channels calculations based on chiral $SU(3)$ dynamics. Among the computed energies of $1s$, $2s$ and $2p$ states in kaonic deuterium the $\bar{K}N$ strong interaction affects only the s states, inducing energy shifts from the levels characteristic of the pure Coulomb and point charge limit of the K^-d atomic system. No energy shift is found for the $2p$ state, so that the $1s$ level shift can be directly associated with the transition energy from the $2p$ to the $1s$ state. The calculated $1s$ level shift and width of kaonic deuterium is $\Delta E - i\Gamma/2 = (670 - i\,508)$ eV, corresponding to a $2p \to 1s$ transition energy of 7.134 keV. Following discussions in Ref. [10] we assign uncertainties of about 10% to Γ and less than 10% to ΔE (not counting the approximately 20% uncertainties in the empirical SIDDHARTA kaonic hydrogen constraints).
- In view of upcoming experimental investigations we have also performed a detailed test of the sensitivity of kaonic deuterium observables with respect to the $I = 1$ component in the $\bar{K}N$ interaction, by varying selectively the real part of the $I = 1$ $\bar{K}N$ potential strength within the uncertainty limits deduced from the kaonic hydrogen data. One can conclude from this test that the $1s$ level shift of kaonic deuterium is indeed expected to provide a significantly improved constraint on the $I = 1$ component, as compared to the SIDDHARTA kaonic hydrogen measurement [1], if the kaonic deuterium level shift can be determined within $\sim 25\%$ accuracy (corresponding to $\sim 2\%$ in the $2p \to 1s$ transition energy). This sets the physics focus on the yet basically unknown K^-neutron sector of the low-energy $\bar{K}N$ interaction.

5.1 Acknowledgments

We thank Tsubasa Hoshino and Shota Ohnishi for their crucial contributions to the computational tasks of this work. The numerical calculations were in part performed on the supercomputer (CRAY XC40) at the Yukawa Institute for Theoretical Physics (YITP), Kyoto University. This work is in part supported by the Grants-in-Aid for Scientific Research on Innovative Areas from MEXT (Grant No. 2404:24105008), by JSPS KAKENHI Grant No. JP16K17694, by the Yukawa International Program for Quark-Hadron Sciences (YIPQS), and by the DFG Cluster of Excellence *Origin and Structure of the Universe*. One of the authors (W. W.) gratefully acknowledges the hospitality of YITP and Kyoto University during his three-months visit in 2017.

References

[1] M. Bazzi *et al.*, Phys. Lett. **B704**, 113 (2011); Nucl. Phys. **A881**, 88 (2012).

[2] C. Curceanu *et al.*, Nucl. Phys. **A914**, 251 (2013);
M. Iliescu *et al.*, J. Phys. Conf. Ser. **770**, 012034 (2016).

[3] J. Zmeskal *et al.*, Acta Phys. Polon. **B46**, 101 (2015).

[4] U. G. Meißner, U. Raha, and A. Rusetsky, Eur. Phys. J. **C47**, 473 (2006).

[5] A. Gal, Int. J. Mod. Phys. **A22**, 226 (2007).

[6] M. Döring and U.-G. Meißner, Phys. Lett. **B704**, 663 (2011).

[7] A. Bahaoui, C. Fayard, T. Mizutani, and B. Saghai, Phys. Rev. **C68**, 064001 (2003).

[8] P. Doleschall, J. Révai, and N.V. Shevchenko, Phys. Lett. **B744**, 105 (2015).

[9] J. Révai, Phys. Rev. **C94**, 054001 (2016).

[10] T. Hoshino, S. Ohnishi. W. Horiuchi, T. Hyodo, and W. Weise, Phys. Rev. **C96**, 045204 (2017).

[11] K. Miyahara and T. Hyodo, Phys. Rev. **C93**, 015201 (2016).

[12] Y. Ikeda, T. Hyodo, and W. Weise, Phys. Lett. **B706**, 63 (2011); Nucl. Phys. **A881**, 98 (2012).

[13] Y. Kamiya, K. Miyahara, S. Ohnishi, Y. Ikeda, T. Hyodo, E. Oset, and W. Weise,
Nucl. Phys. **A954**, 41 (2016).

[14] U. G. Meißner, U. Raha, and A. Rusetsky, Eur. Phys. J. **C35**, 349 (2004).

[15] K. Varga and Y. Suzuki, Phys. Rev. **C52**, 2885 (1995).

[16] J. Mitroy, S. Bubin, W. Horiuchi, Y. Suzuki, L. Adamowicz, W. Cencek, K. Szalewicz,
J. Komasa, D. Blume, and K. Varga, Rev. Mod. Phys. **85**, 693 (2013).

[17] Y. Suzuki, W. Horiuchi, M. Orabi, and K. Arai, Few-Body Syst. **42**, 33 (2008).

[18] A. Doté, T. Hyodo, and W. Weise, Nucl. Phys. **A804**, 197 (2008);
Phys. Rev. **C79**, 014003 (2009).

[19] N. Barnea, A. Gal, and E. Liverts, Phys. Lett. **B712**, 132 (2012).

[20] S. Ohnishi, W. Horiuchi, T. Hoshino, K. Miyahara, and T. Hyodo,
Phys. Rev. **C95**, 044320 (2017).

[21] V. R. Veirs and R. A. Burnstein, Phys. Rev. **D1**, 1883 (1970).

[22] V. Baru, E. Epelbaum, and A. Rusetsky, Eur. Phys. J. **A42**, 111 (2009).

[23] S. S. Kamalov, E. Oset, and A. Ramos, Nucl. Phys. **A690**, 494 (2001).

[24] M. Mai, V. Baru, E. Epelbaum, and A. Rusetsky, Phys. Rev. **D91**, 054016 (2015).

EPJ Web of Conferences **181**, 01007 (2018) https://doi.org/10.1051/epjconf/201818101007
EXA2017

Antiproton–nucleus quasi-bound states within the 2009 version of the Paris $\bar{N}N$ potential

Jaroslava Hrtánková[1,*] and *Jiří* Mareš[1]

[1]Nuclear Physics Institute, 250 68 Řež, Czech Republic

Abstract. We studied the $\bar{p}$ interactions with the nuclear medium within the 2009 version of the Paris $\bar{N}N$ potential model. We constructed the $\bar{p}$–nucleus optical potential using the Paris S- and P-wave $\bar{p}N$ scattering amplitudes and treated their strong energy and density dependence self-consistently. We considered a phenomenological P-wave term as well. We calculated $\bar{p}$ binding energies and widths of the $\bar{p}$ bound in various nuclei. The P-wave potential has very small effect on the calculated $\bar{p}$ binding energies, however, it reduces the corresponding widths noticeably. Moreover, the S-wave potential based on the Paris amplitudes supplemented by a phenomenological P-wave term yields the $\bar{p}$ binding energies and widths in very good agreement with those obtained within the RMF model consistent with $\bar{p}$-atom data.

1 Introduction

The antiproton–nucleus interaction below threshold have been so far studied within phenomenological RMF approaches [1, 2]. The G-parity motivated $\bar{p}$ coupling constants were used to construct the $\bar{p}$–nucleus potential. The absorption of $\bar{p}$ was accounted for in terms of a purely phenomenological optical potential. The $\bar{p}$ optical potential was confronted with $\bar{p}$ atom data. It was found that the $\bar{p}$ coupling constant have to be properly scaled in order to be consistent with the data. Consequently, the $\bar{p}$ potential was applied in the calculations of $\bar{p}$ quasi-bound states in various nuclei [2].

However, it is desirable to study the $\bar{p}$ interactions with the nuclear medium within other theoretical approaches, such as microscopic models of $\bar{N}N$ interaction based on meson-exchange models [3–5] or chiral $\bar{N}N$ interaction models [6, 7]. Comparison between these $\bar{N}N$ interaction models could bring valuable information about in-medium $\bar{p}$ interactions in the direct confrontation with the data from $\bar{p}$ atoms and $\bar{p}$ scattering off nuclei, as well as predictions for $\bar{p}$-nuclear quasi-bound states.

Recently, the 2009 version of the Paris $\bar{N}N$ potential [3] was confronted by Friedman *et al.* with the $\bar{p}$-atom data and antinucleon interactions with nuclei up to 400 MeV/c, including elastic scattering and annihilation cross sections [8]. The analysis revealed the necessity to include the P-wave interaction in order to describe the $\bar{p}$ atom data. The Paris S-wave potential supplemented by a phenomenological P-wave term was found to fit the data on low-density, near-threshold $\bar{p}$-nucleus interaction. This fact stimulated us to apply it in the present calculations of $\bar{p}$-nuclear quasi-bound states and explore the effect of the P-wave interaction on $\bar{p}$ binding energies and widths of $\bar{p}$-nuclear states.

*e-mail: hrtankova@ujf.cas.cz

In Section 2, we briefly introduce the model applied in our calculations. Section 3 presents few representative results together with the discussion of the main findings of our study.

2 Methodology

The binding energies $B_{\bar{p}}$ and widths $\Gamma_{\bar{p}}$ of $\bar{p}$ quasi-bound states in a nucleus are obtained by solving self-consistently the Dirac equation with the optical potential

$$[-i\alpha \cdot \nabla + \beta m_{\bar{p}} + V_{\rm opt}(r)]\psi_{\bar{p}} = \epsilon_{\bar{p}}\psi_{\bar{p}}, \qquad (1)$$

where $m_{\bar{p}}$ is the mass of the antiproton and $\epsilon_{\bar{p}} = -B_{\bar{p}} - {\rm i}\Gamma_{\bar{p}}/2$ ($B_{\bar{p}} > 0$). The S-wave $\bar{p}$–nucleus optical potential $V_{\rm opt}$ enters the Dirac equation as the time component of a 4-vector and is constructed in a '$t\rho$' form as follows:

$$2E_{\bar{p}}V_{\rm opt}(r) = -4\pi\left(F_0\frac{1}{2}\rho_p(r) + F_1\left(\frac{1}{2}\rho_p(r) + \rho_n(r)\right)\right) . \qquad (2)$$

Here, $E_{\bar{p}} = m_{\bar{p}} - B_{\bar{p}}$, F_0 and F_1 are isospin 0 and 1 in-medium amplitudes, and $\rho_p(r)$ [$\rho_n(r)$] is the proton (neutron) density distribution calculated within the RMF NL-SH model [9]. The in-medium amplitudes F_0 and F_1 entering Eq. (2) account for Pauli correlations in the nuclear medium. They are constructed from the free-space $\bar{p}N$ amplitudes using the multiple scattering approach of Wass *et al.* [10] (WRW)

$$F_1 = \frac{\frac{\sqrt{s}}{m_N} f^{\rm S}_{\bar{p}n}(\sqrt{s})}{1 + \frac{1}{4}\xi_k \frac{\sqrt{s}}{m_N} f^{\rm S}_{\bar{p}n}(\sqrt{s})\rho(r)}, \quad F_0 = \frac{\frac{\sqrt{s}}{m_N}[2f^{\rm S}_{\bar{p}p}(\sqrt{s}) - f^{\rm S}_{\bar{p}n}(\sqrt{s})]}{1 + \frac{1}{4}\xi_k \frac{\sqrt{s}}{m_N}[2f^{\rm S}_{\bar{p}p}(\sqrt{s}) - f^{\rm S}_{\bar{p}n}(\sqrt{s})]\rho(r)} . \qquad (3)$$

Here, $f^{\rm S}_{\bar{p}n}$ ($f^S_{\bar{p}p}$) denotes the free-space c.m. $\bar{p}n$ ($\bar{p}p$) S-wave scattering amplitude derived from the Paris $\bar{N}N$ potential as a function of Mandelstam variable $\sqrt{s}$, m_N represents the mass of the nucleon and $\rho(r) = \rho_p(r) + \rho_n(r)$. The factor $\sqrt{s}/m_N$ transforms the amplitudes from the two-body frame to the $\bar{p}$–nucleus frame. The Pauli correlation factor ξ_k is defined as follows

$$\xi_k = \frac{9\pi}{k_{\rm F}^2}\left(4\int_0^\infty \frac{dr}{r}\exp(ikr)j_1^2(k_{\rm F}r)\right) , \qquad (4)$$

where $j_1(k_{\rm F}r)$ is the spherical Bessel function, $k_{\rm F}$ is the Fermi momentum and $k = \sqrt{(\epsilon_{\bar{p}} + m_{\bar{p}})^2 - m_{\bar{p}}^2}$ is the antiproton momentum. The integral in Eq.(4) can be solved analytically. The resulting expression is of the form

$$\xi_k = \frac{9\pi}{k_{\rm F}^2}\left[1 - \frac{q^2}{6} + \frac{q^2}{4}\left(2 + \frac{q^2}{6}\right)\ln\left(1 + \frac{4}{q^2}\right) - \frac{4}{3}q\left(\frac{\pi}{2} - \arctan\left(\frac{q}{2}\right)\right)\right] , \qquad (5)$$

where $q = -ik/k_{\rm F}$.

The analysis of $\bar{p}$ atom data [8] revealed that it is necessary to supplement the Paris S-wave potential by the P-wave interaction to make the real $\bar{p}$ potential attractive in the relevant low-density region of a nucleus. To incorporate the P-wave interaction in our model we supplement the r.h.s. of the S-wave optical potential in Eq. (2) [$2E_{\bar{p}}V^S_{\rm opt} = q(r)$] by a gradient term [8]:

$$2E_{\bar{p}}V_{\rm opt}(r) = q(r) + 3\nabla \cdot \alpha(r)\nabla . \qquad (6)$$

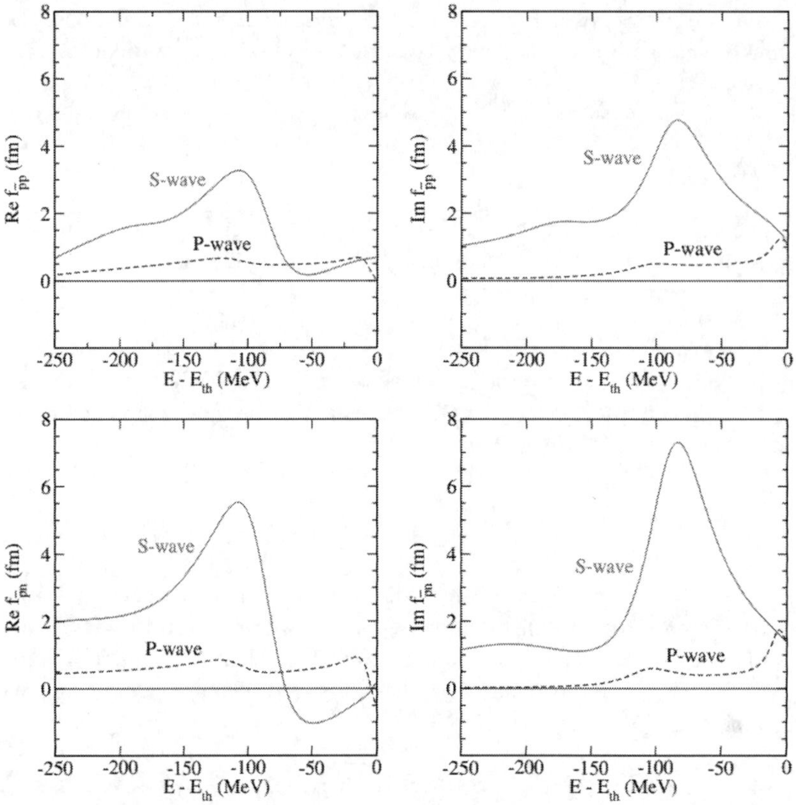

Figure 1. Energy dependence of real (left) and imaginary (right) parts of the Paris 09 $\bar{p}p$ (top) and $\bar{p}n$ (bottom) two-body c.m. scattering amplitudes used in the present calculations: in-medium (Pauli blocked) S-wave amplitudes at $\rho_0 = 0.17$ fm^{-3} and free-space P-wave amplitudes.

The factor $2l + 1 = 3$ in the P-wave part is introduced to match the normalization of the Paris $\bar{N}N$ scattering amplitudes and

$$\alpha(r) = 4\pi \frac{m_N}{\sqrt{s}} \left(f^P_{\bar{p}p}(\sqrt{s})\rho_p(r) + f^P_{\bar{p}n}(\sqrt{s})\rho_n(r) \right) . \tag{7}$$

Here, $f^P_{\bar{p}p}(\sqrt{s})$ and $f^P_{\bar{p}n}(\sqrt{s})$ represent the Paris P-wave $\bar{p}p$ and $\bar{p}n$ free-space c.m. scattering amplitudes, respectively. We do not consider any medium modifications of the P-wave amplitudes since we assume that the P-wave potential should contribute mainly near the surface of the nucleus due to its gradient form.

The analysis of Ref. [8] also revealed that the optical potential constructed from the Paris S- and P-wave amplitudes fails to reproduce the $\bar{p}$ atom data and that it is mainly due to the P-wave amplitude — its real and imaginary parts had to be scaled by different factors to get reasonable fit. On the contrary, the optical potential based on the Paris S-wave potential supplemented by a purely phenomenological P-wave term with $f^P_{\bar{p}N} = 2.9 + i1.8$ fm^3 fits the data well. In our calculations, we adopt both P-wave amplitudes, Paris as well as phenomenological, in order to study their effect on the binding energies and widths of $\bar{p}$-nuclear states.

The Paris amplitudes used in our calculations are shown in Fig. 1. There are $\bar{p}p$ (top) and $\bar{p}n$ (bottom) medium modified S-wave amplitudes (3) at saturation density $\rho_0 = 0.17$ fm^{-3}

and free-space P-wave scattering amplitudes plotted as a function of the energy shift $\delta\sqrt{s} = E - E_{\rm th}$ with $E_{\rm th} = m_{\bar{p}} + m_N$. The S-wave amplitudes vary considerably with energy below threshold. The real in-medium $\bar{p}p$ amplitude is attractive in the entire energy region below threshold. The real part of the in-medium $\bar{p}n$ amplitude is attractive for $\delta\sqrt{s} \leq -70$ MeV with a small repulsive dip near threshold. The imaginary parts of the S-wave amplitudes are comparable or even larger than the corresponding real parts. The energy dependence of the free-space P-wave amplitudes is less pronounced than in the S-wave case. Moreover, the P-wave amplitudes are considerably smaller than the in-medium S-wave amplitudes in the region relevant to $\bar{p}$-nuclear states calculations.

Strong energy dependence of the $\bar{p}N$ amplitudes presented in Fig. 1 requires a proper self-consistent scheme for evaluating the $\bar{p}$ optical potential. The energy argument $\sqrt{s}$ of the amplitudes is expressed in the $\bar{p}$–nucleus frame where the contributions from antiproton and nucleon kinetic energies are not negligible [11]

$$\sqrt{s} = E_{\rm th}\left(1 - \frac{2(B_{\bar{p}} + B_{Nav})}{E_{\rm th}} + \frac{(B_{\bar{p}} + B_{Nav})^2}{E_{\rm th}^2} - \frac{T_{\bar{p}}}{E_{\rm th}} - \frac{T_{Nav}}{E_{\rm th}}\right)^{1/2} . \quad (8)$$

Here, $B_{Nav} = 8.5$ MeV and T_{Nav} are the average binding and kinetic energy per nucleon, respectively, and $T_{\bar{p}}$ represents the $\bar{p}$ kinetic energy. The kinetic energies are evaluated as corresponding expectation values of the kinetic energy operator $\hat{T} = -\frac{\hbar^2}{2m}\Delta$. Since the $B_{\bar{p}}$ appears as an argument in the $\sqrt{s}$, which in turn serves as an argument for $V_{\rm opt}$, $\sqrt{s}$ has to be determined self-consistently. Namely, its value obtained by solving Eq. (8) should agree with the value of $\sqrt{s}$ which serves as input in Eq. (3) and thus Eq. (1), as well.

3 Results

We performed self-consistent calculations of $\bar{p}$-nuclear quasi-bound states in selected nuclei within the model presented in the previous section. We explored the energy and density dependence of the S-wave $\bar{p}$–nucleus potential as well as the role of the $\bar{p}N$ P-wave interaction, and compared the predictions for $\bar{p}$ binding energies and widths with the phenomenological RMF approach [2].

The $\bar{p}N$ amplitudes are strongly energy and density dependent, as was shown in Fig. 1. Consequently, the depth and shape of the $\bar{p}$–nucleus potential depend greatly on the energies and densities pertinent to the processes under consideration. It is demonstrated in Fig. 2 where we present the $\bar{p}$ potential in ^{40}Ca calculated for different energies and densities: i) using the Paris free-space S-wave amplitudes at threshold (denoted by 'th free'), ii) using in-medium Paris S-wave amplitudes at threshold (denoted by 'th medium'), iii) using in-medium Paris S-wave amplitudes at energies relevant to $\bar{p}$ atoms (constructed following Ref. [8]), and iv) using in-medium Paris S-wave amplitudes at energies relevant to $\bar{p}$ nuclei [$\sqrt{s}$ of Eq. (8)]. The $\bar{p}$ potential constructed using the free-space amplitudes has a repulsive real part and fairly absorptive imaginary part. When the medium modifications of the amplitudes are taken into account, the $\bar{p}$ potential becomes attractive and more absorptive. At the energies relevant to $\bar{p}$ atoms, the $\bar{p}$ potential is more attractive and weakly absorptive. Finally, at the energies relevant to $\bar{p}$ nuclei, the $\bar{p}$ potential is strongly attractive, however, also strongly absorptive. The figure clearly shows that proper self-consistent evaluation of the energy $\sqrt{s}$ is essential.

Next, we performed static and dynamical calculations of $\bar{p}$ binding energies and widths using the Paris $\bar{N}N$ potential. In the static calculations, the core nucleus is not affected by the presence of extra $\bar{B}$. In the dynamical calculations, the polarization of the nuclear core due to $\bar{B}$, i.e., changes in the nucleon binding energies and densities, is taken into account. The response of the nuclear core to the extra antiproton is not instant — it could possibly

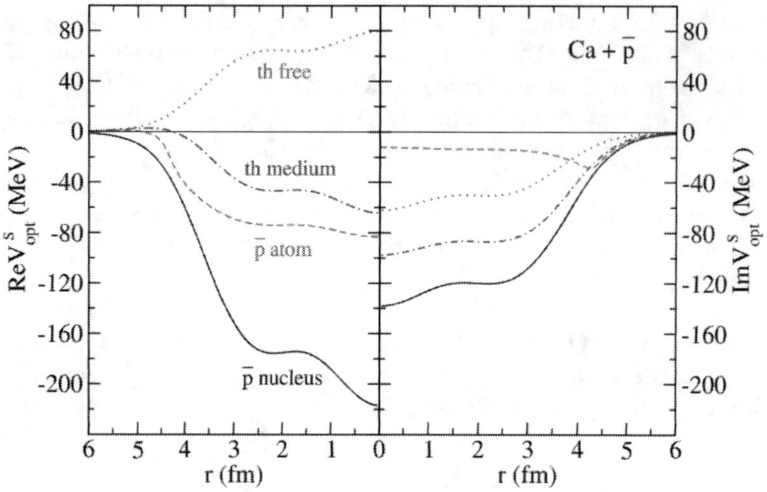

Figure 2. The potential felt by $\bar{p}$ at threshold ('th medium'), in the $\bar{p}$ atom and $\bar{p}$ nucleus, calculated for ^{40}Ca+$\bar{p}$ with in-medium Paris S-wave amplitudes and static RMF densities. The $\bar{p}$ potential calculated using free-space amplitudes at threshold is shown for comparison ('th free').

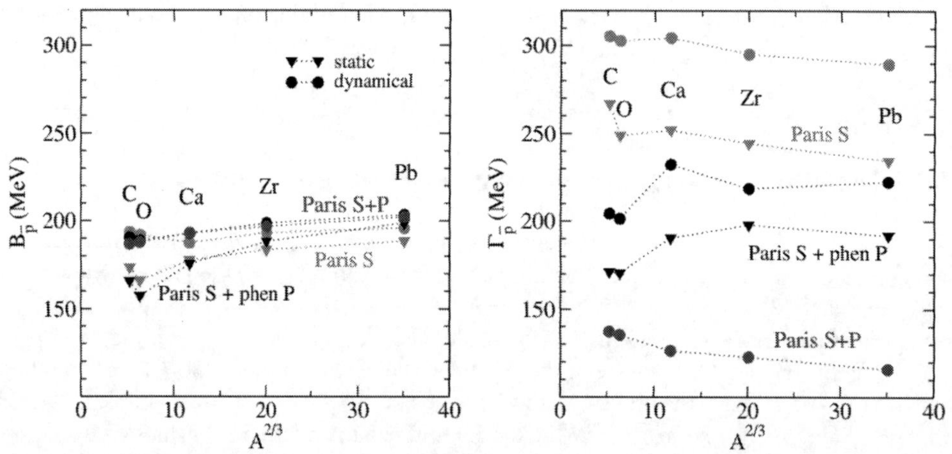

Figure 3. $1s$ $\bar{p}$ binding energies (left panel) and widths (right panel) in various nuclei, calculated statically (triangles) and dynamically (circles) using S-wave Paris potential (red) and including phenomenological P-wave potential (black). The $\bar{p}$ binding energies and widths calculated dynamically using the Paris S + P-wave potential (blue circles) are shown for comparison.

last longer than the lifetime of $\bar{p}$ inside a nucleus [12, 13]. As a result, the antiproton could annihilate before the nuclear core is fully polarized. Our static and dynamical calculations of $\bar{p}$ binding energies and widths may be thus considered as two limiting scenarios.

In Fig. 3, we present $1s$ $\bar{p}$ binding energies (left) and widths (right) as a function of mass number A, calculated statically (triangles) and dynamically (circles) with the Paris S-wave and Paris S-wave + phen. P-wave potentials. We present the $\bar{p}$ binding energies and widths calculated dynamically using the Paris S + P-wave potential for comparison as well.

In dynamical and static calculations alike, the P-wave interaction does not affect much the $\bar{p}$ binding energies — they are comparable with the binding energies evaluated using only the S-wave potential. On the other hand, the $\bar{p}$ widths are reduced significantly when the phenomenological P-wave term is included in the $\bar{p}$ optical potential. The effect is even more pronounced for the Paris P-wave interaction.

The $\bar{p}$ widths calculated dynamically are noticeably larger than the widths calculated statically. It is caused by the increase of the central nuclear density, which exceeds the decrease of the $\bar{p}N$ amplitudes due to the larger energy shift with respect to threshold ($\delta\sqrt{s} \sim -255$ MeV in the dynamical case vs. $\delta\sqrt{s} \sim -200$ MeV in the static case). On the other hand, the $\bar{p}$ binding energies increase only moderately and get closer to each other when the dynamical effects are taken into account. The $\bar{p}$ widths exhibit much large dispersion then the $\bar{p}$ binding energies for the different potentials.

We explored the $\bar{p}$ excited states in selected nuclei as well and compared the results with those obtained within the RMF approach [2]. Fig. 4 shows $\bar{p}$ spectra in ^{40}Ca calculated using the Paris S-wave + phen. P-wave potential and phenomenological RMF approach. The Paris S-wave + phen. P-wave potential yields the $1p$ and $1d$ binding energies slightly larger and thus the s-p and s-d level spacing smaller than the RMF approach. It is an effect of a broader $\bar{p}$ potential well generated by the Paris S-wave + phen. P-wave potential. Nevertheless, both approaches yield comparable $\bar{p}$ widths as well as energies and the overall agreement is surprisingly good.

It is to be noted that there is no spin-orbit splitting of the p and d levels presented in Fig. 4 since the $V_{\rm opt}$ is a central potential constructed from angular momentum-averaged scattering amplitudes. In the RMF approach, the $\bar{p}$ binding energies in $1p$ and $1d$ spin doublets are nearly degenerate, the difference in $\bar{p}$ energies (as well as $\bar{p}$ widths) is up to ~ 1 MeV. This is in agreement with spin symmetry in antinucleon spectra within the RMF approach [14, 15]. In the left panel of Fig. 4 we show the spin-averaged $1p$ and $1d$ $\bar{p}$ binding energies and widths for better comparison with the results obtained with the central Paris potential.

In conclusion, we performed self-consistent calculations of $\bar{p}$-nuclear quasi-bound states using a microscopic potential, namely the Paris $\bar{N}N$ potential, for the first time. We explored

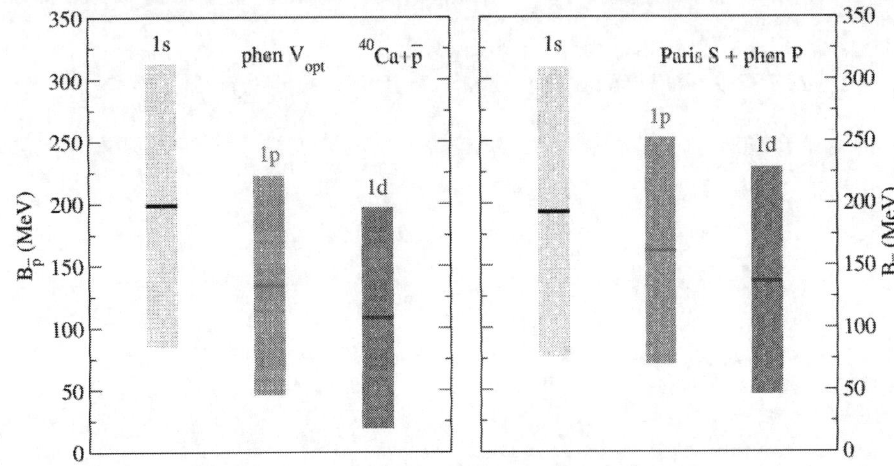

Figure 4. $1s$, $1p$ and $1d$ binding energies (lines) and widths (boxes) of $\bar{p}$ in ^{40}Ca calculated dynamically within the phenomenological RMF $\bar{p}$ optical potential and Paris S-wave + phen. P-wave potential.

the effect of the P-wave interaction on $\bar{p}$ binding energies and widths. We found that the P-wave interaction almost does not affect the binding energies of $\bar{p}$-nuclear states. This is in sharp contrast to the case of $\bar{p}$ atoms where it was found necessary to include the P-wave interaction in order to increase attraction of the $\bar{p}$ optical potential [8]. Moreover, we found good agreement between the results obtained using the phenomenological RMF potential and the Paris S-wave + phenomenological P-wave potential which are the two potentials consistent with antiprotonic atom data and $\bar{p}$ scattering off nuclei at low energies.

Acknowledgements

We thank E. Friedman and A. Gal for valuable discussions, and B. Loiseau for providing us with the $\bar{N}N$ amplitudes. This work was supported by the GACR Grant No. P203/15/04301S.

References

[1] T. J. Bürvenich, W. Greiner, I. N. Mishustin, L. M. Satarov, H. Stöcker, Phys. Rev. **C 71** (2005) 035201.
[2] J. Hrtánková, J. Mareš, Nucl. Phys. **A 945** (2016) 197.
[3] B. El-Bennich, M. Lacombe, B. Loiseau, S. Wycech, Phys. Rev. **C 79** (2009) 054001.
[4] T. Hippchen, K. Holinde, W. Plessas, Phys. Rev. **C 39** (1989) 761.
[5] D. Zhou, R. G. E. Timmermans, Phys. Rev. **C 86** (2012) 044003.
[6] K. W. Kang, J. Haidenbauer, U.-G. Meißner, JHEP **1402** (2014) 113.
[7] L.-Y. Dai, J. Haidenbauer, U.-G. Meißner, JHEP **1707** (2017) 78.
[8] E. Friedman, A. Gal, B. Loiseau, S. Wycech, Nucl. Phys. **A 934** (2015) 101.
[9] M. M. Sharma, M. A. Nagarajan, P. Ring, Phys. Lett. **B 312** (1993) 377.
[10] T. Wass, M. Rho, W. Weise, Nucl. Phys. **A 617** (1997) 449.
[11] A. Cieplý, E. Friedman, A. Gal, D. Gazda, J. Mareš, Phys. Lett. **B 702** (2011) 402.
[12] A. B. Larionov, I. N. Mishustin, L. M. Satarov, W. Greiner, Phys. Rev. **C 78** (2008) 014604.
[13] A. B. Larionov, I. N. Mishustin, L. M. Satarov, W. Greiner, Phys. Rev. **C 82** (2010) 024602.
[14] J. N. Ginocchio, Phys. Rep. **414** (2005) 165.
[15] X. T. He, S. G. Zhou, J. Meng, E. G. Zhao, W. Scheid, Eur. Phys. J. **A 28** (2006) 265.

EPJ Web of Conferences **181**, 01008 (2018) https://doi.org/10.1051/epjconf/201818101008
EXA2017

Single- and multi-nucleon K^- interactions with nuclei near threshold

Eliahu Friedman[1,⋆] and *Avraham* Gal[1,⋆⋆]

[1] *Racah Institute of Physics, the Hebrew University, Jerusalem*

Abstract.
Six recent SU(3) chiral-model EFT approaches to the $\bar{K}$-nucleon interaction near threshold, constrained by K^-p low-energy scattering and reaction data and by the kaonic hydrogen SIDDHARTA experiment, are used as input in kaonic atom calculations. Good agreement with the world-data on kaonic atoms is achieved with optical potentials built on the above models only when $\bar{K}N$ amplitudes are supplemented by a phenomenological multi-nucleon term. Comparing predictions with experimental single-nucleon absorption-at-rest fractions on nuclei, only two of the models together with their associated phenomenological term are acceptable. The information content of K^--nucleus data near threshold is discussed and the topic of deeply-bound kaonic atoms is re-visited.

1 Introduction

The interaction of low energy K^- mesons with nuclei received attention over the years mostly through analyses of experiments with kaonic atoms, see [1, 2] and references therein. Optical-model analyses of strong-interaction observables in kaonic atoms throughout the periodic table achieved fits of χ^2 per point of less that 2 for the whole data put together, which is quite good considering that the data are from different experiments carried out in several laboratories. Advances with theoretical $\bar{K}N$ interaction near threshold, particularly with the K^-p interaction constrained by experiment, raises the possibility of relating the optical potentials to the more fundamental interaction of K^- with nucleons. However, attempts to construct optical potentials from K^--nucleon scattering amplitudes showed that a significant phenomenological term must be included in the potential, which could be related to multi-nucleon processes. Very recently [3] we showed that out of six state-of-the-art models for the K^-N amplitudes only two are able to reproduce experimental single-nucleon absorption fractions [4–6] on nuclei while also forming part of optical potentials that fit kaonic atoms data.

In the present work we extend the discussion of [3] with emphasis on the ratios between the two components of the potential. We discuss the radial sensitivity of kaonic atom experiments, focusing on various extrapolations of the potentials into the nuclear interior presented in earlier works. Finally we discuss the existence of deeply bound K^- *atomic* states [7, 8] and whether they could provide additional information to that available from conventional kaonic atom experiments.

⋆e-mail: elifried@cc.huji.ac.il
⋆⋆e-mail: avragal@savion.huji.ac.il

2 Input amplitudes and optical potentials

Following Cieplý et al. [9] we consider in the present work six sets of K^-N amplitudes derived in various strangeness $S{=}-1$ chiral SU(3) meson-baryon coupled-channel dynamical model approaches at next-to-leading order (NLO). These models are constrained by fits to low-energy K^-p scattering and reaction cross sections, to three precisely determined threshold branching ratios, and to the kaonic hydrogen energy shift and width measured by the SIDDHARTA Collaboration [10, 11]. With such potentials one can take a step forward beyond a purely phenomenological approach and construct an optical model based on a $t\rho$ formulation. All six sets of amplitudes display sharp dependence on energy near threshold making the potential very sensitive to the energy used in its construction. Here, as in recent years, we adopt a 'kinematical' approach where this energy is determined self-consistently [12–14]. In this approach the energy $\sqrt{s}$ and the amplitudes become functions of the nuclear density. Corrections due to Pauli correlations are also included [15]. For attractive optical potentials this energy is below threshold, $\sqrt{s} = M_N + M_{K^-} + \delta\sqrt{s}$, with $\delta\sqrt{s} < 0$.

Tab. 1 shows comparisons between calculation and experiment when the $t\rho$ optical potentials are based only on the K^-N amplitudes as discussed above. Focusing on an average K^--nucleus interaction, our approach is global, handling all the available experiments together, 65 data points from Li to U. As a test the table shows also results for a smaller set, consisting of kaonic atoms of P, S, Cl, Cu, Ag and Pb, where the experimental results are of particularly high quality. It is evident that in all cases there is no agreement between experiment and calculation. Similar conclusions have been obtained in earlier studies of kaonic atoms.

Table 1. χ^2 values for 65 kaonic atom data points and for a sub-set of 18 points, obtained with optical potentials based only on six K^-N amplitudes discussed here (see [3, 9] for notation). The 65 points represent the world data for kaonic atoms. The 18 points are for six species of kaonic atoms where the experimental results are particularly comprehensive, see text.

model	B2	B4	M1	M2	P	KM
$\chi^2(65)$	1174	2358	2544	3548	2300	1806
$\chi^2(18)$	364	733	949	1232	480	449

In order to obtain good fit to experiment, a phenomenological term is added to the single-nucleon based amplitude, namely, we add a term $B(\rho/\rho_0)^\alpha$ and vary the three parameters ReB, ImB and α, with ρ_0 the central nuclear density. The fits obtained in that way are as good as with purely phenomenological potentials. Due to obvious correlations between α and B we eventually grid on the former, and Tab. 2 shows typical results. As noted in the Introduction, a χ^2 per point of less than 2 is considered most acceptable for such global fits to large and diverse data sets.

The additional term $B(\rho/\rho_0)^\alpha$ in the amplitude may be considered as representing multi-nucleon processes. Note that for $\alpha = 1$, a formal dependence of the *potential* on ρ^2 has traditionally been referred to as representing two-nucleon processes. In particular, ImB represents absorptions on two nucleons, which are the pion-less absorptions known well from experiment. Consequently, fits resulting in negative values for ImB must be rejected.

Tab. 3 shows results for the P and KM potentials supplemented by their respective $B(\rho/\rho_0)^\alpha$ term. It is seen that for $\delta\sqrt{s}=0$, i.e. using single-nucleon amplitudes at threshold, the solutions are unacceptable on account of negative ImB. Similar results and significantly larger values of χ^2 are found for the other four sets of amplitudes when $\delta\sqrt{s}=0$. We also note that the results for the P and KM options are essentially the same, within errors. From the results shown in tables 2 and 3 it is not possible to prefer one model to the other, neither on the basis of quality of fits nor from values of the exponent

Table 2. Results of best-fits to 65 and 18 data points when the six amplitudes are supplemented by a complex empirical term $B(\rho/\rho_0)^\alpha$ with α fixed by separate fits, see text.

model	B2	B4	M1	M2	P	KM
$\chi^2(65)$	111	105	121	109	125	123
α	0.3	0.3	0.3	1.0	1.0	1.0
ReB (fm)	2.4±0.2	3.1±0.1	0.3±0.1	2.1±0.2	−1.3±0.2	−0.9±0.2
ImB (fm)	0.8±0.1	0.8±0.1	0.8±0.1	1.2±0.2	1.5±0.2	1.4±0.2
$\chi^2(18)$	17	15	19	17	17	16
α	0.3	0.3	0.3	1.0	1.0	1.0
ReB (fm)	2.7±0.3	2.9±0.1	0.3±0.1	2.2±0.2	−1.3±0.2	−0.9±0.3
ImB (fm)	0.8±0.2	0.8±0.1	0.6±0.1	1.1±0.2	1.4±0.3	1.3±0.3

Table 3. Results of best-fits to the full data set with the P and the KM amplitudes supplemented by a complex empirical term $B(\rho/\rho_0)^\alpha$ with α fixed, using either the sub-threshold prescription for variable $\delta\sqrt{s}$ (var.) or on-threshold values ($\delta\sqrt{s}$=0).

model	P	KM	P	KM	P	KM
$\chi^2(65)$	135	139	125	123	125	122
$\delta\sqrt{s}$	0	0	var.	var.	var.	var.
α	1	1	1	1	2	2
ReB (fm)	−1.9±0.1	−1.8±0.1	−1.3±0.2	−0.9±0.2	−0.5±0.6	0.3±0.7
ImB (fm)	−0.8±0.1	−1.1±0.1	1.5±0.2	1.4±0.2	4.6±0.7	3.8±0.7

α. Additional experimental information is required and that is available as single-nucleon absorption fractions.

3 Single-nucleon absorption fractions

Based on fractions of K^- absorption-at-rest on several nuclear species measured many decades ago, it is possible to distinguish between the six models discussed in the present work [3]. The level width Γ of a kaonic atom state which is usually obtained from the complex eigenvalue $E_{K^-} - i\Gamma/2$ when solving the Klein-Gordon equation with an optical potential V_{K^-}, is also related to the imaginary part of the potential by the overlap integral of Im V_{K^-} and $|\psi|^2$,

$$\Gamma = -2\,\frac{\int \mathrm{Im}\,V_{K^-}\,|\psi|^2\,d\vec{r}}{\int [1-(B_{K^-}+V_C)/\mu_K]\,|\psi|^2\,d\vec{r}}, \tag{1}$$

where B_{K^-}, V_C and μ_K are the K^- binding energy, Coulomb potential and reduced mass, respectively, and ψ is the K^- wave function of the particular state concerned. Partial widths for absorption on a single nucleon and in a multinucleon process are easily calculated for the best fit potentials using the single-nucleon based amplitudes and the corresponding phenomenological additional term.

From the three bubble-chamber experiments mentioned in the introduction [4–6] we adopt a fraction of 0.75±0.05 for the single-nucleon absorption-at-rest of K^- on nuclei heavier than C. Fig. 1 shows very good agreement between calculations and experiment for amplitudes P and KM, supplemented by a phenomenological term. The two models are indistinguishable on this figure. The small https://doi.org/10.1051/epjconf/201818101008

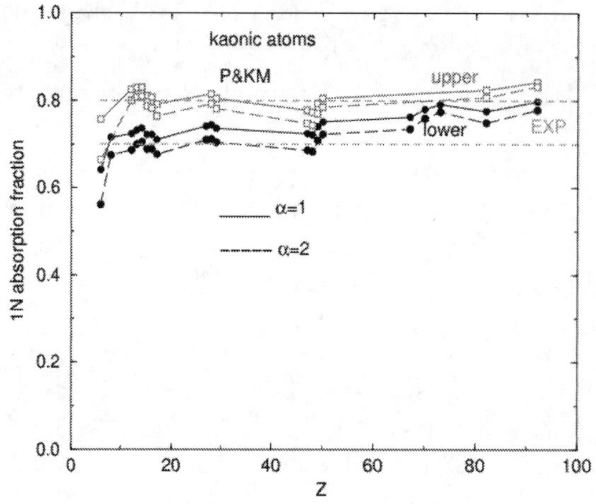

Figure 1. Fractions of single-nucleon absorption for amplitudes P and KM: solid circles are for lower states, open squares for upper states, solid curves use B values from Tab. 3 for α=1, long-dashed curves use B values for α=2. Horizontal dashed lines mark the range of experimental values of the single-nucleon absorption fraction. Models P and KM are indistinguishable from each other on this figure.

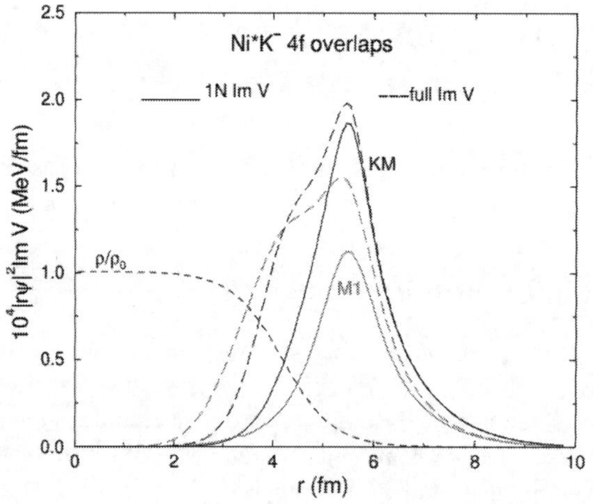

Figure 2. Overlaps of the absolute value squared of the 4f radial wave function in kaonic atoms of Ni with the imaginary part of the optical potential based on the KM amplitude and on the M1 amplitude. Solid lines for potentials due to the K^-N amplitudes; dashed line for the corresponding full potentials including a phenomenological term. Also shown is the relative density for Ni.

dependence on the choice of α is evident. From the radiative widths of the transitions concerned one concludes that about 80% of absorptions take place from the lower states, except for C where most of the absorptions are from the upper state. The other four sets of amplitudes lead to calculated single-nucleon absorption fractions of 0.4-0.6, in disagreement with experiment.

Fig. 2 shows examples for the overlaps of the 4f state in kaonic atoms of Ni. It is easily seen that whereas for the KM amplitudes the area under the single-nucleon curve is about 75% of the area under the full amplitude curve, the corresponding ratio for the M1 amplitudes is only 50%. However, both provide equally good fit to the kaonic atom observables for Ni, and only the absorption fractions distinguish between these models.

4 K^- potentials vs. density

Fig. 3 shows examples of the full amplitudes (single nucleon plus a phenomenological term) for K^- interacting with Ni obtained from global fits to kaonic atoms, based on the P and the KM models. Here the single-nucleon amplitudes are supplemented by a phenomenological term with the exponent α equals 1 (P1 and KM1) or 2 (P2 and KM2), as given in Tab. 3. The full amplitudes are seen to be dominated by the K^-N input at low densities and then the different models produce different amplitudes as the density increases towards the nuclear interior. This is particularly so for the real part of the amplitudes, whereas the imaginary part remains almost model-independent up to about 50% of the central density. The similarity between the P-based and the KM-based amplitudes is clearly seen on this figure.

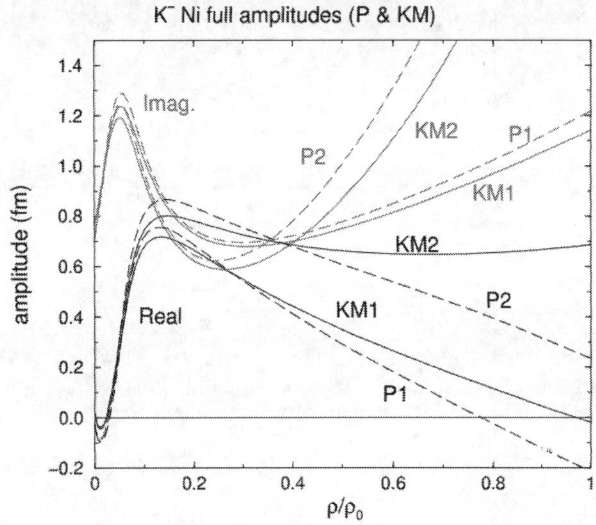

Figure 3. Examples for the full K^-Ni amplitudes (single nucleon plus a corresponding phenomenological term). In black for the real part, in red for the imaginary. P (dashed) and KM (solid lines) are the Prague and Kyoto-Munich models. The 1 and 2 refer to α=1 and 2, respectively, in the phenomenological term.

A more transparent picture emerges from the optical potentials. Fig. 4 shows optical potentials for K^- on Ni from global fits to kaonic atoms data that are based on a theoretical K^-N amplitude plus a phenomenological term and, in addition, agree with single-nucleon absorption fractions. Also shown is a corresponding purely phenomenological optical potential producing similarly good fit to kaonic atom data. Obviously with this type of potential absorption ratios are meaningless and the values at low density are unconstrained by a low density limit. The real parts of the potentials for the four models agree with each other up to about 25% of the central density. The imaginary part is seen to be well-determined up to 50% of the central density, reflecting the observation that strong interaction effects in kaonic atoms are dominated by the imaginary potential where level widths are significantly larger than shifts. It is concluded that at larger densities the various models produce just analytical continuations from values close to the nuclear surface. Note that the present extrapolated imaginary potentials are deeper than the purely phenomenological values.

The depth of the real potential was a topic of some controversy. Fig. 4 suggests that the behaviour at very low densities determines the extrapolation to larger densities. Indeed it is seen that in contrast to the purely phenomenological potential that has a smooth dependence on density, the other potentials have some structure near the surface and hence extrapolate differently to the nuclear interior, see [3] for further details.

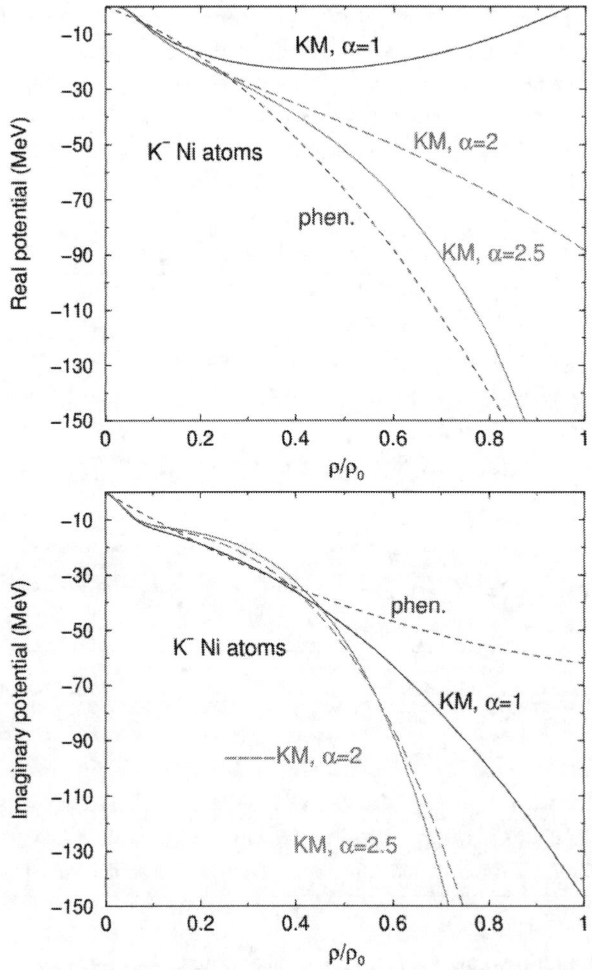

Figure 4. Real part (top) and imaginary part (bottom) of best-fit K^{-}Ni optical potentials based on the KM single nucleon amplitudes plus a phenomenological term with α=1, 2 and 2.5 in the phenomenological term. Also shown is a purely phenomenological potential

5 Deeply bound kaonic atom states

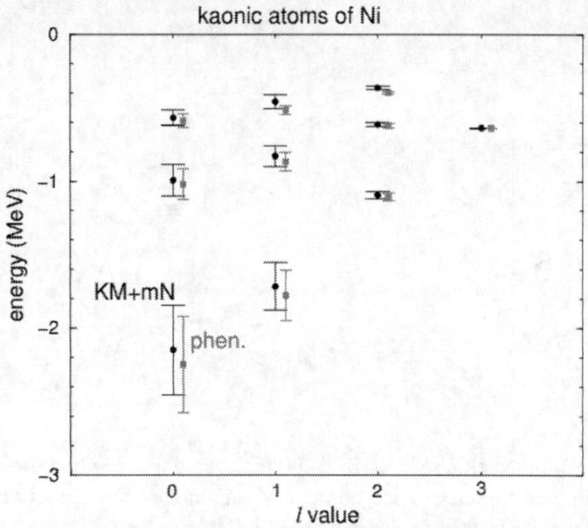

Figure 5. Energy levels of kaonic atom of Ni, showing the energies and the full widths. In red calculated for a purely phenomenological best-fit potential [7, 8], in black for the KM plus phenomenological best-fit potential for the KM model with α=1.

Finally we discuss the possible existence of 'deeply bound' kaonic atom states, i.e. deep atomic states which are not populated by the usual cascade of X-rays. These states have been predicted [7, 8] two decades ago but no attempts have been made to observe them experimentally. It is instructive to follow briefly the history of the deeply bound pionic atoms, observed experimentally for the first time by Yamazaki et al. [16].

The first prediction that the deepest pionic atom states in heavy nuclei are sufficiently narrow to make them observable was by Friedman and Soff in 1985 [17]. The explanation given was the repulsion of the *atomic* wave function from the nucleus due to the strong s-wave repulsive term, thus reducing the overlap with the nucleus and causing the widths of the levels to be small. Three years later Toki and Yamazaki [18] discussed also methods for populating such states, which eventually led to the observation in 1996 [16]. It turned out that the strong interaction level shifts and widths of the deeply bound states were in agreement with predictions made with empirical pionic atom potentials based on the broad data-base of conventional states [19]. In particular, the enhancement within the nucleus of the s-wave isovector term of the pion-nucleus potential was found to agree with what had already been known from global analyses of pionic atoms [20]. We expect similar situation for kaonic atoms.

Fig. 5 shows positions and widths of energy levels in kaonic atom of Ni, calculated from a purely phenomenological potential [7] and from the current KM (α = 1) potential. The results are almost identical for the two models, displaying very small sensitivity to detail, *provided* the potential fits the entire data on kaonic atoms. The mechanism in kaonic atoms is the *effective* repulsion due to strong absorption, which is operative when the absorption is confined to a relatively small region of the atom.

Fig. 6 shows overlaps of atomic wave functions with the Ni nucleus for the observed 4f state and for the predicted 1s state. The absolute scales differ by three orders of magnitudes and so do the widths, about one keV compared to one MeV, but the depth of penetration into the nucleus are the

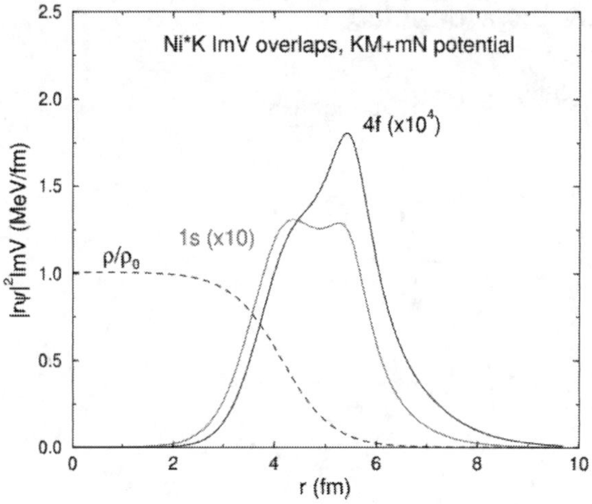

Figure 6. Overlaps of the absolute value squared of the 4f and 1s radial wave function in kaonic atoms of Ni with the imaginary part of the optical potential based on the KM amplitudes plus a phenomenological part with $\alpha=1$. Also shown is the relative density of the nuclear density of Ni.

same. It therefore appears that deeply bound kaonic atom states will not provide information on the interior different from what is known from conventional states.

6 Summary

With six state-of-the-art models providing K^-N amplitudes near threshold it is found, as expected, that empirical terms must be added in each case to produce fit to the world's data on strong interaction effects in kaonic atoms. The six models appear equivalent on this basis, but when the fraction of absorption-at-rest on single nucleons is considered, then only two of the models, the Prague (P) and the Kyoto-Munich (KM) model, pass the test. Comparing amplitudes and potential details from the various acceptable versions, it is concluded that the real potentials are well determined only for densities up to 25% of the central density. Imaginary potentials display better penetration into the nuclear interior, being well determined up to 50% of the central density.

Re-visiting the topic of deeply bound kaonic atoms, we find that such states are well-defined but they are hardly sensitive to details of the potentials, once these are obtained from best-fits to conventional kaonic atoms data. It is concluded that the limited overlap of wave functions with the nucleus, that is the mechanism making these states sufficiently narrow, is the mechanism limiting sensitivities of kaonic atom experiments to only relatively small densities.

References

[1] C.J. Batty, E. Friedman, A. Gal, Phys. Rep. **287** 385 (1997)
[2] E. Friedman, A. Gal, Phys. Rep. **452** 89 (2007)
[3] E. Friedman, A. Gal, Nucl. Phys. A **959** 66 (2017)
[4] H. Davis, F. Oppenheimer, W.L. Knight, F.R. Stannard, O. Treutler, Nuovo Cimento **53A** 313 (1968)

[5] J.W. Moulder, N.E. Garrett, L.M. Tucker, W.M. Bugg, G.T. Condo, H.O. Cohn, R.D. McCulloch, Nucl. Phys. B **35** 332 (1971)
[6] C. Vander Velde-Wilquet, J. Sacton, J.H. Wickens, D.N. Tovee, D.H. Davis, Nuovo Cimento **39A** 538 (1977)
[7] E. Friedman, A. Gal, Phys. Lett. B **459** 43 (1999)
[8] E. Friedman, A. Gal, Nucl. Phys. A **658** 345 (1999)
[9] A. Cieplý, M. Mai, U.-G. Meißner, J. Smejkal, Nucl. Phys. A **954** 17 (2016)
[10] M. Bazzi, et al. (SIDDHARTA Collaboration), Phys. Lett. B **704** 113 (2011)
[11] M. Bazzi, et al. (SIDDHARTA Collaboration), Nucl. Phys. A **881** 88 (2012)
[12] A. Cieplý, E. Friedman, A. Gal, D. Gazda, J. Mareš, Phys. Lett. B **702** 402 (2011)
[13] A. Cieplý, E. Friedman, A. Gal, D. Gazda, J. Mareš, Phys. Rev. C **84** 045206 (2011)
[14] E. Friedman, A. Gal, Nucl. Phys. A **899** 60 (2013)
[15] T. Waas, M. Rho, W. Weise, Nucl. Phys. A **617** 449 (1997)
[16] T. Yamazaki, R.S. Hayano, K. Itahashi, K. Oyama, A. Gillitzer, H. Gilg, M. Knülle, M. Münch, P. Kienle, W. Schott, H. Geissel, N. Iwasa, G. Münzenberg, Z. Phys. A **355** 219 (1996)
[17] E. Friedman, G. Soff, J. Phys. G: Nucl. Phys. **11** L37 (1985)
[18] H. Toki, T. Yamazaki, Phys. Lett. B **213** 129 (1988)
[19] E. Friedman, A. Gal, Phys. Lett. B **432** 235 (1998)
[20] E. Friedman, A. Gal, Nucl. Phys. A **724** 143 (2003)

EPJ Web of Conferences **181**, 01009 (2018) https://doi.org/10.1051/epjconf/201818101009
EXA2017

On the widths and binding energies of K^- nuclear states and the role of K^- multi-nucleon interactions

Jaroslava Hrtánková[1,*] and *Jiří* Mareš[1,**]

[1]Nuclear Physics Institute, 250 68 Řež, Czech Republic

Abstract. We report on our recent self-consistent calculations of K^- nuclear quasi-bound states using K^- optical potentials derived from chirally motivated meson-baryon coupled channels models [1, 2]. The K^- single-nucleon potentials were supplemented by a phenomenological K^- multi-nucleon interaction term introduced to achieve good fits to K^- atom data. We demonstrate a substantial impact of the K^- multi-nucleon absorption on the widths of K^- nuclear states. If such states ever exist in nuclear many-body systems, their widths are excessively large to allow observation.

1 Introduction

The near-threshold K^-N attraction generated by the $\Lambda(1405)$ resonance seems to be strong enough to allow binding of the K^- meson in nuclei. This fact stimulated numerous theoretical studies of K^- nuclear quasi-bound states as well as experimental searches for K^- nuclear clusters [3–5]. However, the question of their binding energies and widths, and even their very existence is still far from being settled. The K^-N interaction near threshold has been recently described within chirally motivated coupled channels meson-baryon interaction models [6–9], parameters of which have been fitted to low-energy experimental data [10, 11]. K^-N scattering amplitudes which enter the construction of K^- optical potentials are strongly energy dependent near and below threshold, i.e., in the region relevant for calculations of kaonic nuclear states. It is important to treat this energy dependence properly [12, 13]. The above chiral meson-baryon interaction models include only K^- absorption on a single-nucleon, $K^-N \to \pi Y$ ($Y = \Sigma, \Lambda$). However, in the nuclear medium, K^- multi-nucleon interactions take place as well [14–16], and have to be considered in realistic evaluations of K^- widths (and to lesser extent also K^- binding energies) in the nuclear medium. Recently, Friedman and Gal [17] supplemented K^- single-nucleon potentials constructed within several chiral coupled channels meson-baryon models by a phenomenological term representing the K^- multi-nucleon interactions. They fitted parameters of this phenomenological term to kaonic data for each meson-baryon interaction model separately. Only the P [7] and KM [6] models were able to reproduce both kaonic data and experimental fractions of K^- absorption at rest [18–20]. In this contribution we apply these two models to calculations of K^- nuclear quasi-bound states in selected nuclei.

In Section 2, we briefly introduce kaon self-energy operator including in-medium modifications of the underlying K^-N scattering amplitudes. We discuss how to incorporate strong

*e-mail: hrtankova@ujf.cas.cz
**e-mail: mares@ujf.cas.cz

energy dependence of the in-medium amplitudes in self-consistent calculations of kaonic nuclear quasi-bound states. In Section 3, we present selected results of our calculations aiming at demonstrating the crucial role played by K^- multi-nucleon interactions in the nuclear medium. A brief summary in Section 4 concludes this contribution.

2 Methodology

The interaction of the K^- meson with a nucleus is described by the Klein-Gordon equation

$$\left[\vec{\nabla}^2 + \tilde{\omega}_{K^-}^2 - m_{K^-}^2 - \Pi_{K^-}(\omega_{K^-}, \rho)\right]\phi_{K^-} = 0\,, \tag{1}$$

where, $\tilde{\omega}_{K^-} = m_{K^-} - B_{K^-} - \mathrm{i}\Gamma_{K^-}/2 - V_C = \omega_{K^-} - V_C$ with B_{K^-} (Γ_{K^-}) being the K^- binding energy (width), m_{K^-} denotes the K^- mass, V_C is the Coulomb potential, and ρ is the nuclear density distribution. The kaon self-energy Π_{K^-} is constructed from in-medium isospin 0 and 1 s-wave amplitudes F_0 and F_1 in a $t\rho$ form as follows:

$$\Pi_{K^-} = 2\mathrm{Re}(\omega_{K^-})V_{K^-}^{(1)} = -4\pi\frac{\sqrt{s}}{m_N}\left(F_0\frac{1}{2}\rho_p + F_1\left(\frac{1}{2}\rho_p + \rho_n\right)\right)\,, \tag{2}$$

where $V_{K^-}^{(1)}$ is the K^--nucleus optical potential, m_N is the nucleon mass, and the kinematical factor $\sqrt{s}/m_N$ transforms the scattering amplitudes from the two-body frame to the K^--nucleus frame. The proton and neutron density distributions, ρ_p and ρ_n, are evaluated within the relativistic mean-field model NL-SH [21].

The in-medium scattering amplitudes were obtained from the free-space amplitudes f_{K^-p} and f_{K^-n}, derived within chiral coupled channels meson-baryon interaction models P and KM, using the multiple scattering approach (WRW) [22] in order to account for Pauli correlations. They are of the following form:

$$F_1 = \frac{f_{K^-n}(\sqrt{s})}{1+\frac{1}{4}\xi_k\frac{\sqrt{s}}{m_N}f_{K^-n}(\sqrt{s})\rho}\,, \quad F_0 = \frac{[2f_{K^-p}(\sqrt{s}) - f_{K^-n}(\sqrt{s})]}{1+\frac{1}{4}\xi_k\frac{\sqrt{s}}{m_N}[2f_{K^-p}(\sqrt{s}) - f_{K^-n}(\sqrt{s})]\rho}\,, \tag{3}$$

where ξ_k is adopted from Ref. [17].

In Fig. 1, we present the K^-p (top) and K^-n (bottom) amplitudes in the P and KM models, modified by the WRW procedure at saturation density $\rho_0 = 0.17$ fm^{-3}, plotted as a function of energy. The corresponding free scattering amplitudes (dotted lines) are presented for comparison. The K^-p in-medium amplitudes are affected significantly by Pauli correlations: the real parts of the amplitudes become attractive in the entire energy region below threshold, and the imaginary parts are considerably lowered. On the other hand, the K^-n amplitudes are modified by Pauli correlations only moderately.

In previous calculations [12, 16], the in-medium K^-N amplitudes in the P model [7] were constructed in a different way. The integration over the intermediate meson-baryon momenta in the underlying Green's function was restricted to a region ensuring the nucleon intermediate energy to be above the Fermi level (denoted further 'Pauli'). In Fig. 2, we compare the Pauli correlated amplitudes with the WRW modified amplitudes in the P model. It is rewarding that both approaches, WRW and Pauli, yield similar K^-N in-medium reduced amplitudes[1] $f_{K^-N}^{\mathrm{r}} = \frac{1}{2}(f_{K^-p}^{\mathrm{r}} + f_{K^-n}^{\mathrm{r}})$ in the subthreshold energy region which is relevant to our calculations of K^- nuclear states. Above threshold, the behavior of Pauli and WRW modified amplitudes is different.

[1] $f_{K^-N}^{\mathrm{r}} = g(p)f_{K^-N}g(p')$, where $g(p)$ is a momentum-space form factor (see Ref. [12])

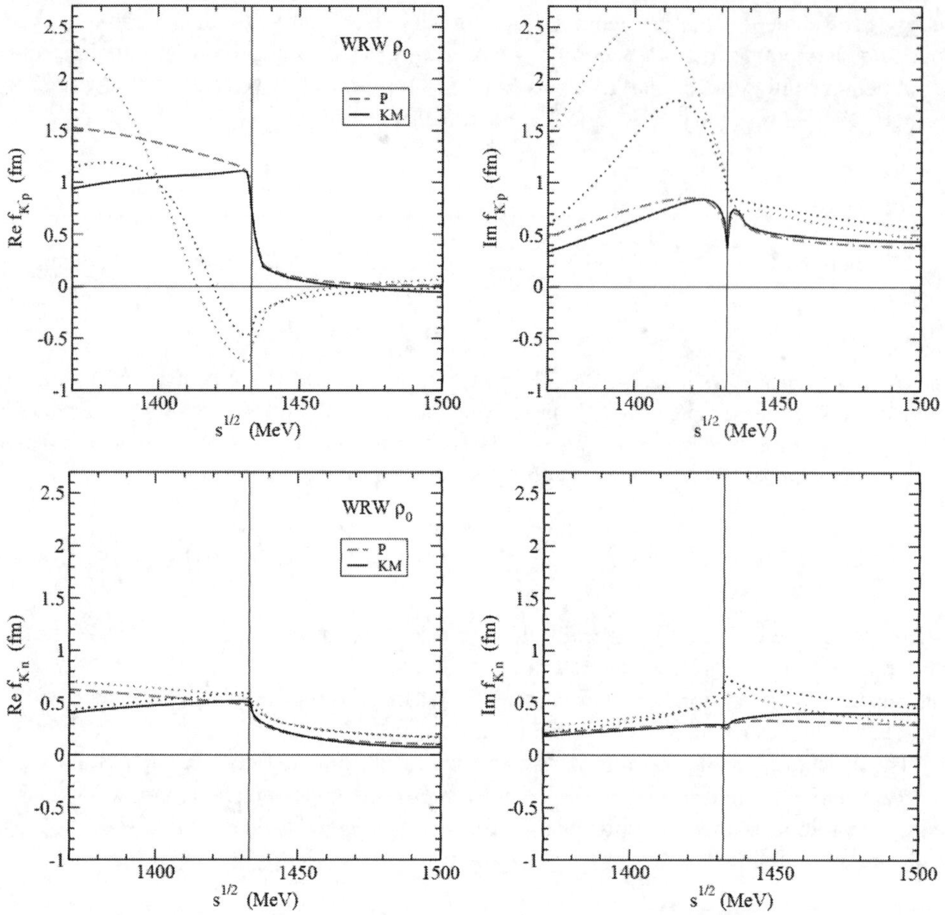

Figure 1. Energy dependence of real (left) and imaginary (right) parts of WRW modified K^-p (top) and K^-n (bottom) amplitudes at $\rho_0 = 0.17$ fm^{-3} in the P and KM models. Corresponding free-space amplitudes are shown for comparison (dotted lines).

Important feature of the K^-p scattering amplitudes shown in Figs. 1 and 2 is their strong energy dependence near threshold caused by the presence of the $\Lambda(1405)$ resonance, dynamically generated in chirally motivated coupled channels models. This feature calls for a proper self-consistent treatment during the evaluation of the K^- optical potential. The amplitudes are a function of energy defined by the Mandelstam variable

$$s = (E_N + E_{K^-})^2 - (\vec{p}_N + \vec{p}_{K^-})^2 , \tag{4}$$

where $E_N = m_N - B_N$, $E_{K^-} = m_{K^-} - B_{K^-}$ and $\vec{p}_{N(K^-)}$ is the nucleon (kaon) momentum. In the K^--nucleus frame the momentum dependent term is no longer zero and provides additional downward energy shift. The energy shift $\delta\sqrt{s} = \sqrt{s} - E_{th} = \sqrt{s} - (m_N + m_K)$ was expanded in terms of binding and kinetic energies as follows [1, 2]:

$$\delta\sqrt{s} = -B_N\frac{\rho}{\bar{\rho}} - \beta_N\left[B_{K^-}\frac{\rho}{\rho_{\max}} + T_N\left(\frac{\rho}{\bar{\rho}}\right)^{2/3} + V_C\left(\frac{\rho}{\rho_{\max}}\right)^{1/3}\right] + \beta_{K^-}\mathrm{Re}V_{K^-}(r) , \tag{5}$$

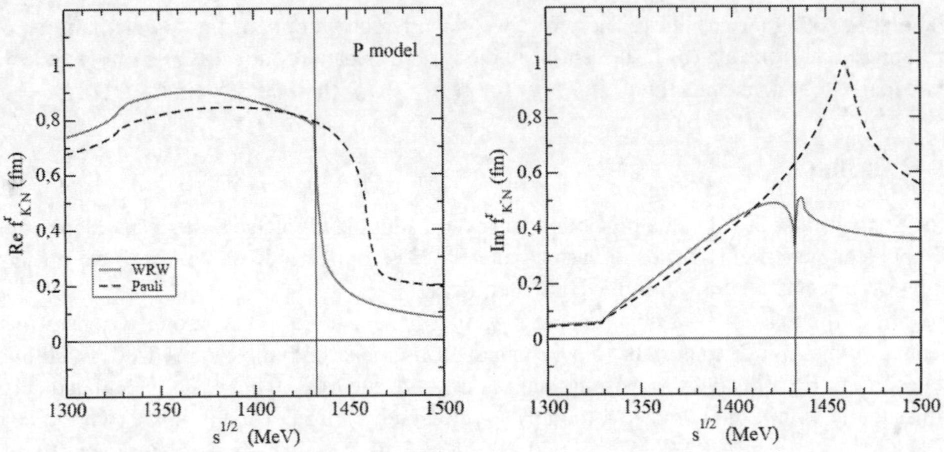

Figure 2. Comparison of real (left) and imaginary (right) parts of WRW (full line) and Pauli (dashed line) modified reduced amplitudes $f^{\mathrm{r}}_{K^-N} = \frac{1}{2}(f^{\mathrm{r}}_{K^-p} + f^{\mathrm{r}}_{K^-n})$ plotted as function of energy $\sqrt{s}$ for $\rho_0 = 0.17\ \mathrm{fm}^{-3}$ in the P model (see text for details).

where $B_N = 8.5$ MeV is the average binding energy per nucleon, $\bar{\rho}$ is the average nuclear density, $\rho_{\max}$ is the maximal value of the nuclear density, $\beta_{N(K^-)} = m_{N(K^-)}/(m_N + m_{K^-})$ and $T_N = 23$ MeV is the average nucleon kinetic energy in the Fermi Gas Model. The energy shift respects the low-density limit, i. e. $\delta\sqrt{s} \to 0$ as $\rho \to 0$.

The above P and KM models describe only the K^- interactions with a single nucleon. However, K^- interactions with two and more nucleons take place in the medium as well [14, 17]. Therefore, we supplemented the K^- single-nucleon potential $V^{(1)}_{K^-}$ from Eq. (2) with a phenomenological optical potential $V^{(2)}_{K^-}$ describing the K^- multi-nucleon interactions:

$$2\mathrm{Re}(\omega_{K^-})V^{(2)}_{K^-} = -4\pi B(\frac{\rho}{\rho_0})^{\alpha}\rho\,. \tag{6}$$

The parameters of the phenomenological potential, complex amplitude B and positive exponent α listed in Table 1, were recently fitted by Friedman and Gal [17] to kaonic atom data for each K^-N interaction model separately. Moreover, the total K^- optical potential, $V_{K^-} = V^{(1)}_{K^-} + V^{(2)}_{K^-}$, was confronted with branching ratios of K^- absorption at rest from bubble chamber experiments [18–20] and the P and KM models discussed in this contribution were found to be the only two interaction models capable of reproducing both experimental constraints simultaneously.

The kaonic atom data probe the K^- optical potential up to at most ~ 50% of the nuclear density [17]. Further inside the nucleus, the shape of the potential is just a matter of ex-

Table 1. Values of the complex amplitude B and exponent α used to evaluate $V^{(2)}_{K^-}$ for chiral meson-baryon interaction models considered in this work.

	P1	KM1	P2	KM2
α	1	1	2	2
ReB (fm)	-1.3 ± 0.2	-0.9 ± 0.2	-0.5 ± 0.6	0.3 ± 0.7
ImB (fm)	1.5 ± 0.2	1.4 ± 0.2	4.6 ± 0.7	3.8 ± 0.7

trapolation to higher densities. Therefore, we considered two options in our calculations: a) we applied the formula (6) in the entire nucleus (full density option - FD) b) we fixed the potential $V^{(2)}_{K^-}$ at constant value $V^{(2)}_{K^-}(0.5\rho_0)$ for $\rho(r) \geq 0.5\rho_0$ (half density limit - HD).

3 Results

The formalism outlined in the previous section was adopted to self-consistent calculations of K^- nuclear quasi-bound states in various nuclei across the periodic table. Here we present only few selected results, for more details see Refs. [1, 2]).

The self-consistently evaluated energy shift $\delta\sqrt{s}$ from Eq (5) is strongly density dependent which plays important role in calculations of kaonic nuclear, as well as atomic states. Figure 3 illustrates the strong density dependence of $\delta\sqrt{s}$ in ^{208}Pb, calculated self-consistently within the P and KM models augmented by the FD variant of $V^{(2)}_{K^-}$ with $\alpha = 1$ (P1, KM1) and $\alpha = 2$ (P2, KM2). The P1 and KM1 models yield smaller energy shift with respect to the K^-N threshold than the original single-nucleon potentials $V^{(1)}_{K^-}$ (P and KM). The P2 and KM2 models yield energy shifts closer to the single-nucleon potentials. The energy shift for the FD option is in any case shallower than for the original K^- single-nucleon potential owing to very strong absorption. It is to be noted that the P and KM models supplemented with various versions of the multi-nucleon interaction term could be regarded as equivalent since they all lie in corresponding uncertainty bands and describe kaonic atom data equally well.

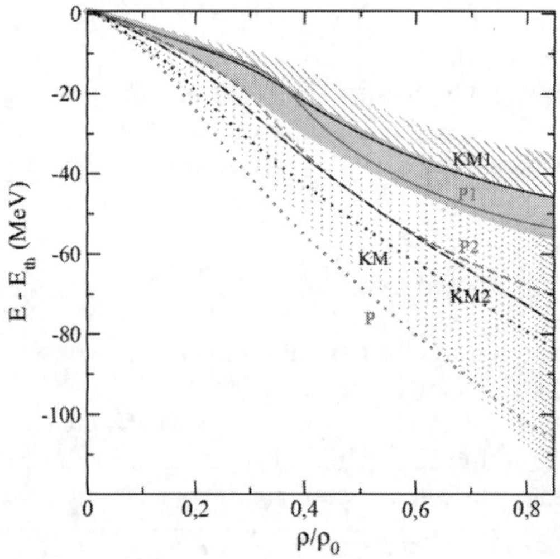

Figure 3. Subthreshold energies probed in the ^{208}Pb+K^- nucleus as a function of relative density ρ/ρ_0, calculated self-consistently in the single-nucleon K^-N amplitude models P and KM (dotted lines), supplemented by the FD variant of $V^{(2)}_{K^-}$ with $\alpha = 1$ (full lines) and $\alpha = 2$ (dashed lines). The shaded areas stand for uncertainties.

In Fig. 4, we present real (left) and imaginary (right) parts of the total K^- potential in ^{208}Pb, calculated self-consistently in the KM1 and KM2 models for the HD and FD options of the multi-nucleon potential $V^{(2)}_{K^-}$. The gray shaded areas stand for uncertainties in $V^{(2)}_{K^-}$ due to the complex parameter B (see Table 1). The underlying chirally-inspired K^- single-nucleon potential (denoted by 'KN') are shown for comparison. The real parts of the K^- optical potential are affected by multi-nucleon interactions markedly less than its imaginary

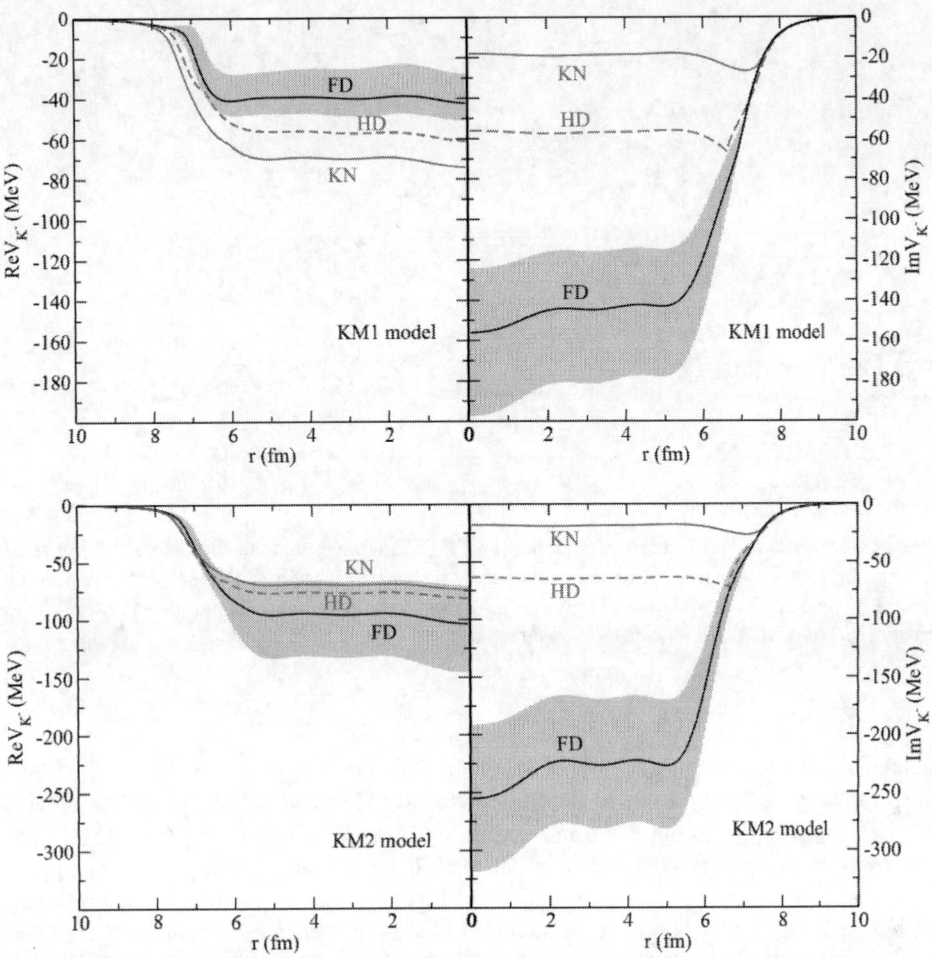

Figure 4. The real (left) and imaginary (right) part of the K^- optical potential in ^{208}Pb, calculated self-consistently in the KM1 (top) and KM2 (bottom) model, for the HD and FD versions of the K^- multi-nucleon potential (see text for details). The shaded areas stand for uncertainties. The single-nucleon K^- potential (KN, green solid lines) calculated in the KM model is shown for comparison.

parts in all considered models. In the KM1 model, the FD and HD options yield ReV_{K^-} shallower than the original single-nucleon $V^{(1)}_{K^-}$ potential. The same holds for the P1 and P2 models (not shown in the figure). In the KM2 model, the overall K^- real potential is deeper than the underlying K^- single-nucleon potential due to the positive sign of ReB. The ReV_{K^-} potentials for HD and FD options differ between each other up to $\approx$ 20 MeV in each interaction model. On the other hand, the imaginary parts of V_{K^-} exhibit much larger dispersion for different versions of $V^{(2)}_{K^-}$, as illustrated in Fig. 4, right panels. The K^- multi-nucleon absorption significantly deepens the imaginary part of the K^- optical potential. For the FD option of $V^{(2)}_{K^-}$, the KM model yields $|\text{Im}V_{K^-}| \gg |\text{Re}V_{K^-}|$ inside the nucleus for both values of α, even when the uncertainties of the K^- multi-nucleon potential are taken into account. The same holds for the P model (not shown in the figure).

The particular role of K^- single- and multi-nucleon absorptions with respect to the nuclear density is illustrated in Fig. 5. Here we compare individual contributions of K^- single

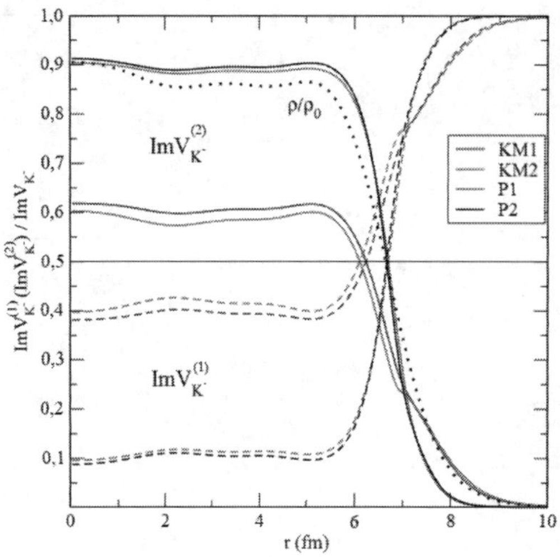

Figure 5. The ratio of $\mathrm{Im}V^{(1)}_{K^-}$ (dashed line) and $\mathrm{Im}V^{(2)}_{K^-}$ (solid line) potentials to the total K^- imaginary potential $\mathrm{Im}V_{K^-}$ as a function of radius, calculated self-consistently for ^{208}Pb+K^- system within different K^-N interaction models and FD option for the K^- multi-nucleon potential. The relative nuclear density ρ/ρ_0 (dotted line) is shown for comparison.

nucleon and multi-nucleon absorptions to the total K^- absorption, expressed as a fraction of $\mathrm{Im}V^{(1)}_{K^-}$ and $\mathrm{Im}V^{(2)}_{K^-}$ with respect to the total imaginary K^- potential $\mathrm{Im}V_{K^-}$ in ^{208}Pb, calculated self-consistently within the P and KM models for the FD option of $V^{(2)}_{K^-}$. The density ρ/ρ_0 (thin dotted line) is shown for comparison. Since the range of the corresponding potentials is different, the relative contributions of $\mathrm{Im}V^{(1)}_{K^-}$ and $\mathrm{Im}V^{(2)}_{K^-}$ to the K^- absorption are changing with radius (density). At the nuclear surface, the K^- absorption on a single nucleon dominates, while it is reduced in the nuclear interior due to vicinity of $\pi\Sigma$ threshold and the multi-nucleon absorption prevails. The models with $\alpha = 2$ yield lower relative fraction of the single-nucleon K^-N absorption in the nuclear medium than the models with $\alpha = 1$. It is due to the self-consistent value of $\sqrt{s}$ at ρ_0 which is closer to the $K^-N \to \pi\Sigma$ threshold in the case with $\alpha = 2$ (see Fig. 3). It is to be noted that both options for multi-nucleon potential, HD and FD, yield similar fraction of the K^- single-nucleon and multi-nucleon absorption inside the nucleus.

Finally, in Table 2 we present $1s$ K^- binding energies B_{K^-} and widths Γ_{K^-}, calculated within the P and KM models. For comparison, we show also K^- binding energies and widths calculated only with the chirally-inspired K^- single-nucleon potentials. The K^- multi-nucleon interactions cause radical increase of K^- widths, while K^- binding energies change much less. The HD option of the $V^{(2)}_{K^-}$ potential yields K^- widths of order $\sim$ 100 MeV and the binding energies much smaller than the corresponding widths in most nuclei. The FD multi-nucleon potential $V^{(2)}_{K^-}$ even does not predict any antikaon bound state in the majority of nuclei. We found $1s$ K^- quasi-bound states in ^{90}Zr and ^{208}Pb but the corresponding K^- binding energies are small and the widths are one order of magnitude larger than the binding energies [2].

[2]For the FD variant of the P2 model, we had to scale huge imaginary part $\mathrm{Im}V_{K^-}$ by factor 0.8 in order to get fully converged self-consistent solution of Eq. 1.

Table 2. 1s K^- binding energies and widths (in MeV) in selected nuclei calculated using the single-nucleon K^-N KM amplitudes (denoted KN); plus a phenomenological amplitude $B(\rho/\rho_0)^\alpha$, where $\alpha = 1$ and 2, for half-density limit (HD) and full density option (FD) (see text for details).

KM model			$\alpha = 1$		$\alpha = 2$	
		KN	HD	FD	HD	FD
^{12}C	B_{K^-}	45	34	not	48	not
	Γ_{K^-}	44	114	bound	125	bound
^{16}O	B_{K^-}	45	34	not	48	not
	Γ_{K^-}	40	109	bound	121	bound
^{40}Ca	B_{K^-}	59	50	not	64	not
	Γ_{K^-}	37	113	bound	126	bound
^{90}Zr	B_{K^-}	69	56	17	72	30
	Γ_{K^-}	36	107	312	120	499
^{208}Pb	B_{K^-}	78	64	33	80	53
	Γ_{K^-}	38	108	273	122	429

P model			$\alpha = 1$		$\alpha = 2$	
		KN	HD	FD	HD	FD
^{12}C	B_{K^-}	64	50	not	64	not
	Γ_{K^-}	28	96	bound	122	bound
^{16}O	B_{K^-}	64	50	not	63	not
	Γ_{K^-}	25	94	bound	117	bound
^{40}Ca	B_{K^-}	81	67	not	82	not
	Γ_{K^-}	14	95	bound	120	bound
^{90}Zr	B_{K^-}	90	74	19	87	not
	Γ_{K^-}	12	88	340	114	bound
^{208}Pb	B_{K^-}	99	82	37	96	47*
	Γ_{K^-}	14	92	302	117	412*

* the solution of Eq. 1 for ImV_{K^-} scaled by factor 0.8

4 Summary

In this contribution, we reported on our most recent self-consistent calculations of K^- nuclear quasi-bound states performed using a K^- single-nucleon potential derived within two chirally motivated coupled channels meson-baryon interaction models P and KM, supplemented by a phenomenological potential representing K^- multi-nucleon interactions. The applied models were recently fitted to kaonic atom data and confronted with the branching ratios of K^- single-nucleon absorption at rest [17]. We demonstrated that the K^- multi-nucleon absorption gives rise to substantial increase of K^- absorption widths. In vast majority of nuclei the widths of K^- nuclear quasi-bound states exceed considerably corresponding K^- binding energies. Identification of such states in experiment seems thus highly unlikely.

Acknowledgements

We thank E. Friedman and A. Gal for valuable discussions. This work was supported by the GACR Grant No. P203/15/04301S.

References

[1] J. Hrtánková, J. Mareš, Phys. Lett. **B 770**, 342 (2017).
[2] J. Hrtánková, J. Mareš, Phys. Rev. **C 96**, 015205 (2017).
[3] W. Weise, Nucl. Phys. **A 835** (2010) 51, and references therein.
[4] A. Gal, E. V. Hungerford, D. J. Millener, Rev. Mod. Phys. **88**, 035004 (2016), and references therein.
[5] N. V. Shevchenko, Few-Body Syst. **58**, 6 (2017), and references therein.
[6] Y. Ikeda, T. Hyodo and W. Weise, Nucl. Phys. **A 881**, 98 (2012).
[7] A. Cieplý, J. Smejkal, Nucl. Phys. **A 881**, 115 (2012).
[8] Z. H. Guo, J. A. Oller, Phys. Rev. **C 87**, 035202 (2013).
[9] M. Mai and U.-G. Meißner, Nucl. Phys. **A 900**, 51 (2013).
[10] A. D. Martin, Nucl. Phys. **B 179**, 33 (1981), and references therein.
[11] M. Bazzi et al (SIDDHARTA Collaboration), Phys. Lett. **B 704**, 113 (2011).
[12] A. Cieplý, E. Friedman, A. Gal, D. Gazda and J. Mareš, Phys. Lett. **B 702**, 402 (2011).
[13] A. Cieplý, E. Friedman, A. Gal, D. Gazda and J. Mareš, Phys. Rev. **C 84**, 045206 (2011).
[14] J. Mareš, E. Friedman, A. Gal, Phys. Lett. **B 606**, 295 (2005).
[15] J. Mareš, E. Friedman, A. Gal, Nucl. Phys. **A 770**, 84 (2006).
[16] D. Gazda, J. Mareš, Nucl. Phys. **A 881**, 159 (2012).
[17] E. Friedman, A. Gal, Nucl. Phys. **A 959**, 66 (2017).
[18] H. Davis, F. Oppenheimer, W. L. Knight, F. R. Stannard, O. Treutler, Nuovo Cimento **53 A**, 313 (1968).
[19] J. W. Moulder, N. E. Garret, L. M. Tucker, W. M. Bugg, G. T. Condo, H. O. Cohn, R. D. McCulloch, Nucl. Phys. **B 35**, 332 (1971).
[20] C. Vander Velde-Wilquet, J. Sacton, J. H. Wickens, D. N. Tovee, D. H. Davis, Nuovo Cimento **39 A**, 539 (1977).
[21] M. M. Sharma, M. A. Nagarajan, and P. Ring, Phys. Lett. **B 312**, 377 (1993).
[22] T. Wass, M. Rho and W. Weise, Nucl. Phys. **A 617**, 449 (1997).

EPJ Web of Conferences **181**, 01010 (2018) https://doi.org/10.1051/epjconf/201818101010
EXA2017

Kaon and Phi Production in Pion-Nucleus Reactions at 1.7 GeV/*c**

Joana Wirth[1,2,**] for the HADES Collaboration

[1]Technische Universität München, Fakultät für Physik, E62
[2]Excellence Cluster Universe, Technische Universität München

Abstract. Nuclear reactions, $\pi^- + A$ ($A = C, W$), at an incident beam momentum of 1.7 GeV/c were measured with the HADES setup at SIS18/GSI. Detailed investigations of the K^+, K^- and ϕ production off nuclei are connected to the study of hadron in-medium properties at nuclear saturation density. A contradictory role is played by the ϕ meson since the ϕN absorption cross-section is assumed to be small due to the OZI suppression which is in contrast to experimental observations. We present the analysis method to identify the ϕ meson exploiting the large K^+K^- branching ratio ($\approx$ 50%). The correction for acceptance and efficiency effects of the detector system will be discussed as well.

1 Introduction

Pion-nucleus reactions allow for a quantitative study of open and hidden strange hadron production and properties at a well defined nuclear density. Of particluar interest is the ϕ meson ($s\bar{s}$) with no-net strangeness. According to the Okubo-Zweig-Iizuka (OZI) rule [1] quark exchange and hence the interaction of the ϕ meson with ordinary (non-strange) baryonic matter is suppressed. Consequently, the ϕN cross-section is assumed to be small.
However, the LEPS Collaboration determined on the basis of the A-dependent ϕ photoproduction yield and a Glauber-type model a large in-medium ϕN absorption cross-section of $\sigma_{\phi N} = 35^{+17}_{-11}$ mb [2]. This extraced absorption cross-section is in agreement with photoproduction measurements performed by the CLAS Collaboration [3]. Here, transperency ratios normalized to carbon where analysed as well within a Glauber model with values of $\sigma_{\phi N}$ in the range of 16 – 70 mb. Also transperency ratios normalized to carbon measured in pA collisions by the ANKE Collaboration point in the same direction with an effective ϕN absorption cross-section in the range of 14 – 21 mb [4]. So far, ϕ absorption in different nuclei has not been studied in pion-induced reactions and should therefore provide complementary information to results obtained with photon and proton beams.

2 The Experiment

Experimental data were obtained with the **H**igh-**A**cceptance **D**i-**E**lectron **S**pectrometer (HADES) [5], currently located at the SIS18 accelerator (GSI, Darmstadt). The detector

*supported by the DFG cluster of excellence "Origin and Structure of the Universe" and SFB 1258
**e-mail: joana.wirth@tum.de

setup comprises six idential sectors surrounding the beam axis covering almost the full azimuthal range and polar angles from 18° to 85°. Two layers of **M**ultiwire **D**rift **C**hambers (MDC), in front and behind a toroidal superconducting magnet, allow for momentum reconstruction with a resolution of $\Delta p/p \approx 3\%$ and particle identification via dE/dx. In this experimental campaign, the first level (LVL1) trigger condition required a T_0 signal in the target detector [6] and a minimum multiplicity of two charged particle hits (M2) in the **M**ultiplicity and **E**lectron **T**rigger **A**rray (META) wall consiting of two time-of-flight detectors, RPC and TOF. In total, 10×10^7 and 13×10^7 events were collected in $\pi^- + C$ and $\pi^- + W$ collisions at $p_{\pi^-} = 1.7$ GeV/c, respectively.

3 Exclusive Phi Production

The ϕ mesons were identified via their dominant decay channel into K^+K^- pairs (BR = 48.9 ± 0.5% [7]). Both charged kaons have been selected employing velocity and momentum cuts[1]. The contamination from other particle species was further reduced by selecting a reconstruced (anti-)kaon mass window of $400 < m$ [MeV/c^2] < 600. Further, the nominal mass value $m_K = 493.677$ MeV/c^2 [7] was then attributed to the identified (anti-)kaon candidates. The resulting K^+K^- invariant mass distribution corresponing to an interval of $750 \leq p$ [MeV/c] < 1250 and $18.8 \leq \theta$ [°] < 27.5 for $\pi^- + C$ collisions is shown in Fig. 1 (left). A clear ϕ peak emerging from the background of non-resonant K^+K^- pairs and track combinations containing misidentified particles is visible. The ϕ signal is described by a double Gaussian accounting for finite resolution effects as well as the collisional broadening due to the scattering of the K^+ and K^- inside the targets, while the background is described by a third-order polynomial together with a Gaussian representing the mass threshold ($2 \times m_K$).
For acceptance and efficiency corrections the Pluto [8] event generator was used. Therefore, the exclusive ϕ production channel $\pi^- + p(^{12}C) \rightarrow \phi[\rightarrow K^+K^-] + n(^{11}B)$ was simulated. The momentum and angular distribution of the generated ϕ mesons is shown in Fig. 2 (left). The K^+K^- pairs originating from ϕ mesons were then processed through simulations modelling the detector response. In Figure 2 (right) the obtained acceptance and efficiency correction matrix is presented. One can see that the acceptance and efficiency corrections strongly depend on phase space, hence a differential correction becomes mandatory. Due the limited ϕ statistic, a double-differential analysis would lead to acceptance and efficiency correction factors averaged over wide and non-constant region in the phase space. This is overcome by weighting each ϕ candidate with its corresponding correction factor in phase space building an integrated and corrected K^+K^- invariant mass distribution (see Fig. 1 (left)).

4 Summary and Outlook

In this contribution, we presented the on-going analysis of ϕ production in $\pi^- + C$ reactions at $p_{\pi^-} = 1.7$ GeV/c measured with the HADES setup. The neutral ϕ is reconstructed via the invariant mass of charged kaon candidates ($M_{K^+K^-}$). On the basis of simulations modelling the detector response in combination with Pluto as an event generator, the acceptance and efficiency corrected ϕ yield can be extracted. In the next step we plan to determine absolute normalized yields. The final goal will be the extraction of an effective ϕN absorption cross-section by comparing the ϕ production in $\pi^- + W$ with respect to $\pi^- + C$ collisions in combination with reaction models providing complementary information to photon- and proton-induced measurements.

[1] $p/\sqrt{p^2 + m_K^2} \pm 0.5 \gtrless \beta$

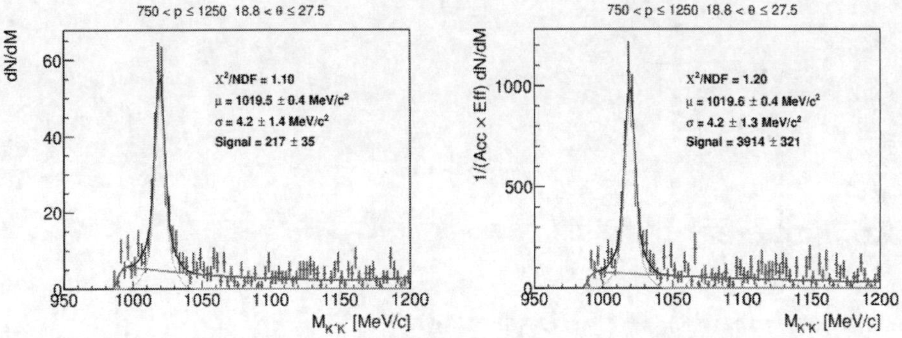

Figure 1. (Color online) Invariant mass distributions for K^+K^- pairs produced in π^-C collisions w/o (left) and with event-by-event acceptance and efficiency corrections (right). The fit consists of two Gaussian for the ϕ signal (light blue line) together with the background described by a polynomial and Gaussian function (pink line). Details see text.

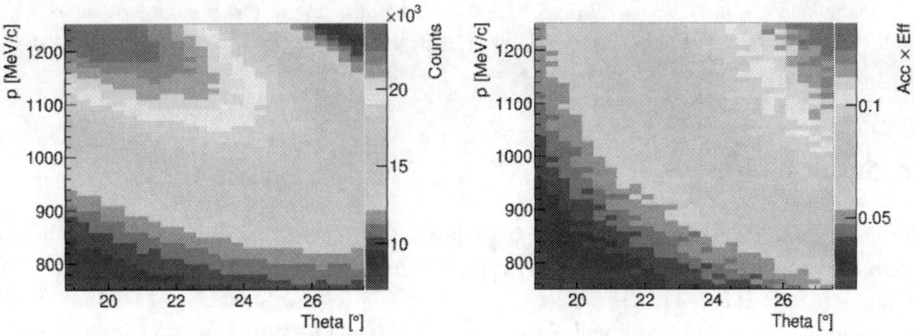

Figure 2. (Color online) The $p-\theta$ distribution of the ϕ mesons generated with the Pluto event generator (left). Combined acceptance and efficiency correction matrix for the ϕ mesons decaying in K^+K^- pairs as a function of $p-\theta$ (right).

References

[1] S. Okubo, Phys. Lett. **5**, 165 (1963); G. Zweig, CERN Report No. TH-401, 1964 (unpublished); J. Iizuka, Prog. Theor. Phys. Suppl. **37**, 21 (1966)
[2] T. Ishikawa, Phys. Lett. B **608**, 215 (2005)
[3] M. H. Wood, Phys. Rev. Lett. **105**, 112301 (2010)
[4] M. Hartmann, Phys. Lett. C **85**, 035206 (2012)
[5] G. Agakishiev *et al.*, Eur. Phys. J. **A41**, 243 (2009)
[6] J. Adamczewski-Musch *et al.*, Eur. Phys. J. **A53** 188 (2017)
[7] C. Patrignani *et al.*, Chin. Phys. **C40**, 100001 (2016)
[8] I. Fröhlich *et al.*, PoS **ACAT2007**, 076 (2007)

Onset of η nuclear binding

Nir Barnea[1], *Betzalel* Bazak[1], *Eliahu* Friedman[1], *Avraham* Gal[1,⋆]
Aleš Cieplý[2], *Jiří* Mareš[2], and *Martin* Schäfer[2]

[1] *Racah Institute of Physics, the Hebrew University, Jerusalem 91904, Israel*
[2] *Nuclear Physics Institute, 250 69 Rez, Czech Republic*

Abstract. Recent studies of η nuclear quasibound states by the Jerusalem-Prague Collaboration are reviewed, focusing on stochastic variational method self consistent calculations of η few-nucleon systems. These calculations suggest that a minimum value $\mathrm{Re}\, a_{\eta N} \approx 1$ fm (0.7 fm) is needed to bind $\eta\,^3$He ($\eta\,^4$He).

1 Introduction

The ηN near-threshold interaction is attractive, owing to the $N^*(1535)$ resonance to which the s-wave ηN system is coupled strongly [1]. This has been confirmed in chiral meson-baryon coupled channel models that generate the $N^*(1535)$ dynamically, e.g. [2]. Hence η nuclear quasibound states may exist [3] as also suggested experimentally by the near-threshold strong energy dependence of the $\eta\,^3$He production cross sections shown in Fig. 1. However, the $\eta\,^3$He scattering length deduced in Ref. [4], $a_{\eta ^3\mathrm{He}} = [-(2.23 \pm 1.29) + i(4.89 \pm 0.57)]$ fm, although of the right sign of its real part, does not satisfy the other necessary condition for a quasibound state pole: $-\mathrm{Re}\, a > \mathrm{Im}\, a$.

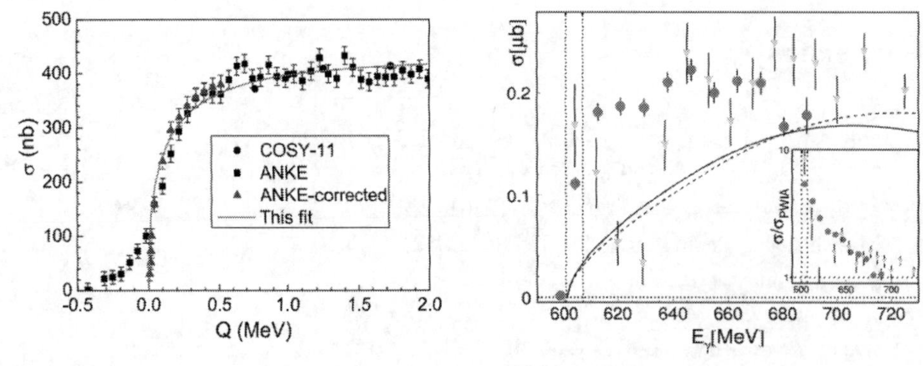

Figure 1. Near-threshold $\eta\,^3$He production cross sections. Left: $dp \rightarrow \eta\,^3$He [4]. Right: $\gamma\,^3\mathrm{He} \rightarrow \eta\,^3$He [5].

⋆Presented by A. Gal (avragal@savion.huji.ac.il) at EXA2017, Vienna, Sept. 2017

Quite generally, experimental searches for η nuclear quasibound states in proton, pion or photon induced η production reactions are inconclusive. Regarding the onset of η nuclear binding, Krusche and Wilkin [6] state: "The most straightforward (but not unique) interpretation of the data is that ηd is unbound, $\eta\,^4$He is bound, but that the $\eta\,^3$He case is ambiguous." Indeed, with $\eta\,^3$He almost bound, one might expect that the denser ^{4}He nucleus should help forming a bound $\eta\,^4$He. Nevertheless, a recent Faddeev-Yakubovsky evaluation [7] of the scattering lengths $a_{\eta^A\mathrm{He}}$ for both He isotopes, $A = 3, 4$, finds this not to be the case, with the denser ^{4}He apparently leading to a stronger reduction of the subthreshold ηN scattering amplitude than in ^{3}He.

The present overview reports and discusses recent few-body stochastic variational method (SVM) calculations of ηNNN and $\eta NNNN$ using several semi-realistic NN interaction models together with two ηN interaction models that, perhaps, provide sufficient attraction to bind η in the ^{3}He and ^{4}He isotopes [8–10].

2 ηN and NN interaction model input

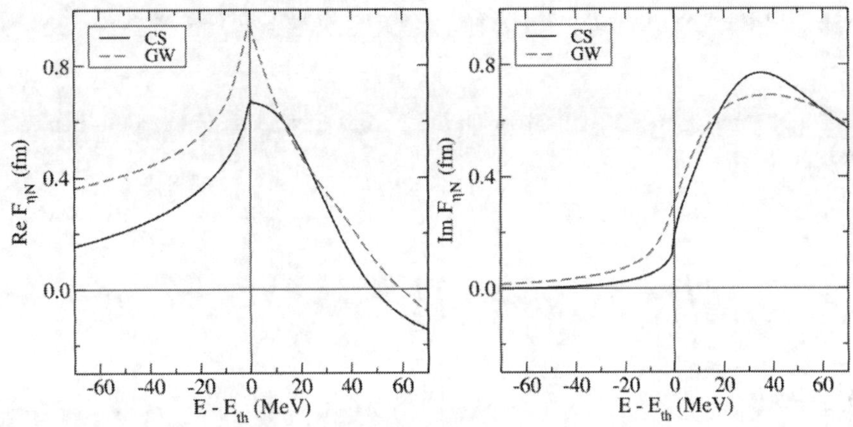

Figure 2. Real and imaginary parts of the ηN cm scattering amplitude near threshold in two meson-baryon coupled channel models: GW [11] and CS [12].

Figure 2 shows ηN s-wave scattering amplitudes $F_{\eta N}(E)$ calculated in two meson-baryon coupled-channel models across the ηN threshold where Re $F_{\eta N}$ has a cusp. These amplitudes exhibit a resonance about 50 MeV above threshold, the $N^*(1535)$. The sign of Re $F_{\eta N}$ below the resonance indicates attraction which is far too weak to bind the ηN two-body system. The threshold values $F_{\eta N}(E_{\rm th})$ are given by the scattering lengths

$$a_{\eta N}^{\rm GW} = (0.96 + i0.26)\ {\rm fm}, \quad a_{\eta N}^{\rm CS} = (0.67 + i0.20)\ {\rm fm}, \tag{1}$$

with lower values below threshold ($E_{\rm th}$ = 1487 MeV). These free-space energy dependent subthreshold amplitudes are transformed to in-medium density dependent amplitudes, in terms of which optical potentials $V_\eta^{\rm opt}(\rho)$ are constructed and used to calculate self consistently η nuclear quasibound states. This procedure was applied in Refs. [13, 14] to several ηN amplitude models, with results for $1s_\eta$ quasibound states in models GW and CS shown in Fig. 3 from ^{12}C to ^{208}Pb.

Figure 3 demonstrates that in both of these ηN amplitude models the $1s_\eta$ binding energy increases with A, saturating in heavy nuclei. Model GW, with larger ηN real and imaginary subthreshold amplitudes than in model CS, gives correspondingly larger values of B_η and Γ_η. While model GW binds

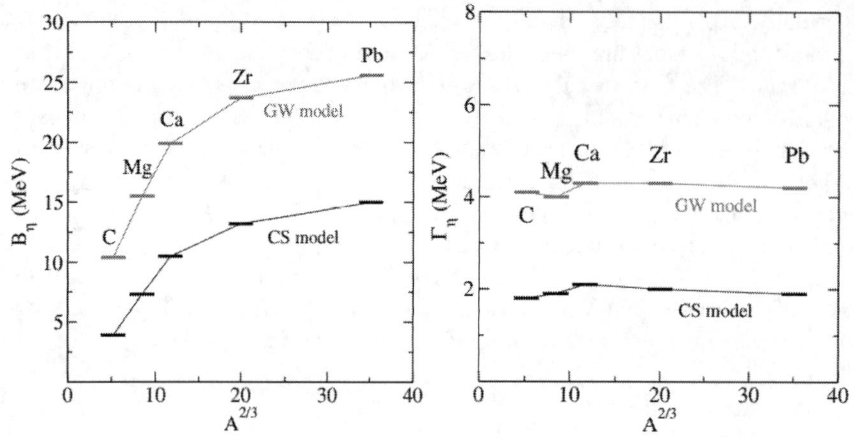

Figure 3. Binding energies B_η (left) and widths Γ_η (right) of $1s_\eta$ quasibound states across the periodic table calculated self consistently [13, 14] using the GW and CS ηN scattering amplitudes of Fig. 2.

η also in nuclei lighter than ^{12}C (not shown in the figure) this needs to be confirmed in few-body calculations.

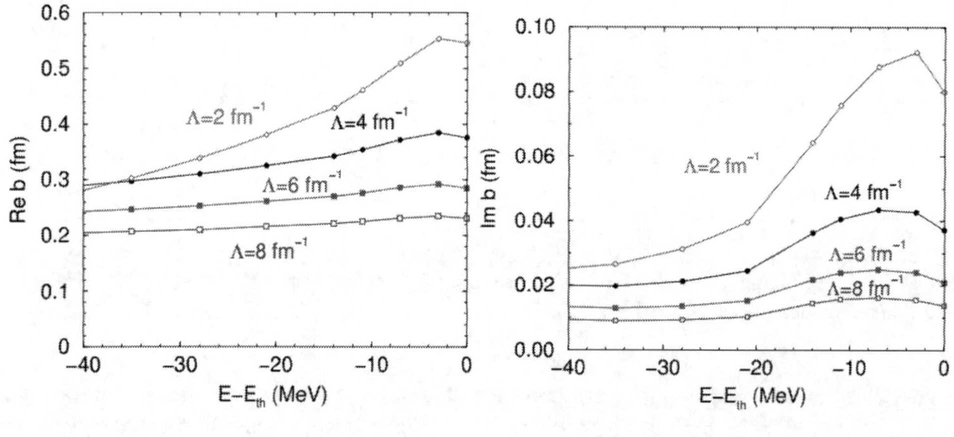

Figure 4. Real and imaginary parts of the strength function $b_\Lambda(E)$ of the effective ηN potential $v^{\rm GW}_{\eta N}(E)$, Eq. (2), obtained from the scattering amplitude $F^{\rm GW}_{\eta N}(E)$ of Fig. 2 below threshold for four values of the scale Λ [9].

Few-body calculations, in distinction from optical model calculations, require the use of effective ηN potentials $v_{\eta N}$ which reproduce the free-space ηN amplitudes below threshold. Fig. 4 shows subthreshold values of the energy dependent strength function $b_\Lambda(E)$ for $v_{\eta N}$ of the form

$$v_{\eta N}(E;r) = -\frac{4\pi}{2\mu_{\eta N}} b_\Lambda(E)\delta_\Lambda(r), \quad \delta_\Lambda(r) = \left(\frac{\Lambda}{2\sqrt{\pi}}\right)^3 \exp\left(-\frac{\Lambda^2 r^2}{4}\right), \tag{2}$$

derived from the scattering amplitude $F^{\rm GW}_{\eta N}(E)$ of Fig. 2 for several choices of inverse range Λ. The normalized Gaussian function $\delta_\Lambda(r)$ is perceived in $\not\pi$EFT (pionless EFT) as a single ηN zero-range

Dirac $\delta^{(3)}(\mathbf{r})$ contact term (CT), regulated by using a momentum-space scale parameter Λ. Regarding the choice of Λ, substituting the underlying short range vector-meson exchange dynamics by a single regulated CT suggests that the scale Λ is limited to values $\Lambda \lesssim m_\rho$ ($\sim$4 fm^{-1}).

Similarly, a $\not{\pi}$EFT energy independent $v_{NN}(r)$ is derived at leading order (LO) by fitting a single regulated CT $\sim \delta_\Lambda(r)$ in each spin-isospin s-wave channel to the respective NN scattering length. A pp Coulomb interaction is included. To avoid NNN and ηNN Thomas collapse in the limit $\Lambda \to \infty$, one introduces a three-body regulated CT for each of these three-body systems [9]:

$$V_{NNN}(r_{ij}, r_{jk}) = d^{\Lambda}_{NNN}\, \delta_\Lambda(r_{ij}, r_{jk}), \qquad V_{\eta NN}(r_{i\eta}, r_{\eta j}) = d^{\Lambda}_{\eta NN}\, \delta_\Lambda(r_{i\eta}, r_{\eta j}), \tag{3}$$

where $\delta_\Lambda(r_{ij}, r_{jk}) = \delta_\Lambda(r_{ij})\delta_\Lambda(r_{jk})$. The three-nucleon CT d^{Λ}_{NNN} is fitted to $B_{\rm exp}(^3{\rm He})$. With no further contact terms, $B_{\rm calc}(^4{\rm He})$ is found in this $\not{\pi}$EFT version [15] to vary moderately with Λ and to exhibit renormalization scale invariance by approaching a finite value $B_{\Lambda\to\infty}(^4{\rm He})$=27.8±0.2 MeV that compares well with $B_{\rm exp}(^4{\rm He})$=28.3 MeV. In contrast, no η-related experimental datum is available for the ηNN CT $d^{\Lambda}_{\eta NN}$ to be fitted to. Two versions for choosing this CT were tested: (i) $d^{\Lambda}_{\eta NN} = d^{\Lambda}_{NNN}$, and (ii) setting $d^{\Lambda}_{\eta NN}$ so that ηd is just bound, i.e. $B_\eta(\eta d) = 0$. Added to $v^{\rm GW}_{\eta N}(E)$, one finds that each of these versions prevents a potential collapse of ηd, with calculated values of $B(\eta\,^4{\rm He})$ that for $\Lambda \geq 4$ fm^{-1} are nearly independent of the adopted version, as shown in Fig. 7 below.

3 Energy independent $\not{\pi}$EFT η nuclear few-body calculations

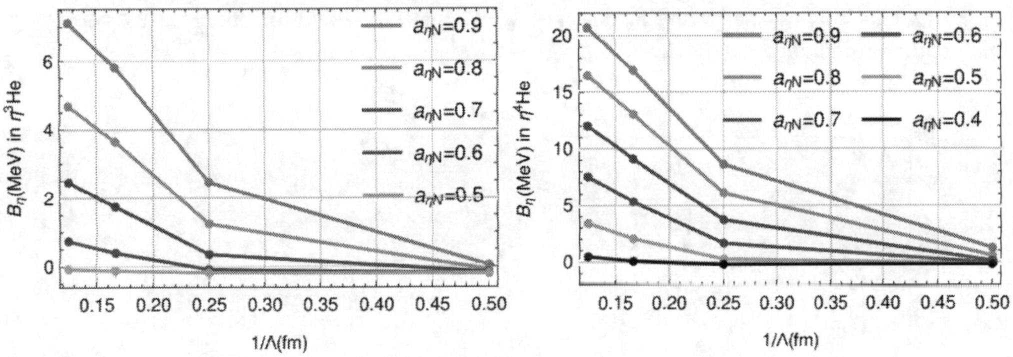

Figure 5. Separation energies B_η obtained in SVM calculations of $\eta\,^3$He (left) and $\eta\,^4$He (right) using $\not{\pi}$EFT NN and ηN real interactions (2) fitted to values of $a_{\eta N} < 1$ fm, plus a universal NNN and ηNN three-body CT (3), $d^{\Lambda}_{\eta NN} = d^{\Lambda}_{NNN}$, as a function of $1/\Lambda$.

Fig. 5 shows η separation energies B_η from $\not{\pi}$EFT SVM calculations of $\eta\,^3$He and $\eta\,^4$He using energy independent ηN potentials $v_{\eta N}(E{=}E_{\rm th}; r)$ fitted to a given real values of $a_{\eta N}$ for a few values of Λ. The figure suggests that binding $\eta\,^3$He ($\eta\,^4$He) requires that $a_{\eta N} \geq 0.55$ fm (0.45 fm), compatible with an effective value Re $a'_{\eta N}$=0.48±0.05 fm derived for a nearly bound $\eta\,^3$He [4]. For input values of $a_{\eta N}$ higher than shown in the figure, beginning at $a_{\eta N}\approx$1.2 fm, the calculated binding energies $B^{A=3,4}_\eta(\Lambda > 4$ fm$^{-1})$ diverge, apparently since ηd becomes bound then at Λ=4 fm^{-1} [8]. Qualitative arguments in support of this ηd onset-of-binding value of $a_{\eta N}$ are given here in Appendix A.

4 Energy dependence in η nuclear few-body systems

Having derived energy dependent ηN potentials $v_{\eta N}(E;r)$, see Eq. (2) and Fig. 4, a two-body subthreshold input energy $\delta\sqrt{s} \equiv E - E_{\rm th}$ needs to be chosen. However, $\delta\sqrt{s}$ is not conserved in the η nuclear few-body problem, so the best one can do is to require that this choice agrees with the expectation value $\langle\,\delta\sqrt{s}\,\rangle$ generated in solving the few-body problem, as given by [10]

$$\langle\delta\sqrt{s}\rangle = -\frac{B}{A} - \xi_N\frac{1}{A}\langle T_A\rangle + \frac{A-1}{A}\mathcal{E}_\eta - \xi_A\xi_\eta\left(\frac{A-1}{A}\right)^2\langle T_\eta\rangle. \tag{4}$$

Here $\xi_{N(\eta)} = m_{N(\eta)}/(m_N + m_\eta)$, $\xi_A = Am_N/(Am_N + m_\eta)$, T_A and T_η denote the nuclear and η kinetic energy operators in appropriate Jacobi coordinates, B is the total binding energy, and $\mathcal{E}_\eta = \langle H - H_N\rangle$ with each Hamiltonian defined in its own cm frame. Self consistency (SC), $\langle\,\delta\sqrt{s}\,\rangle = \delta\sqrt{s}$, is imposed in our calculations, as demonstrated graphically in Fig. 6 (left). Applications of SC to meson-nuclear systems are reviewed in Ref. [16]. For recent K^--atom and nuclear applications see Refs. [17, 18]. More recently, Hoshino et al. [19] argued in a K^-d study that by applying this procedure one violates the requirement of total momentum conservation. In Appendix B here we show specifically for $A = 2$ that our choice of SC Eq. (4) is not in conflict with any conservation law.

Finally, we note that Eq. (4) in the limit $A >> 1$ coincides with the optical model downward energy shift (supplemented by a Coulomb term) used in recent K^- atom and nuclear studies [17, 18]:

$$\langle\delta\sqrt{s}\rangle = -B_N\frac{\rho}{\bar\rho} - \xi_N B_\eta\frac{\rho}{\rho_0} - \xi_N T_N\left(\frac{\rho}{\bar\rho}\right)^{2/3} + \xi_\eta {\rm Re}\,V_\eta^{\rm opt}(\delta\sqrt{s}), \tag{5}$$

where $T_N = \langle T_A\rangle/A = 23.0$ MeV at the average nuclear density $\bar\rho$, $B_N = B_{\rm nuc}/A \approx 8.5$ MeV is an average nucleon binding energy and B_η denotes the calculated η separation energy. All terms here are negative, thereby leading to a downward energy shift.

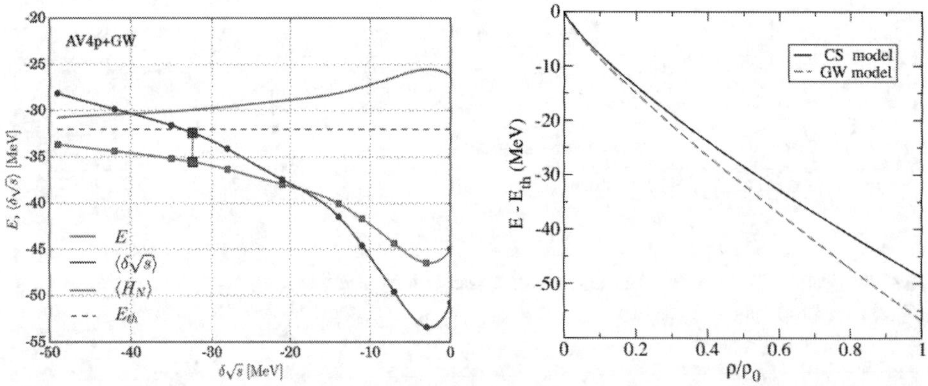

Figure 6. Left: η^4He bound state energy E (red, squares) and the expectation value $\langle\delta\sqrt{s}\rangle$ (blue, circles), calculated using the AV4' NN potential (denoted here AV4p), as a function of the input energy argument $\delta\sqrt{s}$ of the GW ηN potential with $\Lambda = 4$ fm^{-1}. The dotted vertical line marks the self consistent output values of $\langle\delta\sqrt{s}\rangle$ and E. The horizontal dashed line denotes the calculated ^{4}He g.s. energy, marking the threshold of η binding. The green curve shows the expectation value $< H_N >$ of the nuclear core energy. Right: subthreshold ηN energies $\delta\sqrt{s} = E - E_{\rm th}$ probed by the η nuclear optical potential as a function of the relative nuclear RMF density in Ca. Each of the two curves was calculated self consistently for a particular ηN subthreshold amplitude model.

The SC procedure is demonstrated in Fig. 6 (left) for $\eta\,^4$He binding energy calculated using the AV4' NN potential and GW ηN potential with Λ=4 fm^{-1}. The $\eta\,^4$He bound state energy E (excluding rest masses) and the output expectation value $\langle\delta\sqrt{s}\rangle$, where $\delta\sqrt{s}$ stands for the ηN cm energy with respect to its threshold value $E_{\rm th}$, are plotted as a function of the subthreshold input energy argument $\delta\sqrt{s}$ of the potential $v_{\eta N}^{\rm GW}$. The SC condition requires $\delta\sqrt{s} = \langle\delta\sqrt{s}\rangle$ which is satisfied at -32.4 MeV. The corresponding value of $E(\langle\delta\sqrt{s}\rangle)$ then represents the SC energy of $\eta\,^4$He, with $B_\eta^{\rm SC} = 3.5$ MeV, considerably less than the value $B_\eta^{\rm th} = 13$ MeV obtained by disregarding the energy dependence of $v_{\eta N}^{\rm GW}$ and using its threshold value corresponding to $\delta\sqrt{s} = 0$.

In Fig. 6 (right) we present the ηN downward energy shift $\delta\sqrt{s} = E - E_{\rm th}$ as a function of the relative nuclear density ρ/ρ_0 in Ca, evaluated self consistently via Eq. (5) in the CS and GW models. The energy shift at ρ_0 is -55 ± 10 MeV, about twice larger than the SC condition $\delta\sqrt{s} = -B_\eta$ applied in some other works, e.g. [20]. The GW shift exceeds the CS shift owing to the stronger GW amplitude of Fig. 2 and both were incorporated in the calculation of $1s_\eta$ quasibound nuclear states, Fig. 3.

5 Results of η nuclear few-body calculations

Our fully self consistent ηNN, ηNNN and $\eta NNNN$ bound-state calculations [8–10] use the following nuclear core models: (i) $\not\pi$EFT including a three-body contact term [15], (ii) AV4p, a Gaussian basis adaptation of the Argonne AV4' NN potential [21], and (iii) MNC, the Minnesota soft core NN potential [22]. Models GW [11] and CS [12] were used to generate energy dependent ηN potentials which prove too weak to bind any ηNN system when using AV4p or MNC for the nuclear core model. Calculated η separation energies B_η are shown in Figs. 7 and 8.

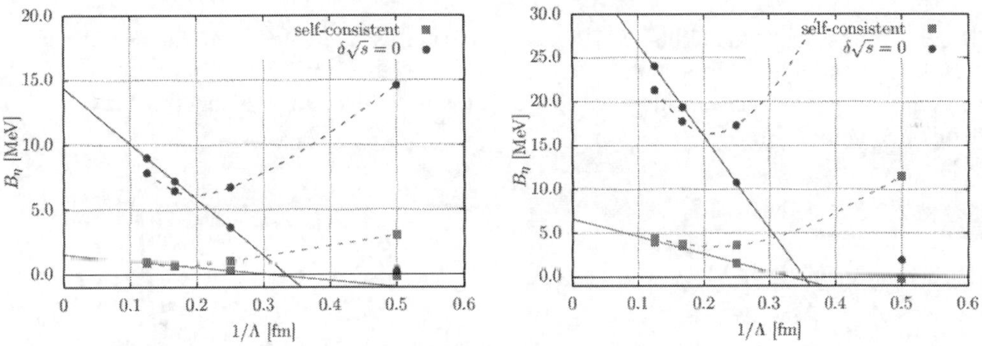

Figure 7. $B_\eta(\eta\,^3\text{He})$ (left) and $B_\eta(\eta\,^4\text{He})$ (right) as a function of $1/\Lambda$ from $\not\pi$EFT few-body calculations [9] using $v_{\eta N}^{\rm GW}$, with (squares) & without (circles) imposing self consistency. Solid lines: $d_{\eta NN}^{\Lambda} = d_{NNN}^{\Lambda}$, dashed lines: $d_{\eta NN}^{\Lambda}$ ensuring that $B_\eta(\eta d) = 0$.

Fig. 7 demonstrates in $\not\pi$EFT the moderating effect that imposing SC (red, squares) by using $v_{\eta N}^{\rm GW}(E_{\rm sc})$, rather than using threshold values $v_{\eta N}^{\rm GW}(E_{\rm th})$ (blue, circles), bears on the calculated B_η values and their Λ scale dependence [9]. Near Λ=4 fm^{-1}, imposing sc lowers $B_\eta(\eta\,^3\text{He})$ by close to 5 MeV and $B_\eta(\eta\,^4\text{He})$ by close to 10 MeV. The figure demonstrates that $B_\eta(\eta\,^4\text{He})$ is always larger than $B_\eta(\eta\,^3\text{He})$. Focusing on scale parameters near Λ=4 fm^{-1} one observes that $\eta\,^3$He is hardly bound by a fraction of MeV, whereas $\eta\,^4$He is bound by a few MeV. The choice of three-body CT $d_{\eta NN}^{\Lambda}$ hardly matters for $\Lambda > 4$ fm^{-1}, becoming substantial at $\Lambda < 4$ fm^{-1}.

Fig. 8 demonstrates in non-EFT calculations the dependence of B_η, calculated self consistently, on the choice of NN and ηN interaction models. Using the more realistic AV4' NN interaction results in less η binding than using the soft-core MNC NN interaction. For $v_{\eta N}^{\rm GW}$ near Λ=4 fm^{-1} the difference amounts to about 0.3 MeV for $\eta\,^3$He and about 1.5 MeV for $\eta\,^4$He; $\eta\,^3$He appears then barely bound whereas $\eta\,^4$He is bound by a few MeV. The weaker $v_{\eta N}^{\rm CS}$ does not bind $\eta\,^3$He and barely binds $\eta\,^4$He using the MNC NN interaction, implying that $\eta\,^4$He is unlikely to bind for the more realistic AV4' NN interaction. For smaller, but still physically acceptable values of Λ down to Λ = 2 fm^{-1}, $\eta\,^3$He becomes unbound and $\eta\,^4$He is barely bound using the AV4' NN and GW ηN interactions.

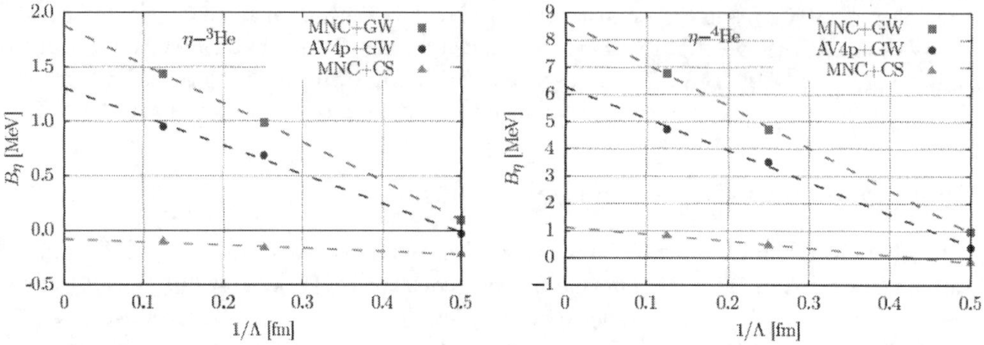

Figure 8. $B_\eta(\eta\,^3$He) (left) and $B_\eta(\eta\,^4$He) (right) as a function of $1/\Lambda$ from few-body calculations [10] using NN and ηN interactions, as marked, and imposing self consistency.

The B_η values calculated in Refs. [8–10] were calculated assuming real Hamiltonians, justified by Im $v_{\eta N} \ll$Re $v_{\eta N}$ from Fig. 4. This approximation is estimated to add near threshold less than 0.3 MeV to B_η. Perturbatively-calculated widths Γ_η of weakly bound states amount to only few MeV, outdating those reported in Ref. [8].

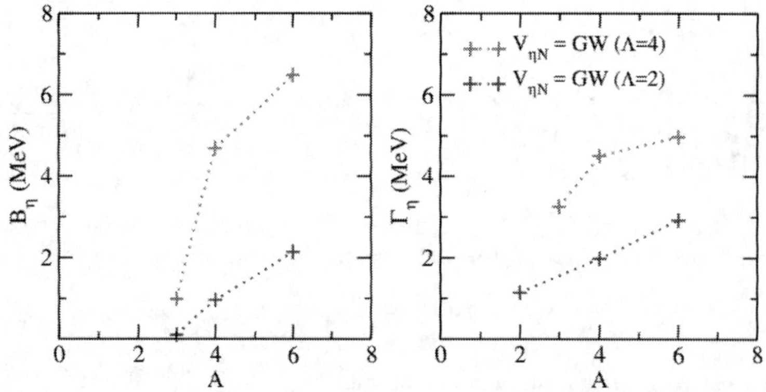

Figure 9. Preliminary SVM results for binding energies B_η (left) and widths Γ_η (right) of $1s_\eta$ quasibound states in ^{3}He, ^{4}He and ^{6}Li, calcualted using the Minnesota NN potential and the GW ηN potential for Λ = 2 and 4 fm^{-1}.

In future work it will be interesting to extend the present SVM few-body calculations to heavier nuclei, beginning with light p-shell nuclei. This represents highly non-trivial task. In Fig. 9 we present preliminary results for $\eta\,^6$Li, using the central Minnesota NN and GW ηN potentials. In this calculation the ^{6}Li nuclear core consisted of a single $S = 1, T = 0$ spin-isospin configuration, yielding $B(^6\text{Li})$=34.66 MeV which is short by almost 2 MeV with respect to a calculation reported in Ref. [24] that used the same NN interaction while including more spin-isospin configurations. The figure suggests that $\eta\,^6$Li is comfortably bound, even for as low value of scale parameter as $\Lambda = 2$ fm^{-1}.

6 Summary

Based mostly on the AV4' results in Fig. 8, which are close to the $\not\pi$EFT results in Fig. 7, we conclude that $\eta\,^3$He becomes bound for Re $a_{\eta N} \sim 1$ fm, as in model GW, while $\eta\,^4$He binding requires a lower value of Re $a_{\eta N} \sim 0.7$ fm, almost reached in model CS. These Re $a_{\eta N}$ onset values, obtained by incorporating the requirements of ηN subthreshold kinematics, are obviously *larger* than those estimated in Sect. 3 upon calculating with $v_{\eta N}(E = E_{\text{th}}; r)$ threshold input. Finally, Re $a_{\eta N} < 0.7$ fm if $\eta\,^4$He is unbound, as might be deduced from the recent WASA-at-COSY search [23].

Appendix A: Onset of ηd binding

Here we apply the Brueckner formula [25], expressing the ηd scattering length in terms of the ηN scattering length, to discuss qualitatively the onset of ηd binding. This formula was originally proposed for a system of a light meson (π meson) and two heavy static nucleons. More recently it was used to estimate the K^-d scattering length (see derivation and discussion in Ref. [26]) where the meson-nucleon mass ratio is similar to that for ηN. For ηd the Brueckner formula assumes the form

$$a_{\eta d} = \int a_{\eta d}(r)|\psi_d(\mathbf{r})|^2 d\mathbf{r}\,, \tag{6}$$

$$a_{\eta d}(r) = \left(1 + \frac{m_\eta}{m_d}\right)^{-1} \frac{\tilde{a}_p + \tilde{a}_n + 2\tilde{a}_p\tilde{a}_n/r}{1 - \tilde{a}_p\tilde{a}_n/r^2}\,, \tag{7}$$

where $\tilde{a} = (1 + m_\eta/m_N)a$, with a_p and a_n standing for $a_{\eta p}$ and $a_{\eta n}$ respectively in the ηN cm system. The numerator in the Brueckner formula consists of single- and double-scattering terms, whereas the denominator provides for the renormalization of these terms by higher-order scattering terms. Since $a_p = a_n$ for the isoscalar η meson, Eq. (7) reduces to a simpler form,

$$a_{\eta d}(r) = \frac{2}{1 + \frac{m_\eta}{m_d}} \frac{\tilde{a}_{\eta N}}{1 - \tilde{a}_{\eta N}/r}\,, \tag{8}$$

which leads to the following approximate expression:

$$a_{\eta d} = \frac{2}{1 + \frac{m_\eta}{m_d}} \frac{\tilde{a}_{\eta N}}{1 - \tilde{a}_{\eta N}\langle 1/r\rangle_d}\,, \tag{9}$$

with expansion parameter $\tilde{a}\langle 1/r\rangle_d$, where $\langle 1/r\rangle_d \approx 0.45$ fm^{-1} for a realistic deuteron wavefunction [27]. Hence, this multiple scattering series faces divergence for sufficiently large ηN scattering length, say $a > 1.4$ fm.

Several straightforward applications of Eq. (9) are as follows:

- For $\mathrm{Re}\, a_{\eta N}^{\mathrm{GW}} = 0.96$ fm, suppressing $\mathrm{Im}\, a_{\eta N}^{\mathrm{GW}}$, one gets $a_{\eta d} = 7.46$ fm. Increasing this GW input value of $a_{\eta N}$, a critical value $a_{\eta N}^{\mathrm{crit}} = 1.40$ fm is reached at which the denominator in Eq. (9) vanishes, signaling the appearance of a zero-energy ηd bound state.
- The LO $\not\pi$EFT nuclear calculations [15] yield a more compact deuteron, $r_{\mathrm{rms}} = 1.55$ fm for $\Lambda \to \infty$ compared to the 'experimental' value $r_{\mathrm{rms}} = 1.97$ fm. Scaling the value $\langle 1/r \rangle_d = 0.45$ used in Eq. (9) by 1.97/1.55, one gets $a_{\eta d} = 18.1$ fm and $a_{\eta N}^{\mathrm{crit}} = 1.10$ fm.
- For the fully complex scattering length $a_{\eta N}^{\mathrm{GW}} = 0.96 + i0.26$ fm, one gets $a_{\eta d} = 4.66 + i4.76$ fm. Increasing $\mathrm{Re}\, a_{\eta N}$ at a frozen value of $\mathrm{Im}\, a_{\eta N}$, $\mathrm{Re}\, a_{\eta d}$ reverses its sign at $\mathrm{Re}\, a_{\eta N}^{\mathrm{crit}} = 1.35$ fm while $\mathrm{Im}\, a_{\eta d}$ keeps positive all through.
- At $\mathrm{Re}\, a_{\eta N}^{\mathrm{crit}} = 1.59$ fm, $|\mathrm{Re}\, a_{\eta d}|$ becomes larger than $\mathrm{Im}\, a_{\eta d}$, which signals a threshold ηd bound state.

Appendix B: ηN subthreshold kinematics

Here we outline the choice of the ηN subthreshold energy shift $\delta\sqrt{s} \equiv \sqrt{s_{\eta N}} - (m_N + m_\eta)$ applied in our η nuclear few-body works [8–10], see Eq. (4), with emphasis on the three-body ηd system. Since the ηN effective potential $v_{\eta N}$ discussed in Sect. 2 is energy dependent, one needs to determine as consistently as possible a fixed *input* value $\delta\sqrt{s}$ at which $v_{\eta N}$ should enter the η nuclear few-body calculation. The two-body Mandelstam variable $\sqrt{s_{\eta N}} = \sqrt{(E_\eta + E_N)^2 - (\vec{p}_\eta + \vec{p}_N)^2}$ which reduces to $(E_\eta + E_N)$ in the ηN two-body cm system is not a conserved quantity in the η nuclear few-body problem since spectator nucleons move the interacting ηN two-body subsystem outside of its cm system. We proceed to evaluate the expectation value of *output* values of $\delta\sqrt{s}$, replacing $\sqrt{s_{\eta N}}$ by $(1/A)\sum_{i=1}^{A} \sqrt{(E_\eta + E_i)^2 - (\vec{p}_\eta + \vec{p}_i)^2}$ due to the antisymmetry of the nuclear wavefunction. Expanding about the ηN threshold, one gets in leading order of p^2

$$\langle \delta\sqrt{s} \rangle \approx \frac{1}{A} \langle \sum_{i=1}^{A} (\mathcal{E}_\eta + \mathcal{E}_i) - \sum_{i=1}^{A} \frac{(\vec{p}_\eta + \vec{p}_i)^2}{2(m_N + m_\eta)} \rangle, \tag{10}$$

where $\mathcal{E}_\eta = E_\eta - m_\eta$ and $\mathcal{E}_i = E_i - m_N$. Since $\sum_{i=1}^{A} \mathcal{E}_i$ is naturally identified with the expectation value of the nuclear Hamiltonian H_N, $\sum_{i=1}^{A} \mathcal{E}_i = \langle H_N \rangle = E_{\mathrm{nuc}} = -B_{\mathrm{nuc}}$, it is natural and also consistent to identify $\mathcal{E}_\eta$ with the expectation value of $(H - H_N)$, $\mathcal{E}_\eta = \langle H - H_N \rangle$. Furthermore, recalling that $\mathcal{E}_\eta + \sum_{i=1}^{A} \mathcal{E}_i = E = -B$, where $E = \langle H \rangle$ is the total η nuclear energy and B is the total binding energy, the sum over the momentum independent part in Eq. (10) gives $[-B + (A-1)\mathcal{E}_\eta]/A$, thereby reproducing two of the four terms in Eq. (4). Note that $\mathcal{E}_\eta$ is negative and its magnitude exceeds the η separation energy B_η. The sum over the momentum dependent part of Eq. (10) yields the other two terms of Eq. (4), which we demonstrate for ηd, $A = 2$.

Since the ηd calculation employes translationally invariant coordinate sets, the total momentum vanishes sharply: $(\vec{p}_\eta + \vec{p}_1 + \vec{p}_2) = 0$. We then substitute $\vec{p}_1^{\,2}$ for $(\vec{p}_\eta + \vec{p}_2)^2$ and $\vec{p}_2^{\,2}$ for $(\vec{p}_\eta + \vec{p}_1)^2$ in the momentum dependent part in Eq. (10), resulting in momentum dependence proportional to $\vec{p}_1^{\,2} + \vec{p}_2^{\,2}$. This is rewritten as

$$\vec{p}_1^{\,2} + \vec{p}_2^{\,2} = \frac{1}{2}[(\vec{p}_1 - \vec{p}_2)^2 + (\vec{p}_1 + \vec{p}_2)^2] = 2\vec{p}_{N:N}^{\,2} + \frac{1}{2}\vec{p}_\eta^{\,2}, \tag{11}$$

where $\vec{p}_{N:N}$ is the nucleon-nucleon relative momentum operator. To obtain the η momentum operator $\vec{p}_\eta$ on the r.h.s. we used again total momentum conservation. Finally, transforming $\vec{p}_{N:N}^{\,2}$ and $\vec{p}_\eta^{\,2}$ to intrinsic kinetic energies, $T_{N:N}$ for the internal motion of the deuteron core and T_η for that of the η meson with respect to the NN cm, one gets for this $A = 2$ special case

$$\langle \delta \sqrt{s} \rangle_{\eta d} \approx -\frac{1}{2}\left(B - \mathcal{E}_\eta + \xi_N \langle T_{N:N} \rangle + \xi_{A=2}\xi_\eta \frac{1}{2}\langle T_\eta \rangle\right), \tag{12}$$

which agrees with Eq. (4) for $A = 2$ upon realizing that $T_{N:N}$ here coincides with $T_{A=2}$ there. To get idea of the relative importance of the various terms in this expression, we assume a near-threshold ηd bound state for which both $\mathcal{E}_\eta$ and $\langle T_\eta \rangle$ are negligible (fraction of MeV each) and $B \to B_d \approx 2.2$ MeV. With $\langle T_{N:N} \rangle \to \langle T_d \rangle$, and with a deuteron kinetic energy $\langle T_d \rangle$ in the range of 10 to 20 MeV, this term provides the largest contribution to the downward energy shift which is then of order −5 MeV for the diffuse deuteron nuclear core.

Acknowledgments

The work of A.C., J.M. and M.S. was supported by the GACR Grant No. P203/15/04301S.

References

[1] R.S. Bhalerao, L.C. Liu, Phys. Rev. Lett. **54**, 865 (1985)
[2] N. Kaiser, P.B. Siegel, W. Weise, Phys. Lett. B **362**, 23 (1995)
[3] Q. Haider, L.C. Liu, Phys. Lett. B **172**, 257 (1986)
[4] J.J. Xie, W.H. Liang, E. Oset, P. Moskal, M. Skurzok, C. Wilkin, Phys. Rev. C **95**, 015202 (2017)
[5] F. Pheron et al., Phys. Lett. B **709**, 21 (2012)
[6] B. Krusche, C. Wilkin, Prog. Part. Nucl. Phys. **80**, 43 (2015)
[7] A. Fix, O. Kolesnikov, Phys. Lett. B **772**, 663 (2017)
[8] N. Barnea, E. Friedman, A. Gal, Phys. Lett. B **747**, 345 (2015)
[9] N. Barnea, B. Bazak, E. Friedman, A. Gal, Phys. Lett. B **771**, 297 (2017), **775**, 364 (2017)
[10] N. Barnea, E. Friedman, A. Gal, Nucl. Phys. A **968**, 35 (2017)
[11] A.M. Green, S. Wycech, Phys. Rev. C **71**, 014001 (2005)
[12] A. Cieplý, J. Smejkal, Nucl. Phys. A **919**, 46 (2013)
[13] E. Friedman, A. Gal, J. Mareš, Phys. Lett. B **725** 334 (2013)
[14] A. Cieplý, E. Friedman, A. Gal, J. Mareš, Nucl. Phys. A **925**, 126 (2014)
[15] J. Kirscher, E. Pazy, J. Drachman, N. Barnea, Phys. Rev. C **96**, 024001 (2017)
[16] A. Gal, E. Friedman, N. Barnea, A. Cieplý, J. Mareš, D. Gazda, Acta Phys. Polon. B **45**, 673 (2014)
[17] E. Friedman, A. Gal, Nucl. Phys. A **959**, 66 (2017)
[18] J. Hrtánková, J. Mareš, Phys. Lett. B **770**, 342 (2017), Phys. Rev. C **96**, 015205 (2017)
[19] T. Hoshino, S. Ohnishi, W. Horiuchi, T. Hyodo, W. Weise, Phys. Rev. C **96**, 045204 (2017)
[20] C. García-Recio, T. Inoue, J. Nieves, E. Oset, Phys. Lett. B **550**, 47 (2002)
[21] R.B. Wiringa, S.C. Pieper, Phys. Rev. Lett. **89**, 182501 (2002)
[22] D.R. Thompson, M. LeMere, Y.C. Tang, Nucl. Phys. A **286**, 53 (1977)
[23] P. Adlarson, et al. (WASA-at-COSY Collab.), Nucl. Phys. A **959**, 102 (2017)
[24] P. Navrátil, E. Caurier, Phys. Rev. C **69**, 014311 (2004)
[25] K.A. Brueckner, Phys. Rev. **89**, 834 (1953)
[26] A. Gal, Int. J. Mod. Phys. A **22**, 226 (2007)
[27] M. Pavón Valderrama, E. Ruiz Arriola, arXiv:nucl-th/0605078

Dynamically generated hadronic states in the $\bar{K}N$ and ηN coupled-channels interactions

Aleš Cieplý[1,*]

[1]Nuclear Physics Institute, 250 68 Řež, Czechia

Abstract. We compare and review the theoretical predictions for the $\bar{K}N$ and ηN elastic amplitudes where sizable variations are found among the considered approaches, especially at subthreshold energies relevant for studies of kaonic atoms and meson-nuclear quasi-bound states. Conditions for an appearance of dynamically generated states in meson-baryon multi-channel interactions are established and discussed for the $\bar{K}N$ and ηN systems.

1 Introduction

A modern treatment of low-energy meson-nucleon interactions is based on effective chiral Lagrangians that implement the QCD dynamics and symmetries in a regime where the large running coupling constant makes the perturbative approach intractable. A good number of chiral approaches have been developed to describe the s-wave interactions of the pseudoscalar meson octet (π, K, η) with the ground state baryon octet (N, Λ, Σ, Ξ), see e.g. [1–7] for papers dealing with the $\bar{K}N$ system and [8–11] for those related to the ηN one. Unfortunately, the presence of resonances hinders the use of perturbative chiral expansion. This problem is standardly overcome by summing the presumably dominant part of the perturbation series by means of either the Lippmann-Schwinger or Bethe-Salpeter coupled-channels equations. The effective potentials that enter these equations are then constructed to match the meson-baryon scattering amplitudes derived from the underlying chiral Lagrangian up to a given chiral order. Free parameters of the models are fixed partly by relating them to meson and baryon properties, and partly by fitting the available experimental data.

In our contribution we reflect briefly on the current status of theoretical approaches to the $\bar{K}N$ and ηN systems and discuss the conditions for an appearance of dynamically generated resonances in such strongly interacting multi-channel systems. Both systems are characterized by well established resonant states observed very close to the channel thresholds. In the strangeness $S = -1$ sector it is the $\Lambda(1405)$ resonance lying about 30 MeV below the $\bar{K}N$ threshold while the low energy ηN interaction representing the $S = 0$ sector is dominated by the $N^*(1535)$ resonance found about 50 MeV above the elastic channel threshold. The resonances have a big impact on dynamics of the $\bar{K}N$ and ηN systems not only in the free space but they also influence significantly the propagation of the $\bar{K}$ and η mesons in nuclear matter at energies close to threshold. In particular, the findings presented in our contribution are also relevant for studies of kaonic atoms and for a possible formation of $\bar{K}$-nuclear and η-nuclear quasi-bound states [12–15].

*e-mail: cieply@ujf.cas.cz

In the next section, we demonstrate the ability of several up-to-date chiral approaches to reproduce selected experimental data and show the ambiguities of the theoretical models when making predictions in areas not so well restricted by the data. We then turn our attention to looking at the origin of the dynamically generated resonances relating them to the poles of the scattering amplitudes and discuss conditions of their existence. The paper is closed with a brief summary.

2 Model predictions and comparisons

2.1 $\bar{K}N$ amplitudes

The meson-baryon channels involved in the $S = -1$ sector ($\pi\Lambda$, $\pi\Sigma$, $\bar{K}N$, $\eta\Lambda$, $\eta\Sigma$ and $K\Xi$) have the channel thresholds spread over a large interval of energies from 1250 to 1810 MeV. However, since we aim at description of $\bar{K}N$ interactions for energies at and below the elastic threshold the relevant experimental data that determine the fitted model parameters are the K^-p threshold branching ratios, low energy elastic and inelastic K^-p cross-sections (see e.g. [1] for references) and 1s level characteristics (the shift and width due to strong interaction) of kaonic hydrogen, measured recently with quite high accuracy [16]. This set of experimental data is reproduced equally well by all current theoretical approaches that include the NLO corrections to the leading order in the chiral expansion. An overview of the models and their direct comparison was discussed in [17]. Here we follow the notation introduced there and label the models by locations of their authors referring to the Kyoto-Munich ($\mathrm{KM_{NLO}}$) [4], Prague ($\mathrm{P_{NLO}}$) [5], Murcia ($\mathrm{M_I}$ and $\mathrm{M_{II}}$) [6] and Bonn ($\mathrm{B_2}$ and $\mathrm{B_4}$) [7] approaches.

In Fig. 1 taken from [17] we demonstrate that the considered approaches lead to very different predictions for the K^-p amplitude extrapolated to sub-threshold energies as well as for the K^-n amplitude. The theoretical ambiguities observed below the $\bar{K}N$ threshold are much larger then those standardly indicated by uncertainty bounds derived from variations of the K^-p scattering length within constraints enforced by the kaonic hydrogen data, see e.g. Ref. [4]. Although the theoretical predictions differ significantly in areas not constrained by experimental data the magnitude of these variations has not been anticipated earlier. We also note that the K^-p amplitude generated by the Bonn models $\mathrm{B_2}$ and $\mathrm{B_4}$ deviates from the other model predictions even at energies above the threshold due to a conceptual difference in a treatment of the s-wave projection. Other differences are caused by different fitting procedures and by inclusion of additional experimental data, the $\pi\Sigma$ mass distributions in the CLAS photo-production measurement [18] in case of the Bonn approach, and more data for processes at higher energies around the $\Lambda(1670)$ resonance in case of the Murcia approach.

2.2 ηN amplitudes

The ηN interactions that represent the $S = 0$ sector were studied in [11]. In close analogy with the $\bar{K}N$ interactions only the two-body meson-baryon channels involving particles from the meson and baryon octets were considered, i.e. the πN, ηN, $K\Lambda$ and $K\Sigma$ channels. The model parameters were fitted to the low energy S_{11} and S_{31} partial waves amplitudes and to the $\pi^-p \to \eta n$ reaction total cross section data, both taken from the SAID database [19]. The quality of the fits was quite good ($\chi^2/dof \approx 1.5$) and in particular, the structure of the $N^\star(1535)$ resonance observed in the ηn production cross sections was nicely reproduced.

In Fig. 2 taken from [13] we compare the elastic ηN amplitude generated by our CS model [11] with predictions by several other authors [8–10, 20]. It should be noted that unlike in the $S = 0$ sector the various theoretical approaches compared here differ conceptually (e.g. the GW model [20] is based on the K-matrix method) as well as in the selection of the fitted

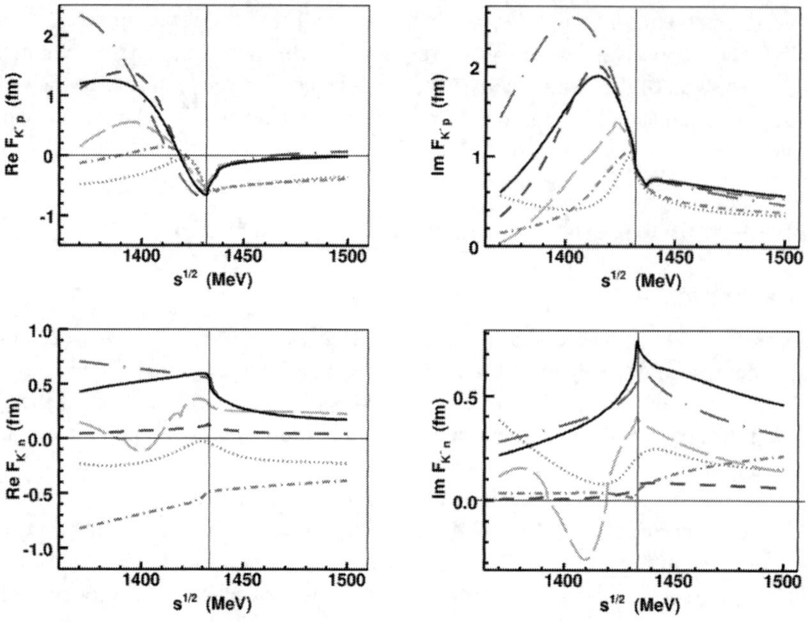

Figure 1. The K^-p (top panels) and K^-n (bottom panels) elastic scattering amplitudes generated by the NLO approaches considered in our work. The various lines refer to the models: B_2 (dotted, purple), B_4 (dot-dashed, red), M_I (dashed, blue), M_{II} (long-dashed, green), P_{NLO} (dot-long-dashed, violet), KM_{NLO} (continuous, black). The thin vertical lines mark the $\bar{K}N$ threshold.

experimental data. As there is no η-hydrogen atom either the lack of any direct data at the ηN threshold energy is reflected by varied predictions for the ηN scattering length $a_{\eta N}$. For the real part of the later the considered models provide us with values in a quite a broad range from $\Re a_{\eta N} = 0.26$ fm up to 0.96 fm, while the imaginary part turns out to be almost model independent, $\Im a_{\eta N} \approx 0.20 - 0.26$ fm. It appears that a sufficiently large value of $\Re a_{\eta N} \gtrsim 0.7$ fm is a prerequisite for forming quasi-bound η-nuclear states as light as η^4He [21]. Our comparison presented in Fig. 2 demonstrates that the current theoretical predictions are not conclusive in this respect with only the models GW [20] and CS [11] meeting the criterion.

3 Dynamically generated poles

The chirally motivated approaches predict two poles assigned to the $\Lambda(1405)$ resonance [22], [3]. The usual assumption is that one of the poles is formed from a resonance in the $\pi\Sigma$ and the other from a quasi-bound state in the $\bar{K}N$ channel [23]. In general, the transition amplitudes matrix F_{ij} (with indexes running over the channel space) has poles for complex energies z (equal to the meson-baryon CMS energy $\sqrt{s}$ on the real axis) if a determinant of the inverse matrix is equal to zero,

$$\det|F^{-1}(z)| = \det|V^{-1}(z) - G(z)| = 0 \quad , \tag{1}$$

where V stands for the potential matrix V_{ij}, that is proportional to a coupling matrix C_{ij} defined by the underlying chiral symmetry, and the intermediate state Green functions are represented by the diagonal G matrix. In the hypothetical zero coupling limit (ZCL), in

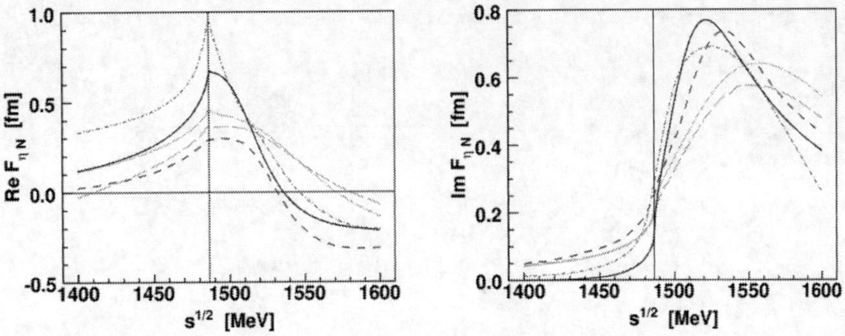

Figure 2. Real (left panel) and imaginary (right panel) parts of the ηN CMS scattering amplitude $F_{\eta N}(\sqrt{s})$ as a function of the total CMS energy $\sqrt{s}$ from five meson-baryon coupled-channel interaction models, in decreasing order of Re $a_{\eta N}$. Dot-dashed curves: GW [20]; solid: CS [11]; dotted: KSW [8]; long-dashed: M2 [10]; short-dashed: IOV [9]. The thin vertical line denotes the ηN threshold.

which the non-diagonal inter-channel couplings are switched off ($V_{ij} = 0$ for $i \neq j$), the poles persist in the decoupled channels and the condition for a pole of the amplitude becomes

$$\prod_n [1/V_{nn}(z) - G_n(z)] = 0 \quad . \tag{2}$$

There will be a pole in channel n at a Riemann sheet (RS) $[+/-]$ (physical/unphysical) if the pertinent n-th factor of the product on the r.h.s. of Eq. (2) equals zero. Then, such ZCL poles can be connected with those found in the physical limit by following their movement upon increasing gradually the inter-channel couplings up to their physical values.

3.1 The strangeness $S = -1$ sector

The concept of relating the poles origins to their existence in the ZCL was used in [17] to analyze and compare the pole contents of the current theoretical approaches to the $\bar{K}N$ interactions that were reviewed in Section 2.1. It follows from Eq. (2) that only states with nonzero diagonal couplings $C_{nn} \neq 0$ can generate poles in the ZCL. If one considers just the leading order Weinberg-Tomozawa interaction only the $\pi\Sigma$, $\bar{K}N$ and $K\Xi$ channels comply with this condition for both, the isoscalar ($I = 0$) and isovector ($I = 1$) states. Though, as some of the models, in particular the Bonn and Murcia ones, have sizable NLO contributions, the ZCL poles may not be restricted to just these three channels. This can be seen in Table 1 that reviews the findings of [17]. There, we show, in both isospin sectors, the channels in which a pole assigned to a given resonance persists when the ZCL is reached. The two poles assigned to the $\Lambda(1405)$ are shown with the indexes 1 and 2. In the complex energy plane the $\Lambda_1(1405)$ pole is usually found at lower energy and further from the real axis than the $\Lambda_2(1405)$ pole. For all models the origin of the $\Lambda_1(1405)$ pole can be traced to a ZCL resonance in the $\pi\Sigma$ channel. The $\Lambda_2(1405)$ pole couples most strongly to the $\bar{K}N$ channel, so it came as a surprise that the pole origin is found in the $\eta\Lambda$ channel for the B_2 and M_{II} models. Most likely, the large NLO couplings occurring in these models are responsible for the appearance of the $\eta\Lambda$ bound state in the ZCL. The other models that have the ZCL pole in the $\bar{K}N$ channel should be preferred if one anticipates a simplified picture of a $\bar{K}N$ bound state submerged in the $\pi\Sigma$ continuum [23].

Table 1. The origins (channels) of the poles generated by the models considered in the $\bar{K}N$ sector for which the poles are found when inter-channel couplings are switched off.

Resonances	Models: P_{NLO}	KM_{NLO}	M_I	M_{II}	B_2	B_4
$\Lambda_1(1405)$	$\pi\Sigma$	$\pi\Sigma$	$\pi\Sigma$	$\pi\Sigma$	$\pi\Sigma$	$\pi\Sigma$
$\Lambda_2(1405)$	$\bar{K}N$	$\bar{K}N$	$\bar{K}N$	$\eta\Lambda$	$\eta\Lambda$	$\bar{K}N$
$\Lambda(1670)$	$K\Xi$	—	$K\Xi$	$K\Xi$	—	$K\Xi$
$\bar{K}N(I=1)$	$\bar{K}N$	$\eta\Sigma$	$\bar{K}N$	$\bar{K}N$	—	—
$\Sigma(1750)$	$K\Xi$	—	$K\Xi$	$K\Xi$	—	$K\Xi$

In the isoscalar sector the models can also account for the $\Lambda(1670)$ resonance that emerges from the $K\Xi$ pole found in the ZCL. It was argued in Ref. [17] that an appearance of such pole is related to a particular condition imposed on a subtraction constant (or an inverse range in case of the Prague approach). If the condition is not met, the pole is missing as happens for the KM_{NLO} and B_2 models. One should note, however, that with an exception of the Murcia approach the other approaches did not aim at describing the experimental data in the $\Lambda(1670)$ energy region, so it is not surprising that the pole is either completely missing or not at an appropriate position in those models.

Similarly, in the isovector sector the models can provide a pole which can be related to the $\Sigma(1750)$ resonance and the origin of this pole can be traced to the $K\Xi$ virtual (or bound) state in the ZCL. Several of the discussed models also predict an isovector $\bar{K}N$ pole located below the $\bar{K}N$ threshold at the Riemann sheet which is physical in the $\pi\Sigma$ and unphysical in the $\bar{K}N$ channel (it would be the third Riemann sheet if only these two channels were coupled). This pole emerges from an isovector $\bar{K}N$ virtual state generated in the ZCL by the Prague and Murcia models, though the Kyoto-Munich model has it in the $\eta\Sigma$ channel. We note that an existence of this pole was already witnessed in Refs. [22], [3] and [12]. It is understood that it relates to the cusp structure in the energy dependence of the elastic K^-n amplitude obtained for both, the P_{NLO} and the KM_{NLO} models as seen in Fig. 1.

3.2 The strangeness $S = 0$ sector

A similar comparative analysis of the poles origins in the ηN related $S = 0$ sector is missing. Among the models discussed in Section 2.2 just the authors of the CS model [11] studied the pole movements when going into the ZCL. For strangeness $S = 0$ and isospin $I = 1/2$ only the πN and $K\Sigma$ channels have nonzero diagonal couplings. Since the πN threshold is too far below the ηN one the only feasible option (barring large NLO contributions) to generate dynamically a pole at energies above 1500 MeV appears to be the $K\Sigma$ channel with a diagonal coupling strong enough to generate a virtual state. This is confirmed in Fig. 3 taken from [11] we show here to demonstrate the movement of the poles on two different Riemann sheets labeled by signs of the imaginary parts of meson-nucleon CMS momenta in all involved channels. In this notation [+,+,+,+] represents the physical RS and the signs are in order of the channel thresholds. Both pole trajectories shown in Fig. 3 emerge from the same $K\Sigma$ virtual state in the ZCL. The trajectory developing on the [+,+,-,-] RS (continuous line, z_1 pole) reaches the position assigned to the $N^\star(1535)$ resonance in the physical limit. The shadow pole that develops on the [+,+,+,-] RS (dashed line, z_2 pole) can be related to the $N^\star(1650)$ resonance. Although, the z_2 pole energy in the physical limit is significantly off the experimental one, the position of the pole can be tuned by varying the model parameters.

Concerning the $N^\star(1535)$ resonance we would also like to stress that it is far from clear that it should be generated dynamically. The fact that the dynamically generated $N^\star(1535)$

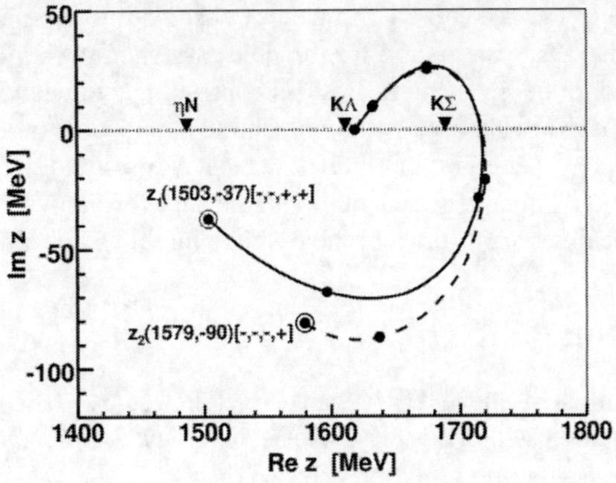

Figure 3. Movement of the poles z_1 and z_2 upon gradually switching off the inter-channel couplings. The positions of the poles in a physical limit are encircled and marked by the labels that also denote the Riemann sheets the poles are located on. The small dots mark positions of the poles for the scaling factors from $x = 0$ (zero coupling limit) to $x = 1$ (physical limit) in steps of 0.2.

pole emerges from the $K\Sigma$ virtual state located at much higher energies indicates that the parameters of the CS model might be stretched beyond their feasible limits. However, the coupled channels model of Ref. [11] has no other choice to accommodate the resonant shape observed in the ηN formation cross section. Thus, one could argue in favor of considering a genuine excited quark state to represent the resonance instead of generating it dynamically. We also note that a satisfactory description of both the $N^\star(1535)$ and $N^\star(1650)$ resonances can be achieved by considering the resonant components of the $\pi\pi N$ system in addition to the pseudoscalar-baryon coupled channels discussed here, see e.g. Refs. [24] and [25].

4 Summary

We have demonstrated that the current theoretical models provide quite varied predictions for the elastic $\bar{K}N$ and ηN amplitudes at subthreshold energies and in case of the ηN system at the threshold too. This observation is intriguing since the same models describe about equally well the available experimental data including the shape of the near threshold $\Lambda(1405)$ and $N^\star(1535)$ resonances observed in the $\pi\Sigma$ and ηN spectra, respectively. Additional experimental constrains on the discussed theoretical models can be provided by studying the impact of nuclear matter on the coupled-channels meson-baryon dynamics. In particular, it was already revelaed in [14] that only the $\bar{K}N$ amplitudes generated by the $\mathrm{KM_{NLO}}$ and $\mathrm{P_{NLO}}$ models are consistent with the systematics of kaonic atom data and reproduce the observed fraction of antikaon absorption on single nucleons. It is worth noting that the Murcia and Bonn approaches that do not pass this additional in-medium test are those characterized by large NLO contributions, a feature that distinguishes them from the Kyoto-Munich and Prague ones.

The origin of resonances generated dynamically in meson-baryon coupled channels systems can be traced to poles emerging due to sufficiently strong interactions in the decoupled channels with inter-channel couplings switched off. These ZCL poles are then moved into their positions found in the physical limit when the inter-channel couplings are gradually turned on. This concept provides us with additional insights on the conditions under which dynamical resonances are formed. The presented contribution reviews our recent findings for the poles assigned to the $\Lambda^\star$ and $\Sigma^\star$ resonances observed in the meson-baryon $S = -1$ channels coupled to the $\bar{K}N$ system, and for the poles assigned to the $N^\star$ resonances in the $S = 0$ channels related to the ηN system. Besides confirming and analyzing the double pole structure of the $\Lambda(1405)$ we also note that some chirally motivated models predict an existence of the isovector pole at energies not far below the $\bar{K}N$ threshold. In the $S = 0$ sector, the chirally motivated coupled-channels dynamics can account for the $N^\star(1535)$ and $N^\star(1650)$ resonances, though the dynamical status of the first one might be questioned.

Acknowledgements

This work was supported by the GACR Grant No. P203/15/04301S. The author also acknowledges a collaboration with M. Mai, U.-G. Meißner and J. Smejkal who contributed to Refs. [17] and [11] the current presentation is based on.

References

[1] N. Kaiser, P. B. Siegel and W. Weise, Nucl. Phys. A **594**, 325–345 (1995)
[2] E. Oset and A. Ramos, Nucl. Phys. A **635**, 99 (1998)
[3] D. Jido, J. A. Oller, E. Oset, A. Ramos and U.-G. Meißner, Nucl. Phys. A **725**, 181–200 (2003)
[4] Y. Ikeda, T. Hyodo and W. Weise, Nucl. Phys. A **881**, 98 (2012)
[5] A. Cieplý and J. Smejkal, Nucl. Phys. A **881**, 115 (2012)
[6] Z. H. Guo and J. A. Oller, Phys. Rev. C **87**, 035202 (2013)
[7] M. Mai and U.-G. Meißner, Eur. Phys. J. A **51**, 30 (2015)
[8] N. Kaiser, P. B. Siegel and W. Weise, Phys. Lett. B **362**, 23–28 (1995)
[9] T. Inoue, E. Oset and M. J. Vicente Vacas, Phys. Rev. C **65**, 035204 (2002)
[10] M. Mai, P. C. Bruns and U.-G. Meißner, Phys. Rev. D **86**, 094033 (2012)
[11] A. Cieplý and J. Smejkal, Nucl. Phys. **A 919**, 46–66 (2013)
[12] A. Cieplý, E. Friedman, A. Gal, D. Gazda and J. Mareš, Phys. Rev. C **84** (2011) 045206
[13] A. Cieplý, E. Friedman, A. Gal and J. Mareš, Nucl. Phys. A **925**, 126–140 (2014)
[14] E. Friedman and A. Gal, Nucl. Phys. A **959**, 66 (2017)
[15] J. Hrtánková and J. Mareš, Phys. Rev. C **96**, 015205 (2017)
[16] M. Bazzi *et al.* [SIDDHARTA Collaboration], Phys. Lett. B **704**, 113 (2011)
[17] A. Cieplý, M. Mai, U.-G. Meißner and J. Smejkal, Nucl. Phys. A **954**, 17 (2016)
[18] K. Moriya *et al.* [CLAS Collaboration], Phys. Rev. C **87**, 035206 (2013)
[19] SAID Partial-Wave Analysis Facility, The George Washington University, URL: http://gwdac.phys.gwu.edu/KWW
[20] A. M. Green and S. Wycech, Phys. Rev. C **71**, 014001 (2005), Erratum: Phys. Rev. C **72**, 029902 (2005)
[21] N. Barnea, E. Friedman and A. Gal, Nucl. Phys. A **968**, 35–47 (2017)
[22] J. A. Oller and U.-G. Meißner, Phys. Lett. B **500**, 263 (2001)
[23] T. Hyodo and W. Weise, Phys. Rev. C **77**, 035204 (2008)
[24] H. Kamano, S. X. Nakamura, T.-S. H. Lee and T. Sato, Phys. Rev. C **88**, 035209 (2013)
[25] E. J. Garzon and E. Oset, Phys. Rev. C **91**, 025201 (2015)

EPJ Web of Conferences **181**, 01013 (2018) https://doi.org/10.1051/epjconf/201818101013
EXA2017

Electromagnetic transition form factors of baryons in a relativistic Faddeev approach

Reinhard Alkofer[1,*], *Christian S.* Fischer[2], and *Hèlios* Sanchis-Alepuz[1]

[1]Institut für Physik, Karl-Franzens–Universität Graz, NAWI Graz, 8010 Graz, Austria

[2]Institut für Theoretische Physik, Justus-Liebig–Universität Giessen, 35392 Giessen, Germany

Abstract. The covariant Faddeev approach which describes baryons as relativistic three-quark bound states and is based on the Dyson-Schwinger and Bethe-Salpeter equations of QCD is briefly reviewed. All elements, including especially the baryons' three-body-wave-functions, the quark propagators and the dressed quark-photon vertex, are calculated from a well-established approximation for the quark-gluon interaction. Selected previous results of this approach for the spectrum and elastic electromagnetic form factors of ground-state baryons and resonances are reported. The main focus of this talk is a presentation and discussion of results from a recent investigation of the electromagnetic transition form factors between ground-state octet and decuplet baryons as well as the octet-only Σ^0 to Λ transition.

1 Introduction

Over the last decades we learned that the structure of baryons, including the nucleon, shows many previously unexpected features. Being composites of three valence quarks, a quark-antiquark sea, and glue, baryons are surprisingly complicated objects. Although there is no doubt that a description of baryons is possible within Quantum Chromo Dynamics (QCD) corresponding investigations are very challenging and computationally demanding. A prominent example is hadron spectroscopy from lattice QCD calculations: It had to overcome many technical obstacles before it reached today's precision, see, *e.g.*, Daniel Mohler's talk at this conference [1].

From those and many other investigations it became evident that the internal structure of mesons and baryons is very sensitive to the way quarks and gluons interact at sub-GeV scales. Even more than confinement the dynamical breaking of chiral symmetry is an important phenomenon shaping the arrangement of the hadrons' constituents. Therefore very likely only the interplay of different theoretical techniques and their confrontation with the available experimental data will allow for an understanding of the intricacies of the hadrons' structure.

In particular, the hadrons' electromagnetic form factors provide information on how the electric charges and electromagnetic multipole moments are distributed. In the spacelike momentum region they carry information about the hadrons' electromagnetic interaction with other charged particles via the exchange of virtual photons. In addition, transition form factors reveal further properties. An important example is the electromagnetic nucleon to Delta

*e-mail: reinhard.alkofer@uni-graz.at

transition for which the experimental results unequivocally demonstrate the existence of deformations. As will be detailed below these deviations from sphericity have a straightforward explanation from the properties of relativistic fermions, see [2, 3] and references therein. This example makes clear that the objective of the approach presented in this talk is much more to gain qualitative insight into the physics governing the structure and dynamics of hadrons than to produce precise numbers for hadronic observables.

The main idea of the here presented relativistic Faddeev approach [4, 5] is to employ the gained knowledge on the QCD Green functions in the Landau gauge and to use the fact that baryons will appear as poles in the six-quark Green funcion. In a similar way as done for the Bethe-Salpeter equations in the relativistic two-body problem (see, *e.g.*, Chapter 6 of ref. [6]) one obtains then fully Poincaré-covariant bound-state equations. It should be emphasized here that there is nothing special about extracting bound state properties in quantum theories from the singularities of amplitudes. On the contrary, the textbook treatment of two-body problems in Quantum Mechanics by mapping the two-body Schrödinger equation to the one for the relative motion and a central potential is the highly exceptional case. But also in Schrödinger theory the scattering amplitudes obtained from such potentials display poles in the complex energy plane at $iE_{B,n}$, *i.e.*, at the bound state energies but on the imaginary axis. It is highly instructive to trace the origin of these poles which is possible already at the level of the one-dimensional Schrödinger equation and the simple potential well: One requires a non-vanishing wave function without an incident wave which leads then to the quantized energy eigenvalue condition $E = iE_{B,n}$, see, *e.g.*, Sect. 3.7.1 of ref. [7].

The solution of the Poincaré-covariant Faddeev equation provides besides the bound-state masses for baryons also the related amplitudes. These serve then in turn as input for the calculation of baryon obervables. In this talk we will focus on baryon form factors, and hereby especially on the transition form factors. The numerical techniques necessary for this kind of calculations — solving first Dyson-Schwinger equations for the Green functions and second Bethe-Salpeter type equations for the bound-state masses and amplitudes, calculating third from them the integrals for the form factors — are quite involved, and correspondingly over the last decades many detailed descriptions of such procedures have been published. For these technical issues we refer here to the most recent publication on these practical matters, ref. [8], as well as references therein.

2 Relativistic Three-Fermion Bound State Equations

As stated in the introduction the here presented Faddeev approach is based on the QCD Green functions in the Landau gauge. As valence quarks are those constituents of hadrons which determine their quantum numbers the most important such function for the description of mesons and baryons within bound state equations is the quark two-point function, respectively, the quark propagator S. Its dressing by gluonic interactions can be expressed via its Dyson–Schwinger equation (DSE)

$$S^{-1}(p) = S_0^{-1}(p) + Z_{1f} \int \frac{d^4q}{(2\pi)^4} \gamma^\mu D_{\mu\nu}(p-q)\, S(q)\, \Gamma^\nu_{gqq}(p,q)\,, \tag{1}$$

see fig. 2 for a pictorial presentation. Hereby, S_0 denotes the renormalized tree-level propagator $S_0^{-1}(p) = Z_2\left(i\not{p} + m\right)$, and Z_2 as well as Z_{1f} are renormalisation constants. m is the bare or current quark mass which, for the different flavours, will be provided as a parameter. (Throughout this study we employ the isospin limit, and set up and down current quark mass to an equal value.)

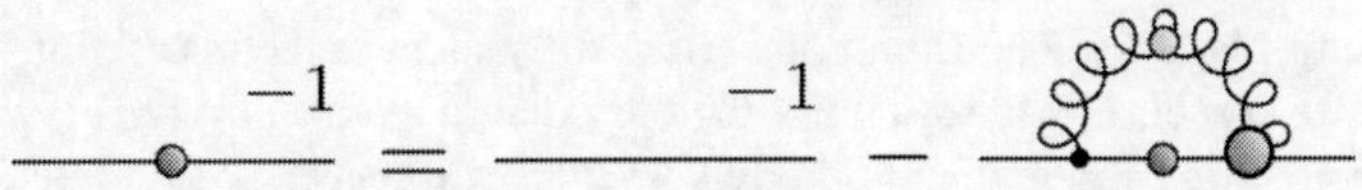

Figure 1. The DSE for the quark propagator.

The solution of this DSE necessitates in principle knowledge about the full quark-gluon vertex Γ^{ν}_{gqq} and the gluon propagator $D^{\mu\nu}$. Although calculations including quite accurate results for both these functions in the determination of the properties of light mesons have been successfully performed [9] we use here a generalized rainbow truncation and approximate the quark-gluon vertex by its bare form multiplied with a suitably adjusted function of the gluon momentum $k = p - q$, $\Gamma^{\nu}_{gqq}(p,q) = \gamma^{\nu} f(k^2)$. The requirement to preserve chiral symmetry leads then to a form of the quark-quark kernel,

$$K^{2-\text{body}} = [\gamma^{\mu} \otimes \gamma^{\nu}]\, f(k^2) D_{\mu\nu}(k)\,, \tag{2}$$

which is considerably easier to implement than in a higher-order chiral-symmetry preserving truncation, see, *e.g.*, ref. [9].

In order to make structures transparent and not to clutter all expressions with their many different indices one needs in an actual calculation we will present the content of the relativistic three-fermion bound state equation in a compact matrix notation. Besides summation over discrete indices it includes also integration over repeated continuous variables. As the three-particle-irreducible interaction within three quarks seems to be of minor importance [3] we are employing the Faddeev approximation and keep only the two-particle-irreducible interaction. The Faddeev equation is then the permuted sum of two-body quark-quark kernels $K^{2-\text{body}}$ [5, 10]:

$$\Psi = \sum_{a=1}^{3} [K^{2-\text{body}}]_{(a)}\, G_0\, \Psi\,. \tag{3}$$

Hereby, the index a determines the interacting pair by labeling the spectator quark. The function G_0 describes the free propagation of three quarks and is given by the disconnected product of three full quark propagators $S(p_i)$.

The interaction of a three-quark system with a single photon is described by the conserved current

$$J^{\mu} = \bar{\Psi}_f \left(G_0^{\mu} - G_0 K^{\mu} G_0 \right) \Psi_i\,, \tag{4}$$

where the incoming and outgoing baryon is represented by the Faddeev amplitudes Ψ_i and Ψ_f, respectively. G_0^{μ} denotes the impulse-approximation diagrams

$$G_0^{\mu} = (S_1 \Gamma^{\mu} S_1) S_2 S_3 + S_1 (S_2 \Gamma^{\mu} S_2) S_3 + S_1 S_2 (S_3 \Gamma^{\mu} S_3)\,, \tag{5}$$

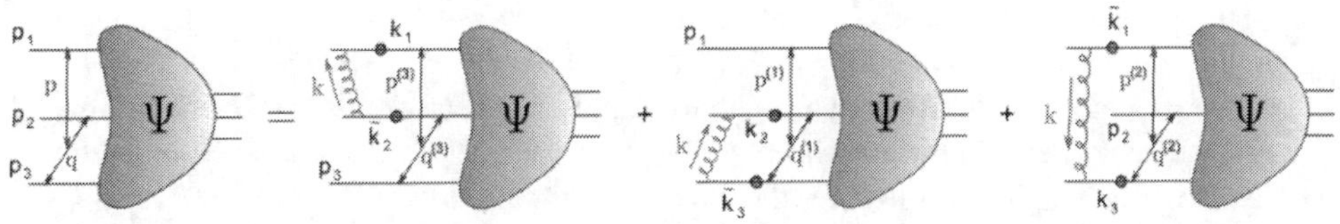

Figure 2. The covariant Faddeev equation in rainbow-ladder approximation.

where Γ^μ is the fully dressed quark-photon vertex. K^μ stands for the interaction of the photon with the Faddeev kernel. For this quantity the only diagrams that survive in rainbow-ladder approximation are those where the two-body kernel is a spectator:

$$K^\mu = \Gamma_1^\mu K_{23} + \Gamma_2^\mu K_{31} + \Gamma_3^\mu K_{12} . \tag{6}$$

For consistency it is important that the quark-photon vertex is calculated from its Dyson-Schwinger equation [11]. The Faddeev amplitudes, including their full Dirac–flavor structure, are determined by solving the three-body Faddeev equation with the interaction kernel K. The numerical solution of this integral equations rests hereby on the use of pseudo-spectral methods, especially on multiple expansions in Chebychev polynomials, see, *e.g.*, [8].

3 Structure of Baryonic Bound State Amplitudes

As this will become important in the interpretation of the results we summarise briefly the structure of the baryonic bound state amplitudes. Coupling three quarks which are spin-1/2 Dirac fermions together in a Poincaré covariant way as well as requiring baryons to be colour singlets and the wanted flavour state determines the elements of a tensorial decomposition of baryons' amplitudes [3, 12]. Phrased otherwise: The structure of the baryonic bound state amplitudes is *independent of any truncation of the bound state equation,* only Poincaré covariance and parity invariance are exploited to derive them. They include all possible internal spin and orbital angular momenta of the considered baryon. For a positive-parity, positive-energy (particle) baryon such an amplitude consists of in a partial-wave decomosition in the baryon's rest frame of 8 s-wave, 36 p-wave and 20 d-wave for spin-1/2 baryons [5] and 4 s-wave, 36 p-wave, 60 d-wave amplitudes and 28 f-wave amplitudes for spin-3/2 [10, 13]. Their relative importance are, of course, determined by the underlying interaction kernel. Although the results presented below indicate that for positive-parity ground-state baryons an s-wave ampitude is the single strongest one typically all others together are of equal importance in practically most observables.

An important exception is given if one considers the deviaton from sphericity for a given baryon. Here, the p-wave amplitudes play a significant role. For a given s-wave contribution which is related to two upper spinor entries of the baryon's four-spinor the p-wave is located in the lower half of the spinor. Those spinors are in the context of Dirac spinors related to anti-fermions, and thus their existence is purely due to the nature of relativistic quantum fields. Therefore, we conclude already at this point that the Poincaré covariant Faddeev approach offers a very natural explanation why the nucleon is not spherically symmetric: It is first a relativistic effect, and it is second a quantum effect due to existence of anti-particles in a relativistically invariant quantum field theory.

Of course, such an explanation does not exclude that dynamically generated d-waves also contribute to the deviation from sphericity, and as a matter of fact, in the results presented below they do. However, the difference is on a qualitative level. Only a fully relativistic approach necessitates to describe baryons and especially the nucleons as spinors being related to a representation of a Clifford algebra. On the other hand, in such an approach such contributions are unavoidable. This leads to the conclusion that in fully relativistic approach based on binding three Dirac-fermionic quarks together spherical baryons are virtually impossible, to obtain them one need an incredible amount of fine-tuning not only in the strength of the interquark interaction but also in its tensorial structure.

4 Earlier selected Results

The electromagnetic structure of baryons is given by several dimensionless and Lorentz-invariant form factors. Hereby, the nucleon as spin-1/2 baryon possesses two form factors which we choose in our calcuation as the two Sachs form factors G_E and G_M. The Δ electromagnetic are G_{E0}, G_{M1}, G_{E2} and G_{M3}: electric monopole, magnetic dipole, electric quadrupole and magnetic octupole. Previously obtained results for the N and Δ electromagnetic form factors are shown in Figs. 3 and 4; details of the respective calculations can be found in Refs. [14, 15]; see also [16] for results on axial form factors.

In all cases the comparison with experimental data and lattice results performs rather well. Note that the only model input in all calculations is the quark-gluon interaction in Eq. (2), and therefore this agreement is quite remarkable. This can be attributed to some extent to the global symmetries of QCD, especially its pattern of spontaneous chiral symmetry breaking, which are respected in every step of these calculations. Furthermore, charge conservation at $Q^2 = 0$ is not imposed but a consequence of the underlying Ward-Takahashi identity which is kept if the quark-photon vertex is calculated consistently. Note, however, that obtaining the quark-photon vertex from a Dyson-Schwinger equation includes automatically the contribution of the vector (ρ) meson as the latter as quark-antiquark bound state contributes to this vertex.

Nevertheless, the neglect of dynamical pions (as Goldstone bosons of spontaneously broken chiral symmetry) in the kernel of the Faddeev equation implies visible defects in the form factors at low Q^2, especially in the electric neutron form factor. Summarizing recent investi-

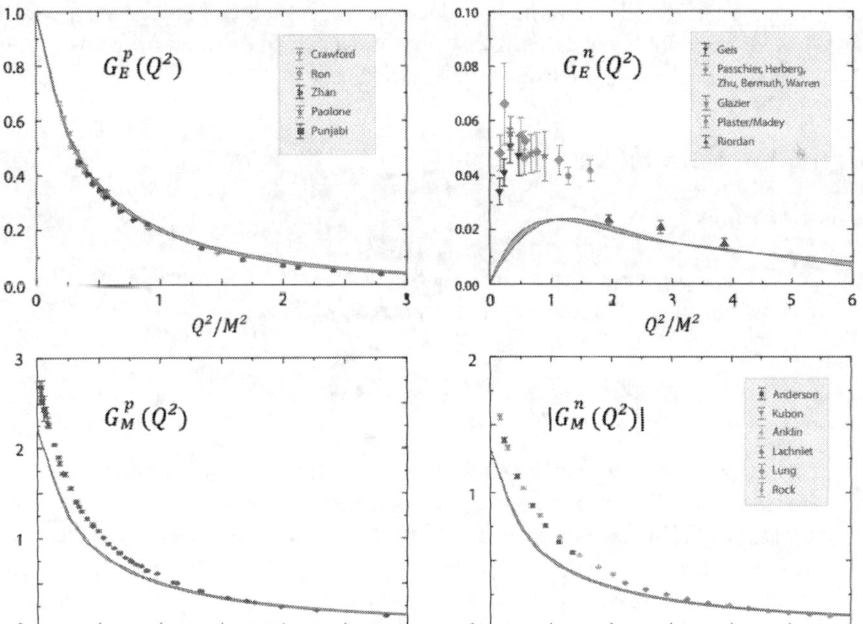

Figure 3. Nucleon electromagnetic form factors G_E and G_M for the proton (left panels) and neutron (right panels) compared to experimental data; see [14] for references.

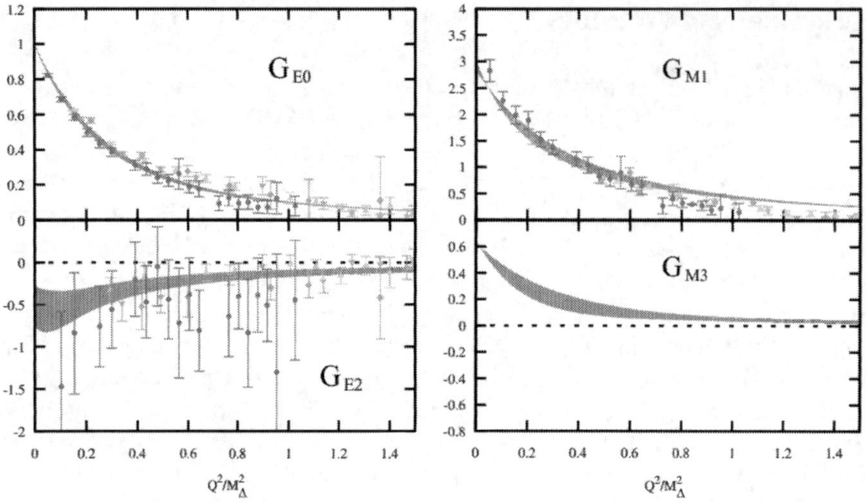

Figure 4. Δ electromagnetic form factors compared to lattice results from [17]; see [15] for references.

gations on this topic, see,*e.g.*, ref. [3] and references therein, these deficiences are symptoms of missing meson-cloud effects, which are known to enhance magnetic moments and charge radii, especially for small pion masses as the physical one. However, the rainbow-ladder truncation employed here does not include such dynamical pion effects, and especially not decays; it produces stable bound states that do not decay and can be viewed as the 'quark core' contribution that is stripped from its pion cloud [18].

5 Transition form factors

$N \to \Delta\gamma$ transition

The $N \to \Delta\gamma$ transition form factors are usually discussed in terms of the magnetic dipole G_M^* and the electric and Coulomb quadrupole ratios R_{EM} and R_{SM}:

$$R_{EM} = -\frac{G_E^*}{G_M^*}, \qquad R_{SM} = -\frac{M_N^2}{M_\Delta^2}\sqrt{\lambda_+\lambda_-}\,\frac{G_C^*}{G_M^*}, \tag{7}$$

which at zero momentum are sensitive to the quadrupole moments of the transition.

We show our results in Fig. 5 compared to experimental data. The results for the magnetic form factor are displayed in the upper panel. Within the theoretical systematic uncertainty estimated by the shaded bands in the plots our results agree well with the experimental data for $Q^2 > 0.8$ GeV^2. Below we observe deviations leading to a systematic underestimation of the form factor. For the electric and scalar form factors G_E and G_C (not shown) we observe a similar trend, the results agreeing with experiment at moderate to high Q^2, and underestimating the data for lower momenta. This trend is in systematic agreement with others for the nucleon electromagnetic and axial form factors as well as the form factors for the Δ [14–16] and indicates, similar as discussed above, the onset of pion cloud effects, which are not included in the present framework (see, however, [28] for pion cloud corrections to baryon masses). For the ratio R_{EM}, displayed in the lower left panel of Fig. 5, we confirm an observation that has

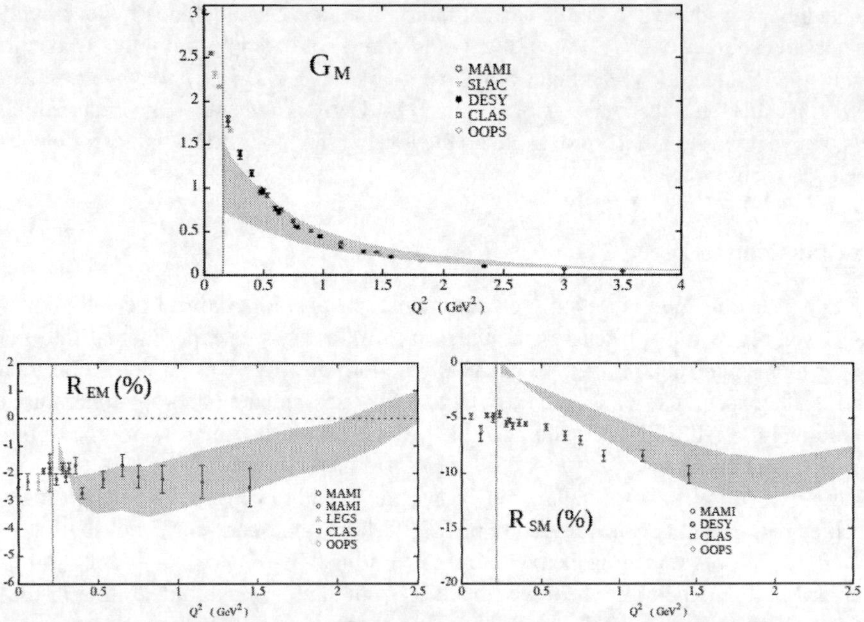

Figure 5. Magnetic form factor G_M, ratio R_{EM} and ratio R_{SM} of the $\gamma^* N \to \Delta$ transition. Vertical dashed line delimit the region below which the singularities of the quark propagator are probed. Experimental data are taken from [19–27].

been previously discussed in the quark-diquark approximation [13]. This quantity is sensitive to the presence of deformations of the nucleon and Δ due to higher angular momentum. In the non-relativistic quark model, these come into play via *d*-wave admixtures to the *s*-wave structure of the octet baryons. In our approach, as discussed above, they are mostly due to the *p*-waves residing in the lower components of the baryon's spinorial amplitudes. This observation also extends to the ratio $R_{SM}(0)$, see the right lower panel of Fig. 5.

Hyperon Octet - Decuplet transition

In the exact $SU(3)$-isospin limit, the transition $\gamma^* \Sigma^{*\,+} \to \Sigma^+$ would be identical to the $\gamma^* N \to \Delta$ studied in the shown in the previous subsection. This is indeed manifest in the magnetic form factor which is comparable in magnitude to the corresponding one in Fig. 5 and qualitatively identical in shape. Similar remarks apply to the $\gamma^* \Sigma^{*\,0} \to \Sigma^0$, see ref. [2] for more details.

Once again, in the exact $SU(3)$-isospin limit the $\gamma^* \Xi^{*\,0} \to \Xi^0$ transition would be identical to the $N\Delta$ or the $\Sigma^{*+}\Sigma^+$ ones. We find that their magnetic form factor is indeed very similar. Since the Ξ is a doubly-strange baryon, this indicates that the isospin-breaking effects are very small.

A good measure of the breaking of $SU(3)$-flavour symmetry is given by the form factors of the $\gamma^* \Sigma^{*\,-} \to \Sigma^-$ and the $\gamma^* \Xi^{*\,-} \to \Xi^-$ transitions, since in the limit of exact symmetry these would vanish identically. Their magnetic form factors indicate a breaking of $SU(3)$ symmetry at the level of a few percent. The smallness of this breaking in the present calculation might result from the fact that it is only generated by the different quark masses of the $s-$ and the $u/d-$quarks, both the current quark mass and the dynamically generated one. Other possible sources of $SU(3)$ breaking would come, for example, from the weakening of the quark-gluon interaction as a function of the quark mass (see [29]). We should mention, however, that such an effect was explored in [30] in relation to the octet and decuplet baryon masses, where it was found to be sizeable on the level of the propagators, but extremely small on the level of observables.

$\Sigma^0 - \Lambda$ transition

The $\gamma^*\Sigma^0 \rightarrow \Lambda$ transition is the only electromagnetic transition allowed between members of the baryon octet. As with other octet hyperon form factors, the experimental information is limited to the magnetic moment [31]. This reaction, however, has attracted considerable theoretical interest. In particular, the form factors for non-vanishing photon momentum have been studied in [32, 33]. The emphasis of [33] was on exploring the role of quark-core and meson-cloud effects on the $\gamma^*\Sigma^0 \rightarrow \Lambda$ transition form factors.

In contrast to the octet-decuplet transitions, our results in Fig. 6 show important dissimilarities with the quark model results of [33]. There, the quark-core contribution to the electric form factor is vanishing, being entirely determined by pion-cloud contributions. This is in contrast to our findings, where we obtain a G_E of similar magnitude to that in [33] but with meson-cloud effects completely absent. This observation is similar to the one discussed above in connection with R_{EM} of the nucleon-Δ transition and the explanation is the same: In quark models with s-wave contributions to the wave functions only, these transitions are zero by default. Thus one need to invoke either unnaturally strong d-wave contributions to the wave-function or attribute the entire non-vanishing form factor to meson cloud effects. In our relativistic framework, however, sizeable p-wave contributions to the baryons wave function appear naturally and thus generate a non-zero result for G_E.

6 Conclusions and Outlook

The here presented results are obtained in the so-called rainbow-ladder truncation with a widely used form of the quark-gluon interaction, the Maris-Tandy model [34]. It allowed for a unified and highly successful description of mesons and baryons, see, *e.g.*, refs. [3, 35] and references therein. The somewhat surprising success of the rainbow-ladder truncation as well as its limitations have been discussed in great detail in the literature, see, *e.g.*, refs. [36, 37] for a recent synopsis. A further and important novel development is the calculation of the

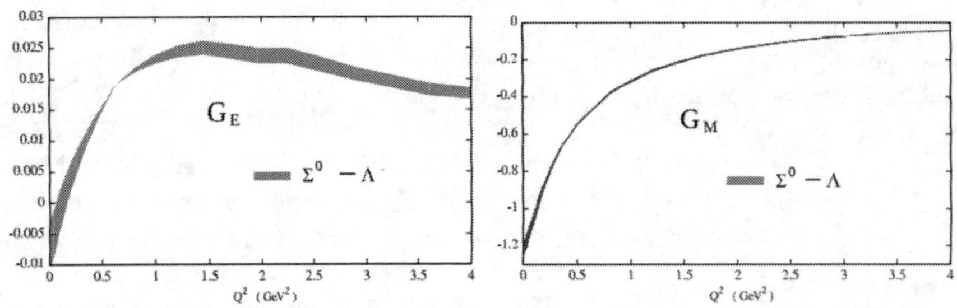

Figure 6. Electric (upper panel) and magnetic (lower panel) form factors of the octet-only $\gamma^*\Sigma^0 \rightarrow \Lambda$ transition.

spectrum of light mesons in a symmetry-preserving truncation based on the three-particle irreducible effective action [9]. The obtained results hereby not only verify the previous arguments on the reliability and the limitations of the rainbow-ladder truncation but mark a very decisive step towards the goal of apparent convergence in the pattern of increasingly sophisticated truncations within the Dyson-Schwinger–Bethe-Salpeter approach. An extension of calculations within this truncation to baryons is in progress, and it will be interesting to see whether and how the obtained results will deepen our understanding of structure of baryons.

Acknowledgements

R.A. is grateful to the organizers of the International Conference on Exotic Atoms and Related Topics - EXA2017 for all their efforts which made this conference possible, and for financial support.

This work has been supported by the project P29216-N36 from the Austrian Science Fund, FWF, by the Helmholtz International Center for FAIR within the LOEWE program of the State of Hesse, and by the DFG grant FI 970/11-1.

References

[1] D. Mohler, these proceedings (2017)

[2] H. Sanchis-Alepuz, R. Alkofer, C.S. Fischer, Eur. Phys. J. **A**, in press (2018), 1707.08463

[3] G. Eichmann, H. Sanchis-Alepuz, R. Williams, R. Alkofer, C.S. Fischer, Prog. Part. Nucl. Phys. **91**, 1 (2016), 1606.09602

[4] G. Eichmann, Ph.D. thesis, Graz U. (2009), 0909.0703, https://inspirehep.net/record/830292/files/arXiv:0909.0703.pdf

[5] G. Eichmann, R. Alkofer, A. Krassnigg, D. Nicmorus, Phys. Rev. Lett. **104**, 201601 (2010), 0912.2246

[6] R. Alkofer, L. von Smekal, Phys. Rept. **353**, 281 (2001), hep-ph/0007355

[7] F. Schwabl, *Quantum Mechanics* (1988)

[8] H. Sanchis-Alepuz, R. Williams (2017), 1710.04903

[9] R. Williams, C.S. Fischer, W. Heupel, Phys. Rev. **D93**, 034026 (2016), 1512.00455

[10] H. Sanchis-Alepuz, G. Eichmann, S. Villalba-Chavez, R. Alkofer, Phys. Rev. **D84**, 096003 (2011), 1109.0199

[11] P. Maris, P.C. Tandy, Phys. Rev. **C61**, 045202 (2000), nucl-th/9910033

[12] C. Carimalo, J. Math. Phys. **34**, 4930 (1993)

[13] D. Nicmorus, G. Eichmann, R. Alkofer, Phys. Rev. **D82**, 114017 (2010), 1008.3184

[14] G. Eichmann, Phys. Rev. **D84**, 014014 (2011), 1104.4505

[15] H. Sanchis-Alepuz, R. Williams, R. Alkofer, Phys. Rev. **D87**, 096015 (2013), 1302.6048

[16] G. Eichmann, C.S. Fischer, Eur. Phys. J. **A48**, 9 (2012), 1111.2614

[17] C. Alexandrou, T. Korzec, G. Koutsou, C. Lorce, J.W. Negele, V. Pascalutsa, A. Tsapalis, M. Vanderhaeghen, Nucl. Phys. **A825**, 115 (2009), 0901.3457

[18] G. Eichmann, A. Krassnigg, M. Schwinzerl, R. Alkofer, Annals Phys. **323**, 2505 (2008), 0712.2666

[19] W. Bartel, B. Dudelzak, H. Krehbiel, J. McElroy, U. Meyer-Berkhout, W. Schmidt, V. Walther, G. Weber, Phys. Lett. **B28**, 148 (1968)
[20] S. Stein, W.B. Atwood, E.D. Bloom, R.L. Cottrell, H.C. DeStaebler, C.L. Jordan, H. Piel, C.Y. Prescott, R. Siemann, R.E. Taylor, Phys. Rev. **D12**, 1884 (1975)
[21] R. Beck et al., Phys. Rev. **C61**, 035204 (2000), nucl-ex/9908017
[22] V.V. Frolov et al., Phys. Rev. Lett. **82**, 45 (1999), hep-ex/9808024
[23] T. Pospischil et al., Phys. Rev. Lett. **86**, 2959 (2001), nucl-ex/0010020
[24] G. Blanpied et al., Phys. Rev. **C64**, 025203 (2001)
[25] N.F. Sparveris et al. (OOPS), Phys. Rev. Lett. **94**, 022003 (2005), nucl-ex/0408003
[26] S. Stave et al. (A1), Phys. Rev. **C78**, 025209 (2008), 0803.2476
[27] I.G. Aznauryan et al. (CLAS), Phys. Rev. **C80**, 055203 (2009), 0909.2349
[28] H. Sanchis-Alepuz, C.S. Fischer, S. Kubrak, Phys. Lett. **B733**, 151 (2014), 1401.3183
[29] R. Williams, Eur. Phys. J. **A51**, 57 (2015), 1404.2545
[30] H. Sanchis-Alepuz, C.S. Fischer, Phys. Rev. **D90**, 096001 (2014), 1408.5577
[31] C. Patrignani et al. (Particle Data Group), Chin. Phys. **C40**, 100001 (2016)
[32] T. Van Cauteren, J. Ryckebusch, B. Metsch, H.R. Petry, Eur. Phys. J. **A26**, 339 (2005), nucl-th/0509047
[33] G. Ramalho, K. Tsushima, Phys. Rev. **D86**, 114030 (2012), 1210.7465
[34] P. Maris, P.C. Tandy, Phys. Rev. **C60**, 055214 (1999), nucl-th/9905056
[35] A. Bashir, L. Chang, I.C. Cloet, B. El-Bennich, Y.X. Liu, C.D. Roberts, P.C. Tandy, Commun. Theor. Phys. **58**, 79 (2012), 1201.3366
[36] H. Sanchis-Alepuz, R. Williams, Phys. Lett. **B749**, 592 (2015), 1504.07776
[37] H. Sanchis-Alepuz, R. Williams, J. Phys. Conf. Ser. **631**, 012064 (2015), 1503.05896

Luminosity determination for the proton-deuteron reaction using $pd \rightarrow {}^3\mathrm{He}\eta$ channel with WASA-at-COSY detector

Magdalena Skurzok[1,*], *Oleksandr* Rundel[1,**], *Aleksander* Khreptak[1,***], and *Paweł* Moskal[1,****]

[1] *Institute of Physics, Jagiellonian University, prof. Stanisława Łojasiewicza 11, 30-348 Kraków, Poland*

Abstract. In this report one of the methods used for the luminosity determination for the experiment performed by WASA Collaboration to search for ^{3}He-η mesic nuclei is presented. The method is based on the analysis of $pd \rightarrow {}^3\mathrm{He}\eta$ process.

1 Introduction

Although strongly bound η- and η'-mesic nuclei, being one of the hottest topics in nuclear and hadron physics, have been postulated many years ago [1], until now their existence has not been experimentally confirmed. Signals measured so far might be only interpreted as an indications of the η- and η'-mesic bound states [2–11]. Recently, several theoretical [12–33] and experimental studies [34–43] are ongoing.

One of the most recent and most promising experiments dedicated to search for η-mesic Helium nuclei has been performed by the WASA-at-COSY Collaboration at the Forschungszentrum Jülich. The measurements were carried out with high statistics and high acceptance with the WASA detection setup in deuteron-deuteron (^{4}He-η) [35, 36] and proton-deuteron (^{3}He-η) [44] fusion reactions. During experiments, the beam momentum was changed slowly and continuously around the η production threshold in each of the COSY acceleration cycles. This allowed us to search for the bound states via the measurement of the excitation functions for chosen decay channels. The analysis dedicated to search for ^{4}He-η mesic nuclei in $dd \rightarrow {}^3\mathrm{He}n\pi^0$ and $dd \rightarrow {}^3\mathrm{He}p\pi^-$ processes results in the upper limits of the total cross section at 90% confidence level equal to about 3 nb and 6 nb, respectively [35].

The search for ^{3}He-η bound states was performed in May 2014 using a ramped proton beam with momentum changing continuously in the range of 1.426-1.635 GeV and pellet deuterium target. The range of the beam momentum corresponds to the range of excess energy $Q_{^3He\eta}$ from -70 to $+30$ MeV. The ^{3}He-η bound states [37, 44] are searched for via studies of the excitation functions for processes corresponding to the three mechanisms: (i) absorption of the η meson by one of the nucleons, which subsequently decays into N^*-π pair e.g.: $pd \rightarrow (^3\mathrm{He}\text{-}\eta)_{bound} \rightarrow ppp\pi^-$, (ii) decay of the η -meson while it is still "orbiting" around a nucleus e.g.: $pd \rightarrow (^3\mathrm{He}\text{-}\eta)_{bound} \rightarrow {}^3\mathrm{He}6\gamma$ reactions and (iii) η

*e-mail: magdalena.skurzok@uj.edu.pl
**e-mail: o.rundel@doctoral.uj.edu.pl
***e-mail: aleksander.khreptak@doctoral.uj.edu.pl
****e-mail: p.moskal@uj.edu.pl

meson absorption by few nucleons e.g.: $pd \to (^3\text{He-}\eta)_{bound} \to ppn$. Almost two weeks of measurement brought the world's largest data sample for ^{3}He-η. The total integrated luminosity and also its dependence on the excess energy above the η production threshold was estimated based on the $pd \to {}^3\text{He}\eta$ process. In this contribution we present the procedure of the luminosity calculation.

2 Luminosity determination based on $pd \to {}^3\text{He}\eta$ with WASA-at-COSY detector

The luminosity determination in the excess energy range is crucial for the normalization of the experimental excitation functions. The luminosity is obtained from the analysis of the reaction that has a known cross section and allows one to calculate cross sections for other measured processes after obtaining the amount of events.

The experimental luminosity can be determined in the whole beam momentum range based on $pd \to ppn_{spec}$ reaction and, in addition, it can be also determined based on the $pd \to {}^3\text{He}\eta$ reaction in the range above the η production threshold ($Q_{^3He\eta} > 0$). The whole $Q_{^3He\eta}$ range from -70 to $+30$ MeV was divided into 40 bins, each one with the width of 2.5 MeV.

Luminosity estimation from the $pd \to {}^3\text{He}\eta$ reaction is the first stage of the data analysis. Events corresponding to this reaction are identified by ^{3}He tracks registered in the forward part of the WASA detector. ^{3}He identification is carried out with the ΔE-E method and kinetic energy reconstruction from the energy deposited in the detector. The angles are reconstructed based on registered tracks with the algorithms implemented by the WASA-at-COSY team and used in the previous analysis.

Kinetic energy ^{3}He is reconstructed based on the value of energy deposited in $FRH1$ layer of the forward detector. For this layer energy deposition by ^{3}He ions is the largest. The kinetic energy is determined from a fit of $\theta(E_{kin})$ spectrum obtained from Monte Carlo simulation with a formula:

$$E_{kin} = f_0(\theta) + f_1(\theta)E_{FRH1}, \tag{1}$$

where f_0 and f_1 are squared functions of the scattering angle θ:

$$f_i(\theta) = a_i + b_i\theta + c_i\theta^2. \tag{2}$$

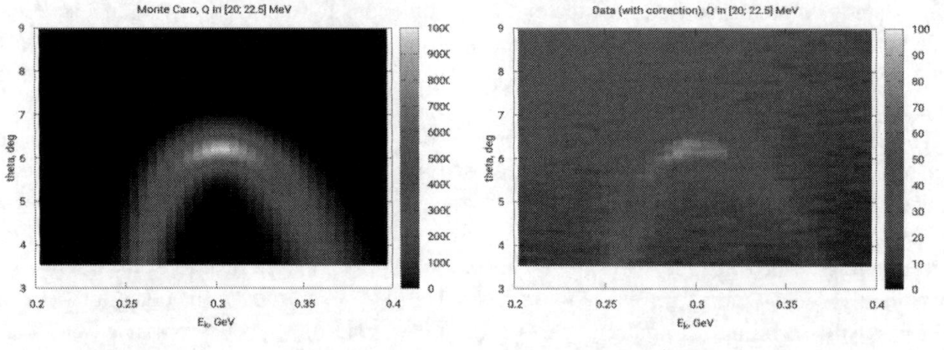

Figure 1. The 2D distribution of ^{3}He energy (horizontal) and θ angle (vertical) for the excess energy interval $Q_{^3He\eta} \in (17.5, 20)MeV$. Left side: reconstruction algorithm run over Monte Carlo simulation. Right side: result of data analysis performed with assuming the correction constants $off_P = 4.6$ MeV/c and $off_E = 2$ MeV.

The fit contains six free parameters. For data, the beam momentum is obtained as a function on accelerator's time-in-cycle that precisely describes the momentum changes but contains unknown offset. It was also assumed that the kinetic energy for data may differ from the reconstructed value by a small correction constant. These two constants were obtained from kinematic distributions (Fig. 1) comparing positions of $E - \theta$ distributions for simulation and data. The values obtained this way are equal to $off_P = 4.6\ MeV/c$ and $off_E = 2\ MeV$.

For obtaining the amount of events corresponding to η creation, the ^{3}He missing mass spectra have been analyzed for each excess energy $Q_{^3He\eta}$ intervals. These events are visible as a peak around the value equal to the mass of the η meson. The value of integrated luminosity for each $Q_{^3He\eta}$ bin was obtained from the formula $L = \frac{N}{\epsilon\sigma}$, where the amount of events N is extracted from the η creation peak area (Fig. 2), the acceptance ϵ is determined based on Monte Carlo simulation (Fig. 3) and the total cross section σ is taken from other experiments [10].

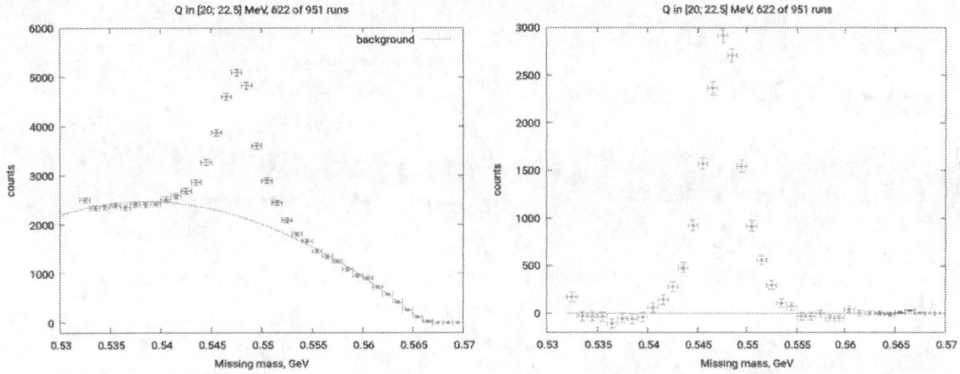

Figure 2. The ^{3}He missing mass spectrum for the excess energy interval $Q_{^3He\eta} \in (20, 22.5)MeV$. Left side: the background around the η creation peak is fit with a polynomial. Right side: missing mass after the background subtraction (for obtaining the amount of η creation events).

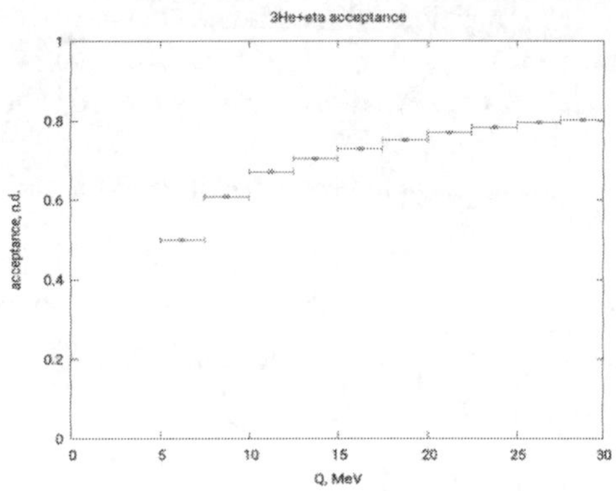

Figure 3. The acceptance values for $pd \rightarrow ^3He\eta$ reaction obtained from Monte Carlo simulations.

The obtained integrated luminosity for each $Q_{^3He\eta}$ bin is shown in Fig. 4. Total integrated luminosity of about 3 pb^{-1} was estimated for whole excess energy range assuming that excess energy intervals for $Q_{^3He\eta} < 0$ have the same values as determined for $Q_{^3He\eta} > 0$ (about 50 nb^{-1}) and taking into account the amount of data that has not been analyzed yet. If we estimate the time of measurement as 10^6 s then obtained value of average luminosity becomes 3 10^{30} $cm^{-2}s^{-1}$. This value is in agreement with value given in the proposal [44].

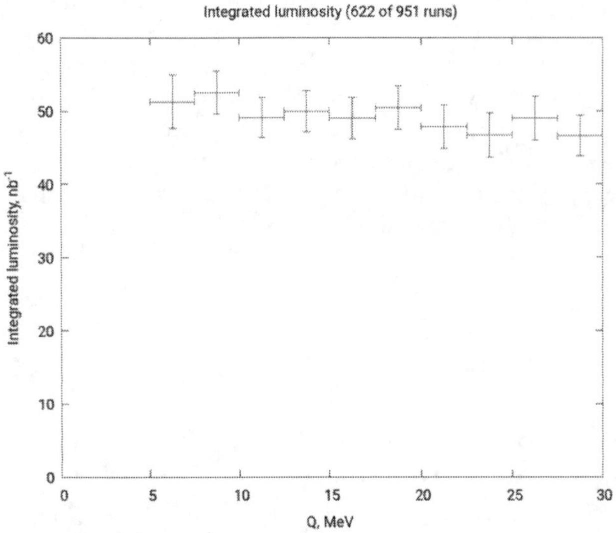

Figure 4. Integrated luminosity determined based on $pd \rightarrow {}^3$Heη reaction for the excess energy range of $Q_{^3He\eta} > 0$. The luminosity was calculated for 65 % of collected data.

3 Summary

One of two planned luminosity determination analyses for the experiment performed with WASA-at-COSY to search the ^{3}He-η bound states in proton-deuteron fusion has been carried out. The analysis was performed for the $pd \rightarrow^3$ Heη reaction. The estimated value of average luminosity is equal to 3 10^{30} $cm^{-2}s^{-1}$ and is in agreement with value presented in proposal [44]. The obtained luminosity will be used for the normalization of the excitation functions for processes in which the ^{3}He-η nuclei are searched for. The data analysis is in progress. The estimated integral luminosity value for this experiment is about 3 pb^{-1} and is the best statistics ever gathered for such experimental conditions.

4 Acknowledgements

We acknowledge the support from the Polish National Science Center through grant No. 2016/23/B/ST2/00784 and Jagiellonian University through grant DSC 2011 for financing young scientist and PhD students No. 142/F/OR/2016.

References

[1] Q. Haider, L. C. Liu, *Phys. Lett.* **B172**, 257 (1986).
[2] G. A. Sokol, L. N. Pavlyuchenko, *arXiv:nucl-ex/0111020*, (2001).
[3] A. Budzanowski *et al., Phys. Rev.* **C79**, 012201 (2009).

[4] J. Berger *et al., Phys. Rev. Lett.* **61**, 919 (1988).
[5] B. Mayer *et al., Phys. Rev.* **C53**, 2068 (1996).
[6] J. Smyrski *et al., Phys. Lett. B* **649**, 258 (2007).
[7] J. Smyrski, P. Moskal, W. Krzemień, *Acta Phys. Polon. Supp.* **2**, 133 (2009).
[8] J. Smyrski, *et al., Nucl. Phys.* **A790**, 438 (2007).
[9] P. Moskal, J. Smyrski, *Acta Phys. Polon.* **B41**, 2281 (2010).
[10] T. Mersmann *et al., Phys. Rev. Lett.* **98**, 242301 (2007).
[11] M. Papenbrock *et al., Phys. Lett.* **B734**, 333 (2014).
[12] N. Ikeno, H. Nagahiro, D. Jido and S. Hirenzaki, Eur. Phys. J. A **53**, 194 (2017).
[13] J. J. Xie *et al., Phys. Rev. C* **95**, 015202 (2017).
[14] A. Fix, O. Kolesnikov, *Phys. Lett.* B **772**, 663 (2017).
[15] N. Barnea, B. Bazak, E. Friedman and A. Gal, *Phys. Lett. B* **771**, 297 (2017).
[16] N. Barnea, E. Friedman, A. Gal, *Nucl. Phys. A* **968**, 35 (2017).
[17] N. Barnea, E. Friedman, A. Gal, *Phys. Lett.* B **747**, 345 (2015).
[18] E. Friedman, A. Gal, J. Mares, *Phys. Lett.* B **725**, 334 (2013).
[19] N. G. Kelkar, *Eur. Phys. J.* A **52**, (2016) 309.
[20] N. G. Kelkar *et al., Rept. Prog. Phys.* **76**, 066301 (2013).
[21] N. G. Kelkar , *Acta Phys. Pol.* B **76**, 113 (2015).
[22] C. Wilkin, Acta Phys. Polon. B **47**, 249 (2016).
[23] S. D. Bass and A. W. Thomas, *Acta Phys. Polon.* B **45**, 627 (2014).
[24] S. D. Bass and A. W. Thomas, *Phys. Lett.* **B634**, 368 (2006).
[25] S. Hirenzaki, H. Nagahiro, Acta Phys. Polon. B **45**, 619 (2014).
[26] H. Nagahiro *et al., Phys. Rev. C* **87**, 045201 (2013).
[27] S. Wycech, W. Krzemien, Acta Phys. Polon. B **45**, 745 (2014).
[28] H. Nagahiro *et al., Phys. Rev. C* **92**, 055205 (2015).
[29] H. Nagahiro, M. Takizawa, and S. Hirenzaki, *Phys. Rev.* **C74**, 045203 (2006).
[30] H. Nagahiro and S. Hirenzaki, *Phys. Rev. Lett.* **94**, 232503 (2005).
[31] K. Tsushima, D-H. Lu, A. W. Thomas and K Saito, *Phys. Lett.* **B443**, 26 (1998).
[32] K. Tsushima, *Nucl. Phys.* **A670**, 198 (2000).
[33] C. Garcia-Recio, J. Nieves, T. Inoue and E. Oset, *Phys. Lett.* **B550** 47 (2002).
[34] Y. K. Tanaka *et al., Phys. Rev. Lett.* **117**, 202501 (2016).
[35] P. Adlarson, *et al.*, Nucl. Phys. A **959**, 102 (2017).
[36] P. Adlarson *et al.*, Phys. Rev. **C87**, 035204 (2013).
[37] M. Skurzok, *et al.*, Acta Phys. Polon. B **47**, 503 (2016).
[38] H. Machner, J. Phys. G **42**, 043001 (2015).
[39] B. Krusche, C. Wilkin, Prog. Part. Nucl. Phys., **80**, 43 (2014).
[40] V. Metag, M. Nanova and E. Ya. Paryev, Prog. Part. Nucl. Phys. **97**, 199 (2017).
[41] P. Moskal, Acta Phys. Polon. **47**, 97 (2016).
[42] P. Moskal, M. Skurzok, W. Krzemien, AIP Conf. Proc. **1753**, 030012 (2016).
[43] P. Moskal, Few Body Syst. **55**, 667 (2014).
[44] P. Moskal, W. Krzemien, M. Skurzok, COSY proposal No. 186.3 (2014).

Search for time-reversal-invariance violation in double polarized antiproton-deuteron scattering

Yuriy Uzikov[1,2,3,★] and *Johann* Haidenbauer[4,★★]

[1] *Joint Institute for Nuclear Research, Dubna 141980, Russia*
[2] *Dubna State University, Dubna 141980, Russia*
[3] *Department of Physics, M.V. Lomonosov Moscow State University, Moscow 119991, Russia*
[4] *Institute for Advanced Simulation and Institut für Kernphysik, Forschungszentrum Jülich GmbH, D-52428 Jülich, Germany*

Abstract.
Apart from the pd reaction also the scattering of antiprotons with transversal polarization p_y^p on deuterons with tensor polarization P_{xz} provides a null-test signal for time-reversal-invariance violating but parity conserving effects. Assuming that the time-reversal-invariance violating $\bar{N}N$ interaction contains the same operator structure as the NN interaction, we discuss the energy dependence of the null-test signal in $\bar{p}d$ scattering on the basis of a calculation within the spin-dependent Glauber theory at beam energies of 50-300 MeV.

1 Introduction

Under CPT symmetry time-reversal-invariance violating but parity conserving (TVPC) forces are considered as a possible source of CP-invariance violation, which is required to account for the matter-antimatter asymmetry in the universe [1]. In contrast to effects from time-reversal-invariance violation together with parity violation such as a permanent electric dipole moment (EDM) of elementary particles, so far much less attention was paid to TVPC effects. The reason why TVPC effects are interesting is that experimental limits on them are still rather weak, in particular, considerably weaker than those for the EDM.

Since the intensity of TVPC interactions within the standard model is extremely small [2], an observation of any effects at the present accuracy level of experiments would be a direct indication of physics beyond the standard model. Indeed a pertinent measurement is planned at the COSY accelerator in the Research Center in Jülich [3]. The observable in question is the integrated cross section for scattering of protons with transversal polarization p_y^p on deuterons with tensor polarization P_{xz}. It provides a null-test signal for TVPC effects [4] and it will be measured in pd scattering at 135 MeV [3]. Theoretical studies of the energy dependence of the expected signal were performed at energies of the planned experiment [5–12] on the basis of the spin-dependent Glauber theory and demonstrate several unexpected effects. Among them are (i) the absense of the contribution from the

★e-mail: uzikov@jinr.ru
★★e-mail: j.haidenbauer@fz-juelich.de

lowest-mass meson-exchange (ρ meson) in the TVPC NN interaction, caused by its specific isospin, spin and momentum dependence; (ii) a strong impact of the deuteron D-wave on the null-test signal due to a destructive interference between the S- and D-wave contributions, even for zero transferred 3-momentum; (iii) oscillating behaviour of the null-test signal as a function of the beam energy, i.e. the vanishing of the TVPC signal at some specific energies is possible even when the TVPC interaction itself is nonzero; (iv) a very small influence of the Coulomb interaction on the TVPC term of the pd forward scattering amplitude $\widetilde{g}$. Furthermore, certain relations between differential observables of elastic pd scattering caused by time-reversal-invariance requirements were obtained and the degree of their violation by TVPC NN forces was studied [13, 14].

Since the spin structure of the amplitude for pd- and $\bar{p}d$ elastic scattering is the same, it is obvious that the integrated cross section for scattering of a polarized ($p_y^{\bar{p}}$) antiproton on tensor polarized (P_{xz}) deuterons also provides a null-test signal for TVPC effects. Furthermore, the TVPC $\bar{N}N$ amplitude for elastic scattering contains the same operator structures as the one for TVPC NN elastic scattering, except for the charge-exchange terms. Therefore, the formalism developed in Refs. [7, 8, 11] within the Glauber theory for the calculation of the null-test signal in pd scattering can be straightforwardly applied to $\bar{p}d$ scattering too. However, due to differences in the hadronic part of the pN and $\bar{p}N$ scattering amplitudes and also in the electromagnetic interactions, the energy dependence of the null test signal in pd and $\bar{p}d$ interaction has to be different. In the present work the energy dependence of the null-test signal in $\bar{p}d$ scattering is studied on the basis of calculations within the spin-dependent Glauber theory using the spin-dependent $\bar{p}N$ amplitudes from a recent partial wave analysis of $\bar{p}p$ scattering [15].

2 Null-test signal for time-reversal-invariance violation

The total cross section for $\bar{p}d$ scattering with TVPC forces included can be written in the same form as for pd scattering [7]

$$\sigma_{tot} = \sigma_0^t + \sigma_1^t \mathbf{p}^{\bar{p}} \cdot \mathbf{p}^d + \sigma_2^t(\mathbf{p}^{\bar{p}} \cdot \mathbf{m})(\mathbf{p}^d \cdot \mathbf{m}) + \sigma_3^t P_{zz} + \widetilde{\sigma} p_y^{\bar{p}} P_{xz}^d \,. \tag{1}$$

Here $\mathbf{p}^{\bar{p}}$ ($\mathbf{p}^d$) is the vector polarization of the initial antiproton (deuteron), P_{zz} and P_{xz} are the tensor polarizations of the deuteron, and $p_y^{\bar{p}}$ is the transversal component of the antiproton vector polarization. The OZ axis is directed along the beam direction $\mathbf{m}$, the OY axis is directed along the vector polarization of the antiproton beam $\mathbf{p}^{\bar{p}}$ and the OX axis is chosen to form a right-handed reference frame. The integrated cross sections σ_i^t ($i = 0, 1, 2, 3$) are those which arise from a standard time-reversal invariant and parity conserving interaction, while the last term $\widetilde{\sigma}$ appears only in the presence of the TVPC interactions and constitutes the TVPC null-test signal. The result (1) can be derived using phenomenological $\bar{p}d$ forward scattering amplitudes and the generalized optical theorem.

The evaluation of the integrated cross sections σ_i^t and $\widetilde{\sigma}$ at beam energies > 100 MeV can be done on the basis of the spin-dependent Glauber theory of $\bar{p}d$ scattering which is formulated similarly to the theory of pd scattering given in Ref. [16]. Indeed, as shown in Ref. [17], this theory allows one to describe rather well available data on differential spin observables of pd scattering in the forward hemisphere at beam energies of 135 – 200 MeV. For the antiproton-deuteron scattering this theory can be applied at even lower energies due to the presence of strong annihilation effects. In the Glauber theory one uses the elastic (on-shell) $\bar{N}N$ scattering amplitudes as input. Hadronic amplitudes of the $\bar{p}N$ scattering are taken here in the same form as for pN scattering [16]

$$\begin{aligned} M_N(\mathbf{p},\mathbf{q};\sigma,\sigma_N) = {} & A_N + C_N\sigma\hat{\mathbf{n}} + C_N'\sigma_N\hat{\mathbf{n}} + B_N(\sigma\hat{\mathbf{k}})(\sigma_N\hat{\mathbf{k}}) + \\ & +(G_N + H_N)(\sigma\hat{\mathbf{q}})(\sigma_N\hat{\mathbf{q}}) + (G_N - H_N)(\sigma\hat{\mathbf{n}})(\sigma_N\hat{\mathbf{n}}) \,, \end{aligned} \tag{2}$$

where $\hat{\mathbf{q}}$, $\hat{\mathbf{k}}$ and $\hat{\mathbf{n}}$ are defined as unit vectors along the vectors $\mathbf{q} = (\mathbf{p}-\mathbf{p}')$, $\mathbf{k} = (\mathbf{p}+\mathbf{p}')$ and $\mathbf{n} = [\mathbf{k}\times\mathbf{q}]$, respectively; $\mathbf{p}$ ($\mathbf{p}'$) is the initial (final) antiproton momentum.

In general, the TVPC NN interaction contains 18 different terms [18]. In the case of the on-shell NN scattering amplitude there are only three terms with different (independent) spin-momentum structures. In the present study we consider the following two terms for the TVPC (on-shell) t-matrix of elastic $\bar{p}N$ scattering which have the same structure as those in TVPC pN scattering

$$t_{\bar{p}N} = h_N[(\boldsymbol{\sigma}\cdot\mathbf{k})(\boldsymbol{\sigma}_N\cdot\mathbf{q}) + (\boldsymbol{\sigma}_N\cdot\mathbf{k})(\boldsymbol{\sigma}\cdot\mathbf{q}) - \frac{2}{3}(\boldsymbol{\sigma}_N\cdot\boldsymbol{\sigma})(\mathbf{k}\cdot\mathbf{q})]/m_p^2 + \\ +g_N[\boldsymbol{\sigma}\times\boldsymbol{\sigma}_N]\cdot[\mathbf{q}\times\mathbf{k}][\boldsymbol{\tau}-\boldsymbol{\tau}_N]_z/m_p^2. \quad (3)$$

Here $\boldsymbol{\sigma}$ ($\boldsymbol{\sigma}_N$) is the Pauli matrix acting on the spin state of the antiproton (nucleon $N = p, n$) and $\boldsymbol{\tau}$ ($\boldsymbol{\tau}_N$) is the isospin matrix acting on the isospin state of the antiproton (nucleon). The momenta $\mathbf{q}$ and $\mathbf{k}$ were already defined above in the context of Eq. (2). Both terms in Eq. (3), h_N and g_N, occur in the TVPC pn interaction. The TVPC $\bar{p}N$ scattering amplitude contains also the charge-exchange term

$$t^{ch} = g'_N(\boldsymbol{\sigma}-\boldsymbol{\sigma}_N)\cdot i\,[\mathbf{q}\times\mathbf{k}][\boldsymbol{\tau}\times\boldsymbol{\tau}_N]_z/m_p^2, \quad (4)$$

which describes the elastic transitions $pn \to np$ and $np \to pn$. Within a picture of one-meson-exchange interaction this g'-term corresponds to the charged ρ-meson exchange [19]. The same term (4) corresponds to the charge-exchange processes $\bar{p}p \to \bar{n}n$ or $\bar{n}n \to \bar{p}p$. However, in contrast to pn scattering these processes are inelastic and therefore the operation of time-reversal invariance transforms, for example, the $\bar{p}p \to \bar{n}n$ amplitude to the $\bar{n}n \to \bar{p}p$ amplitude and does not impose any restrictions on these amplitudes.

The h_N-term in Eq. (3) can be associated with the axial h_1-meson exchange. As shown in Ref. [19], contributions of the π- and σ-meson to the TVPC NN interaction are excluded, which is obviously true for the TVPC $\bar{N}N$ interaction as well.

2.1 TVPC amplitude of $\bar{p}d$ forward scattering

One can write the $\bar{p}d$ forward elastic scattering amplitude in general form taking into account the TVPC $\bar{N}N$ interactions, as it was done for pd elastic scattering [7, 17], and then apply the generalized optical theorem to derive Eq. (1) for the total $\bar{p}d$ scattering cross section. As in Ref. [7], the integrated cross section $\widetilde{\sigma}$ is related to the TVPC term $\widetilde{g}$ of the $\bar{p}d$ forward elastic scattering amplitude by $\widetilde{\sigma} = -4\sqrt{\pi}\,\mathrm{Im}\frac{2}{3}\widetilde{g}$. Furthermore, the TVPC forward amplitude of $\bar{p}d$ elastic scattering $\widetilde{g}$ can be found within the Glauber theory [7]. We consider the h_N- and g_N-terms and take into account both the S- and D-wave components of the deuteron. Taking into account that the g_N-term is excluded in the process $\bar{p}n \to \bar{p}n$ due to the isospin operator in Eq. (3), we obtain the following result for the TVPC forward amplitude from the corresponding equation in Ref. [11]:

$$\widetilde{g} = \frac{i}{4\pi m_p}\int_0^\infty dq q^2\Big[S_0^{(0)}(q) - \sqrt{8}S_2^{(1)}(q) - 4S_0^{(2)}(q) + \sqrt{2}\frac{4}{3}S_2^{(2)}(q) + 9S_1^{(2)}(q)\Big] \\ [-C'_n(q)(h_p+g_p) - C'_p(q)h_n]\,. \quad (5)$$

Here $S_i^{(j)}$ are the elastic form factors of the deuteron defined in Ref. [11]. The first term in the (big) squared brackets in Eq. (5), $S_0^{(0)}(q)$, corresponds to the S-wave approximation, the second term, $S_2^{(1)}(q)$, accounts for the S-D interference, and the last three terms contain the pure D-wave contributions.

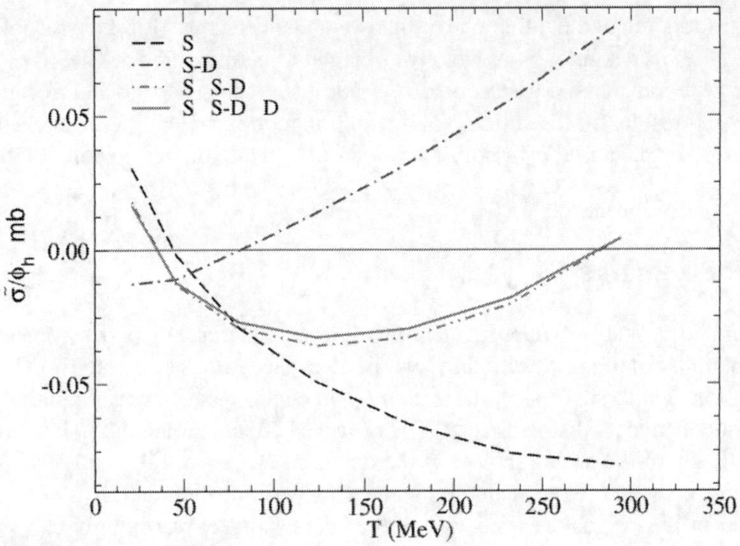

Figure 1. The TVPC signal $\widetilde{\sigma}$ for the h-term, in units of the ratio (ϕ_h, see Ref. [7]) of the TVPC and the strong h_1NN coupling constants, versus the antiproton beam energy T. Results of our calculations accounting for different terms of the deuteron wave function in Eq. (5) are shown, based on the deuteron wave function of the CD Bonn potential and the hadronic $\bar{p}N$ amplitudes from Ref. [15]: S-wave (black), S-D interference (blue), S + S-D waves (green), full result (red).

As was shown in Ref. [11], the contribution of the g'-term to the null-test signal vanishes in *pd* scattering due to the specific spin-isospin structure of the g'-interaction. Formally, for the same reason the charge-exchange g'-term given by Eq. (4) vanishes in the $\bar{p}d$ forward elastic scattering amplitude.

In the first theoretical work [5] where the null-test signal was calculated within the impulse approximation, the Coulomb interaction was not considered. In Ref. [6] Faddeev calculations were performed, but only for *nd* scattering and at rather low energies of ~ 100 keV. The Coulomb interaction was taken into account for the first time in Ref. [7] in a calculation of the null-test signal of *pd* scattering within Glauber theory and found to be negligible. A similar result was found in Ref. [20] using Faddeev calculations.

2.2 Numerical results

Results of numerical calculations of the energy dependence of the null test-signal for the h-term are presented in Fig. 1, in units of the unknown TVPC coupling strength. One can see from this figure that the deuteron S-wave contribution (dashed line) leads to a smooth energy dependence and has a node at an antiproton beam energy of about 50 MeV. The inclusion of the D-wave changes this behaviour considerably (solid line) due to a destructive S-D interference (cf. dash-dotted line). As a result, a second zero of the null-test signal $\widetilde{\sigma}$ appears at higher energies, i.e. at $T \approx 300$ MeV. The maximal value of $\widetilde{\sigma}$ is expected at 100 – 150 MeV. Note that the actual position of the nodes changes only slightly when deuteron wave functions from other NN models are used for the calculation.

Let us consider possible spurious effects that could mimic a TVPC signal. One source for a spurious signal is associated with a nonzero deuteron vector polarization $p_d^y \neq 0$ (in the direction of the incident-proton-beam polarization $\mathbf{p}^p$). In this case, the term $\sigma_1 P_y^{\bar{p}} p_y^d$ in Eq. (1) contributes to the asymmetry corresponding to the difference of the event counting rates for the cases of $p_y^{\bar{p}} P_{xz} > 0$ and $p_y^{p} P_{xz} < 0$ (with the fixed sign of P_{xz}), which is planned to be measured at COSY [3]. According to our calculations, the integrated cross section σ_1 could be equal to zero at antiproton beam energies of $\sim$ 100 MeV (see results for the Jülich $\bar{N}N$ interaction model in Refs. [21, 22]). Therefore, at this energy the spurios signal caused by a nonzero value of the deuteron vector polariziation p_y^d could be minimized.

3 Concluding remarks

We have performed a study of time-reversal-invariance violating but parity conserving effects in antiproton-deuteron scattering. Specifically, we have evaluated the null-test TVPC signal for scattering of antiprotons with transversal polarization p_y^p on deuterons with tensor polarization P_{xz} on the basis of the spin-dependent Glauber theory. The observed effects turned out to be similar to those in *pd* scattering: (i) There is a strong impact of the deuteron D-wave on the null-test signal that arises from a destructive interference between the S- and D-wave contributions; (ii) There is an oscillating behaviour of the null-test signal as a function of the beam energy. Accordingly, it is possible that the signal for TVPC effects is zero at some specific energies, even when the TVPC interaction itself is nonzero.

Acknowledgement. This work was supported in part by the Heisenberg-Landau program.

References

[1] A.D. Sakharov, Pis'ma v ZHETF **5**, 32 (1967).
[2] R.S. Conti, I.B. Khriplovich, Phys. Rev. Lett. **68**, 3262 (1992).
[3] COSY Proposal N 215, Spokespersons: P.D. Eversheim, B. Lorentz, and Yu. Valdau, `http://apps.fz-juelich.de/pax/paxwiki/images/8/8c/215-TRI_Prop_sum.pdf`
[4] H.E. Conzett, Phys. Rev. C **48**, 423 (1993).
[5] M. Beyer, Nucl. Phys. A **560**, 895 (1993).
[6] Y.-H. Song, R. Lazauskas, V. Gudkov, Phys. Rev. C **84**, 025501 (2011).
[7] Yu.N. Uzikov, A.A. Temerbayev, Phys. Rev. C **92**, 014002 (2015).
[8] Yu. N. Uzikov, J. Phys. Conf. Ser. **678**, 012020 (2016).
[9] Yu.N. Uzikov, EPJ Web Conf. **113**, 04027 (2016).
[10] Yu.N. Uzikov, A.A. Temerbayev, Int. J. Mod. Phys. Conf. Ser. **40**, 1660080 (2016).
[11] Yu.N. Uzikov, J. Haidenbauer, Phys. Rev. C **94**, 035501 (2016).
[12] Yu.N. Uzikov, EPJ Web Conf. **138**, 08001 (2017).
[13] A.A. Temerbayev, Yu. N. Uzikov, Izv. Ross. Akad. Nauk Ser. Fiz. **79**, 535 (2015) [Bull. Russ. Acad. Sci. Phys. **79**, 493 (2015)].
[14] A.A. Temerbayev, Yu. N. Uzikov Izv. Ross. Akad. Nauk Ser. Fiz. **80**, 271 (2016) [Bull. Russ. Acad. Sci. Phys. **80** N4, 242 (2016)].
[15] D. Zhou, R.G.E. Timmermans, Phys. Rev. C **86**, 044003 (2012).
[16] M.N. Platonova, V.I. Kukulin, Phys. Rev. C **81**, 014004 (2010).
[17] A.A. Temerbayev, Yu.N. Uzikov, Yad. Fiz. **78**, 38 (2015) [Phys. At. Nucl. **78**, 35 (2015)].
[18] P. Herczeg, Nucl. Phys. **75**, 655 (1966).
[19] M. Simonius, Phys. Lett. **B 78**, 147 (1975).
[20] Y.-H. Song, R. Lazauskas, V. Gudkov, Phys. Rev. C **93**, 065501 (2016).

[21] Yu. N. Uzikov, J. Haidenbauer, Phys. Rev. C **87**, 054003 (2013).
[22] Yu. N. Uzikov, J. Haidenbauer, Phys. Rev. C **88**, 027001 (2013).

Predicting and Discovering True Muonium ($\mu^+\mu^-$)

Henry Lamm[1,*] and *Yao* Ji[2,**]

[1]Department of Physics, University of Maryland, College Park, MD 20742, USA
[2]Institut für Theoretische Physik, Universität Regensburg, Regensburg 93040, Germany

Abstract. The recent observation of discrepancies in the muonic sector motivates searches for the yet undiscovered atom true muonium ($\mu^+\mu^-$). To leverage potential experimental signals, precise theoretical calculations are required. I will present the on-going work to compute higher-order corrections to the hyperfine splitting and the Lamb shift. Further, possible detection in rare meson decay experiments like REDTOP and using true muonium production to constrain mesonic form factors will be discussed.

1 Introduction

Given the dearth of clear signals beyond the Standard Model (BSM) from the LHC, it may prove useful to consider more subtle deviations from well-understood observables. Under this paradigm, one might organize future searches around resolving the "muon problem": the curious coincidence that multiple observables in the muonic sector agree with either theoretical predictions or similar results with other leptonic flavors only at the 2-4σ level.

The most persistent of these is the the anomalous magnetic moment a_μ measured at Brookhaven [1], which differs by $\approx 3\sigma$ from the theoretical predictions. Upcoming experiments at Fermilab [2] and J-PARC [3] are expected to reduce the experimental uncertainty by a factor of four. Simultaneously, the largest two theoretical uncertainties, the hadronic vacuum polarization and hadronic light-by-light, are expected to be reduced sufficiently that if the current mean values persist, the discrepancy would exceed 5σ. Another long-standing discrepancy in the low energy sector, the charge radii from muonic atoms [4, 5], appears to be resolving itself with electronic measurements [6] with near-term experiments to clarify this issue further. If rectified in favor of the muonic results, these observables will put stringent constraints on new physics. At higher energies, the ratio of leptonic decays in D and B mesons have found differences of $2-4\sigma$ with expectations [7–11].

A new class of observables that can shed light on the muon problem are those associated with the bound state ($\mu^+\mu^-$) [12–15]. This state has alternatively been dubbed "true muonium" [16], "bimuonium" [17], and "dimuonium" [18]. Simpler bound states like positronium (e^+e^-), hydrogen, and muonium (μ^+e^-) have attracted significant attention as testing grounds for precision QED studies [19], but are limited in their BSM discovery potential by either the mass suppression $O(m_e/\Lambda_{BSM})$ or large theoretical uncertainties from unknown nuclear structure effects. In contrast, true muonium has a much larger reduced mass ($\mu = \frac{m_\mu}{2}$), and its QCD corrections are limited to the better-understood hadronic loop effects.

*e-mail: hlamm@umd.edu
**e-mail: yao.ji@physik.uni-regensburg.de

Alas, true muonium has yet to be directly observed. The first reason is that it is experimentally difficult to producing low-energy muon pairs, and is exacerbated by the bound state's short lifetime ($\tau \approx 1$ ps), which presents an interesting challenge to experimenters. A second, more prosaic, reason for the neglect is until the a_μ anomaly, it seemed unlikely true muonium would offer any novel physics justifying the large effort.

In this talk, we present state of the art theoretical predictions for key energy splittings and lifetimes. Following this, we discuss the possibilities to observe true muonium in upcoming experiments.

2 Predicting

Resolutions to the muon problem relying upon BSM generically lead to $O(100\text{ MHz})$ corrections to transitions and decay rates of true muonium (e.g, [12]). These $10^{-6}-10^{-9}$ corrections are of plausible size for measurement if the production of large numbers of atoms were possible. By virtue of the annihilation channel, true muonium observables (e.g. transitions, production and decay rates) are sensitive at lower order to new particle content than other atomic transitions where individual measurements can be insensitive to different particle content (i.e. Pseudoscalar contributions to the Lamb shift in (μH) are heavily suppressed). $O(100\text{ MHz})$ is estimated be $O(m\alpha^7)$ in true muonium; therefore, our goal should be to predictions of this level, where hadronic and electroweak effects must be taken into account[13, 20]

The theoretical expression for the energy levels to true muonium from QED can be written

$$E_{n,l,j,s} = -\frac{m_\mu\alpha^2}{4n^2} + m_\mu\alpha^4\left[C_0 + C_1\frac{\alpha}{\pi} + C_{21}\alpha^2\ln\left(\frac{1}{\alpha}\right) + C_{20}\left(\frac{\alpha}{\pi}\right)^2 \right.$$
$$\left. + C_{32}\frac{\alpha^3}{\pi}\ln^2\left(\frac{1}{\alpha}\right) + C_{31}\frac{\alpha^3}{\pi}\ln\left(\frac{1}{\alpha}\right) + C_{30}\left(\frac{\alpha}{\pi}\right)^3 + \cdots\right], \quad (1)$$

where C_{ij} indicate the coefficient of the term proportional to $(\alpha)^i \ln^j(1/\alpha)$. C_{ij} include any dependence on mass scales other than m_μ. The coefficients of single-flavor QED bound states, used in positronium, are known up to $O(m_e\alpha^6)$. Partial results exist for $O(m_e\alpha^7)$ and are an active research area (For updated reviews of the coefficients see [21, 22]).

True muonium has extra contributions that must be considered. Large C_{ij} arise for electronic contributions due to $m_\mu/m_e \approx 200$. In fact, the Lamb shift is dominated by the electronic vacuum polarization [23]. The relative smallness of $m_\tau/m_\mu \approx 17$ and $m_\pi/m_\mu \approx 1.3$ produces contributions to true muonium much larger than analogous contributions to positronium.

Since [24], a number of important contributions have been computed and an update of several important transitions are shown in Table 1. For the HFS, improved calculations of the $O(m_\mu\alpha^5)$ contributions and large $O(m_\mu\alpha^6)$ and $O(m_\mu\alpha^7)$ have reduced the uncertainty be a factor-of-4 [20, 22, 25, 26]. Spin-independent contributions to the spectra are only known partially at $O(m_\mu\alpha^5)$, but some large $O(m_\mu\alpha^6)$ terms have been computed and included the presented results [27].

It should be emphasized that the missing contributions require no new theoretical techniques; positronium, muonium, and muonic hydrogen techniques can be straight-forwardly applied. Most of the unknown corrections arise from virtual electron loops to photon propagators.

Due in part to the difficulty of in-beam laser spectroscopy, consideration of other methods of measuring the Lamb shifts should be considered. An older method utilized for atomic hydrogen [28] has been suggested for true muonium [29] which is similar to the methods

Table 1. Theoretical predictions for key transitions in true muonium. The *had*-lab electron allows for large loop contributionseled uncertainty corresponds to the hadronic contributions, while unlabeled uncertainties indicated estimates of missing contributions

Transition	E_{theory} [MHz]
$1^3S_1 - 1^1S_0$	$42329355(51)_{had}(700)$
$2^3S_1 - 1^3S_1$	$2.550014(16) \times 10^{11}$
$2^3P_0 - 2^3S_1$	$1.002(3) \times 10^7$
$2^3P_1 - 2^3S_1$	$1.115(3) \times 10^7$
$2^3P_2 - 2^3S_1$	$1.206(3) \times 10^7$
$2^1P_1 - 2^3S_1$	$1.153(3) \times 10^7$

pursued currently by the DIRAC experiment for $(\pi^+\pi^-)$ [30]. In this method, a beam of $2S$ state true muonium would be passed through a magnetic field, resulting in level-mixing with $2P$ state which decays by γ emission to the $1S$, decreasing the intensity of the beam. By measuring the beam's intensity as a function of magnetic field, the Lamb shift can be extracted.

Since $m_\mu > m_e$, it is possible for n^3S_1 states to decay into e^+e^- pairs. All currently contemplated searches utilize this decay for discovering true muonium [31–33], so predictions of these rates are desirable. Including all NLO and a large NNLO contribution we find,

$$\Gamma(1^3S_1 \to e^+e^-) = \left(1 + \left[-\frac{221}{36} + \frac{4}{3}\ln\left(2\frac{m_\mu}{m_e}\right) - 0.3899(8)\right]\frac{\alpha}{\pi} + 455(1)\frac{\alpha^2}{\pi^2}\right)\frac{\alpha^5 m_\mu}{6} \tag{2}$$

$$\implies \tau(1^3S_1 \to e^+e^-) = \frac{1}{\Gamma(1^3S_1 \to e^+e^-)} = 1.79560(13) \times 10^{-12}\ \text{s},$$

$$\Gamma(2^3S_1 \to e^+e^-) = \left(1 + \left[-\frac{221}{36} + \frac{4}{3}\ln\left(2\frac{m_\mu}{m_e}\right) - 0.3899(8)\right]\frac{\alpha}{\pi} + 394(1)\frac{\alpha^2}{\pi^2}\right)\frac{\alpha^5 m_\mu}{48} \tag{3}$$

$$\implies \tau(2^3S_1 \to e^+e^-) = \frac{1}{\Gamma(2^3S_1 \to e^+e^-)} = 14.3696(10) \times 10^{-12}\ \text{s},$$

$$\Gamma(3^3S_1 \to e^+e^-) = \left(1 + \left[-\frac{221}{36} + \frac{4}{3}\ln\left(2\frac{m_\mu}{m_e}\right) - 0.3899(8)\right]\frac{\alpha}{\pi} + 400(100)\frac{\alpha^2}{\pi^2}\right)\frac{\alpha^5 m_\mu}{162} \tag{4}$$

$$\implies \tau(3^3S_1 \to e^+e^-) = \frac{1}{\Gamma(3^3S_1 \to e^+e^-)} = 48.5(3) \times 10^{-12}\ \text{s}, \tag{5}$$

where the final NLO term comes from hadronic vacuum polarization [20] and its associated error, and the NNLO coefficient's error is estimated by $O(1)$ for the $n = 1, 2$ because other contributions are not anticipated to be anomalously large. For $n = 3$ the large error is because even the anomalously large NNLO contribution is not known so we estimate based on the lower n and assign a 25% uncertainty. Even at this precision, the theoretical values are lower than the 1%, 5%, and 15% respectively for $n = 1, 2, 3$ that has been suggested as experimental uncertainties attainable at a near-term experiment [33].

Singlet states will predominately decay to two photons, and similar precision is known for these rates [23], but we quote here the leading order values $\tau(n^1S_0 \to \gamma\gamma) = 0.6n^3 \times 10^{-12}$ s.

If very high-intensity true muonium experiments were ever built, it would be possible to measure more exotic decays, including those of the triplet state to neutrinos. The leading order decay rates to mono-energetic neutrinos are known to be $\Gamma(1^3S_1 \to \nu_\mu\bar{\nu}_\mu) \approx 10^{-11}\Gamma_{e^+e^-}$ and $\Gamma(1^3S_1 \to \nu_l\bar{\nu}_l) \approx 10^{-14}\Gamma_{e^+e^-}$. These rates are admittedly small but unlike positronium, due to a $\propto m_\ell^5$ scaling, are around the level of rare mesonic decays. Further, measurement of neutrino decays are related to the general subject of invisible decays. These were shown to constrain a variety of BSM (e.g. extra dimensions, axions, mirror matter, fractional charges, and other low-mass dark matter models) in positronium [34]. In true muonium, these rates are also enhanced due to mass scaling, as well as raising the upper limit on masses from m_e to m_μ.

3 Discovering

The greatest experimental obstacle is producing a sufficiently large number of the bound state. In the past, many production channels have been discussed: $\pi p \to (\mu^+\mu^-)n$ [35], $\gamma Z \to (\mu^+\mu^-)Z$ [35], $eZ \to e(\mu^+\mu^-)Z$ [36, 37], $\mu^+\mu^- \to (\mu^+\mu^-)$ [16], $e^+e^- \to (\mu^+\mu^-)$ [29, 38, 39], $e^+e^- \to (\mu^+\mu^-)\gamma$[29], $\eta \to (\mu^+\mu^-)\gamma$ [40, 41], $K_L \to (\mu^+\mu^-)\gamma$ [42], $Z_1Z_2 \to Z_1Z_2(\mu^+\mu^-)$ [43], and $q^+q^- \to (\mu^+\mu^-)g$ in a quark plasma [44]. Some of the more novel methods of utilizing these production channels considered include: fixed target experiments [45], Fool's Intersection Storage Rings [29], and even from astrophysical sources [46, 47].

Two fixed-target experiments should be highlighted in this discussion. The Heavy Photon Search (HPS) [31] experiment has plans to search for true muonium [45], and DImeson Relativistic Atom Complex (DIRAC) [48] has discussed the possibility in an upgraded run [32]. Additionally, the DIRAC experiment intends to study the Lamb shift in the $(\pi^+\pi^-)$ bound state using a fixed magnetic field and measuring the decay rate as a function of distance [30], and the methods developed could be applied to true muonium.

Brodsky and Lebed's idea to produce true muonium by colliding e^+e^- at an angle with energy near $2m_\mu$ [29] has been developed by a group at BINP into an idea for a low-energy collider [33]. Their proposed physics analysis beyond discovering true muonium could include: production rates, decay lengths, $2P - 1S$ transition probabilities, $2P$ lifetimes, and possibly even transition energies.

Another avenue for discovering true muonium is through a rare meson decay. The proposed REDTOP experiment at Fermilab would search for true muonium in η/η' decays [49]. In anticipation of this, the rates have been recomputed using modern form factors derived from dispersive techniques and applying NLO corrections that will be explained in an upcoming work:

$$\frac{\mathcal{B}(\eta \to \gamma(\mu^+\mu^-))}{\mathcal{B}(\eta \to \gamma\gamma)} = 1.476(5)_{\text{stat}}(4)_{\text{sys}} \times 10^{-9}, \tag{6}$$

$$\frac{\mathcal{B}(\eta' \to \gamma(\mu^+\mu^-))}{\mathcal{B}(\eta' \to \gamma\gamma)} = 1.761(7)_{\text{stat}}(2)_{\text{sys}} \times 10^{-9}, \tag{7}$$

With this branching ratio, about 1000 η events would be produced at REDTOP with their projected luminosity, but this number should be reduced by detector efficiencies and cuts. While no specific experiment has yet been proposed, a similar search could be undertaken in future high-intensity K_L beams being considered at CERN and J-PARC, where the rate has been computed to NLO and including modern experimental form factors [50]:

$$\frac{\mathcal{B}(K_L \to \gamma(\mu^+\mu^-))}{\mathcal{B}(K_L \to \gamma\gamma)} \approx 1.26(2)_{\text{model}} \times 10^{-9}.$$

The authors would like to thank N. Raman for his unpublished calculations. HL is supported by the U.S. Department of Energy under Contract No. DE-FG02-93ER-40762. YJ acknowledges the Deutsche Forschungsgemeinschaft for support under grant BR 2021/7-1.

References

[1] G. Bennett et al. (Muon G-2 Collaboration), Phys. Rev. **D73**, 072003 (2006), hep-ex/0602035
[2] J. Grange et al. (Muon g-2) (2015), 1501.06858
[3] N. Saito (J-PARC g-'2/EDM), AIP Conf. Proc. **1467**, 45 (2012)
[4] A. Antognini et al., Science **339**, 417 (2013)
[5] R. Pohl et al. (CREMA), Science **353**, 669 (2016)
[6] A. Beyer, L. Maisenbacher, A. Matveev, R. Pohl, K. Khabarova, A. Grinin, T. Lamour, D.C. Yost, T.W. Hänsch, N. Kolachevsky et al., Science **358**, 79 (2017)
[7] R. Aaij et al. (LHCb collaboration), Phys. Rev. Lett. **113**, 151601 (2014), 1406.6482
[8] R. Aaij et al. (LHCb), Phys. Rev. Lett. **115**, 111803 (2015), [Addendum: Phys. Rev. Lett.115,no.15,159901(2015)], 1506.08614
[9] G. Ciezarek, M. Franco Sevilla, B. Hamilton, R. Kowalewski, T. Kuhr, V. Luth, Y. Sato (2017), 1703.01766
[10] R. Aaij et al. (LHCb) (2017), 1711.05623
[11] R. Aaij et al. (LHCb), JHEP **08**, 055 (2017), 1705.05802
[12] D. Tucker-Smith, I. Yavin, Phys. Rev. **D83**, 101702 (2011), 1011.4922
[13] H. Lamm, Phys. Rev. D **91**, 073008 (2015)
[14] H. Lamm, Phys. Rev. **D92**, 055007 (2015), 1505.00057
[15] H. Lamm, Phys. Rev. **D94**, 115007 (2016), 1609.07520
[16] V. Hughes, B. Maglic, Bull. Am. Phys. Soc. **16**, 65 (1971)
[17] V.N. Baier, V.S. Synakh, JETP **14**, 1122 (1962)
[18] P. Budini, Tech. rep., CM-P00056754 (1961)
[19] S.G. Karshenboim, Phys. Rept. **422**, 1 (2005), hep-ph/0509010
[20] H. Lamm, Phys. Rev. **A95**, 012505 (2017), 1611.04258
[21] G.S. Adkins, R.N. Fell, Phys. Rev. **A89**, 052518 (2014), 1402.7040
[22] Y. Ji, H. Lamm, Phys. Rev. A **94**, 032507 (2016)
[23] U. Jentschura, G. Soff, V. Ivanov, S.G. Karshenboim, Phys. Rev. **A56**, 4483 (1997), physics/9706026
[24] H. Lamm, *True muonium: the atom that has it all*, in *Twelfth International Conference on the Intersections of Particle and Nuclear Physics, Vail, CO, USA, May 19-24, 2015* (2015), 1509.09306, https://inspirehep.net/record/1395477/files/arXiv:1509.09306.pdf
[25] G.S. Adkins, M. Kim, C. Parsons, R.N. Fell, Phys. Rev. Lett. **115**, 233401 (2015), 1507.07841
[26] Y. Ji, H. Lamm (2017), 1701.04362
[27] H. Lamm, N. Raman, in prep. (2017)
[28] R.T. Robiscoe, Phys. Rev. **138**, A22 (1965)
[29] S.J. Brodsky, R.F. Lebed, Phys. Rev. Lett. **102**, 213401 (2009), 0904.2225
[30] L.L. Nemenov, V.D. Ovsyannikov, Phys. Lett. **B514**, 247 (2001)
[31] A. Celentano (HPS), J. Phys. Conf. Ser. **556**, 012064 (2014)
[32] P. Chliapnikov, DIRAC-NOTE-2014-05 (2014)
[33] A. Bogomyagkov, V. Druzhinin, E. Levichev, A. Milstein, S. Sinyatkin (2017), 1708.05819
[34] A. Badertscher, P. Crivelli, W. Fetscher, U. Gendotti, S. Gninenko, V. Postoev, A. Rubbia, V. Samoylenko, D. Sillou, Phys. Rev. **D75**, 032004 (2007), hep-ex/0609059

[35] S.M. Bilenky, V.H. Nguyen, L.L. Nemenov, F.G. Tkebuchava, Yad. Fiz. **10**, 812 (1969)
[36] E. Holvik, H.A. Olsen, Phys. Rev. **D35**, 2124 (1987)
[37] N. Arteaga-Romero, C. Carimalo, V. Serbo, Phys. Rev. **A62**, 032501 (2000), hep-ph/0001278
[38] V. Baier, V. Synakh, Zhur. Eksptl'. i Teoret. Fiz. **41** (1961)
[39] J. Moffat, Phys. Rev. Lett. **35**, 1605 (1975)
[40] L. Nemenov, Yad. Fiz. **15**, 1047 (1972)
[41] G. Kozlov, Sov. J. Nucl. Phys. **48**, 167 (1988)
[42] J. Malenfant, Phys. Rev. **D36**, 863 (1987)
[43] I. Ginzburg, U. Jentschura, S.G. Karshenboim, F. Krauss, V. Serbo et al., Phys. Rev. **C58**, 3565 (1998), hep-ph/9805375
[44] Y. Chen, P. Zhuang (2012), 1204.4389
[45] A. Banburski, P. Schuster, Phys. Rev. **D86**, 093007 (2012), 1206.3961
[46] S. Ellis, J. Bland-Hawthorn (2015), 1501.07281
[47] S.C. Ellis, J. Bland-Hawthorn (2017), 1712.02022
[48] A. Benelli (DIRAC Collaboration), EPJ Web Conf. **37**, 01011 (2012)
[49] C. Gatto, B. Fabela Enriquez, M.I. Pedraza Morales (REDTOP), PoS **ICHEP2016**, 812 (2016)
[50] Y. Ji, H. Lamm (2017), 1706.04986

The Muonium Antimatter Gravity Experiment

Thomas J. Phillips[1,*] (For the MAGE and muCool Collaborations)

[1]Illinois Institute of Technology, Chicago, IL

Abstract. A direct measurement of the gravitational acceleration of antimatter has the potential to show that we live in a "Dirac-Milne" Universe, which could explain cosmological observations without the need for dark matter, dark energy, inflation, or missing antimatter. Such a measurement would also be sensitive to the possible existence of a fifth force. Cooling antimatter to temperatures where gravitational energies are comparable to thermal energies is challenging for most forms of antimatter, which annihilate upon contact with matter. The exception is the antimuon (μ^+), which is easily cooled by stopping in cold matter, but the short muon lifetime poses challenges. Positive muons that stop in material will combine with free electrons to form muonium, a neutral leptonic atom with most of its mass derived from the 2nd-generation antimuon. We are developing the Muonium Antimatter Gravity Experiment (MAGE) to measure the gravitational force on muonium using a novel, monoenergetic, low-velocity, horizontal muonium beam directed at an ultra-precise atom interferometer. If successful, MAGE will measure for the first time the gravitational coupling to a 2nd-generation particle in a system whose antimatter-dominated mass is not predominantly strong-interaction binding energy. The novel MAGE beam production approach could also have important applications to other muonium experiments as well as to the measurement of $g - 2$.

1 Introduction

Muonium (Mu) is an exotic hydrogen-like atom composed of a positive muon and an electron with unique properties that can be exploited to make numerous fundamental measurements. Since neither the muon nor the electron has any known internal structure, calculations of the bound state are free of the difficulties arising from the finite size of the nucleus in normal atoms. Precision spectroscopic measurements of muonium allow one to determine fundamental constants such as the muon mass, its magnetic moment and magnetic anomaly, and the fine structure constant [1]. The absence of sidereal and diurnal variations in muonium spectroscopic measurements provides stringent limits on CPT violation [1]. Muonium–antimuonium conversion is analogous to $K^0 - \overline{K^0}$, $B^0 - \overline{B^0}$, and neutrino oscillations and appears naturally in many beyond-the-Standard-Model theories [2]. In addition, cold muonium can be ionized to provide a source of cold muons, a method that is currently planned for the muon $g - 2$ experiment under development at J-PARC [3, 4].

The Muonium Antimatter Gravity Experiment (MAGE) Collaboration aims to extend the list of fundamental measurements using muonium by measuring the gravitational force of the

*e-mail: phillips@PHYS.IIT.EDU

earth on muonium [5]. Because the muonium mass is dominated by that of the antimuon, measuring the gravitational force on Mu will provide a direct measurement of the coupling of gravity to antimatter and to a second-generation particle, making the measurement uniquely sensitive to a possible fifth force that couples to the second generation [6].

Muonium experiments are limited by currently available Mu sources and would benefit from the development of better Mu sources in progress at Switzerland's Paul Scherrer Institute (PSI). Indeed, the gravity experiment most likely requires the development of a new Mu source as well as an ultra-precise interferometer. This paper describes these developments as well as the motivation for the gravity measurement.

2 Motivation

At NSF's and DOE's request, the U.S. High Energy Physics Advisory Panel produced a short list of the most important questions in particle physics [7]. It is intriguing that of those nine questions, the measurement of antimatter gravity may have relevance to more than half, and the potential to answer three. While perhaps *a priori* unlikely, an antimatter gravity experiment could show that our Universe is described by "Dirac–Milne" [8, 9] or lattice [10] cosmology, containing equal parts matter and antimatter that repel each other gravitationally. This would explain the mystery of the missing antimatter without the need for additional *CP* violation. With a net gravitational mass of zero, the Universe would be flat and expanding linearly, which fits the Type Ia supernova data with no need of Dark Energy [9–11]. The slower initial expansion allows all parts of the visible Universe to be in thermal contact, resolving the horizon problem with no need for inflation [9, 10] and the age problem with no need for a cosmological constant [9]. Having both positive and negative gravitational mass results in gravitational vacuum polarization [12, 13], which provides a mechanism for Modified Newtonian Dynamics (MOND) [14, 15], and fits galactic rotation curves with no need of Dark Matter. Of course, the Equivalence Principle requires that the acceleration of antimatter on the earth $\bar{g} = g$, and indirect arguments based upon the virtual antimatter component in matter suggest that $\bar{g} = g$ with a high degree of accuracy [16]. However, we already have evidence that virtual particles do not interact gravitationally, for if they did the cosmological constant would be 120 orders of magnitude larger than observed [17]. If antimatter is identified with the $r < 0$ solutions to General Relativity's Kerr–Newman equation (Kerr–Newman singularities have been suggested as a model for electrons [18–22]), then General Relativity would predict $\bar{g} = -g$ and virtual pairs of particles would feel no gravity, so the indirect limits on $\bar{g}$ would not apply. The potential for a single measurement to explain multiple mysteries, with no need to introduce the new physics of cosmic inflation, Dark Energy, and Dark Matter, provides ample motivation for the gravity measurement.

Several collaborations at CERN are conducting experiments that will attempt to measure the gravitational force on antimatter using antihydrogen [23–26], of which one has published broad limits on $\bar{g}$ [26]. However, these experiments are extremely challenging, in some cases requiring decades of work to get to the point where they can make and trap antihydrogen. A primary reason for this difficulty is that antiprotons and antihydrogen annihilate upon contact with ordinary matter. This renders making the cold antihydrogen needed for a gravity measurement particularly challenging. The experiments are made more challenging by the difficulty of getting enough antiproton beam time, a problem about to become worse as antiprotons will be unavailable during CERN's 2-year-long shutdown in 2019–2020. In addition, antihydrogen experiments suffer from the theoretical complication that the antiproton mass is dominated by binding energy so some have argued that a 1% measurement would not be sensitive to the mass of its antiquark constituents [27].

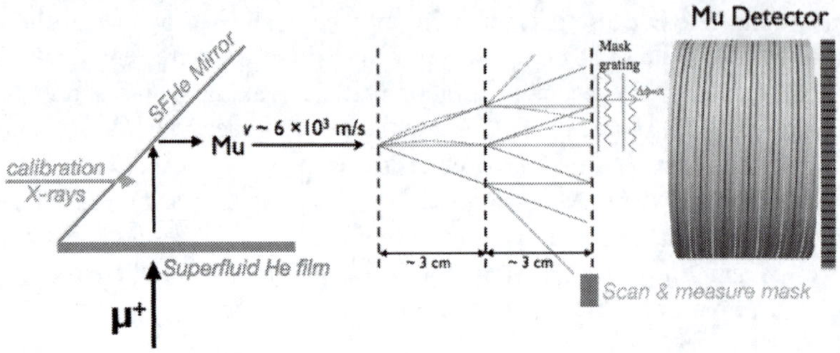

Figure 1. Principle of the MAGE experiment, shown schematically in elevation view (not to scale). Phase shift of π shown for illustrative purposes only; actual phase shift will be small.

Antimuons (μ^+), on the other hand, do not annihilate on contact with ordinary matter, so they are easy to cool by stopping them in a cold material. Once stopped, they can combine with a free electron to form cold muonium. Making a gravity measurement with muonium will be enormously challenging, primarily because of the short (2.2 μs) lifetime of the muon, but the challenges are quite different from the antihydrogen challenges, thus it is possible that the muonium measurement could turn out to be less challenging than the antihydrogen measurement. The Mu mass is 99.5% antimatter thus avoiding antihydrogen's theoretical complications. Additionally, the muonium experiment will be uniquely sensitive to gravitational and fifth-force couplings to second-generation particles [28–30].

3 MAGE: The Muonium Antimatter Gravity Experiment

The principle of MAGE, an application of well-established atom interferometry [31], is illustrated in Fig. 1. A horizontal muonium beam is directed into a three-grating interferometer. The first two gratings set up an interference pattern that has the same period as the gratings. Its phase is determined by scanning the third grating vertically by several grating periods and measuring the resulting sinusoidal variation in detected Mu intensity. Gravity causes a phase shift in the interference pattern which is equivalent to the deflection of an individual muonium atom. The interferometer is aligned using a soft X-ray beam with wavelength comparable to that of the Mu. This X-ray beam will also be used to determine the phase of an undeflected beam. The gratings are able to rotate with respect to the vertical for systematic studies; there is no gravitational phase shift when the grating slits are vertical.

3.1 Muonium Beam

Muonium is produced in many materials when a positive muon is stopped and a weakly bound electron is available to be captured. Since within the material, muonium behaves like a hydrogen atom, its transport is governed mainly by diffusion processes. Therefore in order to reach a high yield in vacuum the stopping target needs to be thin and/or porous, or the muons must be moved to the surface before they form muonium. Vacuum yields from 3 to 40% per stopped muon have been reached with various methods: tungsten foil heated to 2500 K [32], SiO_2 powder [2, 33], laser-ablated SiO_2 aerogel [34] or mesoporous SiO_2 films [35]. For these existing muonium sources, Mu are emitted with a thermal velocity distribution and uniformly in $\cos\theta$ (θ being the angle relative to the surface). This is inefficient for many applications, including the gravity measurement, which accepts Mu only within a small solid angle.

Fortunately, there is an alternative method for making muonium that will produce a beam ideal for the gravity experiment. The basis of this beam is that Mu, which is a light isotope of hydrogen, has a negative chemical affinity in superfluid helium so when it diffuses out of the liquid, it, like hydrogen and deuterium [36], will be ejected with a velocity normal to the surface that is large compared to its thermal velocity.

The concept of forming a Mu beam using a film of superfluid helium (SFHe) was published in 2011 by Taqqu [37]. The formation of Mu in SFHe is a function of charge mobility, which grows exponentially below the Λ point. Muonium formation efficiency in SFHe is near 100% at low temperature [38]. The chemical potential of hydrogen in superfluid helium depends upon the isotopic mass. These potentials have been calculated for hydrogen (37 K), deuterium (14 K), and tritium (7 K) [39], and experimentally verified for deuterium [36]. The corresponding potential for Mu is 270 K [37], so Mu will be ejected from the surface with a kinetic energy of 23 meV, which corresponds to a velocity of 6.3 mm/μs, much larger than its thermal velocity in the superfluid helium. For our application, the superfluid helium will be at 0.2 K, so the energy dispersion will be $\approx 0.2/270 = 0.07\%$, and the 1-dimensional beam divergence $\approx \sqrt{0.2/270} = 27$ mrad.

Muonium that forms in the superfluid helium must reach the surface in order to be ejected into the vacuum. There are two methods that can ensure a large fraction of the Mu reaches the surface: either stop the muons in a very thin film so roughly half the Mu that forms will diffuse to the top surface, or stop the muons in a thick film and use an electric field to drift the muons to the top surface, seeded with electrons, where muons form Mu and are ejected. The MAGE Collaboration plans to investigate both of these methods.

3.1.1 Ultra-Bright Muon Beam

In order to stop muons efficiently in a thin film, they must be extremely low in energy and have a very small velocity dispersion. Conventional surface muon beams are much too hot for this, so the muCool project at PSI is developing a method for compressing the muon phase space by 10 orders of magnitude with an efficiency of $\approx 10^{-3}$, which will allow the muons to be stopped in a 1 μm film of liquid helium. The project is based on ideas published by Taqqu in 2006 [40]: a μ^+ beam is stopped in helium gas where the muons are first compressed transverse to the original beam direction and then in the longitudinal direction before being extracted back into vacuum. A sketch of the setup is shown in Fig. 2. The compression is based on the position-dependent drift-velocity vector of a charged particle under the influence of electric (**E**) and magnetic (**B**) fields in the presence of a gas density gradient. The drift velocity $\mathbf{v}_D$ is given by

$$\mathbf{v}_D = \frac{\mu E}{1+\omega^2\tau_c^2}\left[\hat{\mathbf{E}} + \omega\tau_c\hat{\mathbf{E}}\times\hat{\mathbf{B}} + \omega^2\tau_c^2(\hat{\mathbf{E}}\cdot\hat{\mathbf{B}})\hat{\mathbf{B}}\right]. \tag{1}$$

The effect of the helium gas at 5 mbar is characterized by the mobility of the muons and the mean time between collisions τ_c; $\omega = eB/m_\mu$ is the cyclotron frequency. In the experimental setup shown in Fig. 2 the electric and magnetic fields and the temperature distribution are arranged such that in the transverse compression stage (left side of figure) the muons drift to the right along the bottom wall (colder, higher density, first term of Eq. (1) dominating) and top wall (warmer, lower density, second term of Eq. (1) dominating) of the triangular cell, while in the longitudinal compression stage (center of figure) they follow the magnetic field and are compressed longitudinally (third term of Eq. (1) dominating). Finally the muons with

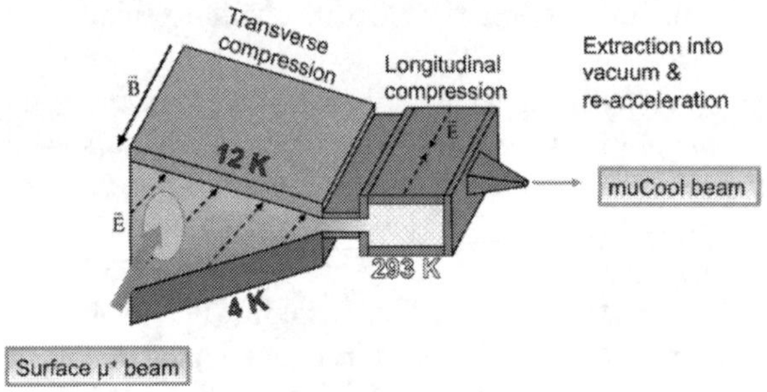

Figure 2. Schematic of the muCool apparatus. The surface muon beam is stopped in cold helium gas with a density gradient designed so that the **E** × **B** drift angle causes transverse compression. Electric field components in the **z** direction cause longitudinal compression. Cooled muons are extracted on the right.

an energy of about 1 eV can be extracted into vacuum through a 1 mm-diameter orifice. Once in the vacuum the muons can be re-accelerated. Extraction of the beam from the magnetic field must be done with care in order to avoid significantly increasing the phase space; an example of how this was done for positrons is described in [41].

The concept of longitudinal compression was demonstrated in an experiment conducted a few years ago [42]. More recently the muCool collaboration has demonstrated longitudinal compression within about 3 μs with an efficiency of approximately 100% (apart from decay losses) [43, 44]. The transverse compression stage is the more complicated as it requires a cryogenic helium gas cell. Transverse compression was demonstrated in 2015 [44].

Having successfully demonstrated both longitudinal and transverse compression, a new target which combines both types of compression at once is currently being tested. If successful, this will improve the efficiency of muCool by reducing the time required for cooling. The next steps will be to extract the cooled muons back into the vacuum, reaccelerate, and extract from the magnetic field.

Ultra-Bright Muonium Beam Production

The muCool μ^+ beam will be directed through a 30 nm-thick silicon-nitride window and stopped inside a 1 μm-thick superfluid helium film at 0.2 K. Roughly half of the formed Mu atoms reach the top surface of the film and are ejected with a velocity of 6.3 mm/μs. Due to the thinness of the helium film the original beam size is preserved and the emission into vacuum takes place with high efficiency. The resulting muonium beam is thus of sub-millimeter beam size, monochromatic ($\Delta E/E \approx 0.07\%$) and with a small divergence (27 mrad). This vertical beam can be reflected to the horizontal by a SFHe-coated mirror [45], making it ideal for the gravity measurements, as the interferometer will be able to accept the entire beam. This beam will also be an ideal source for improving other fundamental measurements using muonium.

Novel E-Field Muonium Source Using a Thick Superfluid Helium Film

While the ultra-bright Mu beam being developed by the muCool program is nearly ideal for the gravity experiment, there is another possibility for developing an even more intense beam suitable for the gravity experiment, and this beam could be available without the technical risk associated with muCool. This is the method described by Taqqu in [37]. It differs from the thin-SFHe-film Mu source just described in that the $\mu^+$are stopped in a thick SFHe film and a surface layer of electrons draws them to the surface before making Mu, substantially raising the efficiency for the Mu to be ejected into the vacuum. This method can utilize an uncompressed surface muon beam by using a thicker window to degrade the muons, and while the beam spot is larger than the nominal area of the gravity experiment, the mismatch for PSI's $\pi E5$ beam is only a factor of 4. This mismatch can be reduced by adjusting the dimensions of the interferometer and by focusing the muonium beam with a curved mirror [45] since the angular divergence of the beam is smaller than the interferometer's acceptance (which is roughly the requirement that the beam pass sequentially through two 1 cm squares separated by ~ 6 cm). The larger number of muons in the uncompressed surface muon beam more than compensates for the lower μ^+ stopping efficiency and interferometer acceptance (see Table 1 and Fig. 5 below), although redirecting the full surface muon beam upward will require more sophisticated beam optics than directing the muCool beam upward, a non-trivial difference.

This thick-film method of making a Mu beam requires preparing a layer of electrons on the upper surface of the superfluid helium. This well-established technique has been demonstrated in [46] with the electrons produced using a tungsten tip in the liquid helium; an external electric field drew them to the surface where they were trapped in a potential well [47]. An electric field of 100 V/cm corresponds to a surface density just above 10^8 e/cm^2, a typical density for these experiments. The electrons are observed to remain at the surface for long times (minutes at low temperatures) after the voltage used to attract the electrons to the surface is removed [47, 48]. An electric field of 100 V/cm, created entirely by the surface electrons, will cause 80% of the stopping muons to drift away from the electrons left by their ionization trail [49]. In liquid helium, positive ions attract a "snowball" of helium atoms which determines their mobility [48, 50]. The drift velocity at 100 V/cm saturates near 42 m/s [51], and at this velocity the muons that stop in the top 100 μm will drift to the surface in 1 muon lifetime, combine with an electron, and be emitted as Mu. A simulation of a simple stainless degrader indicates it is possible to get 5% of a 29.8 MeV/c surface muon beam to stop in SFHe and drift to the surface before decaying, a number that can likely be improved with optimization. Using a surface muon beam such as PSI's $\pi E5$ beam with $1.2 \times 10^8 \mu/s$, we could produce a beam with up to 4×10^6 Mu/s, depending upon the duty cycle required to maintain the surface electrons. For the high-rate beams, electrons may be lost because of recombination with He^+ ions from muon ionization trails as well as Mu formation. In principle it should be possible to continuously resupply the surface electrons and eliminate a duty-cycle factor, but there are too many unknowns about SFHe physics to be confident about this before the development work is done.

3.2 Muonium Interferometer

The MAGE Collaboration plans to use an ultra-precise interferometer to measure the very small deflection the muonium beam experiences in a few muon lifetimes. This interferometer, currently under development at IIT [5] (see Fig. 3), will use a support structure machined from a single silicon block. This is a common technique used in X-ray optics and is particularly effective at cryogenic temperatures where silicon is extremely stable with an extremely small coefficient of thermal expansion.

The interferometer gratings will be mounted in frames that will be moved with piezoelectric actuators for alignment and scanning. The positions of the gratings will be monitored with semiconductor-laser tracking frequency gauges (SL-TFGs), developed by MAGE collaborators J.D. Phillips and R.D. Reasenberg [52]. SL-TFGs have sub-picometer accuracy, already achieved at IIT, more than sufficient for the interferometer.

The baseline MAGE design uses a 100 nm grating period since gratings with this period have been fabricated with adequate fidelity [53]. Gratings with a smaller period would improve the sensitivity of the experiment, so MAGE will use investigate fabricating gratings with finer periods and will use the smallest period that proves practical. With these gratings the interferometer will operate in the quantum regime. The optimum open fraction for the gratings will depend upon several design parameters, but is expected to be near 60% for the first grating and 40% for the second grating. These two gratings set up an interference pattern

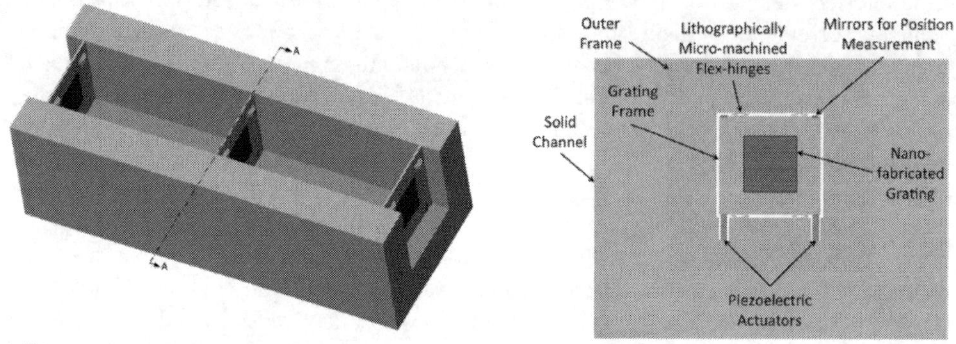

Figure 3. (left) CAD drawing of the muonium interferometer; (right) Section A-A. In blue is the support structure for the gratings, a solid channel machined out of a single silicon block. Each grating is mounted in a silicon frame connected to an outer frame by flex-hinges; a pair of piezoelectric actuators permits small rotations to align the gratings precisely in parallel, as well as scanning of the third grating. Each grating frame has mirrors (or possibly corner-cube retroreflectors) on its top corners that form part of the optical TFGs used to measure their position.

with the same period as the gratings. Since this pattern is too fine to image directly, the phase of the pattern is determined by measuring the transmission of muonium through a third grating of identical period as its position is scanned. An open fraction of 37% is optimum for the third grating [31], and while the maximum theoretical contrast is 20% or more, imperfections in the interferometer will reduce the contrast so a contrast of only 10% is assumed for the sensitivity calculations presented in this paper. The gravitational phase shift depends upon the Mu transit time, so a time-of-flight measurement would be useful for a thermal beam but unnecessary for the mono-energetic Mu beam produced by SFHe.

3.3 Muonium Detector

When Mu decays, the decaying muon produces an energetic positron and a low-energy electron is left behind. The signal for the MAGE Mu detector will be the coincidence of signals from scintillators measuring the positron and a hit in an electron detector. This will allow discrimination of Mu decays occurring before the third interferometer grating from those occurring after the third grating, since the electric field accelerating the electron will be non-zero only after the third grating. Only 1.5% of the muonium entering the interferometer will live long enough to make it past the third grating, so detecting the electron will be important for background rejection for tracks that could have originated before or at the third grating.

A baseline design for the muonium detector's positron tracker is shown in Fig. 4. Hits in the tracker's fibers and paddles localize the Mu decay when combined with the electron location in the transverse plane. The electron from the Mu decay will be accelerated by an electric field onto a position-sensitive detector on the far end. G4beamline simulations [54] indicate that this baseline design would accept 66% of muonium decays from the signal region with only 0.1% background from the much more numerous decays in the interferometer. Simple changes, such as extending the outer tracker 2 cm over the interferometer and adjusting the incident muon's polarization, increase the acceptance to 72%.

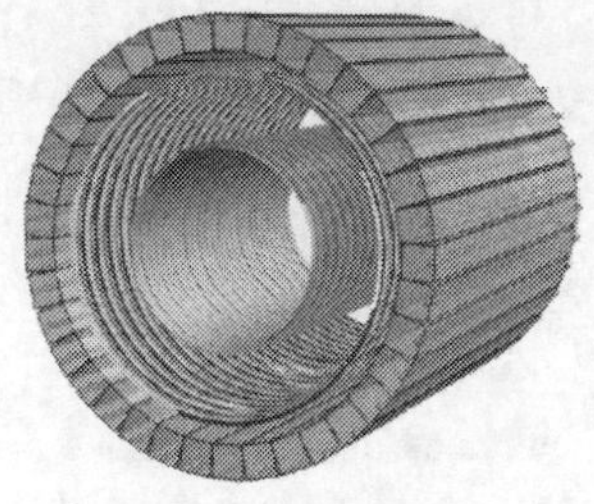

Figure 4. Muonium detector comprised of two cylinders, each having scintillating fibers wound helically around a support tube. The length of each fiber will be approximately one attenuation length, so by reading out both ends of each fiber it will be possible using timing and/or pulse-height differences to determine where the fiber was hit with enough precision to determine the coordinate along the tube with a resolution of a fiber diameter. Azimuth (ϕ) will be measured using scintillating paddles with embedded wave-shifting fibers (read out at one end). Electrodes in the center provide an electric field to accelerate the electron onto a detector (not shown).

4 The MAGE Measurement

The challenge of MAGE is to make a significant gravity measurement within the time permitted by the muon's 2.2 μs lifetime. The phase shift ϕ of the interferometer is given by ([31] Eqn. 52):

$$\phi = [\mathbf{G} \cdot \mathbf{a} + 2\mathbf{G} \cdot (\mathbf{\Omega} \times \mathbf{v})]\, t^2 \tag{2}$$

where $\mathbf{G}$ is the grating vector (in the plane of the grating, perpendicular to its lines, with $|\mathbf{G}| = 2\pi/d$) for a grating with period d, $\mathbf{a}$ is the acceleration of the Mu (e.g. due to gravity), $\mathbf{\Omega}$ is the rotation of the interferometer (e.g. due to the earth's rotation), $\mathbf{v}$ is the velocity of the Mu, and t is the transit time of the Mu between gratings. The sensitivity S of the interferometer to the gravitational acceleration for N_0 detected Mu is [55]:

$$S = \frac{1}{C\sqrt{N_0}} \frac{d}{2\pi} \frac{1}{t^2} \tag{3}$$

where C is the interferometer's contrast. The sensitivity of the gravity measurement is optimized by using finest grating pitch (smallest d) consistent with high fabrication fidelity. The contrast has a theoretical maximum of $C = 29.5\%$ for a three-grating interferometer without separated beams [56], but grating imperfections, misalignments, vibrations, beam coherence characteristics, the lack of collimation, etc., will reduce this contrast so for the calculation (below) we assume $C = 10\%$. We assume 1 cm (5 cm) separation between the Mu source and the interferometer for the thermal (SFHe) beam, and we assume $d = 100$ nm; gratings with this period have been fabricated but not yet with the 1 cm^2 area assumed here.

We use a full calculation (as opposed to a Monte Carlo simulation) of the Mu interferometer in order to optimize the interferometer's sensitivity for the gravity measurement. For a given Mu velocity, a longer interferometer will increase t but reduce N_0 due to the muon's finite lifetime. Figure 5 shows, as a function of the interferometer's separation between gratings, how much beam is required to acquire sufficient statistics to reduce the statistical uncertainty for the gravitational acceleration measurement to 4 m/s^2 (= $0.4g$, to get a 5σ sign measurement assuming $|\bar{g}| = g$) using PSI's $\pi E5$ beam for three different Mu sources: an existing aerogel thermal source, a thin-film SFHe source using the muCool beam, and the thick-film SFHe source. Note that this is actual beam time not including dead time for triggers or any operations that cannot be done while the beam is running, which may include calibrating the interferometer with X-rays, preparing and maintaining the SFHe, etc. Table 1 shows results using the optimum grating separation.

In order to study the systematic uncertainties of the interferometer, it will be important to measure a known signal. It may be possible to do this using the Sagnac effect, measuring the phase shift due to the earth's rotation. This effect results from the interferometer not being in an inertial reference frame and moving slightly during the Mu transit time. The size of the effect is dependent upon the direction of the beam (e.g. relative to north) and the orientation

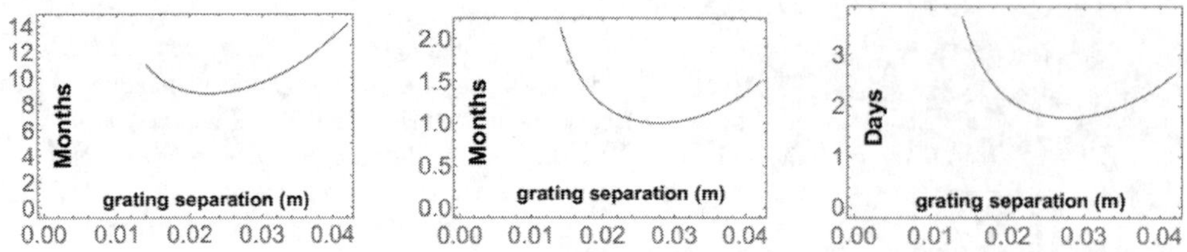

Figure 5. The plot on the left shows beam time required to acquire sufficient statistics to make a measurement with a statistical uncertainty of $0.4g$ using PSI's $\pi E5$ beam for a thermal beam vs. the separation between interferometer gratings. The center plot is the same when using the ultra-bright thin-film superfluid He (muCool) beam. If the novel thick-film Mu source test is successful, the beam time required for a sign determination could be reduced further, as shown on the right plot. (These plots do not include corrections for systematic uncertainties, which will likely be worse for lower Mu rates.) Here the distance from the Mu production target is assumed to be 1 cm for the thermal beam but 5 cm for the SFHe target, since for the SFHe target the Mu beam must be reflected from vertical to horizontal.

Table 1. Estimated performance of Mu sources for the gravity measurement using PSI's $\pi E5$ beam line. Acceptance is for an interferometer with 1 cm^2 gratings, and Beam Time is that required to acquire sufficient statistics for a $\bar{g}$ measurement with a statistical uncertainty of 4 m/s^2, sufficient to determine the sign of $\bar{g}$ with 5σ uncertainty if its magnitude is g. (These times do not include corrections for systematic uncertainties, which will likely be worse for lower Mu rates.) These results indicate that a rate of about 10^4 Mu/s or more in the interferometer's acceptance would enable MAGE, so the SFHe Mu sources need not reach their full potential for this project to succeed.

Mu Source	μ rate	Mu rate	MAGE Acceptance	Beam Time (5σ)
Thermal (300 K)	$1.2\times10^8/s$	$1\times10^6/s$	5×10^{-4}	6400 hours
muCool SFHe	$1.2\times10^5/s$	$5\times10^4/s$	1	720 hours
thick-film SFHe	$1.2\times10^8/s$	$4\times10^6/s$	0.25	43 hours

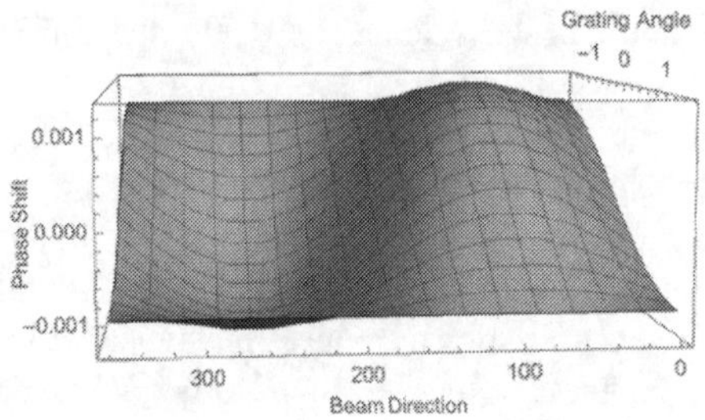

Figure 6. Sagnac effect: Phase shift vs. the (compass) direction of the muonium beam and the angle of the grating relative to horizontal.

of the gratings relative to the rotation axis, as can be seen from the second term in Eqn. 2. This second term is plotted as a function of these two variables in Fig. 6. This effect is small, only about 10% the size of the gravitational phase shift, as can be seen in Fig. 7. However, it may be measurable if systematic uncertainties can be sufficiently controlled and the thick-film SFHe source works at or near its full potential. The statistical uncertainty on a single data point is shown in Fig. 7 for 18 days of beam time using the thick-film SFHe source assuming its full potential. This effect could also be studied using a conventional atomic beam in order to understand systematic effects of the interferometer.

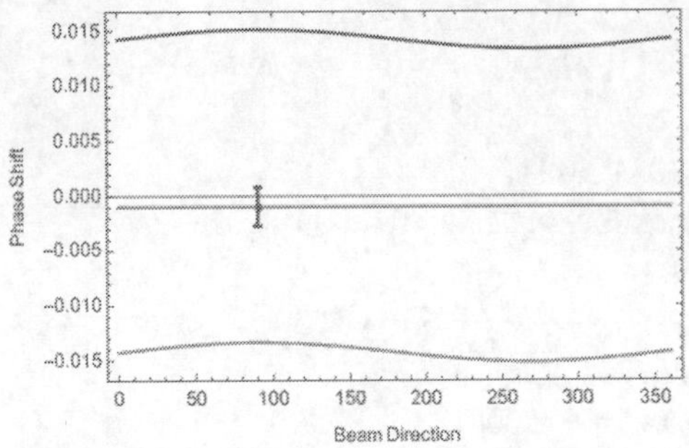

Figure 7. Phase shift vs. the compass direction of the muonium beam including both gravity and the Sagnac effect due to the earth's rotation. The top and bottom curves are for $\bar{g} = \pm g$ with horizontal interferometer gratings, while the center curve is for vertical interferometer gratings where there is no gravitational phase shift. The data point indicates the statistical uncertainty that would be achieved with 18 days of beam time assuming the full potential of the thick-film SFHe muonium source.

5 Conclusion

It is troubling that the sole *raison d'être* for what comprises 95% of the Universe in standard cosmology is to make the 5% we can observe fit the predictions of General Relativity. Despite considerable effort to find it, there is little evidence for Dark Matter, Dark Energy, or inflation beyond the cosmological observations for which their existence is postulated. Antimatter gravity has the potential to explain the biggest mysteries in physics and cosmology without the need to introduce this new physics and without missing antimatter. Efforts to measure antimatter gravity using antihydrogen are underway, but are challenging because antiprotons and antihydrogen annihilate upon contact with matter, making them difficult to cool. An alternate approach to measuring antimatter gravity uses muonium, which can be made cold but decays quickly due to the muon's lifetime. The MAGE collaboration is developing novel muonium beams and an ultra-precise interferometer to enable the measurement of the gravitational force on muonium. This is an extremely challenging experiment, but it has the potential to fundamentally change our understanding of the Universe.

References

[1] K.P. Jungmann, Journal of the Physical Society of Japan **85**, 091004 (2016), `https://doi.org/10.7566/JPSJ.85.091004`

[2] L. Willmann, P.V. Schmidt, H.P. Wirtz, R. Abela, V. Baranov, J. Bagaturia, W. Bertl, R. Engfer, A. Großmann, V.W. Hughes et al., Phys. Rev. Lett. **82**, 49 (1999)

[3] Tech. rep., J-PARC (2011), `https://g2sakura.kek.jp/public/doc/MCDR-submit.pdf`

[4] S. Kanda, Nuclear Physics B - Proceedings Supplements **253**, 212 (2014), the Twelfth International Workshop on Tau-Lepton Physics (TAU2012)

[5] D.M. Kaplan, E. Fischbach, K. Kirch, D.C. Mancini, J.D. Phillips, T.J. Phillips, R.D. Reasenberg, T.J. Roberts, J. Terry, ArXiv **1601**, 07222 (2016)

[6] T.D. Lee, C.N. Yang, Phys. Rev. **98**, 1501 (1955)

[7] P. Drell, A. Albrecht, S. Aronson, K. Baker, J. Bagger, N. Calder, E. Gates, F. Gilman, J. Jackson, S. Kahn et al., Tech. rep., DOE/NSF (2006), `http://www.quantumuniversereport.org/`

[8] A. Benoit-Lévy, G. Chardin, Astronomy & Astrophysics **537**, A78 (2012)

[9] A. Benoit-Lévy, G. Chardin, International Journal of Modern Physics: Conference Series **30**, 1460272 (2014)

[10] M. Villata, Astrophysics and Space Science **345**, 1 (2013)

[11] J.T. Nielsen, A. Guffanti, S. Sarkar, Scientific Reports **6**, 35596 EP (2016)

[12] L. Blanchet, Classical and Quantum Gravity **24**, 3529 (2007)
[13] D.S. Hajdukovic, Astrophys Space Sci **334**, 215 (2011)
[14] J. Bekenstein, M. Milgrom, Astrophys. J. **286**, 7 (1984)
[15] S.S. McGaugh, Phys. Rev. Lett. **106**, 121303 (2011)
[16] D.S.M. Alves, M. Jankowiak, P. Saraswat, ArXiv **0907**, 4110 (2009)
[17] S. Weinberg, Rev. Mod. Phys. **61**, 1 (1989)
[18] G. Chardin, Hyp. Int. **109**, 83 (1997)
[19] H.I. Arcos, J.G. Pereira, Gen. Rel. Grav. **36**, 2441 (2004)
[20] A. Burinskii, Gravitation and Cosmology **14**, 109 (2008)
[21] A. Burinskii, Journal of Physics: Conference Series **361**, 012032 (2012)
[22] M. Villata, Annalen der Physik **527**, 507 (2015)
[23] G.D. *et al.* (57 authors), Tech. Rep. CERN-SPSC-2007-017, CERN (2007), `http://cds.cern.ch/record/1037532`
[24] G. Gabrielse, P. Larochelle, D.L. Sage, B. Levitt, W. Kolthammer, R. McConnell, P. Richerme, J. Wrubel, A. Speck, M. George et al., Phys. Rev. Lett. **100**, 113001 (2008)
[25] G. Dufour, D.B. Cassidy, P. Crivelli, P. Debu, A. Lambrecht, V.V. Nesvizhevsky, S. Reynaud, A.Y. Voronin, T.E. Wall, Adv. High Energy Phys. **2015**, 379642 (2015)
[26] A.E. Charman, C. Amole, M. Ashkezari, M. Baquero-Ruiz, W. Bertsche, E. Butler, A. Capra, C. Cesar, M. Charlton, S. Eriksson et al., Nature Communications **4**, 1785 (2013)
[27] FNAL PAC, *Physics Advisory Committee Meeting March 27-29, 2008 Comments and Recommendations*, (and private communication), `http://www.fnal.gov/directorate/program_planning/Mar2008PACPublic/Mar2008PACReportPublic.pdf`
[28] A. Kostelecky, J. Tasson, ArXiv **1006**, 4106 (2010)
[29] J.D. Tasson, Int. J. Mod. Phys. Conf. Ser. **30**, 1460273 (2014)
[30] J.D. Tasson, Symmetry **8**, 111 (2016)
[31] A.D. Cronin, J. Schmiedmayer, D.E. Pritchard, Rev. Mod. Phys. **81**, 1051 (2009)
[32] A.P. Mills, J. Imazato, S. Saitoh, A. Uedono, Y. Kawashima, K. Nagamine, Physical Review Letters **56**, 1463 (1986)
[33] G. Beer, G. Marshall, G. Mason, A. Olin, Z. Gelbart, K. Kendall, T. Bowen, P. Halverson, A. Pifer, C. Fry et al., Phys. Rev. Lett. **57**, 671 (1986)
[34] G. Beer, Y. Fujiwara, S. Hirota, K. Ishida, M. Iwasaki, S. Kanda, H. Kawai, N. Kawamura, R. Kitamura, S. Lee et al., Progress of Theoretical and Experimental Physics **2014** (2014), `http://ptep.oxfordjournals.org/content/2014/9/091C01.full.pdf+html`
[35] A. Antognini, P. Crivelli, T. Prokscha, K. Khaw, B. Barbiellini, L. Liszkay, K. Kirch, K. Kwuida, E. Morenzoni, F.M. Piegsa et al., Phys. Rev. Lett. **108**, 143401 (2012)
[36] M.W. Reynolds, M.E. Hayden, W.N. Hardy, Journal of Low Temperature Physics **84**, 87 (1991)
[37] D. Taqqu, Phys. Procedia **17**, 216 (2011)
[38] R. Abela, D. Herlach, E. Morenzoni, G. Solt, U. Zimmermann, F. Gygax, A. Schenck, D. Eshchenko, E. Krasnoperov, E. Meillikhov, JETP Lett. **57**, 157 (1993)
[39] M. Saarela, E. Krotscheck, Journal of Low Temperature Physics **90**, 415 (1993)
[40] D. Taqqu, Phys. Rev. Lett. **97**, 194801 (2006)
[41] D. Gerola, W.B. Waeber, M. Shi, S.J. Wang, Review of Scientific Instruments **66**, 3819 (1995), `http://dx.doi.org/10.1063/1.1145443`

[42] Y. Bao, A. Antognini, W. Bertl, M. Hildebrandt, K.S. Khaw, K. Kirch, A. Papa, C. Petitjean, F.M. Piegsa, S. Ritt et al., Phys. Rev. Lett. **112**, 224801 (2014)
[43] A. Eggenberger, I. Belosevic, G. Wichmann, *Development of a low-energy, high-brightness μ^+ beam line*, in *Proceedings of the Seventh Meeting on CPT and Lorentz Symmetry (CPT'16)* (Indiana University, Bloomington, 2016), `http://arxiv.org/abs/1607.06206`
[44] A. Eggenberger, Ph.D. thesis, ETH Zurich (2016)
[45] V.G. Luppov, W.A. Kaufman, K.M. Hill, R.S. Raymond, A.D. Krisch, Physical Review Letters **71**, 2405 (1993)
[46] C. Barenghi, C. Mellor, C. Muirhead, W. Vinen, Journal of Physics C: Solid State Physics **19**, 1135 (1986)
[47] W. Schoepe, G. Rayfield, Z. Naturforsch. **26a**, 1392 (1971)
[48] J. Poitrenaud, F. Williams, Phys. Rev. Lett. **29**, 1230 (1972)
[49] D.G. Eshchenko, V.G. Storchak, J.H. Brewer, G.D. Morris, S.P. Cottrell, S.F.J. Cox, Phys. Rev. B **66**, 035105 (2002)
[50] K. Atkins, Phys. Rev. **116**, 1339 (1959)
[51] M. Kuchnir, J. Ketterson, P. Roach, Phys. Rev. A **6**, 341 (1972)
[52] R. Thapa, J.D. Phillips, E. Rocco, R.D. Reasenberg, Opt. Lett. **36**, 3759 (2011)
[53] R.E. Grisenti, W. Schöllkopf, J.P. Toennies, J.R. Manson, T.A. Savas, H.I. Smith, Phys. Rev. A **61**, 033608 (2000)
[54] T.J. Roberts, *G4beamline: A "swiss army knife" for geant4, optimized for simulating beamlines.*, `http://g4beamline.muonsinc.com/`
[55] M. Oberthaler, Nuc. Inst. Meth. B **192**, 129 (2002)
[56] T.D. Hammond, Ph.D. thesis, MIT (1997), `http://hdl.handle.net/1721.1/46044`

Exotic atoms at extremely high magnetic fields: the case of neutron star atmosphere

Andrea Fontana[1,*], *Alessandro* Colombi[2], *Pietro* Carretta[2], *Alessandro* Drago[3], *Paolo* Esposito[4], *Paola* Gianotti[5], *Carlotta* Giusti[1,2], *Diego* Lonardoni[6], *Alessandro* Lovato[7], *Vincenzo* Lucherini[5], and *Francesco* Pederiva[7]

[1] *INFN, Sezione di Pavia*
[2] *Dipartimento di Fisica, Università degli Studi di Pavia*
[3] *Dipartimento di Fisica e Scienze della Terra, Università degli Studi di Ferrara and INFN, Sezione di Ferrara*
[4] *Anton Pannekoek Institute for Astronomy, University of Amsterdam*
[5] *INFN - Laboratori Nazionali di Frascati*
[6] *National Superconducting Cyclotron Laboratory, Michigan State University and Theoretical Division, Los Alamos National Laboratory*
[7] *Dipartimento di Fisica, Università degli Studi Trento and INFN, Trento Institute for Fundamental Physics and Application*

Abstract. The presence of exotic states of matter in neutron stars (NSs) is currently an open issue in physics. The appearance of muons, kaons, hyperons, and other exotic particles in the inner regions of the NS, favoured by energetic considerations, is considered to be an effective mechanism to soften the equation of state (EoS). In the so-called *two-families scenario*, the softening of the EoS allows for NSs characterized by very small radii, which become unstable and convert into a quark stars (QSs). In the process of conversion of a NS into a QS material can be ablated by neutrinos from the surface of the star. Not only neutron-rich nuclei, but also more exotic material, such as hypernuclei or deconfined quarks, could be ejected into the atmosphere. In the NS atmosphere, atoms like H, He, and C should exist, and attempts to model the NS thermal emission taking into account their presence, with spectra modified by the extreme magnetic fields, have been done. However, exotic atoms, like muonic hydrogen ($p\mu^-$) or the so-called *Sigmium* ($\Sigma^+ e^-$), could also be present during the conversion process or in its immediate aftermath. At present, analytical expressions of the wave functions and eigenvalues for these atoms have been calculated only for H. In this work, we extend the existing solutions and parametrizations to the exotic atoms ($p\mu^-$) and ($\Sigma^+ e^-$), making some predictions on possible transitions. Their detection in the spectra of NS would provide experimental evidence for the existence of hyperons in the interior of these stars.

1 Introduction

The presence of exotic states of matter in the core of neutron stars (NSs) is currently an open problem in physics, and great effort is devoted to its theoretical and experimental investigation [1–3]. The appearance of kaons, hyperons, and other exotic particles is likely to happen in the inner

* e-mail: andrea.fontana@pv.infn.it

regions of a NS due to energetic considerations, and it offers an effective mechanism to soften the equation of state (EoS). This softening affects the entire structure of the star, reducing the pressure and therefore the maximum mass that the star can stably support. The observation of two NSs with masses as high as 2 $M_{\odot}$ [4, 5] seems however to exclude the presence of exotic particles in the core of the star, based on what we know about the interaction between these particles and normal nucleons. This apparent inconsistency, usually referred to as *hyperon puzzle*, is due to the poor knowledge of the interactions involved in these exotic states of matter and to the difficulty of obtaining clear evidence of their presence in NSs from astrophysical observations.

In this work we propose a novel approach to attempt the identification of hyperons and other exotic particles in NSs: the search for signals of the possible formation of exotic atoms in the atmosphere of NS in the process of conversion into a quark star (QS) through the spectroscopic study of their radiation emission. In particular, the spectroscopy of exotic atoms, made in the simplest case by two Fermions like $(p\,\mu^-)$ (muonic hydrogen) or $(\Sigma^+ e^-)$ (Sigmium), can be inferred by the results obtained in the past for hydrogen or hydrogen-like atoms in the extremely strong magnetic field of NSs [6, 7]. It is expected that, under the condition of a strong magnetic field $\boldsymbol{B}$, atoms are of cylindrical shape, and that the traditional level structure observed in terrestrial experiments (*gross*, *fine*, *hyperfine*) is superseded by a much simpler structure of Landau levels with principal quantum number n, with two additional quantum numbers for each level: m, which corresponds to the angular momentum projection on the field axis, and ν, which quantizes the motion along $\boldsymbol{B}$. The evaluation of hypothetical new spectral lines, corresponding to transitions among these levels, could offer alternative interpretations of the anomalies seen in the existing data and/or suggest new observational strategies.

2 Synopsis of neutron stars theory

Neutron stars are very peculiar astrophysical objects. Among their many interesting features, two are relevant for this work: the very high density and the extremely intense magnetic field. Considering nuclear saturation density $\rho_0 \approx 2.5\times10^{14}\ \mathrm{g/cm^3}$, the density of matter in a NS varies from $\approx 3\times10^{-3}\rho_0$ on the surface to $\approx 7\,\rho_0$ in the core, and the magnetic field ranges from 10^7 G on millisecond pulsars to 10^{16} G on magnetars, with an average value of $\approx 10^{12}$ G on most NSs. The extremely high density of NS matter increases towards the center of the star, and it spans more than 10 orders of magnitudes in the range of a dozen kilometers. This fact opens the possibility of extreme physical conditions, where different equilibrium situations are reached as the density increases, from neutronization and hyperon production, to pion (and possibly kaon) condensation, and even to quark deconfinement. Despite the name, NSs indeed contain a rich variety of subatomic particles whose equilibrium is governed by the chemical potential. In general, the baryon chemical potential increases with the density, and the energy of the system is lowered by sharing the conserved baryon number with other baryon species when their chemical potentials exceed their threshold values (i.e. their masses in vacuum). Thus, it is expected that super dense matter is populated by many baryon species, possibly even quarks. The structure and composition of the star and the radial fraction of different baryonic and leptonic species in a NS for a given EoS is shown in Fig. 1. As one can see, there are regions in the NS where different particles (and in particular hyperons) exist and overlap, opening the possibility for their mutual interaction. The reason why hyperons are expected to appear in the interior of NSs is very simple. Let us first we consider a matter made of neutrons only (pure neutron matter). As a first approximation, this system can be cosidered to be a gas of Fermions, with single-particle states occupied up to the Fermi energy, which is determined by the density. If the Fermi energy or, in the most general case, the chemical potential becomes equal to the difference in mass between a neutron and an electrically neutral hyperon, it is energetically convenient for the system to transform a neutron into a Λ particle. The hyperon, being distinguishable from nucleons and not being involved in the

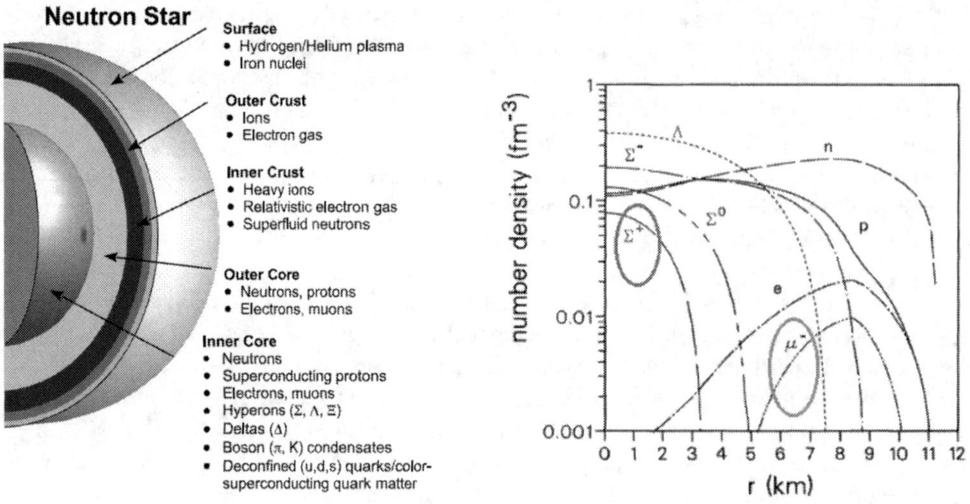

Figure 1. Structure of neutron stars [3] (left) and radial distribution of the different particle fractions [1] (right). Particles of interest for this study, Σ^+ and μ^-, are highlighted in red and green, respectively.

Pauli blocking mechanism, will reduce the total energy of the system, leading to a reduction of the exerted pressure, and thus to a reduced maximum mass. Note that this simple argument does not take into account the effects of the interaction. If the potential between hyperons and nucleons is soft, one can expect that this description is still appropriate. However, in the presence of a strong repulsion, the energetic cost of creating hyperons might become such as to suppress their creation. This is now generally accepted as one of the possible explanations of the hyperon puzzle [8].

Different hypotheses currently exist on the internal structure of a NS, and the most accepted models foresee two typical structures: the so called *traditional neutron stars* and the *quark stars* [3]. A third view is also proposed, the *hybrid neutron stars*, as shown in Ref. [1]. Many EoS have been proposed to describe these hypotheses, from the simple non interacting neutron gas, to more advanced parametrizations that take into account nuclear interactions and causality constraints. The situation is summarized in Fig. 2, where the pressure P in the star is plotted as a function of the energy density ε. The density behavior of the EoS depends on the derivative of P vs ε, and it is generally referred to as the *stiffness* or *softness* of the EoS. The stiffness is directly related to the maximum mass allowable for a given star, and it is therefore connected with astrophysical observations. Typical NS radii are of the order of $11 - 12$ km, and the average mass is $\approx 1.4\,M_{\odot}$, but fluctuations around this value are known [9]. Mass and radius are the main observables in the study of NSs: masses are measured with great precision from binary radio pulsars timing, but radius measurements are much more challenging, and no accurate data currently exist. A new set of constraints could follow from the recent observation of gravitational waves from a binary NS merger, particularly thanks to the analysis of the tidal deformability parameter from the inspiral signal. However, techniques to simultaneously infer both mass and radius have been proposed [10], but not yet successfully applied. Therefore, based on the available information on NS mass and radii, it is currently difficult to discriminate between different EoS models.

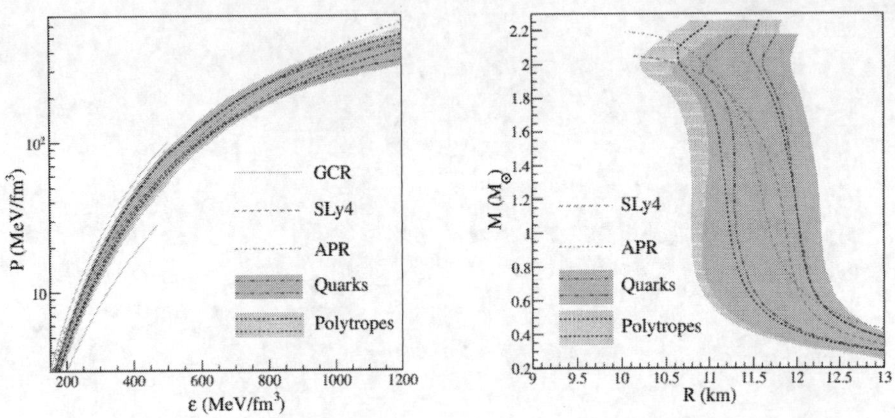

Figure 2. Constraints on the pressure vs. energy density function (left) and mass vs. radius relation (right) as deduced from astrophysical observations, according to the work of Steiner and Gandolfi [11]. The dashed and dot-dashed lines represent the 68% confidence levels, the shadowed areas the 95% confidence level. Curves represent the same quantities from the EoS evaluated in different schemes: GCR - quantum Monte Carlo [12]; SLy4 - Skyrme density functional [13]; APR - variational calculation [14].

3 Neutron star/quark star conversion process

In the so-called *two-families scenario* [15], in which NSs co-exist with QSs, NSs can be very compact and have a maximum mass of about $1.5 - 1.6\,M_\odot$, while QSs can have large radii and be very massive, up to $2.75\,M_\odot$ [15–17]. Concerning the NS branch, with the increase of the central density, hyperons (and/or Δ resonances) will be produced. The increasing softening of the EoS allows the stars in this branch to reach very small radii, even significantly smaller than 11 km (providing a possible strong signature of this scenario). When the strangeness content at the center of the star (most likely due to a non marginal presence of hyperons) reaches a critical value, the star becomes unstable to nucleation of droplets of deconfined strange quark matter. At this point the NS converts into a QS (see Fig. 3). The process of conversion of a NS into a QS has been studied in detail and within different approaches in a series of papers [18–25]. The main result of these analyses is that the combustion proceeds in two phases: a first rapid burning, accelerated by hydro-dynamical instabilities, lasting only a few milliseconds and converting the core of the star, followed by a much slower combustion driven by diffusion of strange quarks from the burnt into the unburnt region, lasting few tens of seconds. During this second phase, material can be ablated by neutrinos from the surface of the star [26]. This is particularly interesting, because it opens the possibility of ejecting in the atmosphere not only neutron-rich nuclei, but also more exotic material, such as hyperons. These particles could interact with the surrounding electrons or protons to form atomic structures, with recombination processes at rest similar to those occurring in laboratory with trapped particles [27], or with in flight capture [28].

Another possible process for generating exotic states of matter in the outer envelope of NSs could be the merger of two QSs, during which nucleosynthesis r-processes are believed to re-ignite, and exotic material could also be ejected and be involved in the process. This topic is of great interest nowadays after the recent multi-messenger observation of a kilonova event in the galaxy NGC 4993 [29], in conjuction with the detection of gravitational waves GW170817 from a binary NS merger with the LIGO/VIRGO observatory [30].

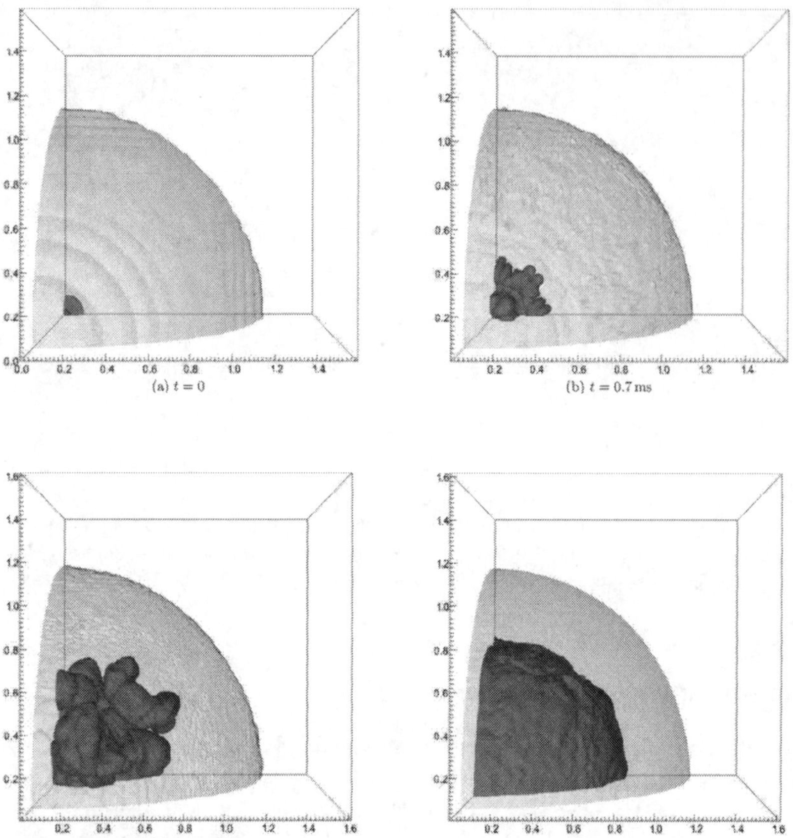

(a) $t = 0$ (b) $t = 0.7$ ms (c) $t = 1.2$ ms (d) $t = 4.0$ ms

Figure 3. NS/QS conversion. Conversion front (red) and surface of the NS (yellow) at different times t: (a) $t = 0$; (b) $t = 0.7$ ms; (c) $t = 1.2$ ms; (d) $t = 4.0$ ms. Spatial units: 10^6 cm [18].

4 Exotic atoms on neutron stars

The formation of exotic structures at the atomic level in the events outlined in the previous section is an intriguing problem that deserves to be investigated theoretically. In fact, given the uncertainties on the EoS, these atoms could provide new observables that enable to distinguish between the two types of compact stars, NSs and QSs.

It is widely accepted that regular atomic structures exist in the outer layers of the star, i.e. in the thin atmosphere. Indeed, results of the spectroscopy of atomic structures in NSs are present in the literature. For instance, hydrogen was investigated in detail after the discovery of pulsars [31], and observational results were obtained in the case of Fe lines in the X-ray spectra of accreting NSs [7, 32, 33]. Theoretical models for H, He, and C atmospheres have been recently proposed, and good fits of observed spectra with C models have been obtained, as in the case of the NS at the centre of the supernova remnant Cassiopeia A with the Chandra X-ray observatory [34]. In addition, the hypothesis for the existence of exotic Coulombic systems, i.e. molecular ions such as (αpe), $(\alpha\alpha e)$, or $(Li^{3+}Li^{3+}e)$, has been also investigated [35]. These atoms, when immersed in the extreme magnetic fields found in the atmosphere of NSs, change dramatically their structure and behavior. Since the magnetic field is so strong that the Coulomb force can be treated as a perturbation, the atoms assume an elongated, cylindrical shape in the direction of the magnetic field (Fig. 4 (left)) [36], and the energy spectrum is much different from the zero—or low—field case, where the traditional quantum numbers n, l, and m are appropriate to describe the spherically symmetric atom. In addition, due to the magnetic confinement, the so-called *decentered* states, shown in Fig. 4 (right), are possible. These are rather unique bound systems, with the source of the field external to the electron cloud and kept together by the magnetic field. The energy levels are distributed according to the quantized cyclotron scheme of Landau levels with quantum number n, and, for each n, additionally described by two new quantum numbers: m, that quantizes the radial wave functions and it is related to the z-component of the angular momentum l_z, and ν, the number of nodes of the longitudinal wave function.

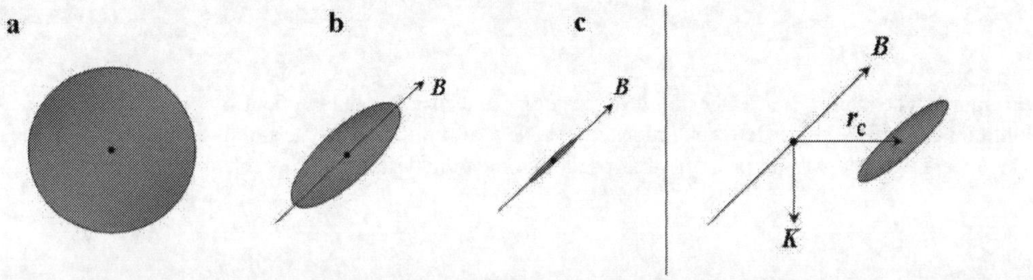

Figure 4. Effects of a strong magnetic field on H atoms: a) $B \ll 10^9$ G, b) $B = 10^9$ G, c) $B = 10^{12}$ G (gray areas: ellipsoids with probability to find an electron $P > e^{-1}$; solid dots: protons) (left). Decentered state, with an H atom confined by the magnetic field, but with the proton external to the electron cloud [37] (right).

The expected modification of the H energy levels for a field $B = 2 \times 10^{12}$ G is shown in Fig. 5 (left). These states are obtained by solving the one-dimensional Schrödinger equation for the wave functions $f_{m\nu}$:

$$-\frac{\hbar^2}{2m_e\rho_0^2} f''_{m\nu} - \frac{e}{\rho_0} V_m(z) f_{m\nu} = E_{m\nu} f_{m\nu} \qquad m, \nu = 0, 1, 2, \ldots, \tag{1}$$

where the average potential $V_m(z)$ is derived from the Landau wave functions radial part (see [7] for details). The eigenvalues of this equation are classified according the quantum numbers m and ν and to the parameter $b = B/B_0$[1] in *tightly-bound* (tb) states ($m \geq 0$, $\nu = 0$) and *weekly-bound* (wb) states ($m \geq 0$, $\nu > 0$). The latter present a doublet sub-structure according to the parity of the solution of the one-dimensional Schrödinger equation. With the definitions $\rho_0 = \left(\frac{\hbar c}{2eB}\right)^{1/2}$ and $\rho_m = (2m+1)^{1/2}\rho_0$, the energy levels are given by the following expressions for the tightly-bound states:

$$E_{m0} \approx -0.16\, A\, l_m^2 \quad \text{a.u.} \qquad (\text{for } 2m+1 \ll b)\,, \tag{2}$$

where:

$$A = \begin{cases} 1 + 1.36 \times 10^{-2}\big[\ln(1000/b)\big]^{2.5} & b < 10^3 \\ 1 + 1.07 \times 10^{-2}\big[\ln(b/1000)\big]^{1.6} & b \geq 10^3 \end{cases},$$

$$l_m = \ln\left(\frac{b}{2m+1}\right), \tag{3}$$

[1] B_0 is a reference field typical for NSs ($B_0 = m_e^2 e^3 c\,\hbar^{-3} = 2.3505 \times 10^9$ G), corresponding to the situation: cyclotron radius ≡ Bohr radius.

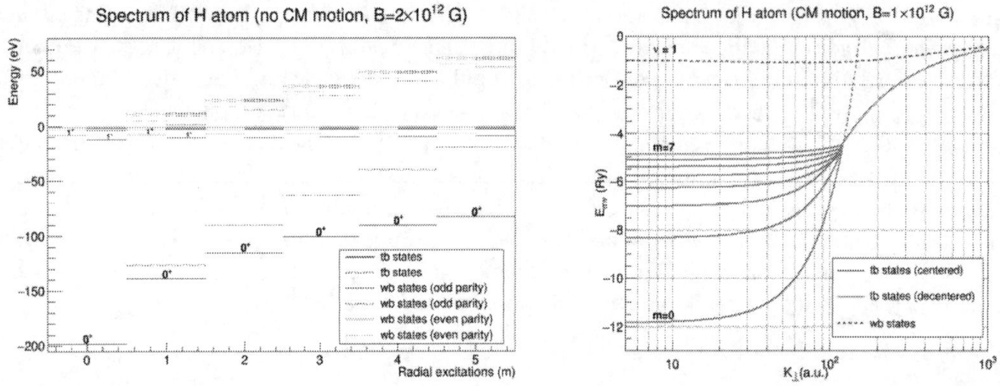

Figure 5. The hydrogen atom energy level structure in a strong magnetic field $B = 2 \times 10^{12}$ G. Static spectrum with infinite proton mass (left, continuous lines) and with finite proton mass (left, dashed lines). Dependence of the lowest energy levels on the transverse pseudomomentum describing the center of mass motion (right).

and for the weakly-bound states:

$$E_{m\nu} = -\frac{1}{(2\nu_1 + \delta)^2} \quad \text{a.u.}, \tag{4}$$

where:

$$\delta = \begin{cases} 2\rho/a_0 & \nu = 2\nu_1 - 1 \\ \left[\ln(a_0/\rho_m)\right]^{-1} & \nu = 2\nu_1 \end{cases} . \tag{5}$$

The effect of the finite proton mass, due to the replacement of the electron mass with the reduced mass, introduces a small correction to each level given by $\Delta E = 29.6\, m \frac{B}{4.7\times 10^{12}\, G}$ eV. The center-of-mass motion introduces an additional degree of freedom, the *pseudomomentum* $\boldsymbol{K}$ [7], which is derived from the canonical momentum $\boldsymbol{\Pi}$:

$$\boldsymbol{K} = \boldsymbol{\Pi} - \frac{e}{c}\boldsymbol{B} \times \boldsymbol{r}, \tag{6}$$

and it is used to define the position of the guiding center $\boldsymbol{r}_c$:

$$\boldsymbol{r}_c = \frac{c\boldsymbol{B} \times \boldsymbol{K}_\perp}{eB^2}, \tag{7}$$

where $\boldsymbol{K}_\perp$ is the pseudomomentum component transverse to the magnetic field direction. In terms of these quantities, the energy levels for the deeply-bound states can be classified as two different solutions, the centered and and the decentered states with energies:

$$\begin{aligned} \text{centered:} \quad & E_{m0}(K_\perp) \approx E_m + \frac{K_\perp^2}{2M_{\perp m}}, \\ \text{decentered:} \quad & E_{m0}(K_\perp) \approx -\frac{b}{K_\perp}, \end{aligned} \tag{8}$$

where E_m is the energy of a bound electron in the fixed Coulomb potential, $M_{\perp m}$ is the effective transverse mass, and all quantities are expressed in atomic units (see [7] for details). Similar parametrizations are also found for the weakly-bound states, resulting in the typical spectrum shown in Fig. 5 (right).

In a typical spectroscopic study of these atoms the wavelength associated to a transition between these levels can be estimated by using the equations above and by taking into account the usual selection rules [31]. For hydrogen, for instance, the $\nu = 1^- \rightarrow \nu = 0^+$ transition (for $m = 0$) produces radiation in the extreme UV regions ($\sim$ 160 eV), but additional transitions are observed for other light atoms (H, He, and C) in the atmospheres of NSs. Moreover, spectral features in the X-ray thermal emission of some isolated pulsars have been observed. While their origin is debated, their interpretation as atomic transitions between different m levels in a magnetized atmosphere is not ruled out [38–40].

It is important to note that, from the theoretical point of view, these results are valid for $2m+1 \ll b$ and $100 \leq b \leq 10^6$ [7]. The latter interval, corresponding to $B_{\min} \approx 3\times10^{11}$ G and $B_{\max} \approx 3\times10^{15}$ G, provides a good coverage of the magnetic field distribution typical of NSs. To generalize these studies to exotic atoms like muonic hydrogen ($p\mu^-$) or Sigmium ($\Sigma^+ e^-$), in first approximation it is possible to rescale the reduced mass and to redefine the parameter b in the equations found for hydrogen. Since the present approach is valid for $100 \leq b \leq 10^6$, one has to consider that a change in the reduced mass would imply a change in the magnetic field range where the solutions are still valid. For example, for muonic hydrogen, for which we have $b = \frac{\hbar^3 B}{m_\mu^2 e^3 c}$, with a ratio $\frac{b}{b^H} \sim 10^{-2}$, an increase of the magnetic field of 10^{16} G is required, while for Sigmium, for which $b = \frac{\hbar^3 B}{m_\Sigma^2 e^3 c}$, with a ratio $\frac{b}{b^H} \sim 1$, no changes in the magnetic field with respect to the hydrogen case are required. The corresponding spectra with the center of mass motion correction are shown in Fig. 6. The full analytical expressions of the wave functions and eigenvalues of the above described levels cannot be easily calculated for exotic atoms. The use of analytical approximations, as shown for example in [41], or ab initio numerical approaches to solve the Schrödinger equation [33], are both possible strategies that require further investigation. However, as an indication of the expected spectral lines, we present in Table 1 an example of possible transitions among these hypothetical states obtained from the above equations. To

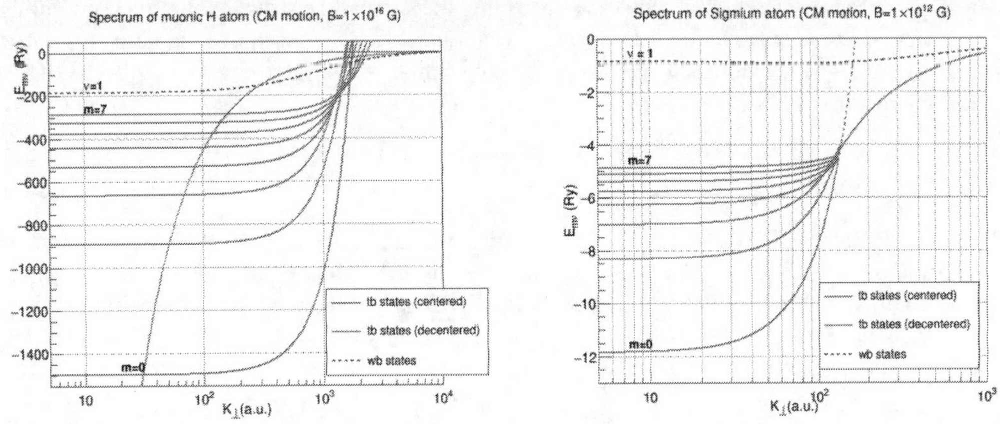

Figure 6. Lowest energy levels for deeply bound states (centered and decentered) and weakly bound states obtained from the hydrogen calculations for the hypothetical muonic hydrogen (left) and Sigmium (right) atoms.

assess the observability of the calculated lines, the knowledge of the formation rate of these systems in the atmosphere of a NS, either at the end of the conversion process into a QS, or in the merger with another NS, is essential. Therefore, future studies on this topic would require a dynamical simulation of the NS evolution and they could include results obtained from other theoretical studies on hypernuclear physics, on exotic atoms formation, and on atmospheric models of NSs [42].

Atom	$K_\perp$ (a.u.)	Energy (eV)	λ (nm)	Spectral region
Hydrogen	0	147.5877	8.4	UV
	10	146.6178	8.5	
	50	129.1634	9.6	
	100	77.6384	16	
Muonic hydrogen	0	17848.5	0.06946	X-ray
	10	17851.5	0.069453	
	50	17870.2	0.069381	
	100	17885.1	0.069323	
Sigmium	0	149.832	8.27	UV
	10	149.029	8.32	
	50	134.488	9.22	
	100	91.7076	13.52	

Table 1. Examples of possible transitions for H, muonic hydrogen, and Sigmium for different values of $K_\perp$.

5 Conclusions

Hydrogen-like atoms, involving μ^-, Σ^+, or other exotic constituents could be formed in the NS/QS conversion. In this work we have investigated the energy level structure of these hypothetical exotic atoms, and made some predictions on transitions that could help to discriminate among different EoS. As a further step in this direction, it would be interesting to estimate the radial exotic fractions during the conversion process, and to numerically solve the one-dimensional Schrödinger equation for exotic atoms. If the results are confirmed, the proposed calculations could give a direct evidence of the presence of hyperons in NSs.

6 Acknowledgments

We thank I. Bombaci, A. Gal, D. Logoteta, A. Ramos, and I. Vidaña for insightful comments and suggestions. The work of D.L. was was supported by the U.S. Department of Energy, Office of Science, Office of Nuclear Physics, under the FRIB Theory Alliance Grant Contract No. DE-SC0013617 titled "FRIB Theory Center - A path for the science at FRIB", and by the NUCLEI SciDAC program. P.E. acknowledges funding in the framework of the NWO Vidi award A.2320.0076.

References

[1] N.K. Glendenning, *Compact stars: Nuclear physics, particle physics, and general relativity* (Springer, 2012)

[2] A. Reisenegger, S.F. Zepeda, Eur. Phys. J. A **52**, 1 (2016)
[3] F. Weber et al., Mod. Phys. Lett. A **29**, 1430022 (2014)
[4] P.B. Demorest et al., Nature **467**, 1081 (2010)
[5] J. Antoniadis et al., Science **340**, 1233232 (2013)
[6] V. Canuto, D.C. Kelly, Astrophysics and Space Science **17**, 277 (1972)
[7] D. Lai, Rev. Mod. Phys. **73**, 629 (2001)
[8] D. Lonardoni, A. Lovato, S. Gandolfi, F. Pederiva, Phys. Rev. Lett. **114**, 092301 (2015)
[9] Kiziltan et al., Astrophys. J. **778**, 66 (2013)
[10] M. Kramer, N. Wex, Class. Quantum Grav. **26**, 073001 (2009)
[11] A.W. Steiner, S. Gandolfi, Phys. Rev. Lett. **108**, 081102 (2012)
[12] S. Gandolfi et al., Phys. Rev. C **85**, 032801 (2012)
[13] E. Chabanat et al., Phys. Scr. **1995**, 231 (1995)
[14] A. Akmal, V.R. Pandharipande, D.G. Ravenhall, Phys. Rev. C **58**, 1804 (1998)
[15] A. Drago, G. Pagliara, Eur. Phys. J. A **52**, 1 (2016)
[16] A. Drago, A. Lavagno, G. Pagliara, Phys. Rev. D **89**, 043014 (2014)
[17] A. Drago, A. Lavagno, G. Pagliara, D. Pigato, Eur. Phys. J. A **52**, 1 (2016)
[18] G. Pagliara, M. Herzog, F.K. Röpke, Phys. Rev. D **87**, 103007 (2013)
[19] A. V. Olinto , Phys. Lett. B **192**, 71 (1987)
[20] I. Bombaci et al., Astrophys. J. **530**, L69 (2000)
[21] I. Bombaci et al., Eur. Phys. J. A **52**, 58 (2016)
[22] A. Drago, A. Lavagno, I. Parenti, Astrophys. J. **659**, 1519 (2007)
[23] M. Herzog, F.K. Röpke, Phys. Rev. D **84**, 083002 (2011)
[24] A. Drago, G. Pagliara, Phys. Rev. C **92**, 045801 (2015)
[25] P. Keränen et al., Astrophys. J. **618**, 485 (2005)
[26] A. Drago, A. Lavagno, B.D. Metzger, G. Pagliara, Phys. Rev. D **93**, 103001 (2016)
[27] C. Amsler et al., Int. J. Mod. Phys. A **29**, 1430035 (2014)
[28] B. Adeva et al., Phys. Rev. D **96**, 052002 (2017)
[29] V. Savchenko et al., Astrophys. J. Lett. **848**, L15 (2017)
[30] Abbott, B. P. et al. (LIGO/Virgo Collaboration), Phys. Rev. Lett. **119**, 161101 (2017)
[31] E.R. Smith et al., Phys. Rev. D **6**, 3700 (1972)
[32] H. Ruder, G. Wunner, H. Herold, J. Trümper, Phys. Rev. Lett. **46**, 1700 (1981)
[33] K. Mori, C.J. Hailey, Astrophys. J. **564**, 914 (2002)
[34] W.C.G. Ho, C.O. Heinke, Nature **462**, 71 (2009)
[35] A.V. Turbiner, J.C.L. Vieyra, Int. J. Mod. Phys. A **22**, 1605 (2007)
[36] P. Mészáros, *High-Energy Radiation from Magnetized Neutron Stars* (Chicago Press, 1992)
[37] P. Haensel, A.Y. Potekhin, D.G. Yakovlev, *Neutron stars 1: Equation of state and structure* (Springer, 2007)
[38] H. Ruder, G. Wunner, H. Herold, F. Geyer, *Atoms in Strong Magnetic Fields* (Springer, 2007)
[39] M. van Kerkwijk, D. Kaplan, Astrophys Space Sci **308**, 191 (2007)
[40] A. Borghese et al., Monthly Notices of the Royal Astronomical Society **468**, 2975 (2017)
[41] A.Y. Potekhin, J. Phys. B: At. Mol. Opt. Phys. **31**, 49 (1998)
[42] A.Y. Potekhin and G. Chabrier and W.C.G. Ho, A&A **572**, A69 (2014)

Studies of discrete symmetries in decays of positronium atoms

Eryk Czerwiński[1,*], *Catalina* Curceanu[2], *Kamil* Dulski[1], *Aleksander* Gajos[1], *Marek* Gorgol[3], *Andrzej* Heczko[1], *Beatrix C.* Hiesmayr[4], *Bożena* Jasińska[3], *Daria* Kisielewska[1], *Grzegorz* Korcyl[1], *Bartłomiej* Korzeniak[1], *Paweł* Kowalski[5], *Tomasz* Kozik[1], *Wojciech* Krzemień[6], *Ewelina* Kubicz[1], *Wojciech* Migdał[1], *Muhsin* Mohammed[1,7], *Szymon* Niedźwiecki[1], *Marek* Pałka[1], *Monika* Pawlik-Niedźwiecka[1], *Lech* Raczyński[5], *Juhi* Raj[1], *Zbigniew* Rudy[1], *Sushil* Sharma[1], *Shivani* Shivani[1], *Roman Y.* Shopa[5], *Michał* Silarski[1], *Magdalena* Skurzok[1], *Wojciech* Wiślicki[5], *Bożena* Zgardzińska[3], *Marcin* Zieliński[1], and *Paweł* Moskal[1]

[1]Faculty of Physics, Astronomy and Applied Computer Science, Jagiellonian University, 30-348 Cracow, Poland
[2]INFN, Laboratori Nazionali di Frascati, CP 13, Via E. Fermi 40, I-00044, Frascati, Italy
[3]Department of Nuclear Methods, Institute of Physics, Maria Curie-Skłodowska University, 20-031 Lublin, Poland
[4]Faculty of Physics, University of Vienna, Boltzmanngasse 5, 1090 Vienna, Austria
[5]Świerk Computing Center, National Centre for Nuclear Research, 05-400 Otwock-Świerk, Poland
[6]High Energy Physics Division, National Center for Nuclear Research, 05-400 Otwock-Świerk, Poland
[7]Department of Physics, College of Education for Pure Sciences, University of Mosul, Mosul, Iraq

Abstract. A positronium - a bound state of electron and positron - is an eigenstate of parity and charge conjugation operators which decays into photons. It is a unique laboratory to study discrete symmetries whose precision is limited, in principle, by the effects due to the weak interactions expected at the level of 10^{-14} and photon-photon interactions expected at the level of 10^{-9}.
The Jagiellonian Positron Emission Tomograph (J-PET) is a detector for medical imaging as well as for physics studies involving detection of electron-positron annihilation into photons. The physics case covers the areas of discrete symmetries studies and genuine multipartite entanglement. The J-PET detector has high angular and time resolution and allows for determination of spin of the positronium and the momenta and polarization vectors of annihilation quanta. In this article, we present the potential of the J-PET system for studies of discrete symmetries in decays of positronium atoms.

1 Introduction

Various systems, like the accelerator neutrinos [1], mesons [2] or nuclei [3], can be used as probes for investigations of charge conjugation (C), parity (P) and time reversal (T) discrete symmetries. The CP symmetry violation plays a key role in explanation of the matter and anti-matter abundance asymmetry in the Universe, the CPT symmetry is conserved in all quantum field theories preserving Lorentz invariance, unitarity and locality, while uniqueness

*e-mail: eryk.czerwinski@uj.edu.pl

of time itself motivates the T symmetry studies. The search for CP and CPT violation was also performed in the decays of a bound state of electron and positron (positronium) [4, 5]. This bound state is also considered for observation of the gravitational quantum states [6]. All of these systems, among others, are used in the Standard Model Extension framework for Lorentz and CPT tests [7–9].

A recent development of the Jagiellonian Positron Emission Tomograph (J-PET) [10–12] provides a new tool for medical applications as well as for discrete symmetries studies in decays of positronium [13]. Test of the C symmetry will be performed as a search for forbidden decays: o-Ps $\rightarrow 4\gamma$ of the positronium triplet state (ortho-positronium) and p-Ps $\rightarrow 3\gamma$ of singlet state (para-positronium). Measurement of the expectation values of symmetry-odd operators for the o-Ps $\rightarrow 3\gamma$ process will test the other fundamental symmetries and their combinations. Such operators can be constructed from o-Ps spin ($\vec{S}$) and momentum of annihilation quantum ($\vec{k}_i$). Additionally, the operators with annihilation quantum polarization ($\vec{\epsilon}_i$) are uniquely available at the J-PET system [13].

2 The J-PET detector

The Jagiellonian Positron Emission Tomograph consists of three cylindrical layers constructed with plastic scintillator strips. Each of such strip made of EJ-230 material is $500 \times 19 \times 7$ mm^3. Single detection module consists of one scintillator strip and the R9800 Hamamatsu photomultiplier connected at each 19×7 mm^2 side. 48 modules placed on radius 425 mm compose 1st layer of the J-PET detector, 48 modules on radius 467.5 mm - 2nd layer and 96 modules on radius 575 mm - 3rd layer. An interaction place and time of γ quantum with scintillator can be determined from the difference and the sum of times of signals from both photomultipliers attached to a given strip [10]. Therefore there are in total 384 analog signals to process. Each of them is probed in time domain at 4 different amplitude thresholds resulting in up to 8 time measurements (leading and trailing edge of the signal at each threshold) [14, 15]. The 1536 Time-to-Digital Converter (TDC) channels are trigger-lessly distributed on the 8 Trigger Readout Boards (TRBs). Apart of interaction place and time [16–18], the energy deposited by gamma quantum is also determined from the measurement of the Time-Over-Threshold (TOT) value, which depends on the amount of light registered by a photomultiplier. The gathered data are offline processed with a dedicated analysis framework [19].

3 The experimental technique

The measurements with a point-like ^{22}Na source placed inside the J-PET system were already performed for the preliminary determination of the detector time and spatial resolution [20, 21]. For physics studies a point-like source covered by the XAD-4 porous polymer [22, 23] was placed in the center of the detector. The XAD-4 material is used to increase the probability of positronium formation. In addition, a test experiment with a source surrounded by aluminum cylinder was also performed [24], while the future measurements will be performed with a cylinder with the inner wall covered by the porous material currently developed at the Maria Curie-Sklodowska University in Lublin. The aim of measurement with a cylinder is to separate in space the positronium formation and annihilation (cylinder wall) from positron emission (source).

The following reactions chain is considered:

$$
\begin{aligned}
{}^{22}\mathrm{Na} &\rightarrow {}^{22}\mathrm{Ne}^{*} + e^{+} + \nu_e \\
{}^{22}\mathrm{Ne}^{*} &\rightarrow {}^{22}\mathrm{Ne} + \gamma \\
e^{+} + e^{-} &\rightarrow \text{o-Ps} \rightarrow 3\gamma
\end{aligned} \tag{1}
$$

The detection of 1274 keV prompt γ from $^{22}\mathrm{Ne}^{*}$ deexcitation is a start signal for the positronium lifetime measurement. The emitted positrons are stopped in the cylinder wall producing o-Ps which may undergo annihilation to 3γ. Due to the coplanarity of annihilation gamma quanta momentum vectors ($\vec{k_i}$) the annihilation point and time is reconstructed with a dedicated method based on the trilateration [25]. Known directions of photons from annihilation of positronium into 3γ enables to calculate the energy of each photon as described in Ref. [26].

The location of positron emission is known since the point-like source is placed in the center of the detection system. Therefore the reconstructed annihilation place and time together with a start signal from prompt gamma allow for e^{+} velocity ($\vec{v}$) direction determination. Due to the parity violation in β-decay the positrons are longitudinally polarized along their direction of motion. The determined spin direction of positron provides the spin of o-Ps as well [27], since the polarization of positron is to large extent preserved during the thermalization process [28, 29].

At J-PET the polymer scintillators are used, therefore the gamma quanta are registered by means of Compton scattering. This scattering is most likely to occur in the plane perpendicular to the electric vector of the photon [30, 31]. Therefore registration of annihilation ($\vec{k_i}$) and scattered ($\vec{k'_i}$) quanta pairs allows to determine its linear polarization $\vec{\epsilon_i} = \vec{k_i} \times \vec{k'_i}$, and hence enables to study the multi-partite entanglement of photons originating from the decays of positronium atoms [32].

The potential of the J-PET detector for studies of discrete symmetries in decays of positronium atoms is described in Ref. [13], while the commissioning of J-PET for such studies is presented Ref. [24].

4 Acknowledgments

The authors acknowledge the financial support from the Polish National Center for Research and Development through grant INNOTECH-K1/IN1/64/159174/NCBR/12, the National Science Centre, Poland through grant No. 2016/21/B/ST2/01222 and by the Polish Ministry for Science and Higher Education through grants Nos. 6673/IA/SP/N/2016 and 7150/E-338/SPUB/2017/1. B.C.H. acknowledges gratefully the Austrian Science Fund FWF-P26783.

References

[1] T. Wąchała, *Acta Phys. Pol.* B **48**, 1969 (2017).

[2] G. Amelino-Camelia *et al., Eur. Phys. J.* C **68**, 619 (2010) [arXiv:1003.3868 [hep-ex]].

[3] A. Aksentyev *et al., Acta Phys. Pol.* B **48**, 1925 (2017).

[4] T. Yamazaki *et al., Phys. Rev. Lett.* **104**, 083401 (2010) [arXiv:0912.0843 [hep-ex]].

[5] P. A. Vetter, S. J. Freedman, *Phys. Rev. Lett.* **91**, 263401 (2003).

[6] P. Crivelli, V. V. Nesvizhevsky and A. Y. Voronin, Adv. High Energy Phys. **2015** 173572 (2015) [arXiv:1408.7051 [physics.atom-ph]].
[7] V. A. Kostelecký and A. J. Vargas, Phys. Rev. D **92** no.5, 056002 (2015) [arXiv:1506.01706 [hep-ph]].
[8] V. A. Kostelecky and N. Russell, Rev. Mod. Phys. **83** 11 (2011) [arXiv:0801.0287 [hep-ph]].
[9] G. S. Adkins, doi:10.1142/9789814327688_0050 (2010), arXiv:1007.3909 [hep-ph].
[10] P. Moskal *et al., Nucl. Instr. and Meth.* A **764**, 317 (2014). [arXiv:1407.7395 [physics.ins-det]].
[11] P. Moskal *et al., Nucl. Instr. and Meth.* A **775**, 54 (2015). [arXiv:1412.6963 [physics.ins-det]].
[12] P. Moskal *et al., Phys. Med. Biol.* **61**, 2025 (2016). [arXiv:1602.02058v2 [physics.ins-det]].
[13] P. Moskal *et al., Acta Phys. Pol.* B **47**, 509 (2016). [arXiv:1602.05226 [nucl-ex]].
[14] M. Pałka *et al., JINST* **12**, P08001 (2017). [arXiv:1707.03565 [physics.ins-det]].
[15] G. Korcyl *et al., Acta Phys. Pol.* B **47**, 491 (2016). [arXiv:1602.05251 [physics.ins-det]].
[16] L. Raczyński *et al., Nucl. Instr. and Meth.* A **764**, 186 (2014). [arXiv:1407.8293 [physics.ins-det]].
[17] L. Raczyński *et al., Nucl. Instr. and Meth.* A **786**, 105 (2015). [arXiv:1503.05188 [physics.ins-det]].
[18] L. Raczyński *et al., Phys. Med. Biol.* **62**, 5076 (2017). [arXiv:1706.00924v2 [physics.ins-det]]
[19] W. Krzemień *et al., Acta Phys. Pol.* B **47**, 561 (2016). [arXiv:1508.02451 [physics.ins-det]].
[20] M. Pawlik-Niedźwiecka *et al., Acta Phys. Pol.* A **132**, no. 5, 1645 (2017).
[21] Sz. Niedźwiecki *et al., Acta Phys. Pol.* B **48**, 1567 (2017).
[22] B. Jasińska *et al., Acta Phys. Pol.* B **47**, 453 (2016). [arXiv:1602.05376 [nucl-ex]].
[23] R. Zaleski, M. Gorgol, A. Kierys, J. Goworek, *Adsorption* **22** 745 (2016).
[24] E. Czerwiński *et al., Acta Phys. Pol.* B **48** no. 10, 1961 (2017).
[25] A. Gajos *et al., Nucl. Instr. and Meth.* A **819**, 54 (2016). [arXiv:1602.07528 [physics.ins-det]].
[26] D. Kamińska *et al., Eur. Phys. J.* C **76**, 445 (2016). [arXiv:1607.08588 [physics.ins-det]].
[27] M. Mohammed *et al., Acta Phys. Pol.* A **132**, 1486 (2017).
[28] J. Van House, P. W. Zitzewitz, *Phys. Rev.* A **29**, 96 (1984).
[29] P. W. Zitzewitz *et al., Phys. Rev. Lett.* **43**, 1281 (1979).
[30] O. Klein, T. Nishina, *Z. Phys.* **52**, 853 (1929).
[31] R. D. Evans, Corpuscles and Radiation in Matter II, Springer Berlin Haidelberg, 218-298 (1958).
[32] B.C. Hiesmayr, P. Moskal, *Scientific Reports* **7**, 15349 (2017).

Interaction of real and virtual $N\bar{N}$ pairs in J/ψ decays

Alexander Milstein[1] and *Sergey* Salnikov[1,2,3,*]

[1]Budker Institute of Nuclear Physics, 630090, Novosibirsk, Russia
[2]Novosibirsk State University, 630090, Novosibirsk, Russia
[3]L.D. Landau Institute for Theoretical Physics, 142432, Chernogolovka, Russia

Abstract. The differential decay rates of the processes $J/\psi \to p\bar{p}\pi^0$, $J/\psi \to p\bar{p}\eta$, $J/\psi \to p\bar{p}\omega$, $J/\psi \to p\bar{p}\rho$, and $J/\psi \to p\bar{p}\gamma$ close to the $p\bar{p}$ threshold are calculated with the help of the $N\bar{N}$ optical potential. We use the potential which has been suggested to fit the cross sections of $N\bar{N}$ scattering together with all other $N\bar{N}$ experimental data available. The $p\bar{p}$ invariant mass spectra of J/ψ decays are in agreement with the available experimental data. The anisotropy of the angular distributions of the decays $J/\psi \to p\bar{p}\pi^0(\eta)$, which appears due to the tensor forces in the $N\bar{N}$ interaction, is predicted close to the $p\bar{p}$ threshold. This anisotropy is large enough to be investigated experimentally. Such measurements would allow one to check the accuracy of the model of $N\bar{N}$ interaction. Using our potential and the Green's function approach we also describe the peak in the $\eta'\pi^+\pi^-$ invariant mass spectrum in the decay $J/\psi \to \gamma\eta'\pi^+\pi^-$ in the energy region near the $N\bar{N}$ threshold.

1 Introduction

Investigation of the nucleon-antinucleon interaction in the low-energy region is an actual topic today. Unusual behavior of the cross sections of several processes has been discovered in recent years. For instance, the cross sections of the processes $e^+e^- \to p\bar{p}$ and $e^+e^- \to n\bar{n}$ reveal an enhancement near the threshold [1–4]. The enhancement near the $p\bar{p}$ threshold is also observed in the decays $J/\psi(\psi') \to p\bar{p}\pi^0(\eta)$ [5–7], $J/\psi(\psi') \to p\bar{p}\omega(\gamma)$ [5, 8–11]. The sharp peak in the vicinity of $N\bar{N}$ threshold has been observed in the cross sections of several processes, i.e., $e^+e^- \to 6\pi$ [12–16] and $J/\psi \to \gamma\eta'\pi^+\pi^-$ [17]. These observations led to numerous speculations about a new resonance [5], $p\bar{p}$ bound state [18–20], or even a glueball state [21–23] with the mass about double proton mass. Another possibility, which we are studying, is the nucleon-antinucleon interaction in the final or intermediate states.

We describe the nucleon-antinucleon interaction by means of an optical potential model. Several optical nucleon-antinucleon potentials [24–26] are usually used to describe the interaction in the low-energy region. All these nucleon-antinucleon potentials have been proposed to fit the nucleon-antinucleon scattering data. These data include elastic, charge-exchange, and annihilation cross sections of $p\bar{p}$ scattering, as well as some single-spin observables. There were attempts to describe the processes of $N\bar{N}$ production in e^+e^- annihilation using these potential models. For instance, using the Paris [27] and Jülich [28] models, it has been shown that the near-threshold enhancement of the cross sections of these processes can be explained by the final-state nucleon-antinucleon interaction. The strong dependence of the ratio of electromagnetic form factors of the proton on the energy in the timelike region near the threshold has been explained by the influence of the tensor part of the nucleon-antinucleon interaction.

In our recent papers [29, 30], to fit the parameters of the potentials, we have suggested to include all available experimental data in addition to the nucleon-antinucleon scattering data. A simple potential model of $N\bar{N}$ interaction in the partial waves $^3S_1 - {}^3D_1$, coupled by the tensor forces, has been suggested [29]. The parameters of this model has been obtained by fitting simultaneously the nucleon-antinucleon scattering data, the cross sections of $p\bar{p}$ and $n\bar{n}$ production in e^+e^- annihilation, and the ratio of electromagnetic form factors of the proton in the timelike region. This model has allowed us to calculate also the contribution of virtual $N\bar{N}$ intermediate state to the processes of meson production in e^+e^- annihilation and to describe the sharp dip in the cross section of 6π production in the vicinity of the $N\bar{N}$ threshold [29]. Similar results have also been obtained in Ref. [31] within the chiral model [26] but without the tensor $N\bar{N}$ interaction taken into account.

The potential [29] has also been used to explain the enhancement observed in the $p\bar{p}$ invariant mass spectra of the decays $J/\psi(\psi') \to p\bar{p}\pi^0(\eta)$ near the $p\bar{p}$ threshold [32]. Note that in these decays in the near-threshold region the most important contribution is also given by the partial waves $^3S_1 - {}^3D_1$. The spectra of these decays, as well as the decays $J/\psi(\psi') \to p\bar{p}\omega(\rho,\gamma)$, have also been studied in Refs. [33, 34] using the chiral model [26].

*e-mail: S.G.Salnikov@inp.nsk.su

In the paper [30] we follow our idea and construct a simple optical potential model of the $N\bar{N}$ interaction in the $^1S_0$ partial wave. This partial wave gives the most important contribution to the final-state $p\bar{p}$ interaction in the decays $J/\psi(\psi') \to p\bar{p}\omega(\rho,\gamma)$ in the energy region close to the $p\bar{p}$ threshold. The parameters of this potential has been obtained by fitting simultaneously the nucleon-antinucleon scattering data and the $p\bar{p}$ invariant mass spectra of the decays $J/\psi(\psi') \to p\bar{p}\gamma$ and $J/\psi \to p\bar{p}\omega$. We have showed that it is possible to describe the pronounced peak in the $p\bar{p}$ invariant mass spectrum of the decay $J/\psi \to p\bar{p}\gamma$ using a simple model of the $N\bar{N}$ interaction. Moreover, in contrast to the results of Ref. [33], our model doesn't predict such peak in the spectrum of the decay $J/\psi \to p\bar{p}\rho$ which has not been observed yet.

We have used our model to calculate the contribution of virtual $N\bar{N}$ pair to the $J/\psi \to \gamma\eta'\pi^+\pi^-$ decay rate in the energy region near the $N\bar{N}$ threshold. Our model describes a peak in the $\eta'\pi^+\pi^-$ invariant mass spectrum. It has been pointed out in Ref. [17] that a contribution of virtual $p\bar{p}$ state may be one of possible origins of the peak in the spectrum. However, in Ref. [17] any models of the $N\bar{N}$ interaction have not been applied.

The present paper is a compilation of the results of Refs. [30, 32], and some parts are identical to the corresponding sections of these papers.

2 Decay amplitudes

We give only the resulting formulas here and the reader is referred to the Refs. [30, 32] for more details.

At first let us discuss the J/ψ decay into a $p\bar{p}$ pair and a pseudoscalar meson. Possible states for a $p\bar{p}$ pair in the decays $J/\psi \to p\bar{p}\pi^0$ and $J/\psi \to p\bar{p}\eta$ have quantum numbers $J^{PC} = 1^{--}$ and $J^{PC} = 1^{+-}$. The dominating mechanism of the $p\bar{p}$ pair creation is the following. The $p\bar{p}$ pair is created at small distances in the $^3S_1$ state and acquires an admixture of $^3D_1$ partial wave at large distances due to the tensor forces in the nucleon-antinucleon interaction. The $p\bar{p}$ pairs have different isospins for the two final states under consideration ($I = 1$ for the $p\bar{p}\pi^0$ state, and $I = 0$ for the $p\bar{p}\eta$ state), that allows one to analyze two isospin states independently. Therefore, these decays are easier to investigate theoretically than the process $e^+e^- \to p\bar{p}$, where the $p\bar{p}$ pair is a mixture of different isospin states.

We derive the formulas for the decay rate of the process $J/\psi \to p\bar{p}x$, where x is one of the pseudoscalar mesons π^0 or η. The following kinematics is considered: $\boldsymbol{k}$ and ε_k are the momentum and the energy of the x meson in the J/ψ rest frame, $\boldsymbol{p}$ is the proton momentum in the $p\bar{p}$ center-of-mass frame, M is the invariant mass of the $p\bar{p}$ system. The following relations hold:

$$p = |\boldsymbol{p}| = \sqrt{\frac{M^2}{4} - m_p^2},$$
$$k = |\boldsymbol{k}| = \sqrt{\varepsilon_k^2 - m^2},$$
$$\varepsilon_k = \frac{m_{J/\psi}^2 + m^2 - M^2}{2m_{J/\psi}}, \qquad (1)$$

where m is the mass of the x meson, $m_{J/\psi}$ and m_p are the masses of a J/ψ meson and a proton, respectively, and $\hbar = c = 1$. Since we consider the $p\bar{p}$ invariant mass region $M - 2m_p \ll m_p$, the proton and antiproton are non-relativistic in their center-of-mass frame, while ε_k is about 1 GeV.

The spin-1 wave function of the $p\bar{p}$ pair in the center-of-mass frame has the form [35]

$$\psi_\lambda^I = \mathbf{e}_\lambda u_1^I(0) + \frac{u_2^I(0)}{\sqrt{2}}\left[\mathbf{e}_\lambda - 3\hat{\boldsymbol{p}}(\mathbf{e}_\lambda\cdot\hat{\boldsymbol{p}})\right], \qquad (2)$$

where $\hat{\boldsymbol{p}} = \boldsymbol{p}/p$, $\mathbf{e}_\lambda$ is the polarization vector of the spin-1 $p\bar{p}$ pair, $u_1^I(r)$ and $u_2^I(r)$ are the components of two independent solutions of the coupled-channels radial Schrödinger equations for $^3S_1 -^3 D_1$ partial waves (see Ref. [32] for more details). The dimensionless amplitude of the decay with the corresponding isospin of the $p\bar{p}$ pair can be written as

$$T_{\lambda\lambda'}^I = \frac{\mathcal{G}_I}{m_{J/\psi}}\psi_\lambda^I\left[\boldsymbol{k}\times\boldsymbol{\epsilon}_{\lambda'}\right]. \qquad (3)$$

Here $\mathcal{G}_I$ is an energy-independent dimensionless constant and $\boldsymbol{\epsilon}_{\lambda'}$ is the polarization vector of J/ψ.

The decay rate of the process $J/\psi \to p\bar{p}x$ can be written in terms of the dimensionless amplitude $T_{\lambda\lambda'}^I$ as (see, e.g., [36])

$$\frac{d\Gamma}{dM\,d\Omega_p\,d\Omega_k} = \frac{pk}{2^9\pi^5 m_{J/\psi}^2}\left|T_{\lambda\lambda'}^I\right|^2, \qquad (4)$$

where Ω_p is the proton solid angle in the $p\bar{p}$ center-of-mass frame and Ω_k is the solid angle of the x meson in the J/ψ rest frame.

Substituting the amplitude (3) in Eq. (4) and averaging over the spin states, we obtain the $p\bar{p}$ invariant mass and angular distribution for the decay rate

$$\frac{d\Gamma}{dM\,d\Omega_p\,d\Omega_k} = \frac{\mathcal{G}_I^2 pk^3}{2^{11}\pi^5 m_{J/\psi}^4}\Big\{\left|u_1^I(0) + \tfrac{1}{\sqrt{2}}u_2^I(0)\right|^2 + \left|u_1^I(0) - \sqrt{2}u_2^I(0)\right|^2(\boldsymbol{n}\cdot\hat{\boldsymbol{k}})^2 + \frac{3}{2}\left[\left|u_2^I(0)\right|^2 - 2\sqrt{2}\,\mathrm{Re}\left(u_1^I(0)u_2^{I*}(0)\right)\right]\times \left[(\boldsymbol{n}\cdot\hat{\boldsymbol{p}})^2 - 2(\boldsymbol{n}\cdot\hat{\boldsymbol{k}})(\boldsymbol{n}\cdot\hat{\boldsymbol{p}})(\hat{\boldsymbol{p}}\cdot\hat{\boldsymbol{k}})\right]\Big\}. \qquad (5)$$

The invariant mass distribution can be obtained by integrating Eq. (5) over the solid angles Ω_p and Ω_k:

$$\frac{d\Gamma}{dM} = \frac{\mathcal{G}_I^2 pk^3}{2^5\,3\pi^3 m_{J/\psi}^4}\left(\left|u_1^I(0)\right|^2 + \left|u_2^I(0)\right|^2\right). \qquad (6)$$

The sum in the brackets is the so-called enhancement factor which equals to unity if the $p\bar{p}$ final-state interaction is turned off.

More information about the properties of $N\bar{N}$ interaction can be extracted from the angular distributions. Integrating Eq. (5) over Ω_p we obtain

$$\frac{d\Gamma}{dM\,d\Omega_k} = \frac{\mathcal{G}_I^2 pk^3}{2^9\pi^4 m_{J/\psi}^4}\left(\left|u_1^I(0)\right|^2 + \left|u_2^I(0)\right|^2\right)\left[1 + \cos^2\vartheta_k\right], \qquad (7)$$

where ϑ_k is the angle between $\boldsymbol{n}$ and $\boldsymbol{k}$. However, the angular part of this distribution does not depend on the features of the $p\bar{p}$ interaction. The proton angular distribution in the $p\bar{p}$ center-of-mass frame is more interesting. To obtain this distribution we integrate Eq. (5) over the solid angle Ω_k:

$$\frac{d\Gamma}{dMd\Omega_p} = \frac{\mathcal{G}_I^2 pk^3}{2^7\, 3\pi^4 m_{J/\psi}^4}\left(\left|u_1^I(0)\right|^2 + \left|u_2^I(0)\right|^2\right)\times \times\left[1+\gamma^I P_2(\cos\vartheta_p)\right], \quad (8)$$

where ϑ_p is the angle between $\boldsymbol{n}$ and $\boldsymbol{p}$, $P_2(x) = \frac{3x^2-1}{2}$ is the Legendre polynomial, and γ^I is the parameter of anisotropy:

$$\gamma^I = \frac{1}{4}\frac{\left|u_2^I(0)\right|^2 - 2\sqrt{2}\,\mathrm{Re}\left[u_1^I(0)u_2^{I*}(0)\right]}{\left|u_1^I(0)\right|^2 + \left|u_2^I(0)\right|^2}. \quad (9)$$

Averaging (5) over the direction of $\boldsymbol{n}$ gives the distribution over the angle ϑ_{pk} between $\boldsymbol{p}$ and $\boldsymbol{k}$:

$$\frac{d\Gamma}{dMd\Omega_{pk}} = \frac{\mathcal{G}_I^2 pk^3}{2^7\, 3\pi^4 m_{J/\psi}^4}\left(\left|u_1^I(0)\right|^2 + \left|u_2^I(0)\right|^2\right)\times \times\left[1-2\gamma^I P_2(\cos\vartheta_{pk})\right]. \quad (10)$$

Note that this distribution can be written in terms of the same anisotropy parameter (9). The mass spectrum (6) and the anisotropy parameter (9) are sensitive to the tensor part of the $N\bar{N}$ potential and, therefore, give the possibility to verify the potential model.

Now let us discuss other J/ψ decays where a $p\bar{p}$ pair and a vector particle (ρ, ω, γ) are produced. Due to the C-parity conservation law, possible states for a $p\bar{p}$ pair in such decays are ${}^1S_0$ and ${}^3P_J$. The S-wave state dominates in the near-threshold region where the relative velocity of the nucleons is small. The $p\bar{p}$ pairs have different isospins for the final states containing a vector meson ($I = 1$ for the $p\bar{p}\rho$ state, and $I = 0$ for the $p\bar{p}\omega$ state). In the case of $p\bar{p}\gamma$ final state, the $p\bar{p}$ pair is a mixture of two isospin states.

The same kinematics relations (1) hold, as well as the relation between the decay rate and the amplitude (4). The dimensionless amplitude of these decays can be written via the radial wave function of the $p\bar{p}$ pair corresponding to the ${}^1S_0$ wave, $\psi_R^I(r)$, as

$$T_{\lambda\lambda'}^I = \frac{\mathcal{G}_I}{m_{J/\psi}}\mathbf{e}_\lambda\left[\boldsymbol{k}\times\boldsymbol{\epsilon}_{\lambda'}\right]\psi_R^I(0). \quad (11)$$

Here $\mathcal{G}_I$ is an energy-independent dimensionless constant, $\mathbf{e}_\lambda$ and $\epsilon_{\lambda'}$ are the polarization vectors of the final vector particle and J/ψ, respectively.

Substituting the amplitude (11) in Eq. (4), averaging over the spin states and integrating over the solid angles, we obtain the $p\bar{p}$ invariant mass distribution

$$\frac{d\Gamma}{dM} = \frac{\mathcal{G}_I^2 pk^3}{2^4\, 3\pi^3 m_{J/\psi}^4}\left|\psi_R^I(0)\right|^2. \quad (12)$$

The wave function module squared is the so called enhancement factor which equals to unity if the $p\bar{p}$ final-state interaction is turned off.

The optical $N\bar{N}$ potential can also be used to calculate the decay rates of the processes with a virtual $N\bar{N}$ pair in the intermediate state. In Ref. [29] it is shown that the total cross section of $N\bar{N}$ production, which is a sum of the cross section of real $N\bar{N}$ pair production (the elastic cross section) and the cross section of the meson production via annihilation of a virtual $N\bar{N}$ pair (the inelastic cross section), can be written in terms of the Green's function of the $N\bar{N}$ pair. According to Ref. [29], in order to switch from the elastic cross section to the total one, we should replace $\left|\psi_R^I(0)\right|^2$ by $\left(-\,\mathrm{Im}\,\mathcal{D}^I(0,0|E)/m_p p\right)$, where $\mathcal{D}^I(r,r'|E)$ is the Green's function of the Schrödinger equation. Therefore, the contribution of the $N\bar{N}$ intermediate state with quantum numbers ${}^1S_0$ to the decay rate of the process $J/\psi \to N\bar{N}x \to particles + x$ (particles in the final state can be nucleons or mesons) has the form

$$\frac{d\Gamma_{\mathrm{tot}}}{dM} = -\frac{\mathcal{G}_I^2 k^3}{2^4\, 3\pi^3 m_p m_{J/\psi}^4}\,\mathrm{Im}\,\mathcal{D}^I(0,0|E), \quad (13)$$

where M is the invariant mass of the mesons, $E = M/2 - m_p$. The Green's function can be written in terms of regular, $\psi_R^I(r)$, and non-regular, $\psi_N^I(r)$, solutions of the Schrödinger equation:

$$\mathcal{D}^I(r,r'|E) = -m_p p\left\{\theta(r'-r)\,\psi_R^I(r)\psi_N^I(r') + +\theta(r-r')\,\psi_N^I(r)\psi_R^I(r')\right\}, \quad (14)$$

where $\theta(x)$ is the Heaviside function.

3 Results and Discussion

3.1 The decays $J/\psi \to p\bar{p}\pi^0(\eta)$

In the present work we use the potential model suggested in Ref. [29] to describe the $p\bar{p}$ interaction in the ${}^3S_1 - {}^3D_1$ coupled channels. The parameters of this model have been fitted using the $p\bar{p}$ scattering data, the cross section of $N\bar{N}$ pair production in e^+e^- annihilation near the threshold, and the ratio of the electromagnetic form factors of the proton in the timelike region. We have slightly refitted the parameters of the model in order to achieve a better description of the invariant mass spectra of the decays considered. By means of this model and Eq. (6), we predict the $p\bar{p}$ invariant mass spectra in the processes $J/\psi \to p\bar{p}\pi^0$ and $J/\psi \to p\bar{p}\eta$. The isospin of the $p\bar{p}$ pair is $I = 1$ and $I = 0$ for, respectively, a pion and η meson in the final state. The model [29] predicts the enhancement of the decay rates of both processes near the threshold of $p\bar{p}$ pair production (see Fig. 1). The invariant mass spectra predicted by our model are similar to those predicted in Ref. [33] with the use of the chiral model.

An important prediction of our model is the angular anisotropy of these J/ψ decays. This anisotropy is the result of D-wave admixture due to the tensor forces in $N\bar{N}$ interaction. The anisotropy (see Eqs. (8) and (10)) is characterized by the parameters γ^1 and γ^0 (9) for the $p\bar{p}\pi^0$ and $p\bar{p}\eta$ final states, respectively. The dependence of the parameters γ^I on the invariant mass of the $p\bar{p}$ pair is shown

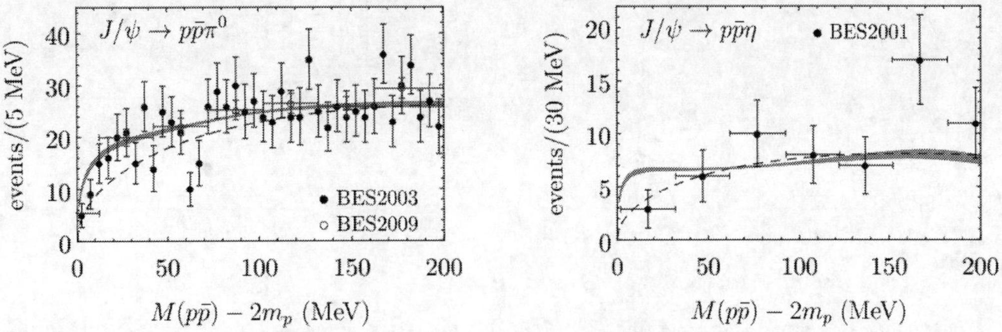

Figure 1. The invariant mass spectra of J/ψ decays to $p\bar{p}\pi^0$ (left) and $p\bar{p}\eta$ (right). The phase space behavior is shown by the dashed curve. The experimental data are taken from Refs. [5–7]. The measurement of Ref. [5] is adopted for the scale of the left plot.

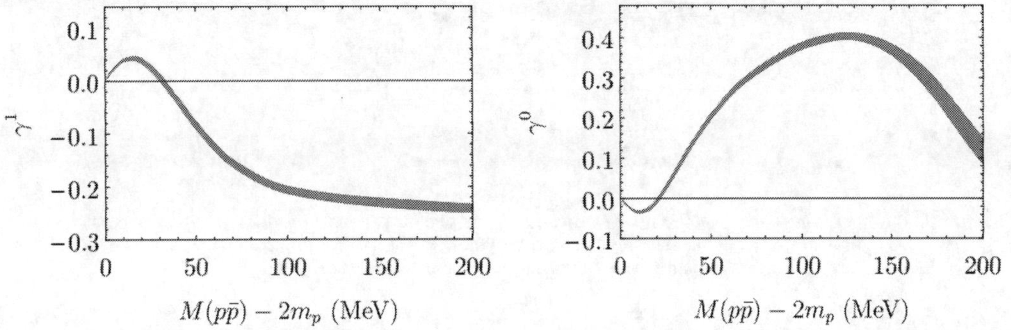

Figure 2. The dependence of the anisotropy parameters γ^I in the decays $J/\psi \to p\bar{p}\pi^0$ (left) and $J/\psi \to p\bar{p}\eta$ (right) on the $p\bar{p}$ invariant mass.

in Fig. 2. For $p\bar{p}$ invariant mass about 100 – 200 MeV above the threshold, significant anisotropy of the angular distributions is predicted. Note that the anisotropy in the distribution over the angle ϑ_{pk} is expected to be two times larger than in the distribution over the angle ϑ_p (compare Eqs. (8) and (10)).

There are some data on the angular distributions in the decays $J/\psi \to p\bar{p}\pi^0$ [6] and $J/\psi \to p\bar{p}\eta$ [7]. However, these distributions are obtained by integration over the whole $p\bar{p}$ invariant mass region. Unfortunately, our predictions are valid only in the narrow energy region above the $p\bar{p}$ threshold. Therefore, we cannot compare the predictions with the available experimental data. The measurements of the angular distributions at $p\bar{p}$ invariant mass close to the $p\bar{p}$ threshold would be very helpful. Such measurements would provide another possibility to verify the available models of $N\bar{N}$ interaction in the low-energy region.

3.2 The decays $J/\psi \to p\bar{p}\gamma(\rho, \omega)$

In order to describe the $p\bar{p}$ interaction in the decays $J/\psi \to p\bar{p}\rho(\omega, \gamma)$ we use an $N\bar{N}$ optical potential for the $^1S_0$ partial wave proposed in Ref. [30]. The data used for fitting the parameters of the potential include the partial contributions of $^1S_0$ wave to the elastic, charge-exchange, and total cross sections of $p\bar{p}$ scattering, and the $p\bar{p}$ invariant mass spectra of the decays $J/\psi \to p\bar{p}\omega$, $J/\psi \to p\bar{p}\gamma$, and $\psi(2S) \to p\bar{p}\gamma$. The partial cross sections of $p\bar{p}$ scattering are calculated from the Nijmegen partial wave S-matrix (Table V of Ref. [25]).

By means of the potential model and Eq. (12), we calculate the $p\bar{p}$ invariant mass spectra in the processes $J/\psi \to p\bar{p}\omega$ and $J/\psi \to p\bar{p}\rho$ (see Fig. 3). The isospin of the $p\bar{p}$ pair is $I = 0$ and $I = 1$ for ω meson and ρ meson in the final state, respectively. Therefore, the decay rates for these processes are given by Eq. (12) with the corresponding constants $\mathcal{G}_I$ and wave functions $\psi^I_R(0)$. Our model fits the experimental data for the decay $J/\psi \to p\bar{p}\omega$ quite well. There are no experimental data for the decay $J/\psi \to p\bar{p}\rho$, therefore, the predictions for the invariant mass spectrum are especially important. The $p\bar{p}$ spectrum in the decay $J/\psi \to p\bar{p}\rho$, calculated in Ref. [33] with the use of the chiral model [26], has a pronounced peak close to the $p\bar{p}$ threshold, while our model predicts a monotonically increasing spectrum without any peak.

The decay amplitude of the process $J/\psi \to p\bar{p}\gamma$ is a sum of two isospin contributions. Therefore, the decay

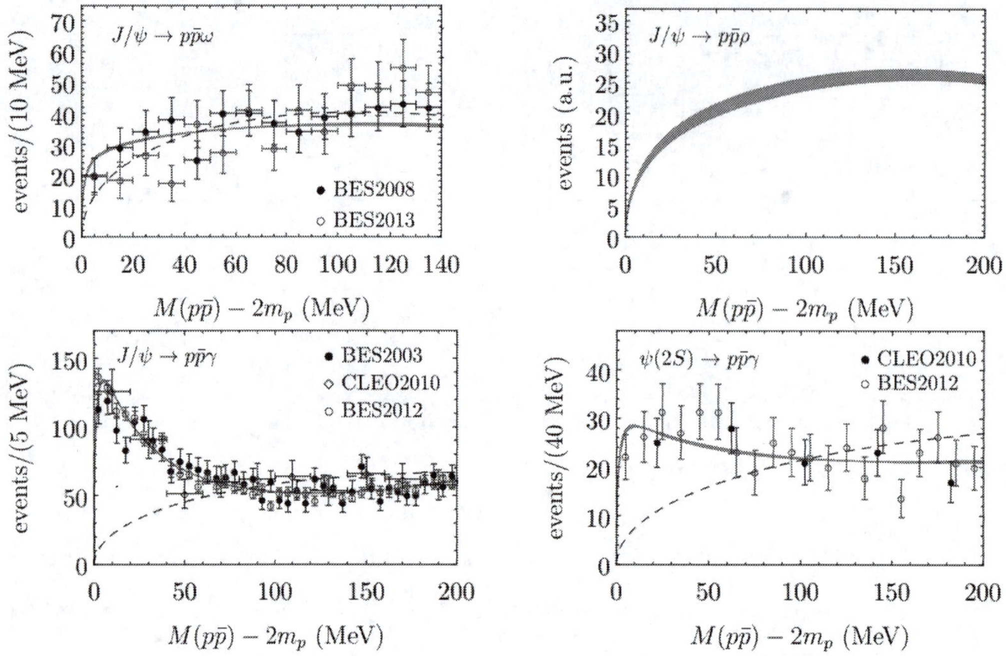

Figure 3. The invariant mass spectra of J/ψ decays to $p\bar{p}\omega$, $p\bar{p}\rho$, $p\bar{p}\gamma$, and $\psi(2S)$ decay to $p\bar{p}\gamma$. The invariant mass spectra without $N\bar{N}$ interaction taken into account are shown by the dashed curves. The experimental data are taken from Refs. [5, 8–11]. The earliest measurements are adopted for the scale of the plots. This Figure has been copied from Ref. [30].

rate reads

$$\frac{d\Gamma_{p\bar{p}\gamma}}{dM} = \frac{pk^3}{2^4\,3\pi^3 m_{J/\psi}^4}\left|\mathcal{G}_{\gamma 0}\psi_R^0(0) + \mathcal{G}_{\gamma 1}\psi_R^1(0)\right|^2 . \quad (15)$$

Our model describes with good accuracy the pronounced peak, seen fairly well in the experimental data for the decay $J/\psi \to p\bar{p}\gamma$ (see Fig. 3). For the best fit, the ratio of the constants is $\mathcal{G}_{\gamma 1}/\mathcal{G}_{\gamma 0} = -1.29 - 0.6\,i$. We have investigated in details the origin of this peak and found out that it arises because of a significant compensation of two isospin amplitudes at energy above 10 MeV per nucleon, though each isospin amplitude has no peak. This leads to another interesting prediction. The decay rate of the process $J/\psi \to n\bar{n}\gamma$, given by the formula

$$\frac{d\Gamma_{n\bar{n}\gamma}}{dM} = \frac{pk^3}{2^4\,3\pi^3 m_{J/\psi}^4}\left|\mathcal{G}_{\gamma 0}\psi_R^0(0) - \mathcal{G}_{\gamma 1}\psi_R^1(0)\right|^2 , \quad (16)$$

should be much larger than that for the process $J/\psi \to p\bar{p}\gamma$. The experimental investigation of the decay $J/\psi \to n\bar{n}\gamma$ would provide important information about the $N\bar{N}$ interaction. For completeness, we also consider the decay $\psi(2S) \to p\bar{p}\gamma$ (the corresponding ratio of the constants is $\mathcal{G}_{\gamma 1}/\mathcal{G}_{\gamma 0} = -1.06 - 0.09\,i$), see Fig. 3.

Making use of our potential model and Eq. (13), we obtain also the predictions for the decay rates of the processes with the interaction of virtual nucleon-antinucleon pairs in the intermediate state (see Fig. 4). A peak in the total and inelastic invariant mass spectra exists near the $p\bar{p}$ threshold, especially in the isoscalar channel. This behavior seems to be the consequence of the existence of a quasi-bound state near the $p\bar{p}$ threshold. Our analysis shows that such state does exist in the isoscalar channel, and its energy is $E_B = (22 - 33\,i)$ MeV. This is an unstable bound state in the classification of Ref. [37] because its energy moves to $E_B = -3$ MeV when the imaginary part of the $N\bar{N}$ potential is turned off.

Let us discuss the exotic behavior of the decay rate of the process $J/\psi \to \gamma\eta'\pi^+\pi^-$ near the $N\bar{N}$ threshold observed in Ref. [17]. An approximate description of the spectrum of this decay was suggested in Ref. [38]. However, due to poor accuracy of the experimental data available at that time, only general resonance structure of the spectrum has been discussed in Ref. [38], but not the sharp dip in the vicinity of $N\bar{N}$ threshold. The new experimental data [17] allows us to investigate this spectrum in more details. The G-parity of the intermediate $N\bar{N}$ state, $G_{N\bar{N}} = C_{N\bar{N}}(-1)^I$, should be equal to that of the final $\eta'\pi^+\pi^-$ state, $G_{\eta'\pi^+\pi^-} = 1$. Taking into account C-parity conservation we obtain $C_{N\bar{N}} = 1$, thus the isospin of the $N\bar{N}$ pair is $I = 0$. Possible $N\bar{N}$ states with positive C-parity are ${}^1S_0$ and ${}^3P_j$, and the former one is expected to dominate in the near-threshold region. Therefore, we believe that the peak in the $\eta'\pi^+\pi^-$ invariant mass spectrum could occur because of the interaction of virtual nucleons in the isoscalar ${}^1S_0$ intermediate state. The contribution

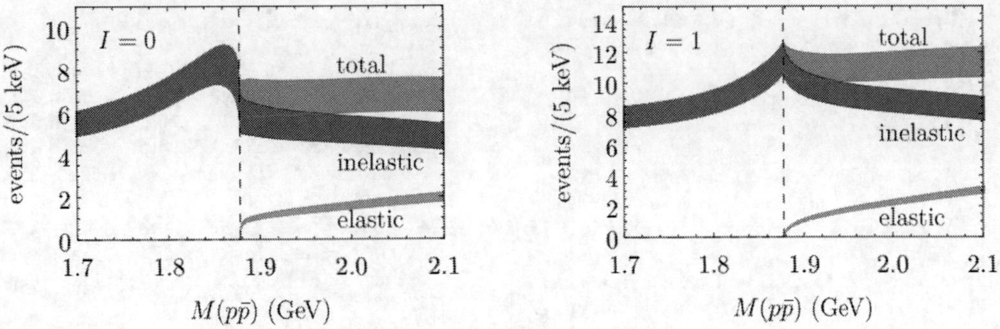

Figure 4. The $p\bar{p}$ invariant mass spectra for the isospin components in the decays $J/\psi \to \gamma p\bar{p}$ (elastic), $J/\psi \to \gamma + p\bar{p} \to \gamma + mesons$ (inelastic), and total. Vertical dashed line is the $p\bar{p}$ threshold. This Figure has been copied from Ref. [30].

of non-$N\bar{N}$ channels should be a smooth function in the vicinity of the $N\bar{N}$ threshold. Therefore, we approximate the invariant mass spectrum of the decay $J/\psi \to \gamma\eta'\pi^+\pi^-$ by the function

$$A \cdot d\Gamma^0_{\rm inel}/dM + B \cdot E + C\,, \qquad (17)$$

where A, B and C are some fitting parameters. The comparison of the experimental data and our fitting formula in Fig. 5 demonstrates good agreement in the near-threshold region.

One can expect that the $\gamma p\bar{p}$ intermediate state plays an important role in the decay $J/\psi \to \gamma\eta'\pi^+\pi^-$ near the threshold of $p\bar{p}$ pair production. Therefore, the branching ratio of this channel should be related to the branching ratio for the $p\bar{p}$ annihilation at rest into the $\eta'\pi^+\pi^-$ final state. To check the validity of this statement, we should take into account that the scale of Fig. 4 for the decay $J/\psi \to p\bar{p}\gamma$ differs from that of Fig. 5 for the decay $J/\psi \to \gamma\eta'\pi^+\pi^-$ because of different total numbers of J/ψ events in the experiments [5] and [17]. After that we find the absolute value of the coefficient A in Eq. (17). The coefficient $A \approx 4 \cdot 10^{-3}$ can be considered as the estimation of the branching ratio of the decay $p\bar{p} \to \eta'\pi^+\pi^-$ at rest. This value is very close to the branching ratio $3.46 \cdot 10^{-3}$ measured in the experiment [39].

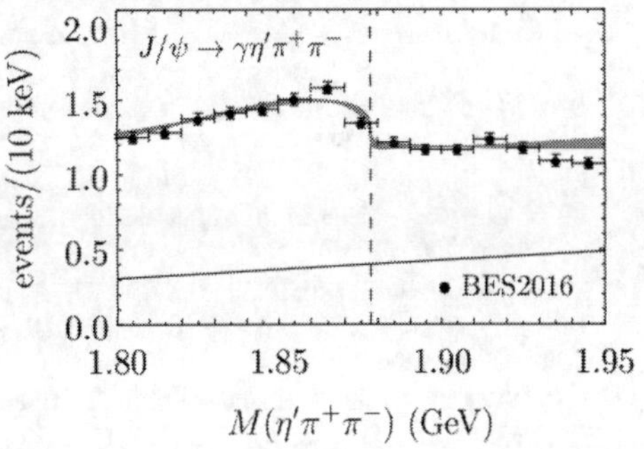

Figure 5. The $\eta'\pi^+\pi^-$ invariant mass spectrum for the decay $J/\psi \to \gamma\eta'\pi^+\pi^-$. The thin line shows the contribution of non-$N\bar{N}$ channels. Vertical dashed line is the $N\bar{N}$ threshold. The experimental data are taken from Ref. [17]. This Figure has been copied from Ref. [30].

4 Conclusions

Using the model proposed in Ref. [29], we have calculated the effects of $p\bar{p}$ final-state interaction in the decays $J/\psi \to p\bar{p}\pi^0(\eta)$. Our results for the $p\bar{p}$ invariant mass spectra close to the $p\bar{p}$ threshold are in agreement with the available experimental data. The tensor forces in the $p\bar{p}$ interaction result in the anisotropy of the angular distributions. The anisotropy in the decay $J/\psi \to p\bar{p}\pi^0$ and especially in the $J/\psi \to p\bar{p}\eta$ decay is large enough to be measured. The observation of such anisotropy close to the $p\bar{p}$ threshold would allow one to refine the model of $N\bar{N}$ interaction.

Using the model of $N\bar{N}$ interaction in $^1S_0$ partial wave we have calculated the effects of $p\bar{p}$ final-state interaction in other J/ψ decays. Our model describes the $p\bar{p}$ invariant mass spectra of the decays $J/\psi \to pp\omega$, $J/\psi \to pp\gamma$, and $\psi(2S) \to p\bar{p}\gamma$ with good precision. We have also obtained the predictions for the $p\bar{p}$ invariant mass spectrum in the decay $J/\psi \to p\bar{p}\rho$ which has not been measured yet. Our prediction for this spectrum differs from the theoretical results obtained earlier. Therefore, the experimental study of the decay rate of this process would help to discriminate different models of the nucleon-antinucleon interaction.

We have used the Green's function approach to calculate the contribution of the interaction of virtual $N\bar{N}$ pairs in the $^1S_0$ state to the cross sections of the processes. In particular we have calculated the contribution of the $N\bar{N}$ intermediate state to the $\eta'\pi^+\pi^-$ invariant mass spectrum for the decay $J/\psi \to \gamma N\bar{N} \to \gamma\eta'\pi^+\pi^-$ in the energy region near the $N\bar{N}$ threshold. Our results are in good agreement with the available experimental data and describe the peak in the invariant mass spectrum just below the $N\bar{N}$ threshold.

We are thankful to V. F. Dmitriev for useful discussions. The work of S. G. Salnikov has been supported by the RScF grant 16-12-10151.

References

[1] B. Aubert, R. Barate, D. Boutigny, F. Couderc, Y. Karyotakis, J.P. Lees, V. Poireau, V. Tisserand, A. Zghiche, E. Grauges et al., Phys. Rev. D **73**, 012005 (2006)

[2] J.P. Lees, V. Poireau, V. Tisserand, E. Grauges, A. Palano, G. Eigen, B. Stugu, D.N. Brown, L.T. Kerth, Y.G. Kolomensky et al., Phys. Rev. D **87**, 092005 (2013)

[3] M.N. Achasov, A.Y. Barnyakov, K.I. Beloborodov, A.V. Berdyugin, D.E. Berkaev, A.G. Bogdanchikov, A.A. Botov, T.V. Dimova, V.P. Druzhinin, V.B. Golubev et al., Phys. Rev. D **90**, 112007 (2014)

[4] R. Akhmetshin, A. Amirkhanov, A. Anisenkov, V. Aulchenko, V. Banzarov, N. Bashtovoy, D. Berkaev, A. Bondar, A. Bragin, S. Eidelman et al., Phys. Lett. B **759**, 634 (2016)

[5] J. Bai, Y. Ban, J. Bian, X. Cai, J. Chang, H. Chen, J. Chen, J. Chen, Y. Chen, S. Chi et al., Phys. Rev. Lett. **91**, 022001 (2003)

[6] M. Ablikim, J. Bai, Y. Bai, Y. Ban, X. Cai, H. Chen, J. Chen, J. Chen, X. Chen, Y. Chen et al., Phys. Rev. D **80**, 052004 (2009)

[7] J. Bai, Y. Ban, J. Bian, J. Chang, A. Chen, H. Chen, H. Chen, J. Chen, X. Chen, Y. Chen et al., Phys. Lett. B **510**, 75 (2001)

[8] J.P. Alexander, D.G. Cassel, S. Das, R. Ehrlich, L. Fields, L. Gibbons, S.W. Gray, D.L. Hartill, B.K. Heltsley, D.L. Kreinick et al., Phys. Rev. D **82**, 092002 (2010)

[9] M. Ablikim, M.N. Achasov, D. Alberto, D.J. Ambrose, F.F. An, Q. An, Z.H. An, J.Z. Bai, R.B.F. Baldini Ferroli, Y. Ban et al., Phys. Rev. Lett. **108**, 112003 (2012)

[10] M. Ablikim, J. Bai, Y. Ban, X. Cai, H. Chen, H. Chen, H. Chen, J. Chen, J. Chen, Y. Chen et al., Eur. Phys. J. C **53**, 15 (2008)

[11] M. Ablikim, M.N. Achasov, O. Albayrak, D.J. Ambrose, F.F. An, Q. An, J.Z. Bai, R. Baldini Ferroli, Y. Ban, J. Becker et al., Phys. Rev. D **87**, 112004 (2013)

[12] B. Aubert, R. Barate, D. Boutigny, F. Couderc, Y. Karyotakis, J.P. Lees, V. Poireau, V. Tisserand, A. Zghiche, E. Grauges et al., Phys. Rev. D **73**, 052003 (2006)

[13] B. Aubert, M. Bona, D. Boutigny, Y. Karyotakis, J.P. Lees, V. Poireau, X. Prudent, V. Tisserand, A. Zghiche, J. Garra Tico et al., Phys. Rev. D **76**, 092005 (2007)

[14] R. Akhmetshin, a.V. Anisenkov, S.A. Anokhin, V.M. Aulchenko, V.S. Banzarov, L.M. Barkov, N.S. Bashtovoy, D.E. Berkaev, A.E. Bondar, a.V. Bragin et al., Phys. Lett. B **723**, 82 (2013)

[15] P.A. Lukin, A.V. Anisenkov, V.M. Aulchenko, R.R. Akhmetshin, V.S. Banzarov, N.S. Bashtovoy, D.E. Berkaev, a.V. Bragin, a.I. Vorobiov, S.E. Gayazov et al., Phys. At. Nucl. **78**, 353 (2015)

[16] A.E. Obrazovsky, S.I. Serednyakov, JETP Lett. **99**, 315 (2014)

[17] M. Ablikim, M.N. Achasov, S. Ahmed, X.C. Ai, O. Albayrak, M. Albrecht, D.J. Ambrose, A. Amoroso, F.F. An, Q. An et al., Phys. Rev. Lett. **117**, 042002 (2016)

[18] A. Datta, P.J. O'Donnell, Phys. Lett. B **567**, 273 (2003)

[19] G.J. Ding, M.L. Yan, Phys. Rev. C **72**, 015208 (2005)

[20] M.L. Yan, S. Li, B. Wu, B.Q. Ma, Phys. Rev. D **72**, 034027 (2005)

[21] N. Kochelev, D.P. Min, Phys. Lett. B **633**, 283 (2006)

[22] B.A. Li, Phys. Rev. D **74**, 034019 (2006)

[23] X.G. He, X.Q. Li, X. Liu, J.P. Ma, Eur. Phys. J. C **49**, 731 (2007)

[24] B. El-Bennich, M. Lacombe, B. Loiseau, S. Wycech, Phys. Rev. C **79**, 054001 (2009)

[25] D. Zhou, R.G.E. Timmermans, Phys. Rev. C **86**, 044003 (2012)

[26] X.W. Kang, J. Haidenbauer, U.G. Meißner, J. High Energy Phys. **2014**, 113 (2014)

[27] V.F. Dmitriev, A.I. Milstein, Phys. Lett. B **658**, 13 (2007)

[28] J. Haidenbauer, X.W.W. Kang, U.G.G. Meißner, Nucl. Phys. A **929**, 102 (2014)

[29] V.F. Dmitriev, A.I. Milstein, S.G. Salnikov, Phys. Rev. D **93**, 034033 (2016)

[30] A.I. Milstein, S.G. Salnikov, Nucl. Phys. A **966**, 54 (2017)

[31] J. Haidenbauer, C. Hanhart, X.W. Kang, U.G. Meißner, Phys. Rev. D **92**, 054032 (2015)

[32] V.F. Dmitriev, A.I. Milstein, S.G. Salnikov, Phys. Lett. B **760**, 139 (2016)

[33] X.W. Kang, J. Haidenbauer, U.G. Meißner, Phys. Rev. D **91**, 074003 (2015)

[34] Y.F. Liu, X.W. Kang, Symmetry (Basel). **8**, 14 (2016)

[35] V.F. Dmitriev, A.I. Milstein, S.G. Salnikov, Phys. At. Nucl. **77**, 1173 (2014)

[36] A. Sibirtsev, J. Haidenbauer, S. Krewald, U.G. Meißner, A.W. Thomas, Phys. Rev. D **71**, 054010 (2005)

[37] A. Badalyan, L. Kok, M. Polikarpov, Y. Simonov, Phys. Rep. **82**, 31 (1982)

[38] J.P. Dedonder, B. Loiseau, B. El-Bennich, S. Wycech, Phys. Rev. C **80**, 045207 (2009)

[39] P. Weidenauer, K.D. Duch, M. Heel, H. Kalinowsky, F. Kayser, E. Klempt, B. May, J. Reifenröther, O. Schreiber, M. Ziegler et al., Zeitschrift für Phys. C Part. Fields **47**, 353 (1990)

EPJ Web of Conferences **181**, 01021 (2018) https://doi.org/10.1051/epjconf/201818101021
EXA2017

Dalitz Plot Analysis of Three-body Charmonium Decays at *BABAR*

Antimo Palano[1,3,⋆]
[1] *INFN and University of Bari, Via Orabona 4, 70125 Bari, Italy*

Abstract. We present results on the measurement of the I=1/2 $K\pi$ S-wave through a model independent partial wave analysis of η_c decays to $K^0_S K^+\pi^-$ and $K^+K^-\pi^0$ produced in two-photon interactions. We also perform Dalitz plot analyses of the J/ψ decays to $\pi^+\pi^-\pi^0$, $K^+K^-\pi^0$ and $K^0_S K^\pm\pi^\mp$ produced in the initial state radiation process.

1 Introduction

Charmonium decays can be used to obtain new information on light meson spectroscopy. In e^+e^- interactions, samples of charmonium decays can be obtained using different processes.

- In two-photon interactions we select events in which the e^+ and e^- beam particles are scattered at small angles and remain undetected. Only resonances with $J^{PC} = 0^{\pm+}, 2^{\pm+}, 3^{++}, 4^{\pm+}$.... can be produced.
- In the Initial State Radiation (ISR) process, we reconstruct events having a (mostly undetected) fast forward γ_{ISR} and, in this case, only $J^{PC} = 1^{--}$ states can be produced.

2 Study of $\eta_c \to K\bar{K}\pi$

The BaBar Dalitz plot analysis of the $\eta_c \to K^+K^-\eta$ and $\eta_c \to K^+K^-\pi^0$ has provided the unexpected observation of $K^*_0(1430) \to K\eta$ [1]. We also find that the η_c three-body hadronic decays proceed almost entirely through the intermediate production of scalar meson resonances.

We study the reactions [2]

$$\gamma\gamma \to K^0_S K^+\pi^-, \qquad \gamma\gamma \to K^+K^-\pi^0.$$

In the following, details on events reconstruction will be given only for the $K^0_S K^+\pi^-$ final state. We select events having only four tracks. Since two-photon events balance the transverse momentum, we require p_T, the transverse momentum of the $K^0_S K^+\pi^-$ system with respect to the beam axis, to be $p_T < 0.08\ GeV/c$. We also define $M^2_{\rm rec} \equiv (p_{e^+e^-} - p_{\rm rec})^2$, where $p_{e^+e^-}$ is the four-momentum of the initial state and $p_{\rm rec}$ is the four-momentum of the $K^0_S K^+\pi^-$ system and remove ISR events requiring $M^2_{\rm rec} > 10\ GeV^2/c^4$. The $K^0_S K^+\pi^-$ mass spectrum in the η_c mass regions is shown in Fig. 1(left) and the η_c Dalitz plot is shown in Fig. 1(right). The Dalitz plot is dominated by the presence of horizontal and vertical uniform bands at the position of the $K^*_0(1430)$ resonance. The corresponding distributions for $\eta_c \to K^+K^-\pi^0$ can be found in Ref. [1] and present similar features.

⋆e-mail: antimo.palano@ba.infn.it

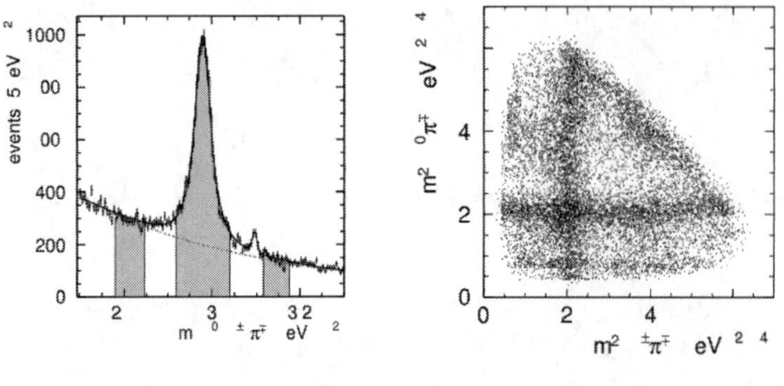

Figure 1. (left) $K_S^0K^+\pi^-$ mass spectrum in two-photon interactions and (right) the $\eta_c \to K_S^0K^+\pi^-$ Dalitz plot.

The η_c signal regions contain 12849 events with (64.3 ± 0.4)% purity for $\eta_c \to K_S^0K^+\pi^-$ and 6494 events with (55.2±0.6)% purity for $\eta_c \to K^+K^-\pi^0$. The backgrounds below the η_c signals are estimated from the sidebands. We observe asymmetric K^*'s in the background to the $\eta_c \to K_S^0K^+\pi^-$ final state due to interference between I=1 and I=0 contributions.

3 Dalitz plot Analysis of $\eta_c \to K\bar{K}\pi$

We perform unbinned maximum likelihood fits using the Isobar model [3] and Model Independent Partial Wave Analysis (MIPWA) [4]. In the MIPWA the $K\pi$ mass spectrum is divided into 30 equally spaced mass intervals 60 MeV/c^2 wide and for each bin we add to the fit two new free parameters, the amplitude and the phase of the $K\pi$ S-wave (constant inside the bin). We also fix the amplitude to 1.0 and its phase to $\pi/2$ in an arbitrary interval of the mass spectrum (bin 11 which corresponds to a mass of 1.45 GeV/c^2). The number of additional free parameters is therefore 58. Due to isospin conservation in the decays, amplitudes are symmetrized with respect to the two $K\pi$ decay modes. The $K_2^*(1420)$, $a_0(980)$, $a_0(1400)$, $a_2(1310)$, ... contributions are modeled as relativistic Breit-Wigner functions multiplied by the corresponding angular functions. Backgrounds are fitted separately and interpolated into the η_c signal regions. The fits improves when an additional high mass $a_0(1950) \to K\bar{K}$ I=1 resonance is included with free parameters in both η_c decay modes. The weighted average of the two measurement is: $m(a_0(1950)) = 1931 \pm 14 \pm 22$ MeV/c^2, $\Gamma(a_0(1950)) = 271 \pm 22 \pm 29$ MeV. The statistical significances for the $a_0(1950)$ effect (including systematics) are 2.5σ for $\eta_c \to K_S^0K^+\pi^-$ and 4.0σ for $\eta_c \to K^+K^-\pi^0$.

The Dalitz plot projections with fit results for $\eta_c \to K_S^0K^+\pi^-$ and $\eta_c \to K^+K^-\pi^0$ are shown in Fig. 2. The fitted fractions and phases are given in Table 1. We observe a good description of the data. We note that the $K^*(892)$ contributions arise entirely from background.

In comparison, the isobar model gives a worse description of the data, with $\chi^2/N_{cells} = 457/254 = 1.82$ and $\chi^2/N_{cells} = 383/233 = 1.63$ respectively for the two η_c decay modes. The resulting $K\pi$ S-wave amplitude and phase for the two η_c decay modes is shown in Fig. 3. We observe a clear $K_0^*(1430)$ resonance signal with the corresponding expected phase motion. At high mass we observe the presence of the broad $K_0^*(1950)$ contribution with good agreement between the two η_c decay modes. Comparing with LASS [5] and E791 [4] experiments we note that phases before the $K\eta'$ threshold are similar, as expected from Watson [6] theorem but amplitudes are very different.

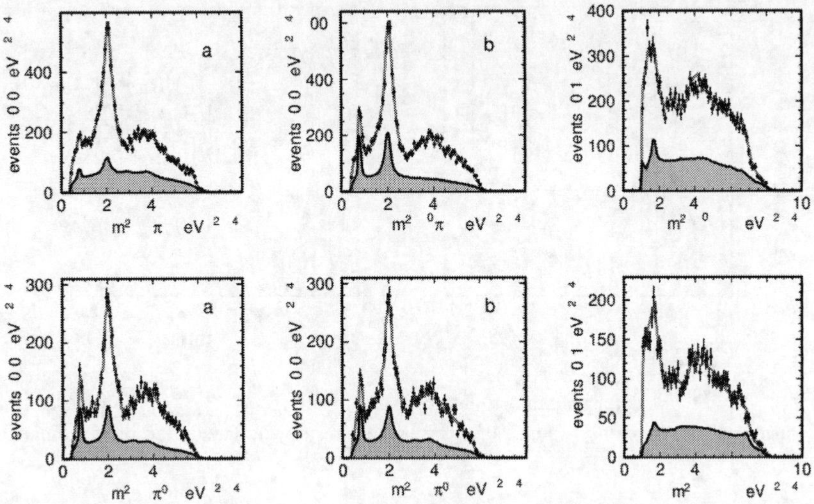

Figure 2. (top) $\eta_c \to K^0_S K^+\pi^-$ and (bottom) $\eta_c \to K^+K^-\pi^0$ Dalitz plots projections. The superimposed curves are from the fit results. Shaded is contribution from the interpolated background.

Table 1. Results from the $\eta_c \to K^0_S K^\pm\pi^\mp$ and $\eta_c \to K^+K^-\pi^0$ MIPWA. Phases are determined relative to the ($K\pi$ $\mathcal{S}$-wave) $\overline{K}$ amplitude which is fixed to $\pi/2$ at 1.45 GeV/c^2.

	$\eta_c \to K^0_S K^\pm\pi^\mp$		$\eta_c \to K^+K^-\pi^0$	
Amplitude	Fraction (%)	Phase (rad)	Fraction (%)	Phase (rad)
($K\pi$ $\mathcal{S}$-wave) $\overline{K}$	107.3 ± 2.6 ± 17.9	fixed	125.5 ± 2.4 ± 4.2	fixed
$a_0(980)\pi$	0.8 ± 0.5 ± 0.8	1.08 ± 0.18 ± 0.18	0.0 ± 0.1 ± 1.7	-
$a_0(1450)\pi$	0.7 ± 0.2 ± 1.4	2.63 ± 0.13 ± 0.17	1.2 ± 0.4 ± 0.7	2.90 ± 0.12 ± 0.25
$a_0(1950)\pi$	3.1 ± 0.4 ± 1.2	−1.04 ± 0.08 ± 0.77	4.4 ± 0.8 ± 0.8	−1.45 ± 0.08 ± 0.27
$a_2(1320)\pi$	0.2 ± 0.1 ± 0.1	1.85 ± 0.20 ± 0.20	0.6 ± 0.2 ± 0.3	1.75 ± 0.23 ± 0.42
$K^*_2(1430)\overline{K}$	4.7 ± 0.9 ± 1.4	4.92 ± 0.05 ± 0.10	3.0 ± 0.8 ± 4.4	5.07 ± 0.09 ± 0.30
Total	116.8 ± 2.8 ± 18.1		134.8 ± 2.7 ± 6.4	
$-2\log\mathcal{L}$	−4314.2		−2339	
χ^2/N_{cells}	301/254=1.17		283.2/233=1.22	

4 Dalitz plot analysis of $J/\psi \to \pi^+\pi^-\pi^0$, $J/\psi \to K^+K^-\pi^0$ and $J/\psi \to K^0_S K^\pm\pi^\mp$

Only a preliminary result exists, to date, on a Dalitz-plot analysis of J/ψ decays to $\pi^+\pi^-\pi^0$ [7]. While large samples of J/ψ decays exist, some branching fractions remain poorly measured. BES III experiment has performed an angular analysis of $J/\psi \to K^+K^-\pi^0$. The analysis requires the presence of a broad $J^{PC} = 1^{--}$ state in the K^+K^- threshold region, which is interpreted as a multiquark state [8].

We study the following reactions:

$$e^+e^- \to \gamma_{\text{ISR}}\,\pi^+\pi^-\pi^0, \qquad e^+e^- \to \gamma_{\text{ISR}}\,K^+K^-\pi^0, \qquad e^+e^- \to \gamma_{\text{ISR}}\,K^0_S K^\pm\pi^\mp$$

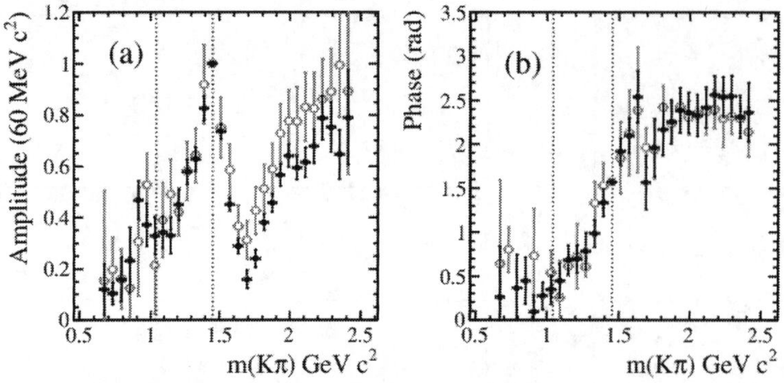

Figure 3. The $I = 1/2$ $K\pi$ S-wave amplitude (a) and phase (b) from $\eta_c \to K^0_S K^+\pi^-$ (solid (black) points) and $\eta_c \to K^+K^-\pi^0$ (open (red) points); only statistical uncertainties are shown. The dotted lines indicate the $K\eta$ and $K\eta'$ thresholds.

where $\gamma_{\rm ISR}$ indicate the (mostly undetected) ISR photon. We compute $M^2_{\rm rec} \equiv (p_{e^-} + p_{e^+} - p_{h^+} - p_{h^-} - p_{h^0})^2$, where h indicates the three hadrons in the final states. This quantity should peak near zero for ISR events. We select events in the ISR region by requiring $|M^2_{\rm rec}| < 2\ GeV^2/c^4$ for the reactions involving a π^0 and $|M^2_{\rm rec}| < 1.5\ GeV^2/c^4$ for the reaction with a K^0_S. We fit the mass spectra using the Monte Carlo resolution functions described by a Crystal Ball+Gaussian functions. We obtain 19560 $\pm$ 164 events for $J/\psi \to \pi^+\pi^-\pi^0$ with (91.3 $\pm$ 0.2)% purity, 2002 $\pm$ 48 for $J/\psi \to K^+K^-\pi^0$ with (88.8 $\pm$ 0.7)% purity, and 3694 $\pm$ 64 with 93.1 $\pm$ 0.4 purity for $J/\psi \to K^0_S K^\pm\pi^\mp$. The efficiency is mapped and fitted on the $(m(h^+h^-), \cos\theta_h)$ plane, where θ_h is the h^+ helicity angle in the J/ψ rest frame. To obtain the measurements of the relative branching fractions, we correct yields by weighting each event by the inverse of the efficiency and perform background subtraction by assigning negative weights to events the J/ψ sidebands regions. We measure the ratio

$$\mathcal{R}_1 = \frac{\mathcal{B}(J/\psi \to K^+K^-\pi^0)}{\mathcal{B}(J/\psi \to \pi^+\pi^-\pi^0)} = 0.120 \pm 0.003(\text{stat}) \pm 0.009(\text{sys}), \quad (1)$$

where many systematic uncertainties cancel out due to the similar event topology of the J/ψ decay modes. The PDG reports $\mathcal{B}(J/\psi \to \pi^+\pi^-\pi^0) = (2.11 \pm 0.07) \times 10^{-2}$, while the branching fraction $\mathcal{B}(J/\psi \to K^+K^-\pi^0)$ has been measured by Mark II [9] using 25 events, to be $(2.8 \pm 0.8) \times 10^{-3}$. These values give a ratio $\mathcal{R}_1^{PDG} = 0.133 \pm 0.038$, in agreement with our measurement. Using a similar procedure and correcting correcting for unseen K^0_S decay modes, we compute the relative branching fraction

$$\mathcal{R}_2 = \frac{\mathcal{B}(J/\psi \to K^0_S K^\pm\pi^\mp)}{\mathcal{B}(J/\psi \to \pi^+\pi^-\pi^0)} = 0.265 \pm 0.005(\text{stat}) \pm 0.021(\text{sys}). \quad (2)$$

The branching fraction $\mathcal{B}(J/\psi \to K^0_S K^\pm\pi^\mp)$ has been measured by Mark I [10], using 126 events, to be $(26 \pm 7) \times 10^{-4}$. Using the above measurements we obtain an estimate of $\mathcal{R}_2^{PDG} = 0.123 \pm 0.033$, which deviates by 3.6σ from our measurement.

5 $J/\psi \to \pi^+\pi^-\pi^0$ Dalitz plot analysis

The Dalitz plot for $J/\psi \to \pi^+\pi^-\pi^0$ is shown in Fig. 4 and is dominated by three $\rho(770)\pi$ contributions. We perform a Dalitz plot analysis using the isobar model with amplitudes described by Zemach

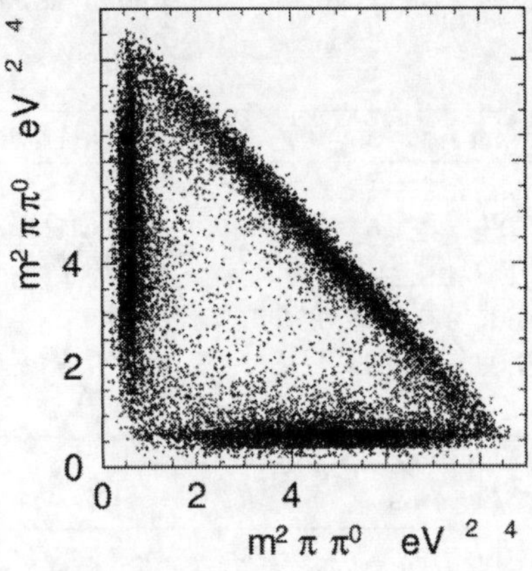

Figure 4. $J/\psi \rightarrow \pi^+\pi^-\pi^0$ Dalitz plot.

tensors [11, 12] and the Veneziano model [13]. The results from the Dalitz analysis are tabulated in Table 2 and fit projections are shown in fig. 5. The Veneziano model deals with trajectories rather

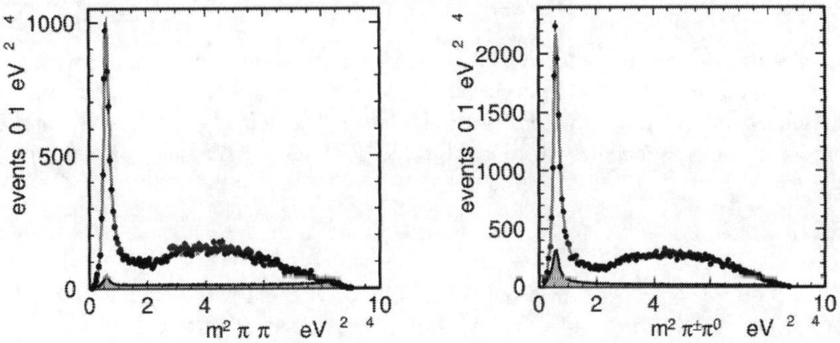

Figure 5. (left) $m^2(\pi^+\pi^-)$ and (right) $m^2(\pi^\pm\pi^0)$ for $J/\psi \rightarrow \pi^+\pi^-\pi^0$. Shaded is the background interpolated from J/ψ sidebands.

than single resonances. The complexity of the model is related to n, the number of Regge trajectories included in the fit which requires n=7, described by 19 free parameters.

Figure 8(a) shows the combinatorial $\cos\theta_\pi$ helicity angle vs. $m(\pi\pi)$. Figure 8 also shows the $m(\pi\pi)$ mass projection for $|\cos\theta_\pi| < 0.2$ for the isobar model fit (b) and Veneziano model (c). The helicity cut removes the ρ reflections enhancing the true ρ signals. The fitted fractions and phases from the Dalitz analyses are summarized in Table 2. The two models give almost similar data representation, but different fractions.

Table 2. Results from the Dalitz-plot analysis of the $J/\psi \to \pi^+\pi^-\pi^0$ channel. When two uncertainties are given, the first is statistical and the second systematic. The error on the amplitude is only statistical.

Final state	Amplitude	Isobar fraction (%)	Phase (radians)	Veneziano fraction (%)
$\rho(770)\pi$	1.	$114.2 \pm 1.1 \ \pm 2.6$	0.	133.1 ± 3.3
$\rho(1450)\pi$	0.513 ± 0.039	$10.9 \pm 1.7 \ \pm 2.7$	$-2.63 \pm 0.04 \pm 0.06$	0.80 ± 0.27
$\rho(1700)\pi$	0.067 ± 0.007	$0.8 \pm 0.2 \ \pm 0.5$	$-0.46 \pm 0.17 \pm 0.21$	2.20 ± 0.60
$\rho(2150)\pi$	0.042 ± 0.008	$0.04 \pm 0.01 \pm 0.20$	$1.70 \pm 0.21 \pm 0.12$	6.00 ± 2.50
$\omega(783)\pi^0$	0.013 ± 0.002	$0.08 \pm 0.03 \pm 0.02$	$2.78 \pm 0.20 \pm 0.31$	
$\rho_3(1690)\pi$				0.40 ± 0.08
Sum		$127.8 \pm 2.0 \pm 4.3$		142.5 ± 2.8
χ^2/ν		$687/519 = 1.32$		$596/508 = 1.17$

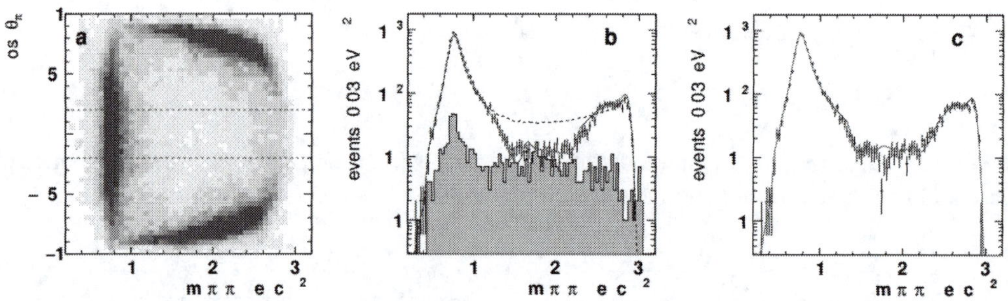

Figure 6. (a) Binned scatter diagram of $\cos\theta_{\pi_3}$ *vs* $m(\pi_1\pi_2)$. (b), (c) $\pi\pi$ mass projection in the $|\cos\theta_\pi| < 0.2$ region for all the three $\pi\pi$ charge combinations. The horizontal lines in (a) indicate the $\cos\theta_\pi$ selection. The dashed line in (b) is the result from the fit with only the $\rho(770)\pi$ amplitude. The fit in (b) uses the isobar model and the shaded histogram shows the background distribution estimated from the J/ψ sidebands. The fit in (c) uses the Veneziano model.

6 $J/\psi \to K^+K^-\pi^0$ and $J/\psi \to K^0_S K^\pm\pi^\mp$ Dalitz plot analyses

We perform a Dalitz-plot analysis of $J/\psi \to K^+K^-\pi^0$ in the J/ψ signal region. Figure 7(left) shows the Dalitz plot for the J/ψ signal region and Fig. 8(left) shows the Dalitz plot projections. We observe that the decay is dominated by the $K^*(892)^\pm K^\mp$ amplitude. We also observe a diagonal band which we tentatively attribute to the $\rho(1450)^0\pi^0$ amplitude. We fit the $J/\psi \to K^+K^-\pi^0$ Dalitz plot using the isobar model. The results from the fit are given in Table 3 and fit projections are shown in Fig. 8(left). The Dalitz analysis is describing well the K^+K^- threshold region and therefore we assign the broad enhancement in the K^+K^- mass spectrum to the presence of the $\rho(1450)$ resonance: the present data do not require the presence of an exotic contribution.

We perform a similar Dalitz plot analysis of $J/\psi \to K^0_S K^\pm\pi^\mp$ in the J/ψ signal region. Figure 7(right) shows the Dalitz plot for the J/ψ signal region and Fig. 8(right) shows the Dalitz plot projections. We fit the $J/\psi \to K^0_S K^\pm\pi^\mp$ Dalitz plot using the isobar model. The results from the best fit are summarized in Table 4. The decay is dominated by the $K^*(892)\bar{K}$, $K^*_2(1430)\bar{K}$ and $\rho(1450)^\pm\pi^\mp$

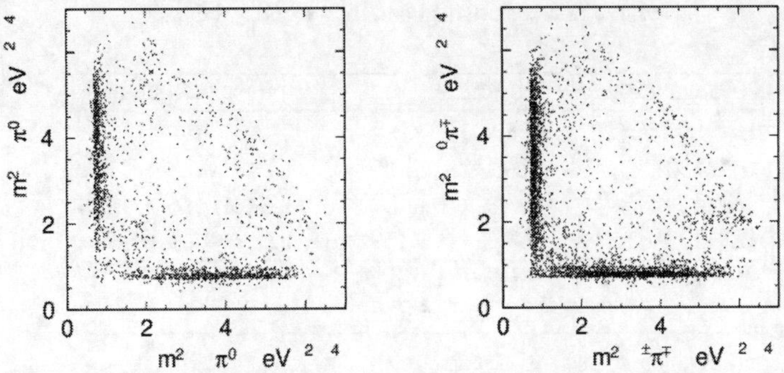

Figure 7. (left) $J/\psi \rightarrow K^+K^-\pi^0$ and (right) $J/\psi \rightarrow K^0_S K^\pm\pi^\mp$ Dalitz plots.

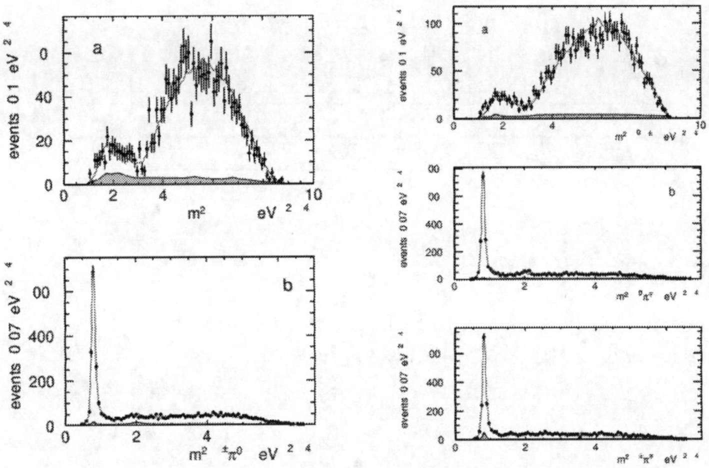

Figure 8. (left) Dalitz plot projections with fit results for $J/\psi \rightarrow K^+K^-\pi^0$. (right) Dalitz plot projections with fit results for $J/\psi \rightarrow K^0_S K^\pm\pi^\mp$. Shaded is the background interpolated from J/ψ sidebands.

amplitudes with a smaller contribution from the $K^*_1(1410)\bar{K}$ amplitude. Also in this case we assign the broad enhancement observed in the $K^0_S K^\pm$ mass projection to the presence of the $\rho(1450)^\pm$ resonance.

In the Dalitz-plot analysis of $J/\psi \rightarrow K^+K^-\pi^0$, the data are consistent with the observation of the decay $\rho(1450)^0 \rightarrow K^+K^-$. This allows a measurement of its relative branching fraction to $\rho(1450)^0 \rightarrow \pi^+\pi^-$. We obtain:

$$\frac{\mathcal{B}(\rho(1450)^0 \rightarrow K^+K^-)}{\mathcal{B}(\rho(1450)^0 \rightarrow \pi^+\pi^-)} = 0.307 \pm 0.084(\text{stat}) \pm 0.082(\text{sys}). \quad (3)$$

Table 3. Results from the Dalitz-plot analysis of the $J/\psi \to K^+K^-\pi^0$ signal region. When two uncertainties are given, the first is statistical and the second systematic.

Final state	fraction (%)	phase (radians)
$K^*(892)^{\pm}K^{\mp}$	$92.4 \pm 1.5 \pm 3.4$	0.
$\rho(1450)^0\pi^0$	$9.3 \pm 2.0 \pm 0.6$	$3.78 \pm 0.28 \pm 0.08$
$K_1^*(1410)^{\pm}K^{\mp}$	$2.3 \pm 1.1 \pm 0.7$	$3.29 \pm 0.26 \pm 0.39$
$K_2^*(1430)^{\pm}K^{\mp}$	$3.5 \pm 1.3 \pm 0.9$	$-2.32 \pm 0.22 \pm 0.05$
Total	107.4 ± 2.8	
χ^2/ν	$132/137 = 0.96$	

Table 4. Results from the Dalitz-plot analysis of the $J/\psi \to K_S^0K^{\pm}\pi^{\mp}$ signal region. When two uncertainties are given, the first is statistical and the second systematic.

Final state	fraction (%)	phase (radians)
$K^*(892)\bar{K}$	$90.5 \pm 0.9 \pm 3.8$	0.
$\rho(1450)^{\pm}\pi^{\mp}$	$6.3 \pm 0.8 \pm 0.6$	$-3.25 \pm 0.13 \pm 0.21$
$K_1^*(1410)\bar{K}$	$1.5 \pm 0.5 \pm 0.9$	$1.42 \pm 0.31 \pm 0.35$
$K_2^*(1430)\bar{K}$	$7.1 \pm 1.3 \pm 1.2$	$-2.54 \pm 0.12 \pm 0.12$
Total	105.3 ± 3.1	
χ^2/ν	$274/217 = 1.26$	

7 Ackowledgements

This work was supported (in part) by the U.S. Department of Energy, Office of Science, Office of Nuclear Physics under contract DE-AC05-06OR23177.

References

[1] J. P. Lees *et al.* [BaBar Collaboration], Phys. Rev. D **89**, no. 11, 112004 (2014)
[2] Charge conjugation is implied through all this work.
[3] S. Eidelman *et al.* [Particle Data Group], Phys. Lett. B **592**, no. 1-4, 1 (2004).
[4] E. M. Aitala *et al.* [E791 Collaboration], Phys. Rev. D **73**, 032004 (2006) Erratum: [Phys. Rev. D **74**, 059901 (2006)]
[5] D. Aston *et al.*, Nucl. Phys. B **296**, 493 (1988).
[6] K. M. Watson, Phys. Rev. **88**, 1163 (1952).
[7] L. Chen *et al.* [MARK-III Collaboration], SLAC-PUB-5674.
[8] M. Ablikim *et al.*, (BESIII Collaboration), Phys. Lett. B **710**, 594 (2012).
[9] M.E.B. Franklin *et al.*, (Mark II Collaboration) Phys. Rev. Lett. **51**, 963 (1983).
[10] F. Vannucci *et al.*, (Mark I Collaboration) Phys. Rev. D **15** 1814 (1977).
[11] C. Zemach, Phys. Rev. **133**, B1201 (1964).
[12] C. Dionisi *et al.*, Nucl. Phys. B **169**, 1 (1980).
[13] A. P. Szczepaniak, M.R. Pennington, Phys. Lett. B **737**, 283 (2014).

EPJ Web of Conferences **181**, 01022 (2018) https://doi.org/10.1051/epjconf/201818101022
EXA2017

Exotic and Conventional Bottomonium Physics Prospects at Belle II

Todd K. Pedlar[1,*]

[1]Luther College Department of Physics
700 College Drive
Decorah, Iowa 52101 USA

Abstract. The Belle II experiment, being constructed at the KEK laboratory in Japan, represents a substantial upgrade of both the Belle detector and the KEKB accelerator. It is expected that Belle II will collect 50 times more data than existing B-Factory samples beginning in 2019. Belle II is uniquely capable of studying the so-called "XYZ" particles: heavy exotic hadrons consisting of more than three quarks. First discovered by Belle, these now number in the dozens, and represent the emergence of a new category within Quantum Chromodynamics. In this article we summarize the capabilities of Belle II to explore both exotic and conventional bottomonium physics, with a particular focus on the physics reach of the first data, where opportunities exist to make an immediate impact in this area.

1 Introduction

While the Belle experiment's principal focus has been on the observation of CP violation and rare decays of B mesons, it has since its inception been the source of a significant number of important observations in both conventional and exotic quarkonia. These results include several first observations, and a number of measurements which are the most precise to date of several conventional bottomonium resonance parameters and branching fractions. The results of our study have also left a number of open questions which Belle II uniquely can answer. Here at the end of 2017, we are now are rapidly approaching the beginning of the Belle II experiment. Currently, the experiment is between two periods of phased commissioning of both detector and accelerator, the second of which is set to begin in early 2018, with full physics data taking with the completed Belle II detector planned for early 2019.

In what follows, we will introduce briefly the changes made between Belle I and Belle II, and upgrades to the KEKB accelerator which has now become SuperKEKB. We then discuss a number of observations made by the Belle Collaboration, and use them as a foundation from which to point out the prospects for new studies of both conventional and exotic bottomonia which are planned for Belle II.

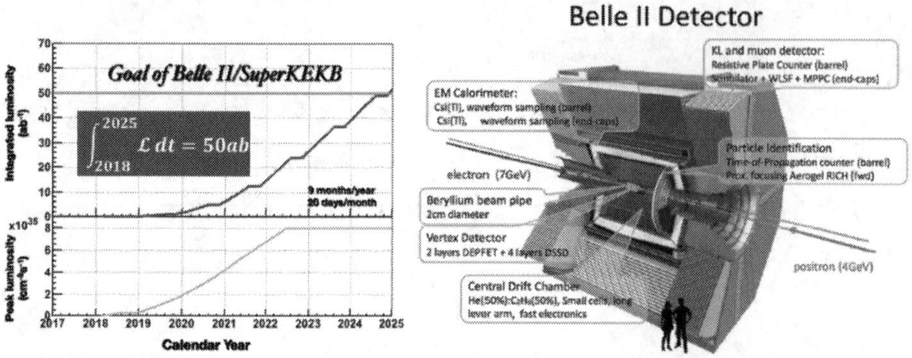

Figure 1. (left) Current luminosity estimates for Belle II. (right) Schematic of the Belle II detector.

2 The Belle II Experiment

2.1 The SuperKEKB Accelerator

The KEKB accelerator, which provided e^+e^- collisions for the Belle Experiment, produced the highest instantaneous luminosity seen in any experiment, reaching a peak value of $2.11 \times 10^{34} cm^{-2} s^{-1}$, and enabling Belle to integrate nearly 1 ab^{-1} over its running period, mostly collecting data at $\Upsilon(4S)$, but also collecting the world's largest samples to date at the $\Upsilon(1S)$ and $\Upsilon(2S)$ resonances, and in the $\Upsilon(5S)$ and $\Upsilon(6S)$ regions.

With upgrades to the accelerator[1], chiefly those which enable SuperKEKB to utilize the nanobeam scheme pioneered by P. Raimondi[2], it is expected that instantaneous luminosities for Belle II will be larger by approximately a factor of 40. Over the course of the running period planned for Belle II, a target of approximately 50 ab^{-1} in total is expected to be integrated (See Fig. 1).

2.2 The Belle II Detector

In order to take advantage of the much increased luminosity afforded by SuperKEKB, we have upgraded the entire Belle detector system. The new detector sits inside the same superconducting solenoid magnet with the iron return yoke, and all components have either been replaced with different detectors or considerably upgraded. The complete Belle II detector system is discussed in full in the Belle II Technical Design Report [3].

One of the most significant impacts of the greater instantaneous luminosity Belle II will see is substantially increased background levels, which will lead to an increase in occupancy and radiation damage, fake hits and pile-up noise in the electromagnetic calorimeter, and neutron induced hits in the muon detection system. New readout electronics for most of the detector and a new trigger system have been introduced in order to compensate for the higher backgrounds. In addition, much improved hadron identification systems have been built, and, as the previous Belle detector, cover nearly the full 4π steradians.

2.3 Early Belle II Run Plan

The commissioning phases of Belle II prior to commencement of physics data runs are denoted *Phase I* and *Phase II*. Phase I concluded in 2017, and involved only the commissioning

*e-mail: pedlto01@luther.edu

Table 1. Summary of data samples collected by CLEO, Belle, and BaBar, together with potential samples collected by Belle II. The numbers reported are luminosities in fb^{-1}, and the number of decays (in millions) of the produced bottomonium state is given in the parentheses adjacent to the reported luminosity.

Expt	Υ(1S)	Υ(2S)	Υ(3S)	Υ(4S)	Υ(5S)	Υ(6S)
CLEO	1.2(21)	1.2(10)	1.2 (5)	16(17.1)	0.1(0.4)	-
BaBar	-	14(99)	30(122)	433(471)	R_b scan	R_b scan
Belle	6(102)	25(158)	3(12)	771(772)	121(36)	5.5
Belle II	-	-	300(1200)	50000(54000)	2000 (600)	100+400(scan)

of SuperKEKB running in single-beam mode, and a specially-built commissioning detector known as BEAST II [4]. Phase II running is expected to commence in January, 2018, and run through summer 2018, and will involve the commissioning of the fully instrumented complement of outer detectors and a partially instrumented inner detector system. During Phase II, the SuperKEKB final focusing magnets will be commissioned and the accelerator will move from single-beam to colliding beam operations for the first time. In this phase, a special configuration of the Beast II commissioning detectors, which will include partial instrumentation of the vertex region by pixel and silicon vertex components like those in the final Belle II vertex system, will be used to study beam-related backgrounds in colliding beam operations.

It is expected that despite the fact that beam collisions will not be optimal in Phase II, and the full inner detectors are not present, Belle II may be able to produce some data useful for physics analyses. Physics running in earnest in Phase III, however, will not begin until after the inner detector region is fully instrumented following the summer shutdown in 2018, and is expected to begin in early 2019. It is during the early running periods of Belle II, when the instantaneous luminosity of SuperKEKB has yet to reach the design specifications, that uniquely large samples of Υ(3S), Υ(6S), and large fine scans in the Υ(6S) region may be accumulated and useful for bottomonium studies - the primary Υ(4S) data sample and a large Υ(5S) sample that may be collected later will also make possible some of the studies discussed below. A table illustrating potential Belle II data samples is presented as Table 1.

3 Studies of the Exotic $Z_b(10610)$ and $Z_b(10650)$ States

Among the most exciting observations we have made at Belle in bottomonium spectroscopy, in addition to our discovery of the singlet-P h_b(nP) states[5] in Υ(5S) dipion transitions (See Fig. 2), were the manifestly exotic Z_b states lying above open-bottom threshold which are responsible for the large rates of production of the singlet-P states [6]-[8]. The presence of these intermediate states in the transitions $\Upsilon(5S) \to \pi^+\pi^- h_b(nP)$ and $\Upsilon(6S) \to \pi^+\pi^- h_b(nP)$ enables the vector-to-pseudoscalar transitions without suppression which would otherwise occur due to the necessary b–quark spin flip in these transitions.

The $Z_b(10610)$ and $Z_b(10650)$ have masses very close to the thresholds for $B\bar{B}^*$ and $B^*\bar{B}^*$, respectively, which indicates possible molecular structure for these states, or at the very least a very substantial molecular composition in the Z_b wave functions. The fact that these states have been observed principally to decay to these B–meson pair combinations strengthens this conclusion [8]. Furthermore, spin-parity measurement, total widths, and production rates are also consistent with expectations for molecular states. The only possible contraindication for molecular identity is the fact that the Z_b states lie slightly *above* the mass thresholds (though not significantly). Molecular states' masses should be below threshold - although it must be

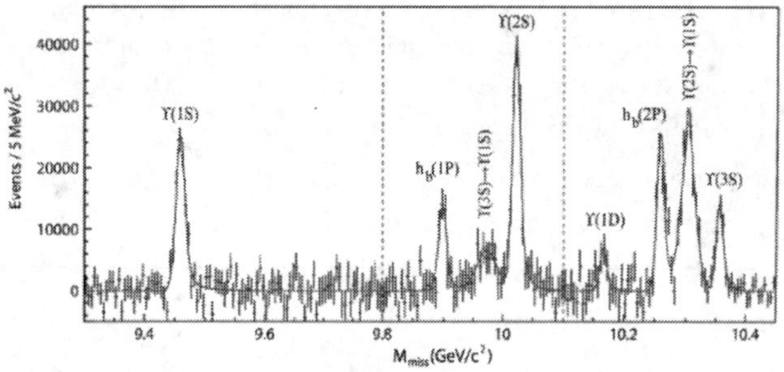

Figure 2. Dipion missing mass spectrum from Belle, for data taken at Υ(5S) [5]. Indicated in this spectrum are transitions both to vector Υ(nS) states, singlet h_b(nP) states, and reflections of dipion transitions among lower Υ(nS) states.

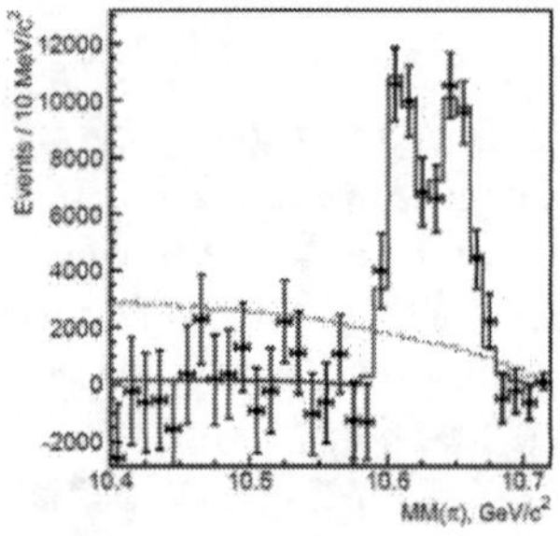

Figure 3. Spectrum of the yield of h_b(1P) versus the single charged pion missing mass from Belle data taken at Υ(5S). The two peaks indicate the production in these transitions of the two charged states Z_b(10610) and Z_b(10650), which saturate the rates for h_b(1P) - that is there is no evidence for continuum three-body process Υ(5S) → $\pi^+\pi^- h_b$(1P).

Table 2. Branching fractions for observed products of Z_b(10610) and Z_b(10650) decays [8]. Branching fractions reported assume that these transitions saturate the total decay rates of the two states.

Channel	Branching Fraction (%)	
	Z_b(10610)	Z_b(10650)
$\Upsilon(1S)\pi^+$	$0.54^{+0.16}_{-0.13}{}^{+0.11}_{-0.08}$	$0.17^{+0.07}_{-0.06}{}^{+0.03}_{-0.02}$
$\Upsilon(2S)\pi^+$	$3.61^{+0.76}_{-0.59}{}^{+0.79}_{-0.53}$	$1.39^{+0.48}_{-0.38}{}^{+0.34}_{-0.23}$
$\Upsilon(3S)\pi^+$	$2.15^{+0.55}_{-0.42}{}^{+0.60}_{-0.48}$	$1.63^{+0.53}_{-0.42}{}^{+0.39}_{-0.28}$
$h_b(1P)\pi^+$	$3.45^{+0.87}_{-0.71}{}^{+0.86}_{-0.63}$	$8.41^{+2.43}_{-2.12}{}^{+1.49}_{-1.06}$
$h_b(2P)\pi^+$	$4.67^{+1.24}_{-1.00}{}^{+1.18}_{-0.89}$	$14.7^{+3.2}_{-2.8}{}^{+2.8}_{-2.3}$
$B^+\bar{B}^{*0} + \bar{B}^0 B^{*+}$	$85.6^{+1.5}_{-2.0}{}^{+1.5}_{-2.1}$	-
$B^{*+}\bar{B}^{*0}$	-	$73.7^{+3.4}_{-4.4}{}^{+2.7}_{-3.5}$

acknowledged that masses of states may shift significantly due to coupled channel effects. Further study of these states is both warranted and eagerly anticipated in Belle II.

While we have learned a great deal about these states in our studies, there is much more to be learned - in particular, we would like to study the Z_b production fractions as a function of energy across the Υ(5S) and Υ(6S) region. Also, while we have some understanding of the principal decay modes of the Z_b states (see Table 2), a healthy 1 – 2 ab^{-1} sample of Υ(5S) expected for Belle II would give the opportunity both to determine whether other channels have significant contribution to the Z_b total widths, and to determine more precisely the various decay branching fractions and angular distributions.

4 Studies of h_b(nP) and η_b(nS) States

The h_b(nP) states were discovered as noted previously in $\pi^+\pi^-$ transitions from Υ(5S), in a sample of 121.4 fb^{-1}. Subsequently transitions Υ(5S) → ηh_b(1P) were also observed and therefore a large sample of Υ(5S), of order ten times the previous size, is something we'd like to accumulate to study this process. As shown in Figure 4, Belle made the observation of

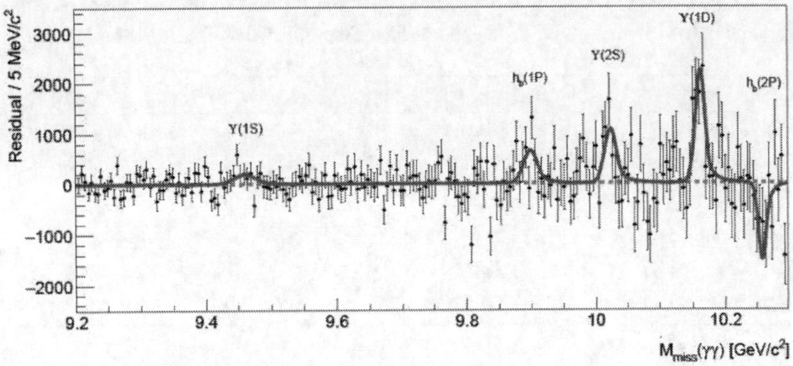

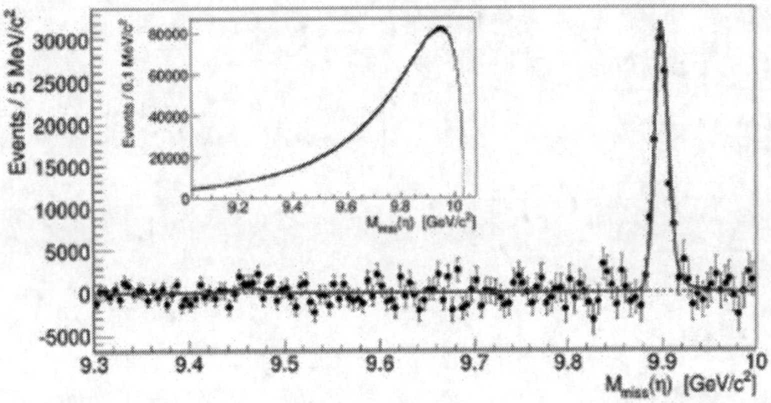

Figure 4. (top) Evidence for η transitions from $\Upsilon(5S)$ seen in η missing mass from the Belle sample of 121.4 fb^{-1}. Peaks are observed for transitions to $\Upsilon(1S)$, $h_b(1P)$, $\Upsilon(2S)$, and $\Upsilon(1D)$. (bottom) Observation of $\Upsilon(4S) \to \eta h_b(1P)$ in the η missing mass spectrum.

a very large branching fraction for $\Upsilon(4S) \to \eta h_b(1P)$, and this offers yet another avenue for accumulation of large $h_b(1P)$ data sample. Naturally, then, there is a reasonable expectation that η transitions from $\Upsilon(6S)$ may also be large enough to provide a useful sample of both $h_b(1P)$ and $h_b(2P)$.

As a result, data samples taken in Belle II at all three of the $\Upsilon(4S)$, $\Upsilon(5S)$, and $\Upsilon(6S)$ states have the potential to serve as a source of $h_b(\mathrm{nP})$, and with such samples, we can study both the decays of $h_b(\mathrm{nP})$ and those of $\eta_b(\mathrm{nS})$ by tagging $h_b(\mathrm{nP})$ in $\pi^+\pi^-$ or η transitions, and subsequently $\eta_b(\mathrm{nS})$ in hard radiative decays of higher $h_b(\mathrm{nP})$ which decay prolifically to the $\eta_b(\mathrm{nS})$. The branching fractions of $\Upsilon(4S)$ and $\Upsilon(5S)$ to produce $h_b(\mathrm{nP})$ are given in Table 3, and using them, and efficiencies from Belle II Monte Carlo study, we can expect usable samples (assuming 50 ab^{-1} at $\Upsilon(4S)$ and 2 ab^{-1} at $\Upsilon(5S)$) of millions of $h_b(\mathrm{nP})$ and $\eta_b(1S)$, and a few hundred thousand $\eta_b(2S)$. With such samples, we will have the opportunity to make precise measurements of both the lineshape for the M1 photons which are produced in the transitions, the $\eta_b(\mathrm{nS})$ masses and widths, and potentially some important decay branching fractions (*e.g.* $\eta_b(1S) \to \gamma\gamma$, which is of fundamental importance as it is related directly to the $b\bar{b}$ wave function at the origin, and $\eta_b(2S) \to \pi^+\pi^-\eta_b(1S)$).

It is likely that in Belle II, additionally, that a sample of order 300 fb^{-1} will be taken at $\Upsilon(3S)$. As such, we can expect with such a sample to have another 600k $\eta_b(1S)$ for study

Table 3. Known branching fractions and expected yields in Belle II for the production of the singlet states h_b(nP) and η_b(nS) in decays of Υ(4S) and Υ(5S), assuming 50 ab^{-1} at Υ(4S) and 2 ab^{-1} at Υ(5S).

Process	Branching Fraction (%)	expected observed yield ($\times 10^6$)
$\Upsilon(4S) \to \eta h_b(1P)$	0.183 ± 0.023	4
$\Upsilon(5S) \to \pi^+\pi^- h_b(1P)$	0.35 ± 0.11	8
$\Upsilon(5S) \to \pi^+\pi^- h_b(2P)$	$0.60 \pm 0.20)$	1.4
$h_b(1P) \to \gamma\eta_b(1S)$	$49.2 \pm 5.7^{+5.6}_{-3.3}$	3.7
$h_b(2P) \to \gamma\eta_b(1S)$	$22.3 \pm 3.8^{+3.1}_{-3.3}$	0.3
$h_b(2P) \to \gamma\eta_b(2S)$	$47.5 \pm 10.5^{+6.8}_{-7.7}$	0.6

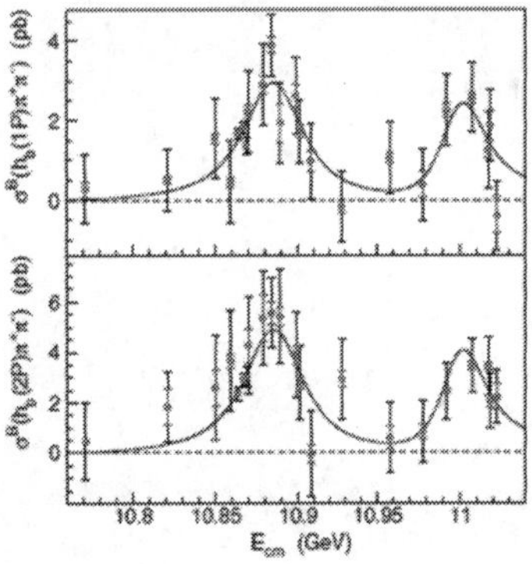

Figure 5. h_b(nP) cross section as a function of center of mass energy, showing the resonance curves of both Υ(5S) and Υ(6S).

of η_b(1S). One of the more important reasons to look particularly at this sample is the difference in mass measurements for η_b(1S) that persists between the analyses of M1 transitions $\Upsilon(nS) \to \gamma\eta_b(1S)$ by BaBar [9] and CLEO [10], and the analysis of E1 transitions $h_b(nP) \to \gamma\eta_b(1S)$ by Belle [11].

5 Studies of Transitions from the $\Upsilon(6S)$

While the h_b(nP) states can be studied in significant detail in large data samples Belle II expects to accumulate at Υ(4S) and Υ(5S), it is also likely that we can add to our knowledge of these states (and possibly the 3S singlet states) by studying transitions from Υ(6S) as well. In Belle's small data sample of a 5.6 fb^{-1} in the vicinity of Υ(6S), transitions to both $\pi^+\pi^-\Upsilon(1S)$ and $\pi^+\pi^- h_b(nP)$ were observed [12]. The yield of h_b(nP) states is shown in Fig. 5. While production of both h_b(nP) states had been previously established in data taken at the Υ(5S) resonance [5], this study revealed 3.5σ and 5.3σ signals for h_b(1P) and h_b(2P), respectively at the Υ(6S). Furthermore, the Υ(5S) also has been observed to decay to $\eta h_b(1P)$, so a search for similar transitions to both h_b(nP) states would prove interesting. For these reasons, then, any appreciable data sample taken at Υ(6S), can serve as an additional and revealing source of h_b(nP) decays.

What is of more interest, however, is the resonant substructure in these transitions from Υ(6S) as compared to transitions from Υ(5S). In Belle, the resonant substructure of these transitions was investigated, and, as shown in Fig. 6, one very interesting fact stands out: while transitions through at least one of the charged Z_b states can be established, it cannot be definitively stated whether such transitions take place through both (rather than just one) state.

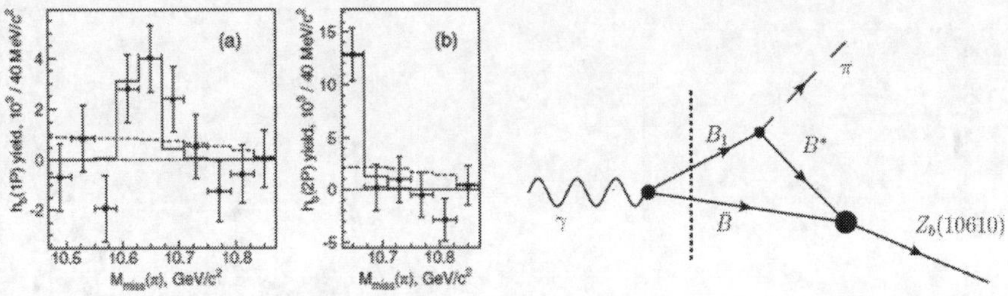

Figure 6. (left) Evidence for Z_b produced as intermediate states in $\Upsilon(6S) \to \pi^+\pi^- h_b(nP)$ transitions. The statistics are insufficient to determine the relative population of the two Z_b states in these decays. (right) Potential decay diagram for $\Upsilon(6S)$ if its wave function has a sizeable $B_1 B$ component, as suggested in Ref. [14].

The $\Upsilon(6S)$, like the other vector states above open bottom threshold, is almost certainly not a simple $b\bar{b}$ bound state; at the very least, coupled channel considerations play a large role in its decays. Along these lines, a very interesting comparison that has been suggested is between the $\Upsilon(6S)$ and the $Y(4260)$. Both these states transition to lower narrow heavy quarkonium states by dipion transitions, and do so through charged quarkonium-like states. The mass of $\Upsilon(6S)$ is very near the threshold for the production of a $B_1 B$ pair, just as the $Y(4260)$ is near the threshold for the charm analog, $D_1 D$. Because of its unusual properties, and the proximity to the $D_1 D$ threshold, the $Y(4260)$ has been interpreted as a possible molecule [13] - and the $\Upsilon(6S)$ might likewise be interpreted. One important indicator for the possibility of $\Upsilon(6S)$ as a $B_1 B$ molecule would be the exclusive transition through the lower Z_b state, $Z_b(10610)$ [14] via the diagram in Fig.6. As already noted, the current small data sample near $\Upsilon(6S)$ is insufficient to rule out or require $Z_b(10610)$ dominance in its transitions to lower bottomonia. More data then are needed to settle this very interesting question.

A data sample of $\sim 20\,\text{fb}^{-1}$ at the peak of $\Upsilon(6S)$ would enable more detailed studies of the $\Upsilon(6S) \to \pi^{\pm} Z_b^{\mp}$ process and determine the question of relative production of the $Z_b(10650)$ and $Z_b(10610)$. Such a sample at $\Upsilon(6S)$ would also make possible searches for expected Z_b partners [15] in the decays $\Upsilon(6S) \to \gamma W_b$ and possibly $\Upsilon(6S) \to \pi^+\pi^- W_b$. These transitions would allow searches for transitions to lower narrow bottomonia by analogy to our observations of states in $\Upsilon(5S)$ decays, including possibly some of the as-yet observed narrow states $\Upsilon(2D)$, $h_b(1P)(3P)$, and $\Upsilon(1F)$. Further, a fine grained search for $\pi^+\pi^-\Upsilon(nS)$ with high statistics (say, $1 - 2\,\text{fb}^{-1}$ per data point) in the region above 10.7 GeV could potentially add insight as the thresholds for production of the Z_b and W_b states are crossed.

6 Studies of the $\Upsilon(1D, 2D)$ States

Another of the narrow sub-threshold states of bottomonium that can be studied with Belle II data is the $\Upsilon(1D)$ and $\Upsilon(2D)$ states - knowledge of these L=2 D-wave states of bottomonium is relatively limited. The triplet of 1D states (having total angular momentum $J = 1, 2$, and 3 has thus far only been observed as an undifferentiated conglomerate in experimental papers from CLEO [16] in four-photon cascades in data taken at $\Upsilon(3S)$ and BaBar [17] in their $\Upsilon(3S)$ data sample by reconstruction of the $\Upsilon(1D) \to \pi^+\pi^-\Upsilon(1S) \to \pi^+\pi^-\ell^+\ell^-$ final state (See Fig. 7). The $\Upsilon(2D)$ levels are completely unknown at present.

In data taken at Belle, we have seen an indication (though not statistically significant) of the process $\Upsilon(5S) \to \pi^+\pi^-\Upsilon(1D)$ in our $h_b(nP)$ discovery paper [5], and most recently we have discovered the transition $\Upsilon(5S) \to \eta\Upsilon(1D)$ (See Fig. 4) but again in these cases, we only

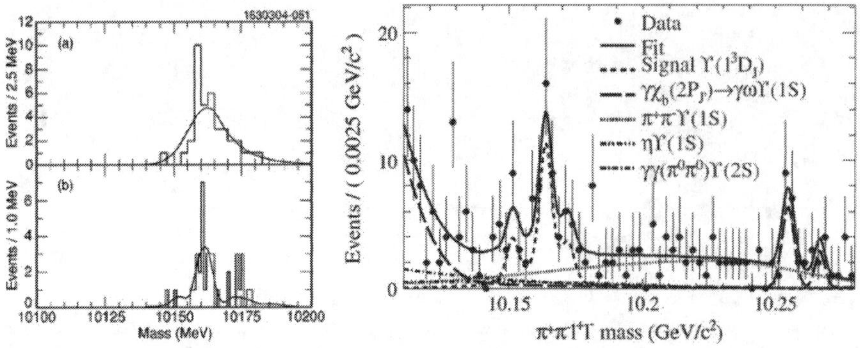

Figure 7. Observation of the ϒ(1D) states by (left) CLEO [16] and (right) BaBar [17].

have a broad observation of a state or states at the expected ϒ(1D) mass. For the purposes of understanding these high angular momentum states, it will be essential that we have the ability to differentiate the members of the triplet.

We also would like to observe decays of the ϒ(1D) states in addition to the radiative transitions to lower $\chi_{bJ}(nP)$ states that have already been observed. Examples of the most likely transitions include $\Upsilon(1D) \to \pi^+\pi^-\Upsilon(1S)$ or $\eta\Upsilon(1S)$. The much larger data samples of Belle II should allow us a good opportunity both to isolate one or more of the ϒ(1D) states from the others and possibly to observe the $\Upsilon(2D)$ triplet expected to lie near $10.45\mathrm{GeV}/c^2$.

The BaBar observation [17] suggests the possibility of resolution of the three states with an ϒ(3S) data sample approaching $250 - 300\,\mathrm{fb}^{-1}$, which we expect to collect sometime early in Phase III of Belle II operations. Furthermore the $J^{PC} = 1^{--}$ member of the ϒ(1D) and $\Upsilon(2D)$ triplets could also be reached directly by scanning the e^+e^- beams across each resonance. The beauty of this potential search is that by it, both the mass and possibly the full width of the 1^{--} members of the triplets can be established. With predicted Γ_{ee} partial widths of order $1eV$, a $2\,\mathrm{fb}^{-1}$ per point scan would result in a signal of greater than 5σ significance.

7 Conclusions

In addition to its ambitious plans in CP violation and searches for extremely rare decays of B mesons which will shed light on new physics, the Belle II experiment has the opportunity and indeed excellent prospects to shed light on the nature of strongly interacting systems such as conventional states of bottomonium and the exotic bottomonium-like states recently discovered.

Acknowledgments

The author thanks the organizers of the 6th International Conference on Exotic Atoms and Related Topics for the invitation to speak, his Belle Collaboration colleagues for their various contributions of analyses presented, and the National Science Foundation for their support of this work through Grant #1506412.

References

[1] Y. Ohnishi et al., "Accelerator design at SuperKEKB," PTEP 2013, 03A011 (2013)

[2] M. Baszczyk et al. (SuperB Collaboration), INFN-13-01-PI, LAL-13-01, SLAC-R-1003, arXiv:1306.5655.
[3] T. Abe *et al.* (Belle Collab.) 2, KEK-REPORT-2010-1, arXiv:1011.0352.
[4] M. Nayak *et al.* , PoS BEAUTY **2016** 059 (2016).
[5] A. Bondar *et al.* (Belle Collab.) ,Phys. Rev. Lett. **108**, 122001 (2001).
[6] I. Adachi *et al.* (Belle Collab.) , Phys. Rev. Lett. **108**, 032001 (2001).
[7] P. Krokovny *et al.* (Belle Collab.) , Phys. Rev. D **88**, 052016 (2013).
[8] A. Garmash *et al.* (Belle Collab.) , Phys. Rev. Lett. **116**, 212001 (2016).
[9] B. Aubert *et al.* (BaBar Collab.) , Phys. Rev. Lett. **101**, 071801 (2008); Phys. Rev. Lett. **103**, 161801 (2009).
[10] G. Bonvicini *et al.* (CLEO Collab.) , Phys. Rev. D **81**, 031104(R) (2010).
[11] R. Mizuk *et al.* (Belle Collab.) Phys. Rev. Lett. **109**, 232002 (2012).
[12] R. Mizuk *et al.* (Belle Collab.) Phys. Rev. Lett. **117**, 142001 (2016).
[13] Q. Wang, C. Hanhart, and Q. Zhao, Phys. Rev. Lett. **111**, 132003 (2013).
[14] M. Voloshin, B2TiP Pittsburgh 2016.
[15] A. E. Bondar *et al.* Phys. Rev. D **84**, 054010 (2011).
[16] G. Bonvicini *et al.* (CLEO Collab.) , Phys. Rev. D **70**, 032001 (2004).
[17] P. del Amo Sanchez *et al.* (BaBar Collab.) , Phys. Rev. D **82**, 11102 (2010).

EPJ Web of Conferences **181**, 01023 (2018) https://doi.org/10.1051/epjconf/201818101023
EXA2017

Results on b hadron properties in CMS

Thomas Madlener[1,⋆] on behalf of the CMS collaboration

[1] *Institute of High Energy Physics (HEPHY), Vienna*

Abstract. The lifetime is among the fundamental properties of particles and is one of the important observables in heavy hadrons allowing to test the theoretical tools describing their physics. This report presents lifetime measurements of B^0, B_s^0, Λ_b^0 and B_c^+ hadrons performed on data, corresponding to 19.7 fb^{-1}, collected by the CMS experiment at the LHC at a center-of-mass energy of $\sqrt{s} = 8$ TeV. The B^0 lifetime is measured to be $453.0 \pm 1.6\,(\text{stat}) \pm 1.5\,(\text{syst})\,\mu\text{m}$ in $J/\psi K_S^0$ and $457.8 \pm 2.7\,(\text{stat}) \pm 2.7\,(\text{syst})\,\mu\text{m}$ in $J/\psi K^*(892)^0$. The B_s^0 meson lifetime is measured in two decay modes with differing contributions from the heavy and light eigenstates, resulting in two different measured lifetimes: $504.3 \pm 10.5\,(\text{stat}) \pm 3.7\,(\text{syst})\,\mu\text{m}$ in $J/\psi\pi^+\pi^-$ and $443.9 \pm 2.0\,(\text{stat}) \pm 1.2\,(\text{syst})\,\mu\text{m}$ in $J/\psi\phi(1020)$ respectively. The lifetime of the Λ_b^0 hadron is found to be $443.1 \pm 8.2\,(\text{stat}) \pm 2.7\,(\text{syst})\,\mu\text{m}$. The B_c^+ lifetime is measured with respect to the B^+ lifetime and is found to be $162.3 \pm 8.2\,(\text{stat}) \pm 4.7\,(\text{syst}) \pm 0.1\,(\tau_{B^+})\,\mu\text{m}$.

1 Introduction

Precise lifetime measurements of heavy hadrons play an important role in the study of non-perturbative Quantum Chromodynamics (QCD). The QCD inspired Heavy Quark Expansion (HQE) model describes the weak decays of hadrons containing a heavy quark and provides accurate estimates of the ratio of their lifetimes [1]. The large production rate of b hadrons at the LHC and the detection capabilities of the CMS detector allow to make precise b hadron lifetime measurements, which in turn make it possible to test these predictions.

In this report the measurement of the B^0, B_s^0, B_c^+ and Λ_b^0 hadrons lifetimes at the center-of-mass energy of $\sqrt{s} = 8$ TeV by the CMS collaboration [2] are discussed.

The lifetime measurement is based on the reconstruction of the transverse decay length L_{xy}, where $\vec{L}_{xy}$ is the flight distance vector from the primary vertex to the decay vertex of the b hadron, projected onto the transverse momentum $\vec{p}_T$ of the b hadron momentum. The proper decay time times the speed of light c is measured via

$$ct = cL_{xy}\frac{M}{p_T}, \tag{1}$$

where M is the world average value of the b hadron mass [3].

⋆e-mail: thomas.madlener@cern.ch

2 Dataset and event selection

The data used in the analysis has been been collected by the CMS experiment at the LHC in 2012 in proton-proton collisions at a center-of-mass energy of $\sqrt{s} = 8\,\text{TeV}$ and corresponds to an integrated luminosity of 19.7 fb^{-1}.

The analysis is based on a data sample collected with a trigger requiring two oppositely charged muons within the pseudorapidity region $|\eta| < 2.2$. The dimuon is requested to have a transverse momentum, p_T, greater than 7.9 GeV and to have an invariant mass $2.8 < M(\mu^+\mu^-) < 3.35\,\text{GeV}$. Both muons must have a distance of closest approach to the event vertex in the transverse plane smaller than 0.5 cm. A fit to a common vertex must have a χ^2 probability greater than 0.5 %. The offline selection criteria of the dimuon system replicate the trigger selection or impose slightly stricter requirements, such as the invariant mass being within 150 MeV of the world-average value [3] unless stated otherwise.

Muons are identified using the standard CMS muon reconstruction procedure [4], which requires multiple hits in the detector with a consistent trajectory. Tracks are required to fulfill the standard CMS high-purity requirements [5] and have p_T greater than 0.5 GeV.

3 Reconstruction of b hadrons

The offline selection starts by reconstructing J/ψ candidates from pairs of oppositely charged muons. Candidate b hadrons are built by combining a J/ψ candidate with reconstructed neutral particles or track(s) depending on the decay mode under consideration. Intermediate neutral particles are reconstructed by combining pairs of oppositely charged tracks with appropriately assigned masses. A candidate b hadron is fitted to a common vertex constraining the dimuon mass to the world-average J/ψ mass [3]. In fits including a K^0_S or Λ^0 hadron the world-average mass values are used for these particles. Primary Vertices (PV) are fitted from the reconstructed tracks in the event using the beamspot - the estimated proton-proton interaction region - as a constraint. The PV with the smallest pointing angle is used, where the pointing angle is defined as the angle between the reconstructed b hadron momentum and the vector joining the PV and the decay vertex.

3.1 Reconstruction of B^0, B^0_s and Λ^0_b hadrons

The B^0, B^0_s and Λ^0_b hadrons are reconstructed in the following decay channels:

$$
\begin{aligned}
B^0 &\to J/\psi K^0_S &&\to \mu^+\mu^-\pi^+\pi^- \\
B^0 &\to J/\psi K^*(892)^0 &&\to \mu^+\mu^- K^\pm\pi^\mp \\
B^0_s &\to J/\psi\phi(1020) &&\to \mu^+\mu^- K^+K^- \\
B^0_s &\to J/\psi\pi^+\pi^- &&\to \mu^+\mu^-\pi^+\pi^- \\
\Lambda^0_b &\to J/\psi\Lambda^0 &&\to \mu^+\mu^- p\pi^-
\end{aligned}
$$

The B^0, B^0_s and Λ^0_b hadrons must have $p_T > 13$ GeV, except for the $B^0_s \to J/\psi\phi(1020)$ channel, where no such requirement is imposed. The vertex fit χ^2 probability has to be greater than 0.1 % for the latter decay only. For all channels the lifetime measurement is restricted to events for which the ct of the b hadron is larger than 0.02 cm to avoid reconstruction and resolution effects in the low-ct region.

3.2 Reconstruction of $B_c^+ \to J/\psi\pi^+$ and $B^+ \to J/\psi K^+$

Both decays are reconstructed using the same algorithm and selection criteria and as a result the dependence of the efficiencies on the proper decay time is similar.

The dimuon invariant mass must lie in the range $\pm 3\sigma$ from the nominal J/ψ mass, where σ is the average resolution for the J/ψ signal depending on the J/ψ pseudorapidity. The vertex χ^2 probability for the dimuon has to be greater than 1 % and the b hadron vertex χ^2 probability has to greater then 5 %. The p_T of the b hadron and the charged track have to be greater than 10 GeV and 3.3 GeV, respectively. The lifetime measurement is limited to $ct > 0.01$ cm to ensure a constant ratio of the meson efficiencies. Additionally, the angle $\cos\theta$ is required to be greater than 0.98, where $\cos\theta = \vec{L}_{xy} \cdot \vec{p}_{T,B}/(|\vec{L}_{xy}| \cdot |\vec{p}_{T,B}|)$ and $\vec{L}_{xy}$ and $\vec{p}_{T,B}$ refer to the transverse decay length and transverse momentum of the B^+ and B_c^+ mesons.

4 B^0, B_s^0 and Λ_b^0 lifetime measurements

The neutral B_q^0 decay rate is characterized by the average decay width, $\Gamma_q = (\Gamma_L^q + \Gamma_H^q)/2$, and the decay width difference, $\Delta\Gamma_q = \Gamma_L^q - \Gamma_H^q$, where $\Gamma_{L,H}^q$ are the decay widths of the light (L) and heavy (H) mass eigenstates. The B^0 system has a small lifetime difference compared to the average lifetime $\frac{\Delta\Gamma_d}{\Gamma_d} = (-0.3 \pm 1.5)\%$ [3], and thus the the ct distribution can be treated as an exponential. However, in the B_s^0 system, $\frac{\Delta\Gamma_s}{\Gamma_s} = (12.4 \pm 1.1)\%$ [3] and the deviation from an exponential has to be considered for the measurement.

Two lifetimes associated to the B_s^0 are measured in the decay channels $J/\psi\pi^+\pi^-$ and $J/\psi\phi(1020)$. The latter final state is an admixture of two CP-even and one CP-odd state and can be used to measure the effective lifetime of the B_s^0 system. This measurement is complementary to the weak mixing phase analysis in the same channel [6]. The $J/\psi\pi^+\pi^-$ final state is reconstructed in the mass window $0.9240 < M(\pi^+\pi^-) < 1.204$ GeV, dominated by the $f_0(980)$ resonance [7, 8]. Thus, this is a CP-odd final state and the measured lifetime is related to the heavy B_s^0 mass eigenstate, $\tau_{B_s^0}^{\text{CP-odd}} \approx 1/\Gamma_H$.

4.1 Data modelling

For all three hadrons a three-dimensional unbinned maximum-likelihood fit using the b hadron mass, ct and the per-event ct uncertainty, σ_{ct}, is performed. For the B_s^0 modes an extended unbinned maximum-likelihood fit is used.

The invariant mass distribution for the signal is modelled with a sum of two Gaussian functions, except for the B_s^0, where one Gaussian is sufficient for describing the data. The combinatorial background is modelled by a linear polynomial or an exponential in the case of the $B_s^0 \to J/\psi\phi(1020)$. For the $B_s^0 \to J/\psi\pi^+\pi^-$ decay additional terms are added for describing backgrounds from partially reconstructed or misidentified b hadron decays: The decays $B^0 \to J/\psi\pi^+\pi^-$ and $B_{(s)}^0 \to J/\psi h_1^+ h_2^-$, where h_1^+ and h_2^- are charged hadron tracks that are not both pions, are modelled by two separate Gaussian functions while the shape of the $B^+ \to J/\psi K^+$ are taken from simulations.

The ct signal distribution is modelled by an exponential function convolved with the detector resolution and then multiplied by an efficiency function. The backgrounds are described by a superposition of exponential functions likewise convolved with the resolution. The resolution is described by a Gaussian function with the event-per-event width taken from the ct uncertainty distribution and are in the order of some μm.

The signal and background σ_{ct} distributions are modelled with a Gaussian function convolved with either two exponential functions or the sum of two gamma functions.

The invariant mass and ct distributions obtained from data together with the superimposed fit results are shown in Figs. 1–3

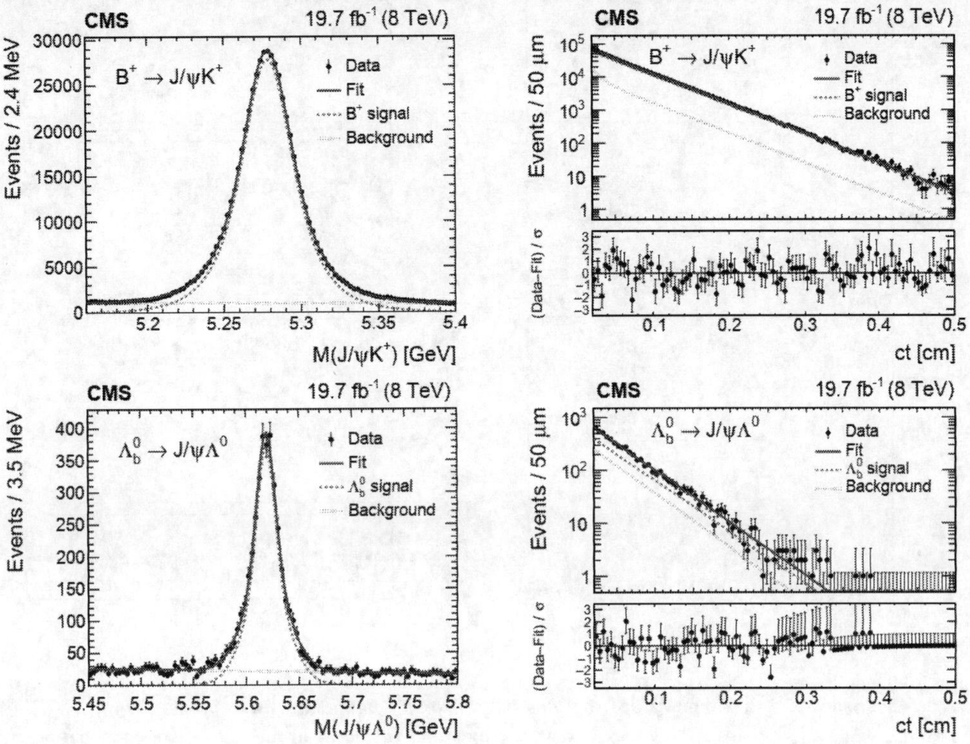

Figure 1. Invariant mass (left) and ct distributions (right) for B^+ (top) and Λ_b^0 candidates (bottom) together with projections of the fit to the data. The contributions from signal (dashed), background (dotted) and the sum of the two (solid) are shown. The bottom panels of the right figures show the differences between the observed data and the fit divided by the data uncertainty [2].

4.2 Efficiency

To determine the reconstruction and selection efficiency as a function of ct, fully simulated MC samples are used for each decay mode. The efficiency is defined as the generated ct distribution of the selected events after reconstruction divided by the ct distribution obtained from an exponential decay with the lifetime set to the value used in the generation of the events. For the $B_s^0 \to J/\psi\phi(1020)$ channel the sum of the two exponentials generated with the theoretical $B_s^0 \to J/\psi\phi(1020)$ decay rate model [9] is used as the denominator. The efficiencies for the different decay modes are shown in Fig. 4.

5 B_c^+ lifetime measurements

In the current B_c^+ decay model different processes contribute to the decay width. The weak decay can either occur through the b or c quark decaying with the other quark as spectator or through an annihilation process, which is predicted to contribute up to 10 % of the decay width [10]. Thus lifetime

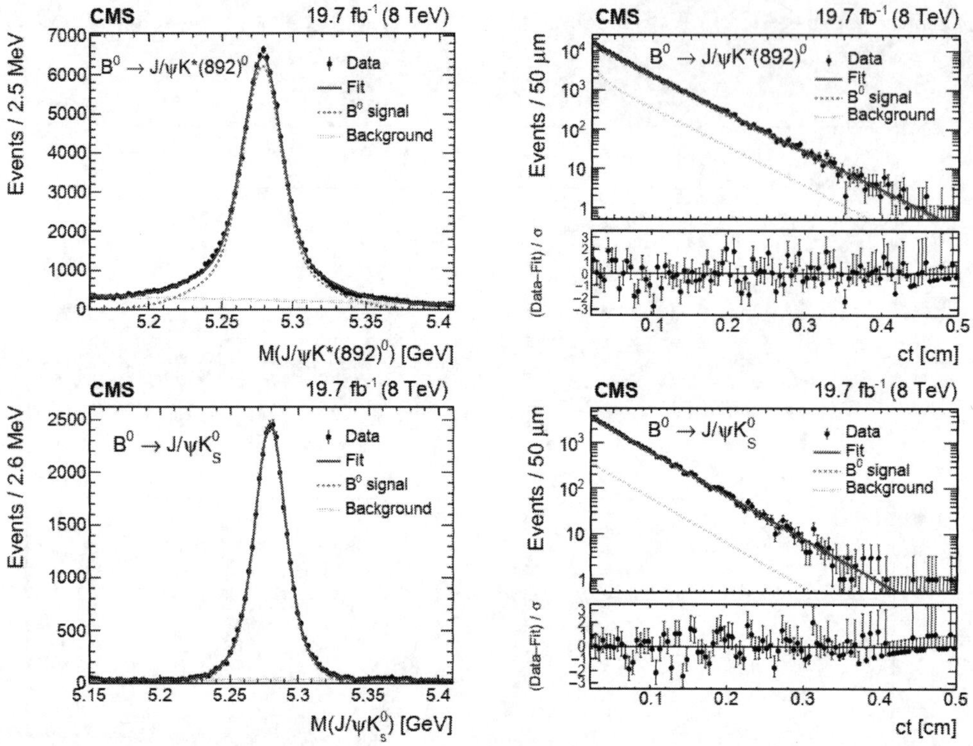

Figure 2. Invariant mass (left) and ct distributions (right) for B^0 candidates reconstructed from $J/\psi K^*(892)^0$ decays (top) and $J/\psi K_S^0$ decays (right) together with projections of the fit to the data. The contributions from signal (dashed), background (dotted) and the sum of the two (solid) are shown. The bottom panels of the right figures show the differences between the observed data and the fit divided by the data uncertainty [2].

measurements can be used to test the B_c^+ decay model. Additionally, recent lifetime measurements reported by the LHCb collaboration [11, 12] are significantly larger than those measured at the Tevatron [13–15]. A measurement by CMS may help to resolve this ambiguity in experimental results.

The B_c^+ lifetime measurement uses the method developed by the LHCb collaboration [11, 12], which exploits the precise knowledge of the B^+ lifetime [3] to obtain the B_c^+ lifetime by measuring the difference in total width between the B_c^+ and the B^+,

$$\Delta\Gamma \equiv \Gamma_{B_c^+} - \Gamma_{B^+} = \frac{1}{\tau_{B_c^+}} - \frac{1}{\tau_{B^+}}. \tag{2}$$

The signal decay time distribution $N_B(ct)$ can be expressed as the product of an exponential decay function $E(ct|\tau_B) = \exp(-ct/\tau_B)$ convolved with the time resolution function of the detector $R(ct)$ and the efficiency function $\epsilon_B(ct)$. The ratio of B_c^+ to B^+ events at any given proper time then becomes

$$\frac{N_{B_c^+}(ct)}{N_{B^+}(ct)} = \mathcal{R}(ct) = \frac{[E(ct|\tau_{B_c^+}) \otimes R(ct)]\epsilon_{B_c^+}(ct)}{[E(ct|\tau_{B^+}) \otimes R(ct)]\epsilon_{B^+}(ct)}. \tag{3}$$

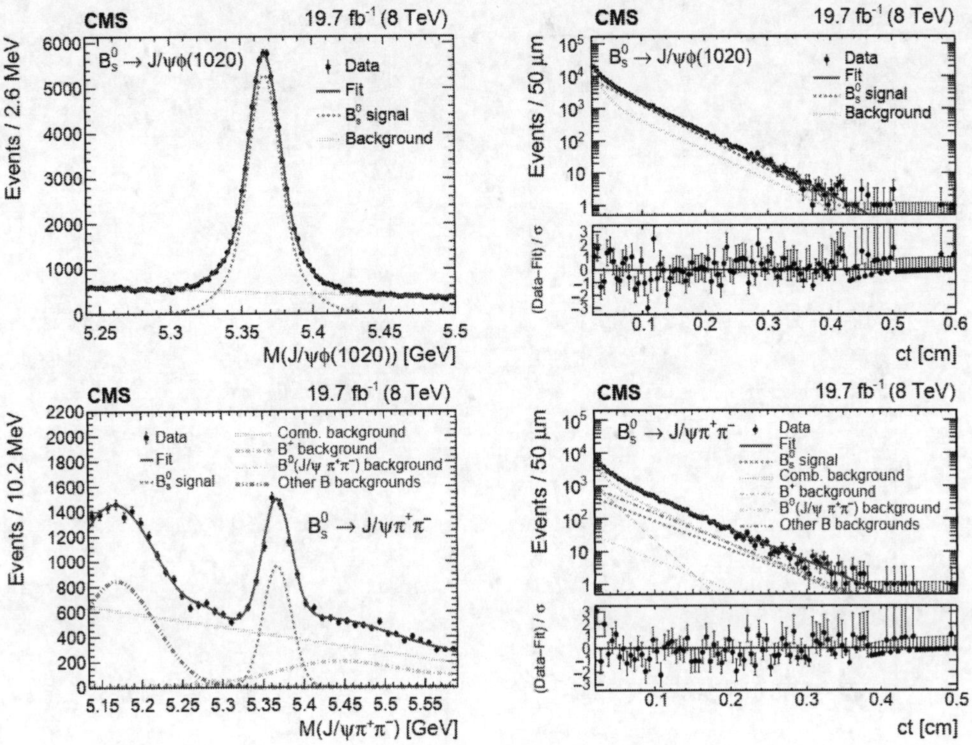

Figure 3. Invariant mass (left) and *ct* distributions (right) for B_s^0 candidates reconstructed from $J/\psi\phi(1020)$ decays (top) and from $J/\psi\pi^+\pi^-$ decays (bottom) together with projections of the fit to the data. The full fit function (solid) and the signal (dashed) are shown for both decays. The total background (dotted) is shown for the top plots. For the bottom plots the combinatorial background (dotted), misidentified $B^+ \rightarrow J/\psi K^+$ (dash-dotted), $B^0 \rightarrow J/\psi\pi^+\pi^-$ (dashed-dotted-dotted-dotted), and partially reconstructed and misidentified B contributions (dashed-dotted-dotted) are shown separately. The bottom panels of the right figures show the differences between the observed data and the fit divided by the data uncertainty [2].

Studies using simulated pseudo events have been used to verify that Eq. 3 is not significantly affected by the time resolution and thus it can be simplified to

$$\mathcal{R}(ct) = R_\epsilon(ct)\exp(-\Delta\Gamma t), \tag{4}$$

where residual effects of the time resolution are absorbed into the ratio of the efficiency functions $R_\epsilon(ct)$, evaluated from MC simulations.

5.1 Fit model and results

To extract the B_c^+ lifetime a binned chi-square fit to the ratio of the efficiency-corrected *ct* distribution of the $B_c^+ \rightarrow J/\psi\pi^+$ and $B^+ \rightarrow J/\psi K^+$ channels is used. The B_c^+ and B^+ signal *ct* distributions from data are obtained by dividing the data sample in different *ct* bins. In each bin an unbinned maximum-likelihood fit to the $J/\psi\pi^+$ and $J/\psi K^+$ invariant mass distribution is performed, where the peak position and resolution is fixed to the values obtained from fits to the full samples, which are

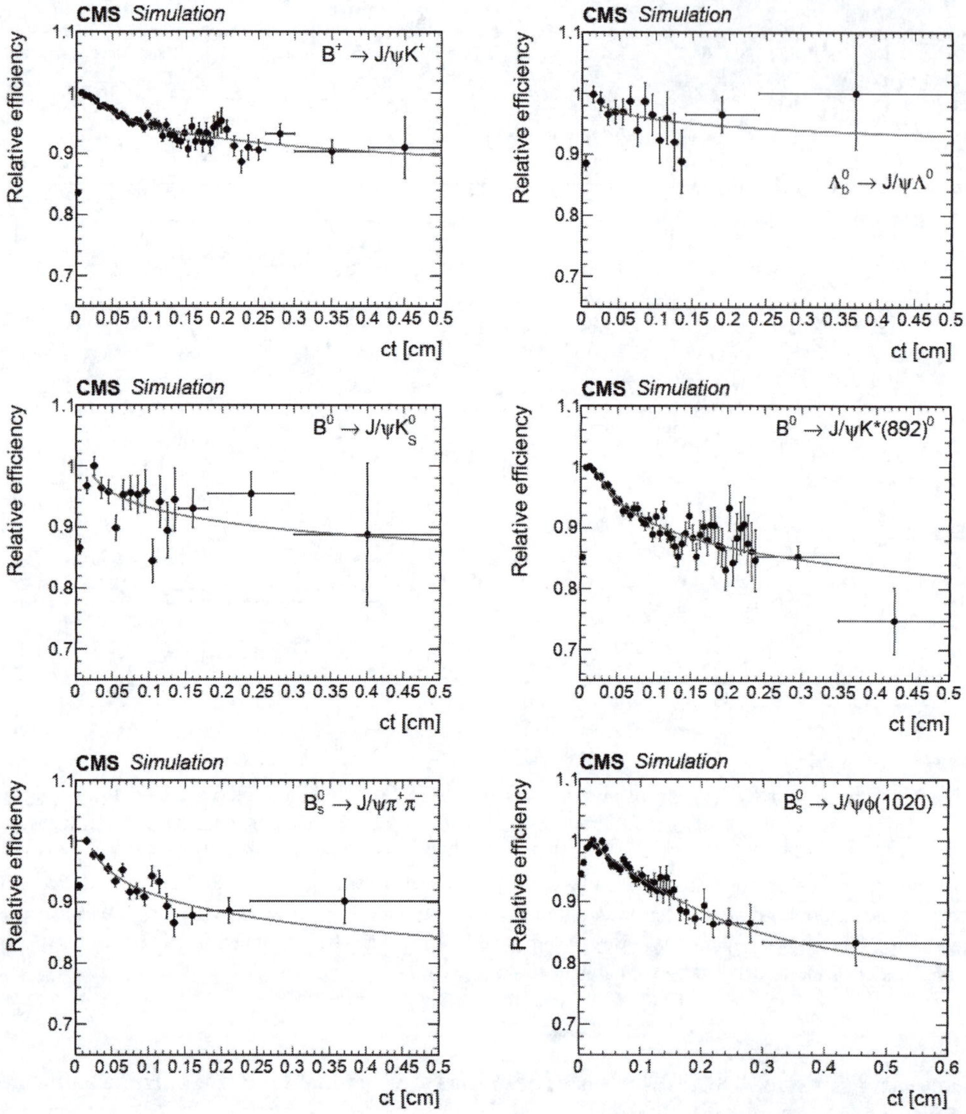

Figure 4. The reconstruction and selection efficiencies from simulation with a superimposed fit to an inverse power function as function of *ct* for the different decay channels, $B^+ \to J/\psi K^+$ (top left), $B^0 \to J/\psi K^*(892)^0$ (top right), $B^0 \to J/\psi K^0_S$ (center left), $\Lambda^0_b \to J/\psi \Lambda^0$ (center right), $B^0_s \to J/\psi \pi^+\pi^-$ (bottom left), and $B^0_s \to J/\psi \phi(1020)$ (bottom right) [2].

shown in Fig. 5. The bin edges, which are equal for the B_c^+ and B^+ *ct* distribution are chosen to ensure similar statistical uncertainties in the B_c^+ signal yields among all bins. The B_c^+ and B^+ yields are shown in the left plot of Fig. 6 and the ratio of the two efficiency functions are shown in the right plot of Fig. 6. Figure 7 shows the ratio of the efficiency-corrected *ct* distributions along with the result of a fit to an exponential function. The fit results in $\Delta\Gamma = 4.12 \pm 0.30$ c/mm.

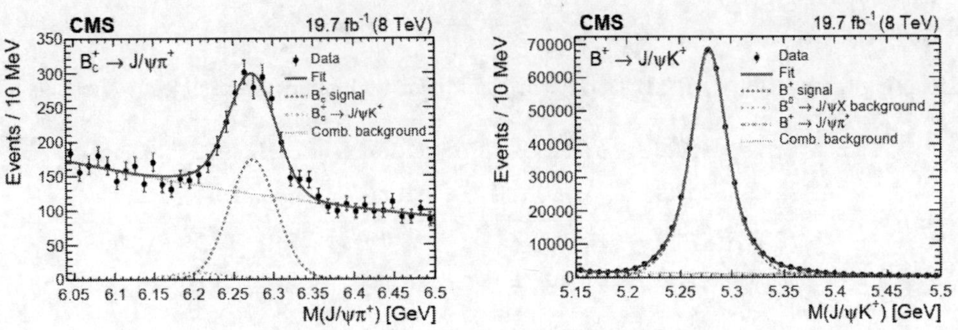

Figure 5. The $J/\psi\pi^+$ (left) and $J/\psi K^+$ (right) invariant mass distribution. The solid line represents the total fit, the dashed line the signal component, and the dotted line the combinatorial background. Contributions from $B_c^+ \to J/\psi K^+$, $B^+ \to J/\psi\pi^+$ and $B^0 \to J/\psi X$ are also shown [2].

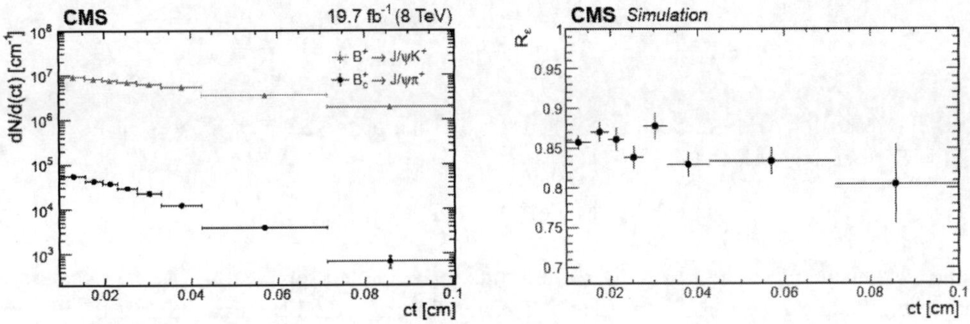

Figure 6. B_c^+ and B^+ signal yields normalized to the bin width (left) and ratio (right) of the B_c^+ and B^+ efficiency distributions as function of *ct* [2].

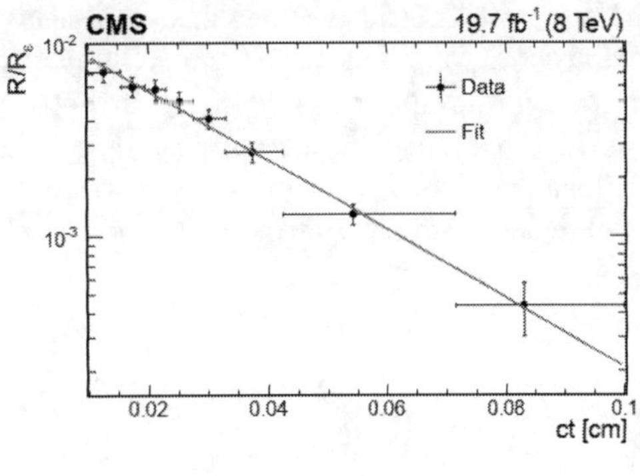

Figure 7. Ratio of the efficiency corrected B_c^+ to B^+ *ct* distributions together with the result of the fit to an exponential function [2].

6 Results

The resulting lifetimes of the B^0, B_s^0 and Λ_b^0 hadrons measured in the different decay channels are:

$$c\tau_{B^0} = 453.0 \pm 1.6\,(\text{stat}) \pm 1.5\,(\text{syst})\,\mu\text{m} \quad \text{in } J/\psi K^*(892)^0, \tag{5}$$
$$c\tau_{B^0} = 457.8 \pm 2.7\,(\text{stat}) \pm 2.7\,(\text{syst})\,\mu\text{m} \quad \text{in } J/\psi K_S^0, \tag{6}$$
$$c\tau_{B_s^0} = 504.3 \pm 10.5\,(\text{stat}) \pm 3.7\,(\text{syst})\,\mu\text{m} \quad \text{in } J/\psi\pi^+\pi^-, \tag{7}$$
$$c\tau_{B_s^0} = 443.9 \pm 2.0\,(\text{stat}) \pm 1.2\,(\text{syst})\,\mu\text{m} \quad \text{in } J/\psi\phi(1020), \tag{8}$$
$$c\tau_{\Lambda_b^0} = 443.1 \pm 8.2\,(\text{stat}) \pm 2.7\,(\text{syst})\,\mu\text{m} \quad \text{in } J/\psi\Lambda^0. \tag{9}$$

Both measurements of the B^0 lifetime are with the current world-average values, which are

$$c\tau_{B^0} = 455.7 \pm 1.2\,\mu\text{m}\ [3],$$
$$c\tau_{B_s^0 \to J/\psi\phi(1020)} = 443.4 \pm 3.6\,\mu\text{m}\ [16],$$
$$c\tau_{B_s^0 \to J/\psi\pi^+\pi^-} = 497.1 \pm 9.6\,\mu\text{m}\ [16],$$
$$c\tau_{\Lambda_b^0} = 440.7 \pm 3.0\,\mu\text{m}\ [3].$$

Using the result for $\Delta\Gamma$ from Section 5.1 and the lifetime of the B^+ meson, $c\tau_{B^+} = 491.1 \pm 1.2$ μm [3], the B_c^+ lifetime is found to be

$$c\tau_{B_c^+} = 162.3 \pm 8.2\,(\text{stat}) \pm 4.7\,(\text{syst}) \pm 0.1\,(\tau_{B^+})\,\mu\text{m}, \tag{10}$$

where the systematic uncertainty from the B^+ lifetime [3] is quoted separately. These results are in agreement with the average value from LHCb (153.4 ± 2.8 μm) [11, 12] and confirm a B_c^+ lifetime value higher than the average value measured at the Tevatron (135.5 ± 9.6 μm) [13–15].

7 Summary

The lifetimes of the B^0, B_s^0, B_c^+ and Λ_b^0 hadrons have been measured using different decay channels containing a J/ψ meson. The data, corresponding to an integrated luminosity of 19.7 fb^{-1}, have been collected by the CMS experiment in proton-proton collisions at a center-of-mass energy of 8 TeV. All measured lifetimes are compatible with the current world-average values. The precision of each channel is comparable or better than previous measurements in the respective channel for the neutral hadrons. The B_c^+ lifetime is in agreement with results from LHCb and confirms a longer lifetime than previously reported by the Tevatron experiments.

References

[1] A. Lenz, Int. J. Mod. Phys. **A30**, 1543005 (2015), `arXiv:1405.3601`
[2] A.M. Sirunyan et al. (CMS collaboration) (2017), CMS-BPH-13-008, CERN-EP-2017-244. Subm. to Eur. Phys. J. C., `arXiv:1710.08949`
[3] C. Patrignani, Particle Data Group, Chinese Physics C **40**, 100001 (2016)
[4] The CMS collaboration, Journal of Instrumentation **7**, P10002 (2012), `arXiv:1206.4071`
[5] V. Khachatryan et al. (CMS collaboration), Eur. Phys. J. **C70**, 1165 (2010), `arXiv:1007.1988`
[6] V. Khachatryan et al. (CMS collaboration), Phys. Lett. **B757**, 97 (2016), `arXiv:1507.07527`

[7] R. Aaij et al. (LHCb), Phys. Rev. **D87**, 052001 (2013), arXiv:1301.5347
[8] R. Aaij et al. (LHCb), Phys. Rev. **D89**, 092006 (2014), arXiv:1402.6248
[9] A.S. Dighe, I. Dunietz, R. Fleischer, Eur. Phys. J. **C6**, 647 (1999), arXiv:hep-ph/9804253
[10] V.V. Kiselev (2002), arXiv:hep-ph/0211021
[11] R. Aaij et al. (LHCb collaboration), Phys. Lett. **B742**, 29 (2015), arXiv:1411.6899
[12] R. Aaij et al. (LHCb collaboration), Eur. Phys. J. **C74**, 2839 (2014), arXiv:1401.6932
[13] A. Abulencia et al. (CDF Collaboration), Phys. Rev. Lett. **97**, 012002 (2006)
[14] T. Aaltonen et al. (CDF Collaboration), Phys. Rev. D **87**, 011101 (2013)
[15] V.M. Abazov et al. (D0 Collaboration), Phys. Rev. Lett. **102**, 092001 (2009)
[16] Heavy Flavor Averaging Group, ArXiv e-prints (2016), arXiv:1612.07233 [hep-ex]

Hadron spectroscopy in LHCb

Antimo Palano[1,3,⋆]

[1] *INFN and University of Bari, Via Orabona 4, 70125 Bari, Italy*

Abstract. The LHCb experiment is designed to study the properties and decays of heavy flavored hadrons produced in pp collisions at the LHC. The data collected in the LHC Run I enables precision spectroscopy studies of beauty and charm hadrons. The latest results on spectroscopy of conventional and exotic hadrons are reviewed. In particular the discovery of the first charmonium pentaquark states in the $J/\psi p$ system, the possible existence of four-quark states decaying to $J/\psi\phi$ and the confirmation of resonant nature of the $Z_c(4430)^-$ mesonic state are discussed. In the sector of charmed baryons, the observation of five new Ω_c states, the observation of the Ξ^+_{cc} and the study of charmed baryons decaying to D^0p are presented.

1 Introduction

In the simple quark model, only two types of quark combinations are required to account for the existing hadrons, i.e. $q\bar{q}$ combinations form mesons, while baryons are made up of three quarks. However, in the quark model proposed by Gell-Mann and Zweig in 1960s [1] other SU(3) color-neutral combinations of quarks and gluons such as gg glueballs, $q\bar{q}g$ hybrids, $q\bar{q}q\bar{q}$ tetraquarks, $q\bar{q}qqq$ pentaquarks etc. are allowed. The world's largest data sample of beauty and charm hadrons collected by LHCb during LHC Run I provides great opportunities for studying the production and properties of heavy hadrons. The present document describes some recent results from LHCb in exotic and standard hadron spectroscopy. For all the studies the LHCb data corresponding to 3 fb^{-1} of integrated luminosity in 7 and 8 TeV pp collisions has been used.

2 The observation of pentaquarks in $\Lambda^0_b \to J/\psi pK^-$

Evidences/observations of pentaquarks candidates were reported around 2005 by several experiments, always with low statistics and approximate statistical treatment of the data. For a summary and references see ref. [2]. High statistics searches gave negative results (see for example ref. [3]).

LHCb experiment studied the decay of $\Lambda^0_b \to J/\psi pK^-$ [4]. A sample of $26,007 \pm 166$ signal candidates, on 5.4% background, within ±15 MeV of the $J/\psi pK^-$ mass peak has been selected using the full Run I dataset. An anomalous peaking structure in the $J/\psi p$ invariant mass spectrum might be observed on the Dalitz plot shown in Fig. 1.

Since the contribution of $\Lambda^* \to K^-p$ resonances corresponding to the vertical bands only dominates the low $m^2(Kp)$ region, it is unlikely that the horizontal band is an effect of interference between

⋆e-mail: antimo.palano@ba.infn.it

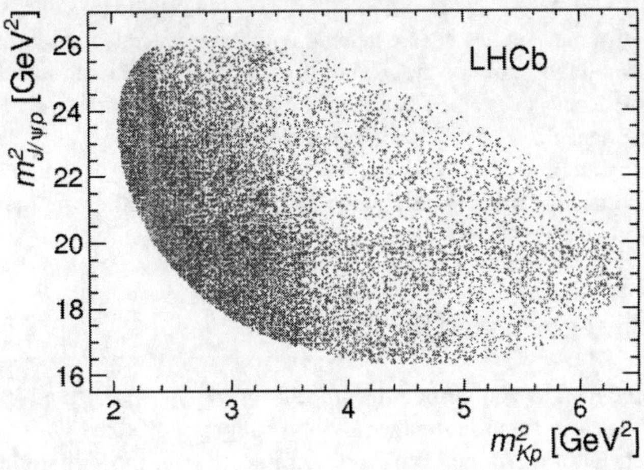

Figure 1. Dalitz plot of $\Lambda_b^0 \to J/\psi p K^-$.

the Λ^* states. This structure can also be seen in the invariant mass projections shown in Fig. 2. If the peak structure observed in Fig. 2(b) represents a resonance strongly decaying into $J/\psi p$, the minimal valence quarks should be $c\bar{c}uud$, a charmonium pentaquark state. A full six-dimensional amplitude

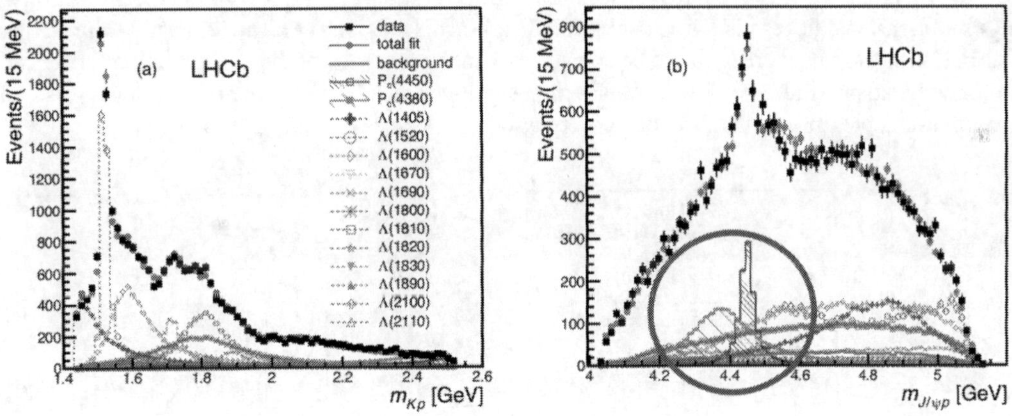

Figure 2. Dalitz plot projections of $\Lambda_b^0 \to J/\psi p K^-$ with superimposed the results from the amplitude analysis.

fit with resonance invariant mass, three helicity angles and two differences between decay planes has been applied to describe the data. It allowed interference between the decays $\Lambda_b^0 \to J/\psi\Lambda^*$ and $\Lambda_b^0 \to P_c^+ K^-$, with $\Lambda^* \to pK^-$ and $P_c^+ \to J/\psi p$, where P_c^+ indicates the pentaquark candidate. The amplitude model in the fit contains 14 well-defined Λ^* states listed by Particle Data Group [6] and two P_c^+ states: $P_c(4380)^+$ and $P_c(4450)^+$. With a full amplitude analysis, the masses of these two states are measured to be $4380 \pm 8(\text{stat}) \pm 29(\text{syst})$ MeV and $4449.8 \pm 1.7(\text{stat}) \pm 2.5(\text{syst})$ MeV, with the widths $205 \pm 18(\text{stat}) \pm 86(\text{syst})$ MeV and $39 \pm 5(\text{stat}) \pm 19(\text{syst})$ MeV, respectively. The preferred spin-parity values are $(3/2^-, 5/2^+)$, $(3/2^+, 5/2^-)$ and $(5/2^+, 3/2^-)$, where the first one is the J^P assignment given by the best fit. This analysis has been extended with a model-independent approach [7] as first proposed by the BaBar collaboration for the $Z_c(4430)^-$ analysis [8]. The model independent analysis gives similar results and allows to exclude that data can be described by the pK^- contributions alone. The evidence of the resonant character of the discovered pentaquark states was obtained observing the evolution of the complex amplitude in the Argand diagram for $P_c(4450)^+$, which is consistent with a rapid counter-clockwise change of the phase when its magnitude reaches the maximum, a behavior characteristic for a resonance.

3 Exotic baryonic resonances in $\Lambda_b^0 \to P_c^+\pi^-$

A possible way for an independent confirmation of the discovered $P_c(4380)^+$ and $P_c(4450)^+$ states is to check whether such states also occur in the the Cabibbo-suppressed decay $\Lambda_b^0 \to J/\psi p\pi^-$. Analyzing the same Run I dataset, but now in the Cabibbo suppressed mode, the signal yield was 1885 ± 50, about 15 times less than that of the Cabibbo-favored channel and with a relatively high background level [5]. The following possible contributions to the decay amplitude were taken into account in the analysis: $\Lambda_b^0 \to J/\psi N^*$ with $N^* \to p\pi^-$, $\Lambda_b^0 \to P_c^+\pi^-$ with $P_c^+ \to J/\psi p$, and $\Lambda_b^0 \to Z_c(4200)^- p$ with $Z_c^- \to J/\psi\pi^-$ where the $Z_c(4200)^-$ has been reported by the Belle collaboration in the decay $B^0 \to J/\psi K^+\pi^-$ [9]. The N^* resonances were modeled using relativistic Breit-Wigners, except the $N(1535)$, which is modeled by a Flatté lineshape, as it couples also to the near-threshold $n\eta$ mode. The full amplitude analysis used in the first pentaquark analysis has been applied to this Cabibbo suppressed Λ_b^0 decay. Both reduced and extended models of the N^* contributions have been tested, the latter including a wider set of resonances and allowed couplings. Figure 3 shows the projections of the six-dimensional fit results to the background-subtracted data onto the $m(p\pi^-)$ and $m(J/\psi p)$ mass distributions. The data were found to be consistent with the presence of the two exotic candidates, but the limited statistics did not allow to obtain high significances (3.1σ) for the two P_c states, also due to the possible presence of the Z_c exotic contribution.

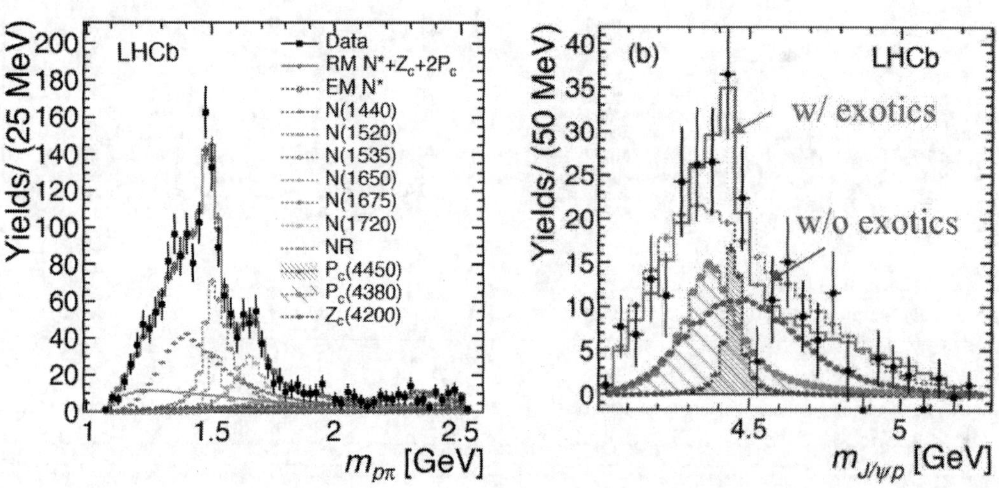

Figure 3. Dalitz plot projections of $\Lambda_b^0 \to J/\psi p\pi^-$ with superimposed the results from the amplitude analysis.

4 Confirmation of resonant nature of $Z_c(4430)^-$

The $Z_c(4430)^-$ state was first reported by the Belle collaboration [10] as a charged resonance structure in the $\psi'\pi^-$ invariant mass distribution in the decay $B^0 \to \psi' K^+\pi^-$. A relatively narrow peak has been found, $\Gamma \approx 45$ MeV, in the $\psi'\pi^-$ mass spectrum with a mass of 4433 ± 5 MeV. The BaBar collaboration could explain the enhancement as a reflection of the known K^* states [8]. In 2013 Belle collaboration, using more data with about 2000 signal events, performed a full amplitude analysis of the same final state. They obtained a significance of 5.2σ for the $Z_c(4430)^-$ state, with a mass $M(Z_c) = 4485 \pm 22 \pm 28$ MeV and width $\Gamma(Z_c) = 200 \pm 41 \pm 26$ MeV, favoring the $J^P = 1^+$ spin-parity assignment by more than 3.4σ.

The LHCb experiment collected about 25,000 B^0 signal events, a yield one order of magnitude larger than the BaBar or Belle datasets, and observed the $Z_c(4430)^-$ with a significance larger than 13.9σ [11]. The LHCb collaboration performed both a model independent analysis as well as a full amplitude analysis. The projection of the fit including the $Z_c(4430)^-$ state is shown in Fig. 4 and evidences a good description of the data. The measured parameters are $M(Z_c) = 4475 \pm 7^{+15}_{-25}$ MeV, $\Gamma(Z_c) = 172 \pm 13^{+37}_{34}$ MeV. The Argand diagram of the $Z_c(4430)^-$ amplitude (Fig. 4) shows the resonance behavior for the first time. The spin-parity was measured to be 1^+, by excluding other hypotheses by at least 9.7σ. For a charged charmonium state, $Z_c(4430)^-$ has a minimum quark content of $c\bar{c}u\bar{d}$ which clearly does not fit into the traditional quark model.

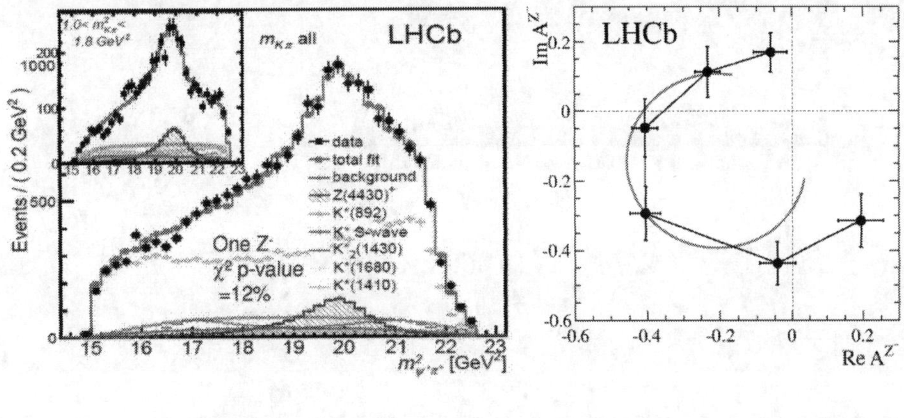

Figure 4. (Left) $\psi'\pi^-$ mass distribution from $\bar{B}^0 \to \psi'\pi^- K^+$ with superimposed the results from the amplitude analysis. (Right) Argand diagram for $Z_c(4430)^-$.

5 Study of $B \to J/\psi\phi K$

There has been a great deal of experimental and theoretical interest in $J/\psi\phi$ mass structures in $B^+ \to J/\psi\phi K^+$ decays since the CDF collaboration presented 3.8σ evidence for a near-threshold $X(4140)$ mass peak, with width $\Gamma = 11.7$ MeV [12]. Much larger widths are expected for charmonium states at this mass, therefore it has been suggested that the $X(4140)$ peak could be a molecular state, a tetraquark state, a hybrid state or a rescattering effect. Subsequent measurements resulted in a confusing experimental situation. LHCb experiment has reconstructed 4289 ± 151 $B^+ \to J/\psi\phi K^+$ decays, with roughly uniform efficiency and a relatively low background across the entire $J/\psi\phi$ mass range [13]. A full amplitude analysis has been performed, including resonant contributions from

Table 1. Parameters of the resonances decaying to $J/\psi\phi$ resulting from the amplitude analysis of $B^+ \to J/\psi\phi K^+$

Resonance	σ	J^{PC}	M (MeV)	$\Gamma(MeV)$
X(4140)	8.4	1^{++}	$4160 \pm 4^{+5}_{-3}$	$83 \pm 21^{+21}_{-14}$
X(4274)	5.8	1^{++}	$4273 \pm 8^{+17}_{-4}$	$56 \pm 11^{+8}_{-11}$
X(4500)	6.1	0^{++}	$4506 \pm 11^{+12}_{-15}$	$92 \pm 21^{+21}_{-20}$
X(4700)	5.6	0^{++}	$4704 \pm 10^{+14}_{-24}$	$120 \pm 31^{+42}_{-33}$

K^* resonances decaying to ϕK^+. Four resonant contributions in the $J/\psi\phi$ final state are found to contribute, whose parameters and quantum numbers are given in Table 1. The fit results are superimposed to the $J/\psi\phi$ mass spectrum in fig. 5. It can be noted that the $X(4140)$ width is substantially larger than

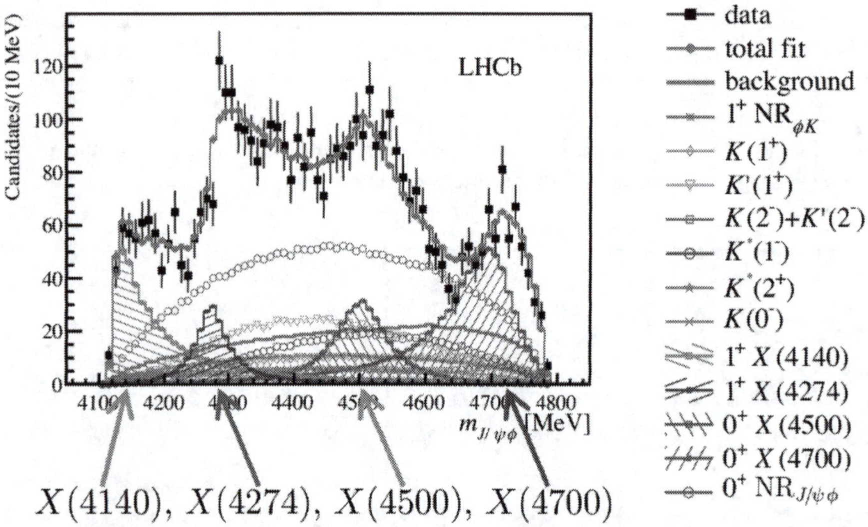

Figure 5. $J/\psi\phi$ mass projection with superimposed the fit result from the amplitude analysis of $B^+ \to J/\psi\phi K^+$

previously determined. The interpretation of these states is still an open issue.

6 Observation of five new Ω_c^0 states

The spectroscopy of singly charmed baryons cqq' is intricate. With three quarks and numerous degrees of freedom, many states are expected. At the same time, the large mass difference between the charm quark and the light quarks provides a natural way to understand the spectrum by using the symmetries provided by Heavy Quark Effective Theory (HQET). In recent years, considerable improvements have been made in the predictions of the properties of these heavy baryons. In many of these models, the heavy quark interacts with a (qq) diquark, which is treated as a single object. These models predict seven states in the mass range 2.9–3.2 GeV. In particular, among the expected charmed baryon states the spectrum of Ω_c baryons, which have quark content css and isospin zero, is largely unknown: only the Ω_c and $\Omega_c(2770)^0$, presumed to be the $J^P = 1/2^+$ and $3/2^+$ ground states, have been observed.

The LHCb experiment has reconstructed a large ($\approx 10^6$ events) and high-purity (83%) sample of Ξ_c^+ baryons, reconstructed in the Cabibbo-suppressed decay mode $pK^+\pi^-$ [14]. Surprisingly, in the $\Xi_c^+ K^-$ mass spectrum, shown in Fig. 6, five new, narrow excited Ω_c states are observed: the $\Omega_c(3000)^0$, $\Omega_c(3050)^0$, $\Omega_c(3066)^0$, $\Omega_c(3090)^0$, and $\Omega_c(3119)^0$. The measured masses and widths are reported in Table 2. The data indicate also the presence of a broad structure around 3188 MeV that is fitted as a single resonance but could be produced in other ways, for example as a superposition of several states. In addition, the partially reconstructed decay $\Omega_c(3066)^0$, $\Omega_c(3090)^0$ and $\Omega_c(3119)^0$ in their possible decay to $\Xi_c'^+ K^-$ are observed via their feed-down in the low mass region. This observation has generated large interest, especially for the presence of very narrow states, and many phenomenological models have been proposed, including the possibility of the presence of pentaquark states.

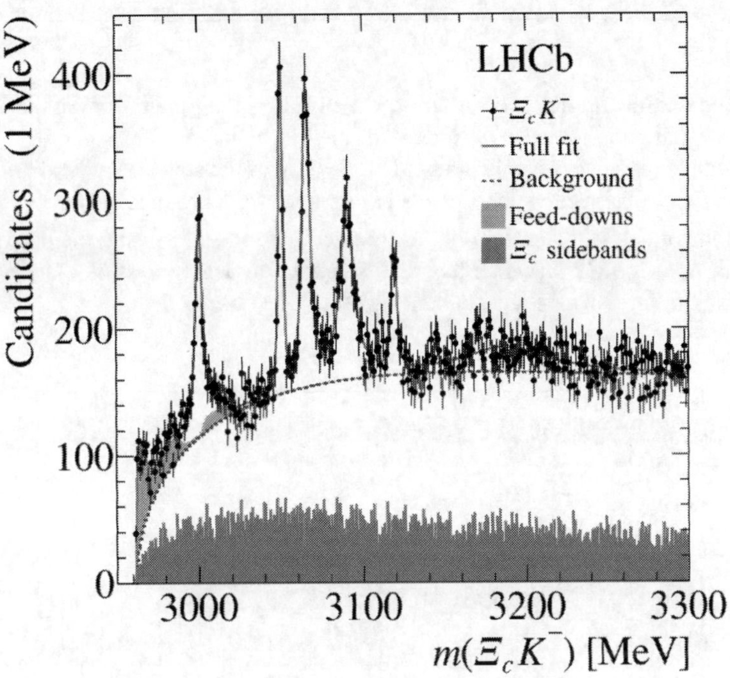

Figure 6. $\Xi_c^+ K^-$ mass spectrum with the observation of five new Ω_c states.

7 Observation of the double charmed baryon Ξ_{cc}^{++}

The constituent quark model of hadrons predicts the existence of three weakly decaying ground state $J^P = 1/2^+$ baryons with two charmed valence quarks: Ξ_{cc}^+ (ccd) and Ξ_{cc}^{++} (ccu), which form an isospin doublet, and Ω_{cc}^+ (ccs). Several QCD computational formalisms have been used to predict their masses, lifetimes, relative production rates, and the spectroscopy of their excited states. The only previously published evidence of their existence comes from the observations of the Ξ_{cc}^+ baryon claimed by the SELEX collaboration [15]. They observed narrow peaking structures at a mean mass of 3518.7 ± 1.7 MeV in the mass spectra of $\Lambda_c^+ K^- \pi^+$ (15.9 signal events over 6.1 ± 0.5 background events, 6.3σ significance) and pD^+K^- (5.62 signal events over 1.38 ± 0.13 background events, 4.8σ

Table 2. Measured parameters of the five new Ω_c states.

Resonance	Mass (MeV)	Γ (MeV)
$\Omega_c(3000)^0$	$3000.4 \pm 0.2 \pm 0.1^{+0.3}_{-0.5}$	$4.5 \pm 0.6 \pm 0.3$
$\Omega_c(3050)^0$	$3050.2 \pm 0.1 \pm 0.1^{+0.3}_{-0.5}$	$0.8 \pm 0.2 \pm 0.1$
		$< 1.2 MeV$, 95% CL
$\Omega_c(3066)^0$	$3065.6 \pm 0.1 \pm 0.3^{+0.3}_{-0.5}$	$3.5 \pm 0.4 \pm 0.2$
$\Omega_c(3090)^0$	$3090.2 \pm 0.3 \pm 0.5^{+0.3}_{-0.5}$	$8.7 \pm 1.0 \pm 0.8$
$\Omega_c(3119)^0$	$3119.1 \pm 0.3 \pm 0.9^{+0.3}_{-0.5}$	$1.1 \pm 0.8 \pm 0.4$
		$< 2.6 MeV$, 95% CL
$\Omega_c(3188)^0$	$3188.1 \pm 4.8 \pm 12.7$	$60 \pm 15 \pm 11$

significance). The lifetime of the state was unexpectedly short (less than 33 fs at the 90% confidence level) and had a production rate several orders of magnitude larger than theoretical predictions for their hyperon-beam on fixed target collisions. Identification of the state as the Ξ_{cc}^+ baryon implied that 20% of all Λ_c^+ baryons observed by the SELEX experiment were produced in decays of Ξ_{cc}^+ baryons.

LHCb experiment has searched for double charmed baryon states using the decay mode $\Xi_{cc}^{++} \to \Lambda_c K^- \pi^+ \pi^+$, with $\Lambda_c \to pK^-\pi^+$ (which has a 10% branching fraction) [16]. The $\Lambda_c K^- \pi^+ \pi^+$ mass spectrum is shown in Fig. 7 where a prominent signal can be observed.

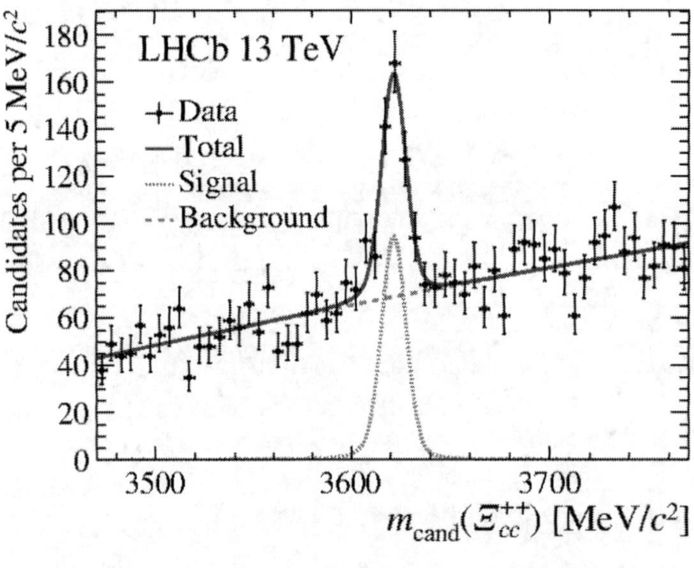

Figure 7. $\Lambda_c K^- \pi^+ \pi^+$ mass spectrum with the observation of Ξ_{cc}^{++}.

LHCb observes 313 ± 33 decays with a significance of $> 12\sigma$ of a Ξ_{cc}^{++} state that has a mass of $m(\Xi_{cc}^{++}) = 3621.40 \pm 0.72(\text{stat}) \pm 0.27(\text{syst}) \pm 0.14(\Lambda_c)$ MeV and a width that is much smaller than experimental resolution and that is consistent with a particle that flies a significant distance before decaying.

This is the first published observation of the Ξ_{cc}^{++} baryon. The published SELEX Ξ_{cc}^{+} state has a mass that is 103 ± 2 MeV smaller than the Ξ_{cc}^{++} state, so the two observations are inconsistent with being isospin partners. A consistent interpretation of the two observations may prove challenging.

8 Charmed baryon spectroscopy using Λ_b decays

LHCb experiment has performed a full amplitude analysis of the Cabibbo-favoured decay $\Lambda_b \to D^0 p \pi^-$ [17]. This Λ_b decay mode allows to study baryon spectroscopy in the $D^0 p$ final state and the determination of the quantum numbers of the observed states. The analysis was possible due to the large statistics and the clean Λ_b reconstructed signal (11,200 events, 86% purity).

Figure 8 shows the fit projection on the $D^0 p$ mass spectrum, where three Λ_c^+ states are observed, with parameters given in Table 3.

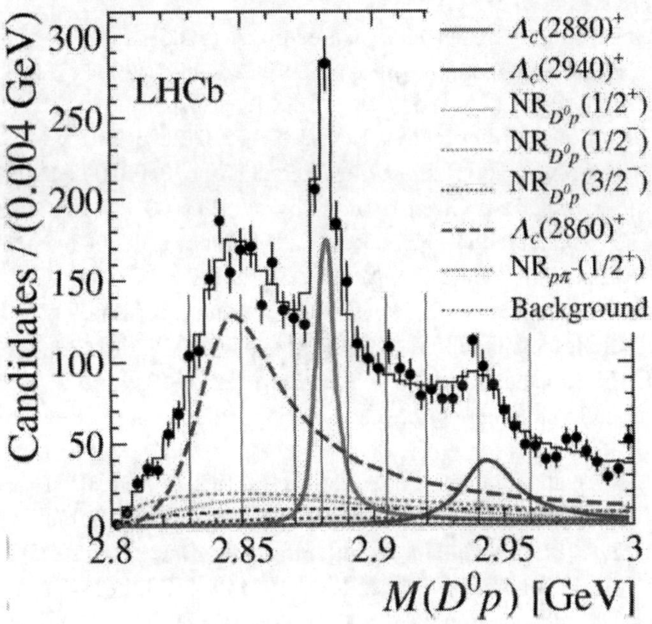

Figure 8. $D^0 p$ mass projection with the results from the amplitude analysis of $\Lambda_b \to D^0 p \pi^-$

The two $\Lambda_c(2880)^+$ and $\Lambda_c(2940)^+$ states have been previously observed by BaBar [18] experiment, but without a spin-parity determination. For $\Lambda_c(2860)^+$ the present analysis represents a first observation.

Table 3. Parameters of resonances decaying to $D^0 p$ from the amplitude analysis of $\Lambda_b \to D^0 p \pi^-$. The tabulated errors are statistic, systematic and related to the fitting model.

Resonance	J^{PC}	M (MeV)	$\Gamma(MeV)$
$\Lambda_c(2860)^+$	$3/2^+$	$2856.1^{+2.0}_{-1.7} \pm 0.5^{+1.1}_{-5.6}$	$67.6^{+10.1}_{-8.1} \pm 1.4^{+5.9}_{-20.0}$
$\Lambda_c(2880)^+$	$5/2^+$	$2881.75 \pm 0.29 \pm 0.07^{+0.14}_{-0.20}$	$5.43^{+0.77}_{-0.71} \pm 0.29^{+0.75}_{-0.00}$
$\Lambda_c(2940)^+$	$3/2^-$	$2944.8^{+3.5}_{-2.5} \pm 0.4^{+0.1}_{-4.6}$	$27.7^{+8.2}_{-6.0} \pm 0.9^{+5.2}_{-10.4}$

9 Conclusions

LHCb is a flavor factory, exploring a large set of physics topics. In particular, in the spectroscopy field, many new unexplored regions are being studied. These analyses are producing unexpected results, such as the discovery of "exotic" states, or the observation of many unexpected resonances and particles. Basic ingredients of these results are: the high statistics and purity of the final states and highly sophisticated and newly developed full amplitude analyses. This field is in rapid development and much more experimental and theoretical work is needed to understand the full pattern. Many more analyses are underway, making use of the large amount of data which are being collected at LHC.

References

[1] M. Gell-Mann, Phys. Lett. **8**, 214 (1964).

[2] A. Dzierba, C. Mayer and A. Szczepaniak, hep-ex/04120 (2005).

[3] B. Aubert *et al.* [BaBar Collaboration], Phys. Rev. Lett. **95**, 042002 (2005). J. M. Link *et al.* [FOCUS Collaboration], Phys. Lett. B **639**, 604 (2006).

[4] R. Aaij *et al.* [LHCb Collaboration], Phys. Rev. Lett. **115**, 072001 (2015).

[5] R. Aaij *et al.* [LHCb Collaboration], Phys. Rev. Lett. **117**, no. 8, 082003 (2016).

[6] C. Patrignani et al. (Particle Data Group), Chin. Phys. C, **40**, 100001 (2016) and 2017 update.

[7] R. Aaij *et al.* [LHCb Collaboration], Phys. Rev. Lett. **117**, no. 8, 082002 (2016).

[8] B. Aubert *et al.* [BaBar Collaboration], Phys. Rev. D **79**, 112001 (2009).

[9] K. Chilikin *et al.* [Belle Collaboration], Phys. Rev. D **90**, no. 11, 112009 (2014).

[10] S. K. Choi *et al.* [Belle Collaboration], Phys. Rev. Lett. **100**, 142001 (2008).

[11] R. Aaij *et al.* [LHCb Collaboration], Phys. Rev. Lett. **112**, no. 22, 222002 (2014).

[12] CDF collaboration, T. Aaltonen *et al.*, [CDF Collaboration], Phys. Rev. Lett. **102** , 242002 (2009).

[13] R. Aaij *et al.* [LHCb Collaboration], Phys. Rev. Lett. **118**, 022003 (2017).

[14] R. Aaij *et al.* [LHCb Collaboration], Phys. Rev. Lett. **118**, 182001 (2017).

[15] A. Ocherashvili *et al.* [SELEX Collaboration], Phys. Lett. B **628**, 18 (2005).

[16] R. Aaij *et al.* [LHCb Collaboration], Phys. Rev. Lett. **119**, 112001 (2017).

[17] R. Aaij *et al.* [LHCb Collaboration], JHEP **1705**, 030 (2017).

[18] B. Aubert *et al.* [BaBar Collaboration], Phys. Rev. Lett. **98**, 012001 (2007).

EPJ Web of Conferences **181**, 01025 (2018) https://doi.org/10.1051/epjconf/201818101025
EXA2017

Tomography and gravitational radii for hadrons by three-dimensional structure functions

S. Kumano[1,2], *Qin-Tao Song*[1], and *O. V. Teryaev*[3]

[1] *KEK Theory Center, Institute of Particle and Nuclear Studies, KEK and Department of Particle and Nuclear Physics, Graduate University for Advanced Studies (SOKENDAI), Ooho 1-1, Tsukuba, Ibaraki, 305-0801, Japan*

[2] *J-PARC Branch, KEK Theory Center, Institute of Particle and Nuclear Studies, KEK and Theory Group, Particle and Nuclear Physics Division, J-PARC Center, 203-1, Shirakata, Tokai, Ibaraki, 319-1106, Japan*

[3] *Bogoliubov Laboratory of Theoretical Physics, Joint Institute for Nuclear Research, 141980 Dubna, Russia*

Abstract. Three-dimensional tomography of hadrons can be investigated by generalized parton distributions (GPDs), transverse-momentum-dependent parton distributions (TMDs), and generalized distribution amplitudes (GDAs). The GDA studies had been only theoretical for a long time because there was no experimental measurement until recently, whereas the GPDs and TMDs have been investigated extensively by deeply virtual Compton scattering and semi-inclusive deep inelastic scattering. Here, we report our studies to determine pion GDAs from recent KEKB measurements on the differential cross section of $\gamma^*\gamma \to \pi^0\pi^0$. Since an exotic-hadron pair can be produced in the final state, the GDAs can be used also for probing internal structure of exotic hadron candidates in future. The other important feature of the GDAs is that the GDAs contain information on form factors of the energy-momentum tensor for quarks and gluons, so that gravitational form factors and radii can be calculated from the determined GDAs. We show the mass (energy) and the mechanical (pressure, shear force) form factors and radii for the pion. Our analysis should be the first attempt for obtaining gravitational form factors and radii of hadrons by analysis of actual experimental measurements. We believe that a new field of gravitational physics is created from the microscopic level in terms of elementary quarks and gluons.

1 Introduction

Inclusive lepton deep inelastic scattering (DIS) has been investigated since 1970's and it is described by structure functions and parton distribution functions (PDFs) expressed by the Bjorken-scaling variable x. Since it is the longitudinal momentum fraction for a parton in a hadron, the inclusive DIS probes the one-dimensional structure of hadrons. However, it became necessary to understand three-dimensional (3D) structure of hadrons for precisely describing exclusive and semi-inclusive reactions and for finding the origin of nucleon spin including contributions from partonic orbital angular momenta. As the 3D structure functions, generalized parton distributions (GPDs) and transverse-momentum-dependent parton distributions (TMDs) have been investigated both theoretically and experimentally. They are measured by the deeply virtual Compton scattering (DVCS) and semi-inclusive deep inelastic lepton scattering. There is another type of 3D structure functions called generalized distribution amplitudes (GDAs), which can be investigated by the s-t crossed process to the DVCS ($\gamma^* h \to \gamma h$), namely the two-photon process $\gamma^* \gamma \to h\bar{h}$.

The situation is illustrated in Fig. 1. The nucleonic PDFs and form factors have been investigated for a long time, and recent studies focus on the 3D structure functions, GPDs, TMDs, and GDAs. These functions are obtained by integrating the generating functions, generalized transverse-momentum-dependent parton distributions (GTMDs) or Wigner distributions. Although there are much experimental progress on the GPDs and TMDs in the last several years, there was no experimental information on the GDAs until recently. However, the Belle collaboration reported the cross-section data on the two-photon process $\gamma^* \gamma \to \pi^0 \pi^0$ in 2016 [1], so that it became possible to extract the pion GDAs from their data [2].

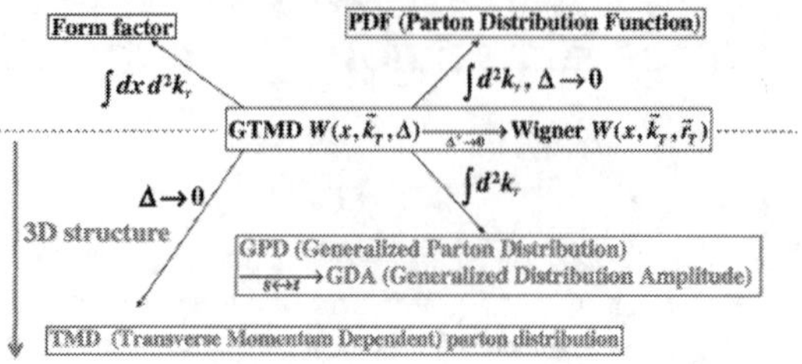

Figure 1. 3D structure functions and Wigner distributions.

The determination of the GDAs is valuable in studying 3D tomography of hadrons for finding the origin of nucleon spin, because the GPDs and GDAs are related by the s-t crossing and because both distributions are obtained from common double distributions by using the Radon transform as explained in Sect. 2. Such tomography studies could be used also for probing internal structure of exotic hadron candidates because an unstable hadron pair could be produced in the two-photon process [3], whereas unstable hadrons cannot be used as fixed targets in measuring the GPDs and TMDs. There is another important advantage in studying the GPDs and GDAs for probing gravitational source by the energy-momentum tensor of quarks and gluons.

The electric charge and magnetic form factors of the nucleons are measured in electron scattering and their radii are determined from them. The charge radius of the pion is measured as 0.672 ± 0.008 fm. The charged particles in the pion, namely quarks, contribute to the charge form factor and the radius. In the same way, it is interesting to measure the gravitational mass distributions and radii for the pion or any hadron. Here, both quarks and gluons contribute to the gravitational distributions. It is not measured in direct scattering experiment like the electron scattering because of the ultra-weak nature of gravitational interactions. However, there is a way to access them by the 3D structure functions, such as the GDAs, because they contain the factors of the energy-momentum tensor for quarks and gluons [2]. We know the the energy-momentum tensor is the source of gravity.

We discuss determination of the pion GDAs and gravitational form factors by analyzing the Belle measurements in this report [2]. First, the GDAs are introduced in Sect. 2, and the cross section of the two-photon process $\gamma^* \gamma \to \pi^0 \pi^0$ is explained with the GDAs in Sect. 3. Analysis results are shown in Sect. 4, and our studies are summarized in Sect. 5

2 Generalized distribution amplitudes and gravitational form factors

We explain the GDAs in comparison with the GPDs, which have been studied extensively. First, the GPDs can be experimentally studied by the DVCS $\gamma^* h \to \gamma h$ in Fig. 2(a) if $Q^2 = -q^2$ is large enough to satisfy the factorization that the process is described by the hard perturbative QCD part and the soft GPD one. Here, q is the initial virtual-photon momentum. The s-t crossed process of the DVCS is the

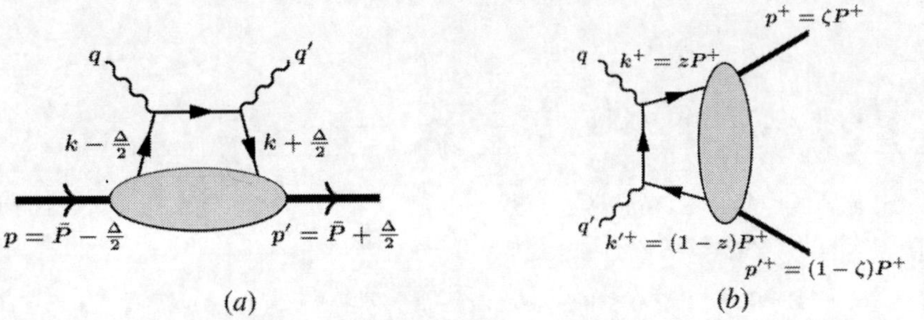

Figure 2. (*a*) Virtual Compton process for GPDs (*b*) Two-photon process for GDAs.

two-photo process to produce a hadron pair $\gamma^*\gamma \to h\bar{h}$ as shown in Fig. 2(*b*). It is also factorized into the hard part and the soft GDA one if Q^2 is large enough.

The pion GPDs $H_q^{\pi^0}$ are defined by off-forward matrix elements of quark and gluon operators with a lightcone separation, and the GDAs $\Phi_q^{\pi^0\pi^0}$ are defined by the same operator between the vacuum and the hadron pair as [4, 5]

$$H_q^{\pi^0}(x,\xi,t) = \int \frac{dy^-}{4\pi} e^{ix\bar{P}^+y^-} \left\langle \pi^0(p') \left| \bar{q}(-y/2)\gamma^+ q(y/2) \right| \pi^0(p) \right\rangle \Big|_{y^+=\vec{y}_\perp=0},$$
$$\Phi_q^{\pi^0\pi^0}(z,\zeta,W^2) = \int \frac{dy^-}{2\pi} e^{i(2z-1)P^+y^-/2} \langle \pi^0(p)\pi^0(p') \,|\, \bar{q}(-y/2)\gamma^+ q(y/2) \,|\, 0 \rangle \Big|_{y^+=\vec{y}_\perp=0}. \quad (1)$$

These GPDs and GDAs are defined for quarks, and similar expressions exist also for gluons. Here, link operators for the color gauge invariance are not explicitly written for simplicity. The PDFs are given by the forward limit of the GPDs: $q^{\pi^0}(x) = H_q^{\pi^0}(x, \xi = 0, t = 0)$ for quarks at $x > 0$ (antiquark distributions at $x < 0$). Using the initial and final pion (photon) momenta p and p' (q and q'), we define average momenta ($\bar{P}$, $\bar{q}$) and momentum transfer Δ as

$$\bar{P} = \frac{p+p'}{2}, \quad \bar{q} = \frac{q+q'}{2}, \quad \Delta = p' - p = q - q'. \quad (2)$$

Then, the Bjorken variable x, the skewdness parameter ξ, and the momentum-transfer squared t are given by

$$x = \frac{Q^2}{2p\cdot q}, \quad \xi = \frac{\bar{Q}^2}{2\bar{P}\cdot\bar{q}}, \quad t = \Delta^2, \quad (3)$$

where $Q^2 = -q^2$ and $\bar{Q}^2 = -\bar{q}^2$. The DVCS process is factorized, if the kinematical condition $Q^2 \gg |t|,\ \Lambda_{\rm QCD}^2$ is satisfied, to express it in terms of the GPDs in Fig. 2(*a*). Here, $\Lambda_{\rm QCD}$ is the QCD scale parameter.

The variables of the GDAs are the momentum fractions z and ζ in Fig. 2(*b*) and the invariant-mass squared W^2, and they are defined by

$$z = \frac{k\cdot q'}{P\cdot q'} = \frac{k^+}{P^+}, \quad \zeta = \frac{p\cdot q'}{P\cdot q'} = \frac{p^+}{P^+} = \frac{1+\beta\cos\theta}{2}, \quad W^2 = (p+p')^2 = (q+q')^2 = s, \quad (4)$$

where β is the pion velocity given by $\beta = |\vec{p}|/p^0 = \sqrt{1-4m_\pi^2/W^2}$, and the scattering angle is θ in the center-of-mass frame of final pions. The two-photon process is factorized and expressed by the GDAs if the condition $Q^2 \gg W^2,\ \Lambda_{\rm QCD}^2$ is satisfied. The GPDs and GDAs are related with each other by the

s-t crossing as

$$\Phi_q^{\pi^0\pi^0}(z',\zeta,W^2) \longleftrightarrow H_q^{\pi^0}\left(x=\frac{1-2z'}{1-2\zeta},\ \xi=\frac{1}{1-2\zeta},\ t=W^2\right). \tag{5}$$

However, the physical regions of the GDAs ($0 \le z \le 1$, $|1-2\zeta| \le 1$, $W^2 \ge 0$) do not necessarily correspond to the physical ones of the GPDs ($|x| \le 1$, $|\xi| \le 1$, $t \le 0$):

$$0 \le |x| < \infty,\quad 0 \le |\xi| < \infty,\quad |x| \le |\xi|,\quad t \ge 0. \tag{6}$$

Therefore, the GDA studies may not be directly utilized for clarifying the GPDs. There is a way to circumvent this issue by using the Radon transform, which is often used in tomography studies in general.

Let us consider possible two-pion states in the reaction $\gamma^*\gamma \to \pi\pi$. The isospin $I=1$ $\pi\pi$ state is antisymmetric under the exchange of the pions, whereas the $I=0$ and $I=2$ $\pi\pi$ states are symmetric. The C parity of the $\pi\pi$ state is $C=(-1)^{L+S}=(-1)^L=1$ with $S=0$ because the C parity of $\gamma^*\gamma$ is even. It means L is even. Then, the Pauli principle $(-1)^L(-1)^I(-1)^S=1$ suggests that the $\pi\pi$ isospin state is $I=0$ or 2. However, the GDAs are defined by the vector-type nonlocal operator and the isospin of $\bar{q}q$ is 0 or 1, so that the $\pi\pi$ states from the two photons are $I=0$ with $L=$ even numbers $(0,2,\cdots)$. Therefore, the GDAs in the process $\gamma^*\gamma \to \pi^0\pi^0$ are C-even functions denoted with (+):

$$\Phi_q^{\pi^0\pi^0}(z,\zeta,W^2)=\Phi^{\pi\pi(I=0)}(z,\zeta,W^2)=\Phi_q^{\pi\pi(+)}(z,\zeta,W^2). \tag{7}$$

The charge-conjugation and isospin symmetries require the relations for the GDAs:

$$\begin{aligned}
&\Phi_q^{\pi^0\pi^0}(z,\zeta,W^2)=-\Phi_q^{\pi^0\pi^0}(1-z,\zeta,W^2)=\Phi_q^{\pi^0\pi^0}(z,1-\zeta,W^2),\\
&\Phi_u^{\pi^0\pi^0}(z,\zeta,W^2)=\Phi_d^{\pi^0\pi^0}(z,\zeta,W^2),
\end{aligned} \tag{8}$$

which are considered as constraints in setting up the parametrization of the π^0 GDAs.

The Radon transform is defined for a function $f(x)$ in n dimensions as [6]

$$\hat{f}(p,\xi)=\int d^n x\, f(x)\,\delta(p-\xi\cdot x), \tag{9}$$

where x is the n-dimensional space coordinate $[x=(x_1,x_2,\cdots,x_n)]$ and ξ is the unit vector in n dimensions $[\xi=(\xi_1,\xi_2,\cdots,\xi_n)]$. We express the GPDs and GDAs by the common double distributions (DDs) F_q, G_q, and D_q with different Radon transforms given by [7]

$$\begin{aligned}
H_q(x,\xi,t)&=\int d\beta\, d\alpha\, \delta(x-\beta-\xi\alpha)\left[F_q(\beta,\alpha,t)+\xi G_q(\beta,\alpha,t)\right],\\
\Phi_q^{h\bar{h}}(z,\zeta,W^2)&=-2(1-2\zeta)\int d\beta\, d\alpha\delta(1-2z-(1-2\zeta)\beta+\alpha)\\
&\qquad\times F_q(1-2z,\alpha,W^2)-2D_q(x/\xi,W^2),\\
q(x)&=\int_{-1+x}^{1-x} d\alpha\, F_q(\beta,\alpha,t=0).
\end{aligned} \tag{10}$$

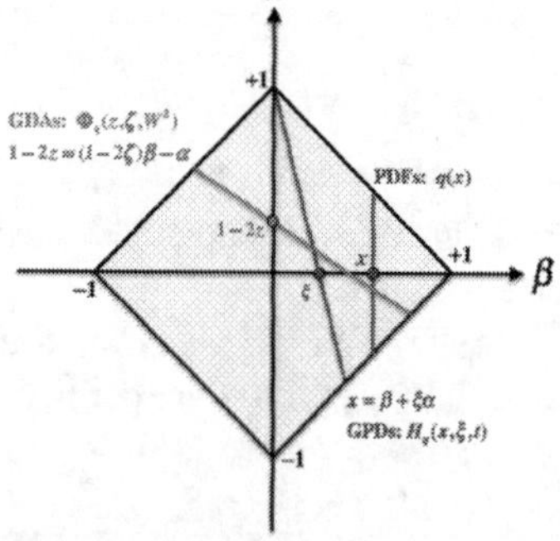

Figure 3. Radon transforms for PDFs, GPDs, and GDAs.

These integral paths are shown in Fig. 3. Namely, the GPDs are obtained by integrating the DDs over the slight line $x=\beta+\xi\alpha$, the PDFs are by the integral over the vertical line with the condition of the forward limit ($t = 0$), and the GDAs are by the Radon transform along the different line $1 - 2z - (1 - 2\zeta)\beta + \alpha = 0$. If the DDs are determined by the GDA studies, they can be used for finding the GPDs and vice verse. Therefore, the GDA studies should be valuable for clarifying the 3D tomography of hadrons and also finding the origin of nucleon spin including orbital-angular-momentum contributions.

There is an important application of the GPD and GDA studies for investigating gravitational aspects of hadrons. The GPDs and GDAs are defined in Eq. (1) with the common nonlocal operator. If the n-th moment of this operator is calculated, we obtain

$$2(P^+/2)^n \int_0^1 dz\,(2z-1)^{n-1} \int \frac{dy^-}{2\pi} e^{i(2z-1)P^+y^-/2} \bar{q}(-y/2)\gamma^+ q(y/2)\Big|_{y^+=\vec{y}_\perp=0} = \bar{q}(0)\gamma^+\left(i\overleftrightarrow{\partial}^+\right)^{n-1} q(0). \quad (11)$$

We notice in this equation that the right-hand side is the usual vector-type current $\bar{q}\gamma^\mu q$ for $n = 1$ and the second ($n = 2$) moment is the energy-momentum tensor $\bar{q}\gamma^\mu i\partial^\nu q$ for quarks as shown in Fig. 4. Actually, the second moment is expressed as [5]

$$\int_0^1 dz\,(2z-1)\,\Phi_q^{\pi^0\pi^0}(z,\zeta,W^2) = \frac{2}{(P^+)^2}\langle\,\pi^0(p)\,\pi^0(p')\,|\,T_q^{++}(0)\,|\,0\,\rangle. \quad (12)$$

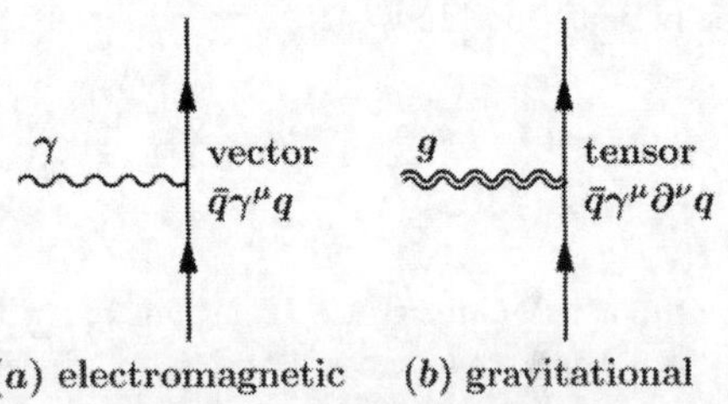

Figure 4. Electromagnetic and gravitational form factors.

In general, the quark energy-momentum tensor is defined by $T_q^{\mu\nu}(x) = \bar{q}(x)\,\gamma^{(\mu}i\overleftrightarrow{D}^{\nu)}\,q(x)$, where D^μ is the covariant derivative $D^\mu = \partial^\mu - ig\lambda^a A^{a,\mu}/2$ with the QCD coupling constant g and the SU(3) Gell-Mann matrix λ^a.

The matrix element of the energy-momentum tensor is expressed in terms of the timelike gravitational form factors $\Theta_1(s)$ and $\Theta_2(s)$ as

$$\langle\,\pi^0(p)\,\pi^0(p')\,|\,T_q^{\mu\nu}(0)\,|\,0\,\rangle = \frac{1}{2}\left[\,(s\,g^{\mu\nu} - P^\mu P^\nu)\,\Theta_{1,q}(s) + \Delta^\mu\Delta^\nu\,\Theta_{2,q}(s)\,\right], \quad (13)$$

where $P = p + p'$ and $\Delta = p' - p$. Using Eqs. (12) and (13), we can calculate the timelike gravitational form factors $\Theta_1(s)$ and $\Theta_2(s)$ if the GDAs are determined by analyzing experimental data on $\gamma^*\gamma \to \pi^0\pi^0$. The spacelike gravitational form factors $\Theta_1(t)$ and $\Theta_2(t)$ are defined by the matrix element as

$$\langle\,\pi^0(p')\,|\,T_q^{\mu\nu}(0)\,|\,\pi^0(p)\,\rangle = \frac{1}{2}\left[\,(t\,g^{\mu\nu} - q^\mu q^\nu)\,\Theta_{1,q}(t) + P^\mu P^\nu\,\Theta_{2,q}(t)\,\right], \quad (14)$$

where $q = p' - p$. The spacelike form factors are calculated by using the dispersion relation from the timelike ones, which are obtained directly from the determined GDAs. As explained in Sect. 4, the form factor Θ_2 indicates the gravitational mass (energy) distribution and Θ_1 is the mechanical (pressure, shear force) distribution.

3 Cross section for two-photon process $\gamma^*\gamma \to \pi^0\pi^0$ and GDAs

We analyze cross-section measurements for the two-photon process $\gamma^*\gamma \to \pi^0\pi^0$ and try to determine the GDAs of the pion. The cross section is expressed by the matrix element $\mathcal{M}(\gamma^*\gamma \to \pi^0\pi^0)$ as

$$d\sigma = \frac{1}{4q\cdot q'}\overline{\sum_{\lambda,\lambda'}}|\mathcal{M}(\gamma^*\gamma \to \pi^0\pi^0)|^2 \frac{d^3p}{(2\pi)^3\,2E_p}\frac{d^3p'}{(2\pi)^3\,2E_{p'}}(2\pi)^4\delta^4(q + q' - p - p'), \quad (15)$$

and it is written by the hadron tensor $\mathcal{T}_{\mu\nu}$ and photon-polarization vectors $\epsilon^\mu(\lambda)$ and $\epsilon^\nu(\lambda')$ as $i\mathcal{M}(\gamma^*\gamma \to \pi^0\pi^0) = \epsilon^\mu(\lambda)\,\epsilon^\nu(\lambda')\,\mathcal{T}_{\mu\nu}$. The hadron tensor is generally defined by the matrix element of the electromagnetic current J^{em}_μ and then by the GDAs for the pion, if the hard scale satisfies the factorization condition $Q^2 \gg W^2,\ \Lambda^2_{\rm QCD}$, in the leading order of the running coupling constant α_s and leading twist [2]:

$$\mathcal{T}_{\mu\nu} = i\int d^4y\, e^{-iq\cdot y}\langle \pi^0(p)\pi^0(p')|T J^{em}_\mu(y) J^{em}_\nu(0)|0\rangle = -g_T^{\mu\nu} e^2 \sum_q \frac{e_q^2}{2}\int_0^1 dz \frac{2z-1}{z(1-z)}\Phi_q^{\pi^0\pi^0}(z,\zeta,W^2), \quad (16)$$

where $g_T^{\mu\nu} = -1$ for $\mu = \nu = 1,\ 2$ and $g_T^{\mu\nu} = 0$ for $\mu,\ \nu =$ others. The helicity amplitude A_{ij} is defined as $A_{ij} = \varepsilon^{(i)}_\mu(q)\,\varepsilon^{(j)}_\nu(q')\,\mathcal{T}^{\mu\nu}/e^2$ ($i = -,\ 0,\ +;\ j = -,\ +$), and the differential cross section is expressed by the helicity amplitude A_{++}:

$$\frac{d\sigma}{d(\cos\theta)} = \frac{\pi\alpha^2}{4(Q^2+s)}\sqrt{1-\frac{4m_\pi^2}{s}}\,|A_{++}|^2, \quad A_{++} = \sum_q \frac{e_q^2}{2}\int_0^1 dz \frac{2z-1}{z(1-z)}\Phi_q^{\pi^0\pi^0}(z,\xi,W^2), \quad (17)$$

where the parity-conservation relation $A_{--} = A_{++}$ is used. The gluon GDA contributes in the higher-order of α_s, and the terms A_{0+} and A_{0-} are higher-twist ones. They are neglected in our analysis.

Because of the page limitation, the details of the GPD parametrization are not discussed in this article and they should be found in Ref. [2]. Only the outline is explained in the following. First, the GPDs should satisfy the symmetry relations in Eq. (8). The asymptotic ($Q^2 \to \infty$) z-dependence form is given by $z(1-z)(2z-1)$ for the π^0 GDAs, so that the parameter α may be assigned to its functional form as $z^\alpha(1-z)^\alpha(2z-1)$. Since there are S- and D-wave contributions to the $\pi^0\pi^0$ state, the quark GPDs are expressed as

$$\Phi_q^{\pi^0\pi^0}(z,\zeta,W^2) = N_\alpha z^\alpha(1-z)^\alpha(2z-1)\,[\widetilde{B}_{10}(W^2) + \widetilde{B}_{12}(W^2)P_2(\cos\theta)], \quad (18)$$

where $P_2(\cos\theta)$ is the Legendre polynomial. The S- and D-wave terms are $\widetilde{B}_{10}(W^2)$ and $\widetilde{B}_{12}(W^2)$, respectively, and they are given by the GDA continuum part and resonance contributions from $f_0(500)$ and $f_2(1270)$ [8]:

$$\widetilde{B}_{10}(W^2) = -\left[\left(1+\frac{2m_\pi^2}{W^2}\right)\frac{10}{9}M^\pi_{2(q)}F^\pi_q(W^2) + \sum_{f_0}\frac{5\,g_{f_0\pi\pi}\bar{f}_{f_0}}{3\sqrt{2}\sqrt{(M_{f_0}^2-W^2)^2+\Gamma_{f_0}^2 M_{f_0}^2}}\right]e^{i\delta_0(W)},$$

$$\widetilde{B}_{12}(W^2) = \left(1-\frac{4m_\pi^2}{W^2}\right)\frac{10}{9}\left[M^\pi_{2(q)}F^\pi_q(W^2) + \frac{g_{f_2\pi\pi}f_{f_2}M_{f_2}^2\beta^2}{\sqrt{2}\sqrt{(M_{f_2}^2-W^2)^2+\Gamma_{f_2}^2 M_{f_2}^2}}\right]e^{i\delta_2(W)}. \quad (19)$$

The decay constants $\bar{f}_{f_0}$ and f_{f_2} ($\equiv f_f$) have Q^2 dependence, and it is expressed as

$$f_f(Q^2) = f_f(Q_0^2)\left[\frac{\alpha_s(Q^2)}{\alpha_s(Q_0^2)}\right]^{\gamma_n/\beta_0}, \quad \gamma_n = C_F\left[1-\frac{2}{(n+1)(n+2)}+4\sum_{j=2}^{n+1}\frac{1}{j}\right], \quad (20)$$

effectively by including the scale dependence of the distribution amplitude part [3]. These decay constants are evaluated by the QCD sum rule often at $Q^2 = 1$ GeV^2, so that they are evolved to the average scale $\langle Q^2\rangle = 16.6$ GeV^2 of the used Belle data. The scale dependence of the decay constants indicates $f_f(Q^2 \to \infty) = 0$, which means the resonance contributions vanish in the scaling limit. In this limit, there are sum rules for the GDAs as

$$\int_0^1 dz\, \Phi_q^{\pi^0\pi^0}(z,\zeta,W^2) = 0, \qquad \int_0^1 dz\,(2z-1)\,\Phi_q^{\pi^0\pi^0}(z,\zeta,0) = -4\, M_{2(q)}^{\pi^0}\,\zeta(1-\zeta). \tag{21}$$

Here, $M_{2(q)}^h$ is the momentum fraction carried by flavor-q quarks and antiquarks in the pion, so that the total quark fraction is $\sum_q M_{2(q)}^{\pi^0}$. The first terms of $\widetilde{B}_{10}$ and $\widetilde{B}_{12}$ in Eq. (19) are constrained at $W^2 = 0$ by the second sum rule. From Eqs. (12), (13), and (18), we obtain the timelike gravitational form factors expressed by the S- and D-wave terms of the GDAs as

$$\Theta_{1,q}(s) = -\frac{3}{5}\widetilde{B}_{10}(W^2) + \frac{3}{10}\widetilde{B}_{20}(W^2), \quad \Theta_{2,q}(s) = \frac{9}{10\beta^2}\widetilde{B}_{20}(W^2), \tag{22}$$

and the total timelike gravitational form factors of the pion is obtained by adding them as $\Theta_n(s) = \sum_{i=q}\Theta_{n,i}(s)$ where $n = 1$ or 2. We notice that the form factor Θ_2 originates from the D-wave part of the GDAs and Θ_1 from both S- and D-wave terms. The overall form factor for the continuum term is taken as $F_q^\pi(W^2) = 1/[1 + (W^2 - 4m_\pi^2)/\Lambda^2]^{n-1}$ with the cutoff Λ and the factor n taken as the constituent-counting value $n = 2$ [9].

The $f_0(980)$ contribution is not included in the analysis because the differential cross-section data of the Belle do not show such contribution in the invariant-mass region at $W \simeq 1$ GeV and because there is no theoretical estimate on the decay constant by reflecting its exotic nature, tetra-quark or $K\bar{K}$ configuration. There is a QCD-sum-rule calculation by assuming an ordinary $q\bar{q}$-type configuration for $f_0(980)$; however, calculated cross sections are totally in contradiction to the Belle data. It means that the $q\bar{q}$ structure is not supported by the differential cross section data of the Belle collaboration for $f_0(980)$ as it has been claimed for a long time [10].

Some explanations are needed for understanding Eq. (19). First, we comment on coupling constants. There are theoretical estimates on the decay constant f_{f_2}, and another one $\bar{f}_{f_0}$ is considered as a parameter in our analysis because there is no theoretical estimate. The coupling constants $g_{f_0\pi\pi}$ and $g_{f_2\pi\pi}$ are determined by the 2π decay widths. Second, there are contributions to the cross section $\gamma^*\gamma \to \pi^0\pi^0$ from the processes with intermediate mesons as shown in Fig. 5. Considering the $\pi\pi$ intermediate state, we need to include the $\pi\pi$ phase shifts $\delta_0(W)$ and $\delta_2(W)$ in Eq. (19). In our analysis, we use the $\pi\pi$ phase shifts by Bydzovsky, Kaminski, Nazari, and Surovtsev [11]. Above the $K\bar{K}$ threshold at $2m_{K^+} = 0.987$ GeV, the $K\bar{K}$ channel opens, and then $\eta\eta$ channel also opens at higher energies. We do not include these effects in our current analysis; however, we introduce additional parameters in the $\pi\pi$ phase shifts above the $K\bar{K}$ threshold and they are determined by fitting the Belle data.

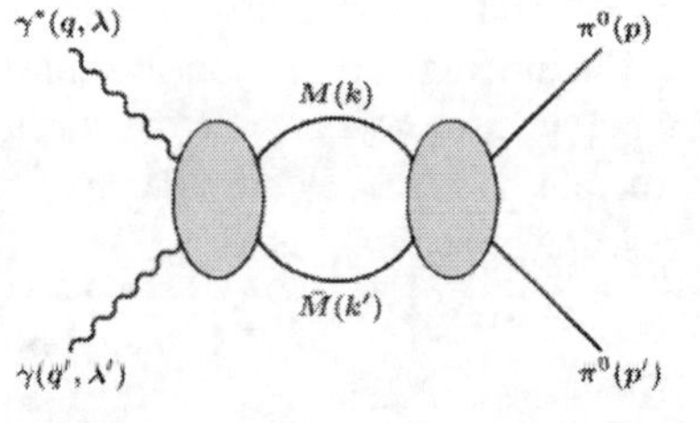

Figure 5. Intermediate-meson contribution.

4 Analysis results

The Belle experimental data for $\gamma^*\gamma \to \pi^0\pi^0$ are analyzed for extracting the pion GDAs. For this purpose, we select the data with large enough Q^2 to satisfy the factorization condition $Q^2 \gg W^2, \Lambda_{QCD}^2$ so that the amplitude is factorized into the hard perturbative QCD part and the soft GDA one. As such a condition, we take the scale $Q^2 \geq 8.92$ GeV^2 for the Belle data. Then, there are 550 points

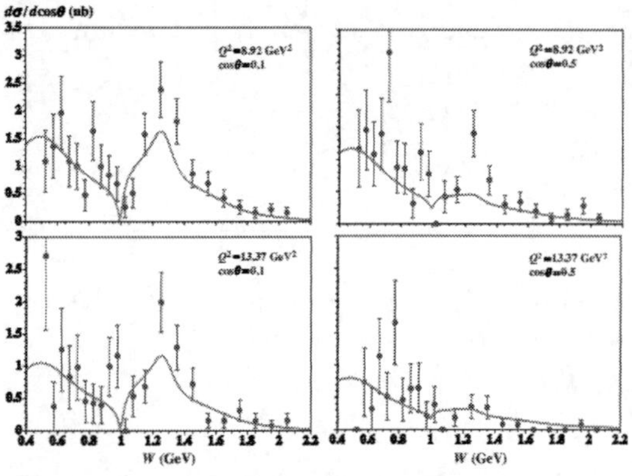

Figure 6. Comparison with Belle data [2].

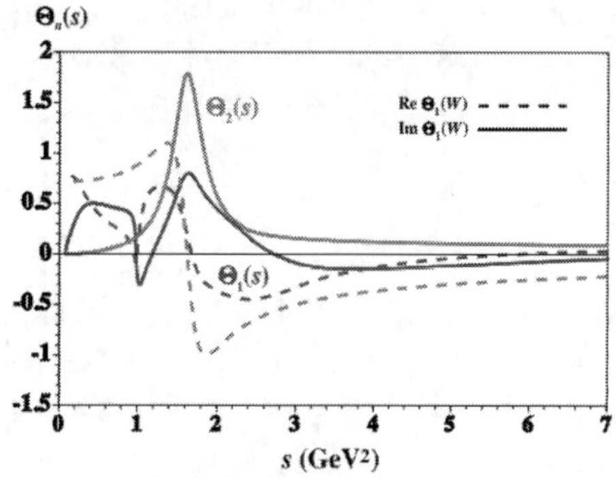

Figure 7. Timelike gravitational form factors for π [2].

of data with the scales Q^2 =8.92, 10.93, 13.37, 17.23, and 24.25 GeV^2. The invariant-mass range is 0.5 GeV $< W <$ 2.1 GeV, and the scattering angles are $\cos\theta = 0.1, 0.3, 0.5, 0.7$, and 0.9 in the data.

The pion GDAs are determined by fitting these cross-section data with the equations in the previous section [2]. The theoretical cross sections are compared with the Belle data, as an example, at Q^2 =8.92, 13.37 GeV^2 and $\cos\theta = 0.1, 0.5$ in Fig. 6. Here, the results are shown by introducing phase parameters for the S-wave phase. We obtained a reasonable result to explain the data. At $\cos\theta = 0.1$, there is a conspicuous peak from $f_2(1270)$ in the cross section; however, it becomes relatively small at $\cos\theta = 0.5$. There is a $f_0(500)$ effect at small W, and it overlaps with the continuum term in Eq. (19). From the determined GDAs, we calculate the timelike gravitational form factors in Fig. 7 by using Eq. (22). Since they are timelike, they contain both real and imaginary parts. In Θ_2, the $f_2(1270)$ resonance feature is clear at $\sqrt{s}$ = 1.27 GeV, whereas Θ_1 has more complicated s dependence with both S- and D-wave contributions.

The timelike form factors are converted to the spacelike ones by using the dispersion integral over the real positive t ($\equiv s$) with the consideration that singularities of the form factor $\Theta_n(t)$ (n = 1, 2) is in the positive real t axis from $4m_\pi^2$:

$$\Theta_n(t) = \int_{4m_h^2}^{\infty} \frac{ds}{\pi} \frac{\operatorname{Im}\Theta_n(s)}{s - t - i\varepsilon}. \tag{23}$$

Then, the space-coordinate densities are calculated by the Fourier transforms of the spacelike form factors:

$$\rho_n(r) = \int \frac{d^3q}{(2\pi)^3}\, e^{-i\vec{q}\cdot\vec{r}}\, \Theta_n(q) = \int_{4m_h^2}^{\infty} \frac{ds}{4\pi^2} \frac{e^{-\sqrt{s}r}}{r} \operatorname{Im}\Theta_n(s). \tag{24}$$

Using these equations, we obtain the form factors and densities in Figs. 8 and 9 [2].

Physics meaning of these form factors and densities is understood in the following way. The static energy-momentum tensor may be defined by the three-dimensional Fourier transform as [12] $T_q^{\mu\nu}(\vec{r}) = \int d^3q/[(2\pi)^3\, 2E] e^{i\vec{q}\cdot\vec{r}} \left\langle \pi^0(p') \middle| T_q^{\mu\nu}(0) \middle| \pi^0(p) \right\rangle$ with the pion energy $E = \sqrt{m_\pi^2 + \vec{q}^{\,2}/4}$. The $\mu\nu = 00$ component satisfies the mass relation $\int d^3r\, T_q^{00}(\vec{r}) = m_\pi \Theta_{2,q}(0)$, which means that Θ_2 and $\rho_2(r)$ indicate the mass (energy) distributions in the pion. At finite t, Θ_1 also contributes to the mass distribution. The $\mu\nu = ij$ (i, j = 1, 2, 3) components are generally written as $T_q^{ij}(\vec{r}) = p_q(r)\,\delta_{ij} + s_q(r)(r_i r_j/r^2 - \delta_{ij}/3)$ in terms of the pressure $p(r)$ and shear force $s(r)$. Since $T_q^{ij}(\vec{r})$ is expressed by

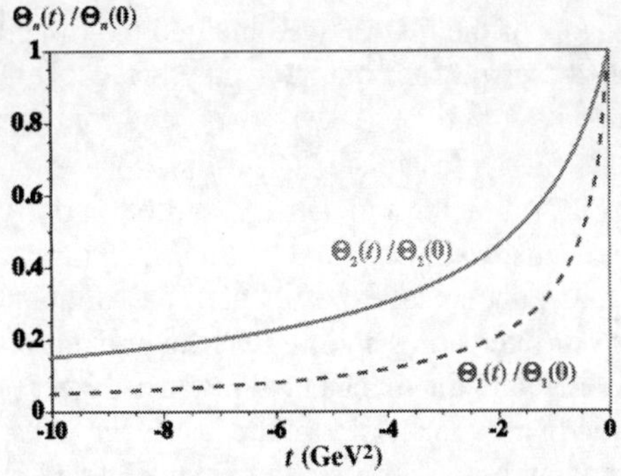

Figure 8. Spacelike gravitational form factors [2].

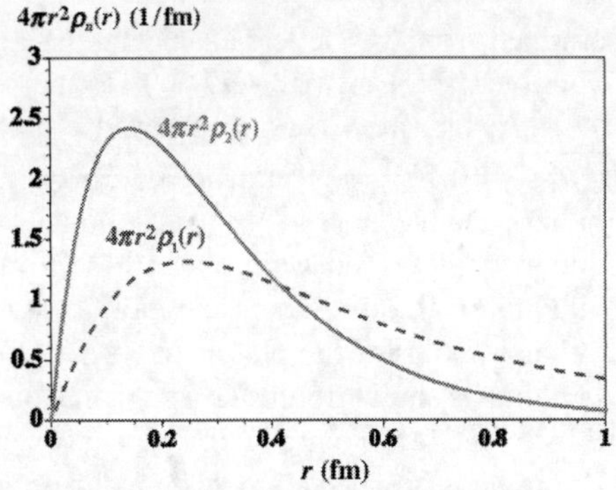

Figure 9. Mass and mechanical densities [2].

only Θ_1, Θ_1 and $\rho_1(r)$ indicate pressure and shear-force distributions in the pion. We may call Θ_2 and $\rho_2(r)$ the gravitational mass (or energy) form factor and density, and Θ_1 and $\rho_1(r)$ may be called the mechanical (pressure, shear force) form factor and density. As shown in Figs. 8 and 9, the mass density has a harder distribution than the mechanical one. From the form factors or densities, we obtain the root-mean-square (rms) radii for both distributions: $\sqrt{\langle r^2\rangle_{\text{mass}}} = 0.69\,\text{fm}$ and $\sqrt{\langle r^2\rangle_{\text{mech}}} = 1.45\,\text{fm}$. We mentioned that the parameters are introduced for the S-wave phase; however, an equally good fit is also obtained by assigning them to the D-wave part. In the D-wave case, our results are slightly different and there are some ambiguities on this assignment. By considering this ambiguity, we obtain the evaluated gravitational radii as [2]:

$$\sqrt{\langle r^2\rangle_{\text{mass}}} = 0.56 \sim 0.69\,\text{fm}, \quad \sqrt{\langle r^2\rangle_{\text{mech}}} = 1.45 \sim 1.56\,\text{fm}. \tag{25}$$

This is the first report on the gravitational radii for a hadron from experimental measurements. It is especially interesting to find that the mass radius is similar or slightly smaller than the charge radius $\sqrt{\langle r^2\rangle_{\text{charge}}} = 0.672 \pm 0.008$ fm and that the mechanical radius is larger. However, the studies of the gravitational form factors and radii for hadrons are still in the beginning stage, and further investigations are needed.

We believe that this new field has bright future to understand gravitational physics from microscopic quark and gluon level. Gravity studies are mainly on macroscopic systems because gravitational interactions are ultra-weak ones and they cannot be detected generally in microscopic particle-physics measurements. However, as we explained in this report, we can find the gravitational source originates from quarks and gluons by using the technique of hadron tomography, namely by the 3D structure functions. The KEKB will produce much accurate cross-section data in the near future by the upgraded super-KEKB, so that the errors in Fig. 6 should become much smaller in a few years. Furthermore, the 3D structure functions can be investigated at various high-energy facilities in the world such as LHC, RHIC, CERN-COMPASS, JLab, Fermilab, J-PARC, GSI, and ILC. Time has come to investigate the 3D tomography including the GDAs for clarifying gravitational properties of hadrons.

5 Summary

We have determined the GDAs, gravitational form factors, and densities for the pion by analyzing the Belle cross-section measurements on the two-photon process $\gamma^*\gamma \to \pi^0\pi^0$. The GDAs are provided with several parameters by considering the continuum term and resonances ones, and they are determined from the Belle data. By taking the first moments of the GDAs, we obtained the timelike gravitational form factors (Θ_1, Θ_2) for the pion. Using the dispersion relation, they are converted to the spacelike form factors. Then, the space-coordinate distributions ($\rho_1(r)$, $\rho_2(r)$) and rms radii, $\sqrt{\langle r^2 \rangle_{\text{mass}}} = 0.56 \sim 0.69$ fm and $\sqrt{\langle r^2 \rangle_{\text{mech}}} = 1.45 \sim 1.56$ fm, are calculated. The functions Θ_2 and $\rho_2(r)$ have the meaning of the gravitational mass (energy) form factor and density, and Θ_1 and $\rho_1(r)$ are the mechanical (pressure, shear force) form factor and density. This should be the first finding on the gravitational form factors and radii for a hadron by analyzing actual experimental measurements. The charge radius of the pion is 0.672 ± 0.008 fm. It is our interesting finding that the gravitational mass radius is similar to this charge radius or slightly smaller, and the mechanical radius is larger.

Since the 3D tomography has been a very popular topic in hadron physics in the last several years, much progress is expected in this novel field of gravitational physics from the fundamental quark and gluon level. Gravitational physics in microscopic systems had been a speculative project for a long time due to ultra-weak interaction nature. However, time has come to investigate it in the microscopic level because the gravitational source from quarks and gluons can be determined, as we showed in this work. Our work is just the beginning of such studies, and much progress is expected in this new research area.

Acknowledgement

This work was supported by Japan Society for the Promotion of Science (JSPS) Grants-in-Aid for Scientific Research (KAKENHI) Grant Number JP25105010.

References

[1] M. Masuda *et al.* (Belle Collaboration), Phys. Rev. D **93**, 032003 (2016).

[2] S. Kumano, Qin-Tao Song, and O. V. Teryaev, arXiv:1711.08088 [hep-ph].

[3] H. Kawamura and S. Kumano, Phys. Rev. D **89**, 054007 (2014).

[4] For a review, see M. Diehl, Phys. Rept. **388**, 41 (2003).

[5] M. Diehl, T. Gousset, B. Pire, and O. Teryaev, Phys. Rev. Lett. **81**, 1782 (1998); M. Diehl, T. Gousset, and B. Pire, Phys. Rev. D **62**, 073014 (2000); M. V. Polyakov, Nucl. Phys. B **555**, 231 (1999); M. V. Polyakov and C. Weiss, Phys. Rev. D **60**, 114017 (1999).

[6] S. R. Deans, *The Radon transform and some of its applications*, (Dover Publications Ins., Mineola, New York, 1992).

[7] O. V. Teryaev, Phys. Lett. B **510**, 125 (2001); C. Mezrag, H. Moutarde, and J. Rodriguez-Quintero, Few Body Syst. **57**, 729 (2016).

[8] I. V. Anikin, B. Pire, L. Szymanowski, O. V. Teryaev, and S. Wallon, Phys. Rev. D **71**, 034021 (2005).

[9] H. Kawamura, S. Kumano, and T. Sekihara, Phys. Rev. D **88**, 034010 (2013). W.-C. Chang, S. Kumano, and T. Sekihara, Phys. Rev. D **93**, 034006 (2016).

[10] S. Kumano, V. R. Pandharipande, Phys. Rev. D **38**, 146 (1988); F. E. Close, N. Isgur, and S. Kumano, Nucl. Phys. B **389**, 513 (1993); T. Sekihara and S. Kumano, Phys. Rev. D **92**, 034010 (2015).

[11] V. Nazari, P. Bydzovsky, and R. Kaminski, Phys. Rev. D **94**, 116013 (2016); P. Bydzovsky, R. Kaminski, and V. Nazari, Phys. Rev. D **90**, 116005 (2014).

[12] M. V. Polyakov, Phys. Lett. B **555**, 57 (2003).

EPJ Web of Conferences **181**, 01026 (2018) https://doi.org/10.1051/epjconf/201818101026
EXA2017

Production of Hypernuclei by Antiprotons

Horst Lenske[1,⋆] and *Alexei* Larionov[1,2,⋆⋆]

[1] *Institut für Theoretische Physik, Justus-Liebig-Universität Gießen, Germany*
[2] *Frankfurt Institute for Advanced Studies (FIAS), D-60438 Frankfurt am Main, Germany*

Abstract. The production of hypernuclei is investigated for $\bar{p}+A \to \bar{\Lambda}+_{\Lambda}A$ reactions in a covariant meson exchange approach. Besides the conventional pseudo-scalar (K) and vector (K^*) channels, we study for the first time also contributions from the correlated πK scalar channel, described by the κ/K_0^* meson. Initial and final state interactions are considered by eikonal theory. The total and angular differential cross sections of the coherent process $\bar{p}+{}^AZ \to {}^A_\Lambda(Z-1)+\bar{\Lambda}$ are evaluated at the beam momenta 1.5...20 GeV/c within the meson exchange model with bound proton and Λ-hyperon wave functions. It is shown that the shape of the beam momentum dependence of the hypernucleus production cross sections with various discrete Λ states is strongly sensitive to the presence of the scalar κ meson exchange in the $\bar{p}p \to \bar{\Lambda}\Lambda$ amplitude. This can be used as a clean test of the exchange by scalar πK correlation in coherent $\bar{p}A$ reactions.

1 Introduction

Hypernuclei are being produced in manifold ways, by photon-, pion-, antikaon-, proton-, antiproton- and nucleus-nucleus interactions [1]. All these types of reactions, except the $\bar{p}A$ one, are rather well studied both experimentally and theoretically. But up to now, there are only few theoretical studies of hypernuclear production in antiproton-nucleus interactions [2–5], and all of them address the incoherent production mechanism. Incoherent hypernucleus production in central collisions is initiated by antikaon production in elementary in-medium $\bar{p}N$ annihilation reactions followed by a sequence of strangeness exchange reactions of the type $\bar{K}N \to Y\pi$. Such reactions are the ideal tool to investigate simultaneously the production of single- and multi-strangeness systems, as discussed e.g. in our recent review article [6]. However, a different approach is required if spectroscopic studies of the final hypernuclei are the aim. For that purpose, the proper method are peripheral reactions by which hypernuclei in bound discrete quantum states are obtained. A comprehensive overview of the status of such studies is found in the recent reviews in [7] and [6]. We have investigated previously coherent hypernuclear production for proton- and pion-induced reactions in [8] and [9], respectively, and for photo-production in [10]. Since such coherent reactions are of a perturbative character, they can be described quantum mechanically by distorted wave methods.

Not much is known, in fact, on the basic $\bar{p}+p \to \bar{\Lambda}+\Lambda$ reaction amplitude. Here, we use a meson exchange model. On the antibaryon side a $\bar{u}$-quark must be changed into a $\bar{s}$ quark while on the baryon side a u-quark has to be transformed into a s-quark. That can be viewed as the propagation of

⋆e-mail: horst.lenske@physik.uni-giessen.de
⋆⋆e-mail: larionov@fias.uni-frankfurt.de

positively charged mesons of a $[u\bar{s}]$ quark structure with strangeness $|S| = 1$ from baryon to antibaryon (or of $[\bar{u}s]$ mesons with strangeness $S = -1$ in the opposite direction). Obvious candidates for such a process are the pseudo-scalar (0^-) kaon K and vector (1^-) K^* mesons. However, we have to expect that also πK correlated exchange in the scalar 0^+-channel may play an important role. The $0^+, |S| = 1$-channel is represented by the $K_0^*(800)$ or κ mesons [11] which may be considered as the $|S| = 1$ members of the (hypothetical) scalar meson octet to which also the $\sigma/f_0(600)$ and the $\delta/a_0(980)$ mesons belong. As for the other 0^+-mesons also the κ meson is characterized by a rather broad spectral distribution with uncertain mass and width. In the present context, the κ/K_0^* meson contributes of course through t-channel exchange processes. In that sense, we consider κ exchange as an economical way to account for the correlated πK channel. The κ exchange channel is of particular interest for multi-strangeness baryonic matter as found in heavy ion collisions and in neutron stars. κ exchange is also an indispensable part of baryon-baryon interaction approaches utilizing the $SU(3)$-flavour group structure. The Nijmegen group was probably the first one to introduce that channel explicitly [12, 13] into their treatment of baryon-baryon scattering while in the Juelich model that channel is treated dynamically as a πK-correlation [14]. We note that in the present context the unnatural parity K-exchange is strongly suppressed for transition involving bound proton and Λ states, respectively, because it is a purely relativistic effect proceeding through the lower wave function components of the Dirac spinors. Thus, coherent hypernucleus production reactions are perfect tools to addressing specifically the exchange of the natural parity K^* and κ mesons. Any other independent source of information allowing to probe hyperon-nucleon and hyperon-nucleus interactions is highly wanted. In this respect, hypernuclear reaction physics may provide important clues.

In section 2 we introduce the Lagrangian describing our covariant production model and the relativistic mean-field approach for bound baryon states. Initial and final state interactions of the scattered antibaryons are taken into account in the eikonal approximation. In section 3 we study the population of discrete bound states in the $^{40}_{\Lambda}$Cl hypernucleus. Angular distributions and total hypernucleus production cross sections are discussed. In section 4 we summarize our results and present our conclusions.

2 Theory of Coherent Antiproton-Nucleus Reactions

Investigations of peripheral reactions with strongly absorbed hadrons require a reliable model for elastic scattering in order to account properly for the extremely important initial (ISI) and final state (FSI) interactions. ISI and FSI will affect both the magnitude and the shape of momentum and angular distributions. Thus, descriptions using lowest order Born approximations as the plane wave approach underlying the Impulse Approximation (IA) do not lead to reliable results. In [15], a Glauber approach was developed for elastic scattering of antiprotons on nuclei, including also investigations of spin-observables. Data of total cross sections as well as angular distributions of cross sections and azimuthal asymmetries and polarization observables are well described over large ranges of energies, corresponding to p_{lab} from a few hundred MeV/c to 20 GeV/c, and over a wide region of momentum transfers. Typical results for elastic scattering angular distributions are displayed in Fig. 1, showing that the LEAR data are well reproduced. A detailed discussion of the dependencies of elastic scattering cross sections and polarization variables on incident energy and target mass and the comparison to proton elastic scattering are found in [15]. For the present case of high energy studies, eikonal theory is used.

The basic $\bar{p} + p \to \bar{\Lambda} + \Lambda$ reaction amplitude is not well explored. Here, we use a meson exchange model. The exchange of K, K^* and κ mesons, respectively, is introduced by the following interaction

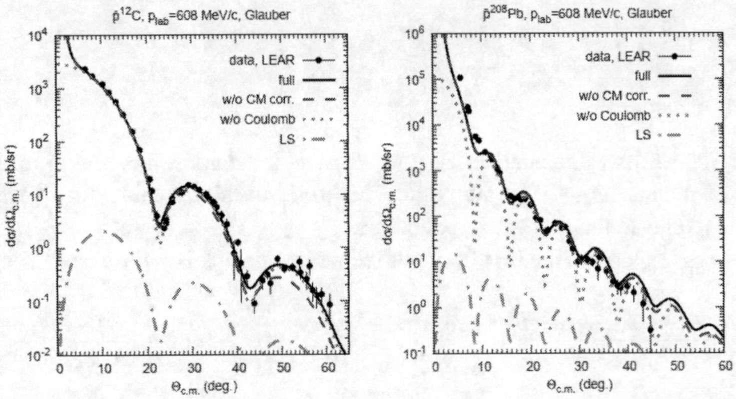

Figure 1. Angular differential cross section of proton elastic scattering on ^{12}C and ^{208}Pb and beam momentum p_{lab} = 608 MeV/c corresponding to T_{lab} = 180 MeV. Contributions of various interaction are shown. For more details see Ref. [15]. The data are from ref. [16].

Lagrangians [17–19]:

$$\mathcal{L}_{KN\Lambda} = -ig_{KN\Lambda}\bar{N}\gamma^5\Lambda K + \text{h.c.}\,, \tag{1}$$

$$\mathcal{L}_{K^*N\Lambda} = \bar{N}\left(G_v\gamma^\mu - \frac{G_t\sigma^{\mu\nu}\partial_\nu^{K^*}}{m_N+m_\Lambda}\right)\Lambda K^*_\mu + \text{h.c.}\,, \tag{2}$$

$$\mathcal{L}_{\kappa N\Lambda} = -g_{\kappa N\Lambda}\bar{N}\Lambda\kappa + \text{h.c.}\,. \tag{3}$$

In the case of the bound proton and Λ we include their wave functions in the field operators of the Lagrangians Eqs. (1)-(3) and calculate the S-matrix in the second order perturbation theory, as discussed in [20]. The coupling constants will be discussed below in section 2.1

The K, K^* and κ exchange (noninvariant) partial matrix elements are expressed as

$$M_K = -g^2_{KN\Lambda}F^2_K(t)\sqrt{\Omega}\,\bar{u}_{-p_1,-\lambda_1}\gamma^5 u_{-p_3,-\lambda_3}\frac{1}{t-m^2_K}\int d^3r e^{-i\boldsymbol{qr}}\bar{\psi}_4(\boldsymbol{r})\gamma^5\psi_2(\boldsymbol{r})\,, \tag{4}$$

$$M_{K^*} = -F^2_{K^*}(t)\sqrt{\Omega}\,\bar{u}_{-p_1,-\lambda_1}\Gamma^\mu(-q)u_{-p_3,-\lambda_3}G_{\mu\nu}(q)\int d^3r e^{-i\boldsymbol{qr}}\bar{\psi}_4(\boldsymbol{r})\Gamma^\nu(q)\psi_2(\boldsymbol{r})\,, \tag{5}$$

$$M_\kappa = g^2_{\kappa N\Lambda}F^2_\kappa(t)\sqrt{\Omega}\,\bar{u}_{-p_1,-\lambda_1}u_{-p_3,-\lambda_3}\frac{1}{t-m^2_\kappa}\int d^3r e^{-i\boldsymbol{qr}}\bar{\psi}_4(\boldsymbol{r})\psi_2(\boldsymbol{r})\,. \tag{6}$$

Here, $\psi_2(\boldsymbol{r})$ and $\psi_4(\boldsymbol{r})$ are the relativistic wave functions of the bound proton and Λ, respectively.

p_i is the four-momentum and $\lambda_i = \pm 1/2$ is the spin magnetic quantum number of an antiproton ($i = 1$) and anti-Lambda ($i = 3$). $q = p_3 - p_1$ is the four-momentum transfer, $t = q^2$. In Eq. (5),

$$G_{\mu\nu}(q) = \frac{-g_{\mu\nu} + q_\mu q_\nu/m^2_{K^*}}{t - m^2_{K^*}} \tag{7}$$

is the K^* meson propagator. We neglected the widths of the K^* and κ mesons in their propagators as the momentum transfers are space-like (e.g. $-t$ = 0.08...1.7 GeV^2 at $p_{\rm lab}$ = 2 GeV/c). The $K^*N\Lambda$ vertex function is defined as

$$\Gamma^\mu(q) = iG_v\gamma^\mu + \frac{G_t}{m_N+m_\Lambda}\sigma^{\mu\nu}q_\nu\,. \tag{8}$$

Monopole vertex form factors $F_{K,K^*,\kappa}$ are used:

$$F_j(t) = \frac{\Lambda_j^2 - m_j^2}{\Lambda_j^2 - t}\,, \quad j = K, K^*, \kappa\,. \tag{9}$$

Similar to refs. [21, 22] we included in Eqs. (4)-(6) the attenuation factor $\sqrt{\Omega}$ to describe the modification of the elementary $\bar{p}p \to \bar{\Lambda}\Lambda$ amplitude due to ISI in the $\bar{p}p$ channel and FSI in the $\bar{\Lambda}\Lambda$ channel, which corresponds to the absorptive potentials.

The differential cross section in the rest frame of the target nucleus is written as

$$d\sigma = \frac{2\pi\delta^{(4)}(p_1 + p_A - p_3 - p_B)}{2p_{\text{lab}}} \frac{1}{2} \sum_{\lambda_1, m_2, \lambda_3, m_4} |M_K + M_{K^*} + M_\kappa|^2 \frac{d^3 p_3}{(2\pi)^3 2E_3} d^3 p_B\,, \tag{10}$$

where p_A and p_B are the four momenta of the initial nucleus (A) and final hypernucleus (B). The δ function in Eq. (10) takes into account the recoil of the hypernucleus. m_2 and m_4 are the total angular momentum magnetic quantum numbers of the occupied proton state from the valence shell and of the Λ hyperon, respectively. The factor of $1/2$ expresses the averaging over λ_1.

2.1 Coupling Constants

The choice of coupling constants is based on $SU(3)$ relations [23]:

$$g_{KN\Lambda} = -g_{\pi NN}\frac{3 - 2\alpha_{PS}}{\sqrt{3}}\,, \tag{11}$$

$$G_{v,t} = -G^{\rho}_{v,t}\frac{3 - 2\alpha_{E,M}}{\sqrt{3}}\,, \tag{12}$$

$$g_{\kappa N\Lambda} = -g_{\sigma NN}\frac{3 - 2\alpha_S}{3 - 4\alpha_S}\,, \tag{13}$$

where α's are the D-type coupling ratios. The πNN coupling constant is very well known, $g_{\pi NN} = 13.4$ [24]. The vector ρNN coupling constant is also fixed, $G^{\rho}_{v} = 2.66$, however, the tensor ρNN coupling constant is quite uncertain, $G^{\rho}_{t} = 10.9...20.6$ [17]. The σNN coupling constant can be estimated either from the Bonn model [25] or from the Walecka-type models (cf. [26]). In both cases one obtains $g_{\sigma NN} \simeq 10$. The α's for the octets of light pseudoscalar and vector mesons are reasonably well determined [17, 19]: $\alpha_{PS} \simeq 0.6$, $\alpha_E \simeq 0$, $\alpha_M \simeq 3/4$. However, there is no any phenomenological information on α_S.

Thus, the coupling constants G_t and $g_{\kappa N\Lambda}$, the cutoff parameters Λ_K, Λ_{K^*} and Λ_κ, and the attenuation factor Ω remain to be determined from comparison with experimental data. We adjusted these parameters to describe the beam momentum dependence of the $\bar{p}p \to \bar{\Lambda}\Lambda$ cross section. The two sets of parameters, (1) without κ meson and (2) with κ meson, are listed in Table 1. In the calculations we used the mass $m_\kappa = 682$ MeV [27].

As we see from Fig. 2, in the calculation with set 1 the peak of the $\bar{p}p \to \bar{\Lambda}\Lambda$ cross section at $p_{\text{lab}} \simeq 2$ GeV/c is exhausted by the K exchange. In contrast, in the case of set 2 the peak is exhausted mostly by the κ exchange. The K^* exchange contribution grows with beam momentum and becomes dominant at $p_{\text{lab}} > 3$ GeV/c. We note that set 2 gives a steeper increasing angular differential cross section towards $\Theta_{\text{c.m.}} = 0$ at $p_{\text{lab}} = 2.060$ GeV/c in a good agreement with experimental data [30].

Table 1. Parameters of the $\bar{p}p \to \bar{\Lambda}\Lambda$ amplitude. The value of $g_{KN\Lambda}$ slightly differs from -13.3 as given by Eq. (11) and is taken from from K^+N scattering analysis of ref. [28]. The cutoff parameters Λ_K, Λ_{K^*} and Λ_κ are in GeV. The attenuation factors are shown in the last column.

Set	$g_{KN\Lambda}$	G_v	G_t	$g_{\kappa N\Lambda}$	Λ_K	Λ_{K^*}	Λ_κ	Ω
1	-13.981	-4.6	-8.5	—	2.0	1.6	—	0.0150
2	-13.981	-4.6	-9.0	-7.5	1.8	2.0	1.8	0.0044

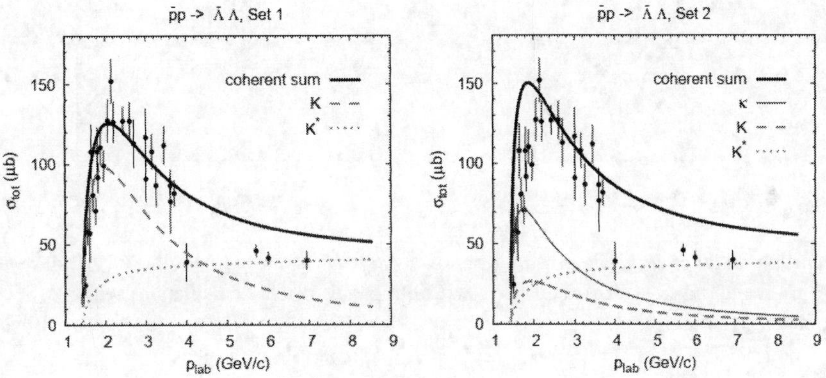

Figure 2. Angle-integrated cross section of the elementary process $\bar{p}p \to \bar{\Lambda}\Lambda$ as a function of the beam momentum calculated without (Set 1) and with (Set 2) inclusion of the κ meson. Experimental data are from ref. [29].

Table 2. Binding energies of the Λ states in the $^{40}_{\Lambda}$Cl nucleus. Empirical Λ binding energies (spin-orbit splitting not resolved) for $^{40}_{\Lambda}$Ca from ref. [31] are given in brackets.

Λ state	B_Λ [MeV]
$1s_{1/2}$	18.55 (18.7 ± 1.1)
$1p_{3/2}$	10.20 (9.9 ± 1.1)
$1p_{1/2}$	9.26 (9.9 ± 1.1)
$1d_{5/2}$	2.14 (1.5 ± 1.1)
$2s_{1/2}$	1.44
$1d_{3/2}$	0.84 (1.5 ± 1.1)

2.2 Nucleon and Hyperon States

The nucleon and hyperon ($B = N, \Lambda$) single particle bound state wave functions are determined as solutions of a static Dirac equation with scalar and vector potentials (cf. refs. [9, 31, 32]):

$$(-i\alpha \cdot \nabla + \beta m^*_B(r) + V_B(r) + V_C(r) - \varepsilon)\,\psi_B(\mathbf{r}) = 0\,, \tag{14}$$

where $m^*_B(r) = m_B + S_B(r)$ is the effective (Dirac) mass. Both the scalar (S_B) and nuclear vector (V_B) potentials are chosen in the form of superpositions of the classical meson fields, $\sigma(I = 0, J^P =$

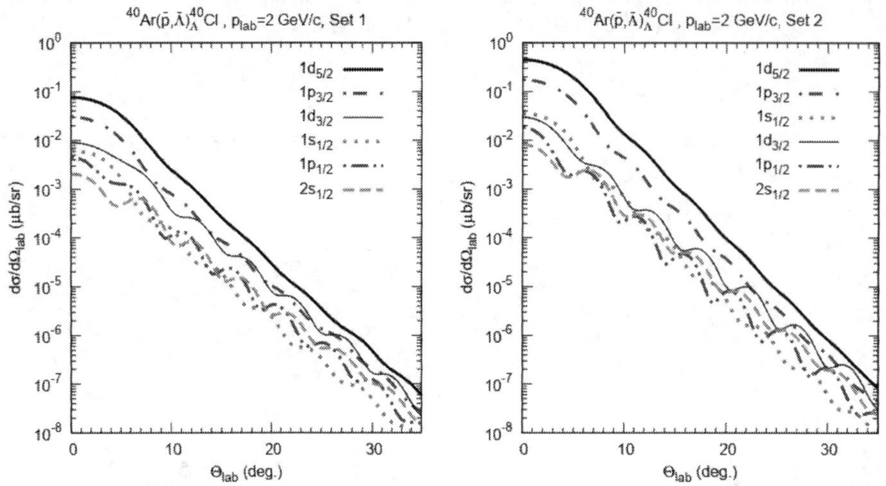

Figure 3. Angular differential cross section of the reaction $^{40}\mathrm{Ar}(\bar{p}, \bar{\Lambda})^{40}_{\Lambda}\mathrm{Cl}$ at $p_{\mathrm{lab}} = 2$ GeV/c. Lines show the calculations for Λ in various states, as indicated. Left and right panels display calculations without (Set 1) and with (Set 2) κ exchange.

$0^+)$, $\omega(0, 1^-)$, $\delta(1, 0^+)$ and $\rho(1, 1^-)$, weighted by the strong interaction coupling constants appropriate for the given baryon. For the present purpose, the meson fields are parameterized by Woods-Saxon form factors. For protons also the static Coulomb potential (V_C) contributes [31]. The obtained Λ energy levels are in good agreement with the empirical systematics and with our previous relativistic mean-field calculations [31], as seen from Table 2. A detailed discussion of the model parameters and spectroscopic results for Λ binding energies and wave functions is found in [20].

3 Results on Hypernuclear Production by Peripheral Antiproton Reactions

An appropriate case for exploratory studies is ^{40}Ar which may be of potential interest as target material for future experiments at PANDA@FAIR. In order to assure that after the reaction the residual core nucleus carries as little excitation energy as possible, we consider only strangeness creation processes on protons in the $1d_{3/2}$ valence shell of ^{40}Ar.

The differential hypernuclear production cross sections with the Λ occupying various shells are compared in Fig. 3. Irrespective of spin-orbit effects, overall the cross sections are larger for larger hyperon orbital angular momentum, i.e. l_Λ. This is a consequence of the interplay of several effects:

- The momentum transfer at $\Theta = 0$ is small (~ 0.3 GeV/c) implying a suppression of the $p \to \Lambda$ transitions with large orbital momentum transfer.
- The number of the spin states of the Λ contributing to the transition probability, i.e. $2(2l_\Lambda + 1)$, grows obviously with l_Λ.
- For Λ states with larger l_Λ the hyperon probability distribution is increasingly shifted to larger radii. Hence, the absorption effects are diminished with increasing l_Λ.

The inclusion of κ exchange leads to significant enhancement of the cross sections at small polar angles for all states of the produced hypernucleus.

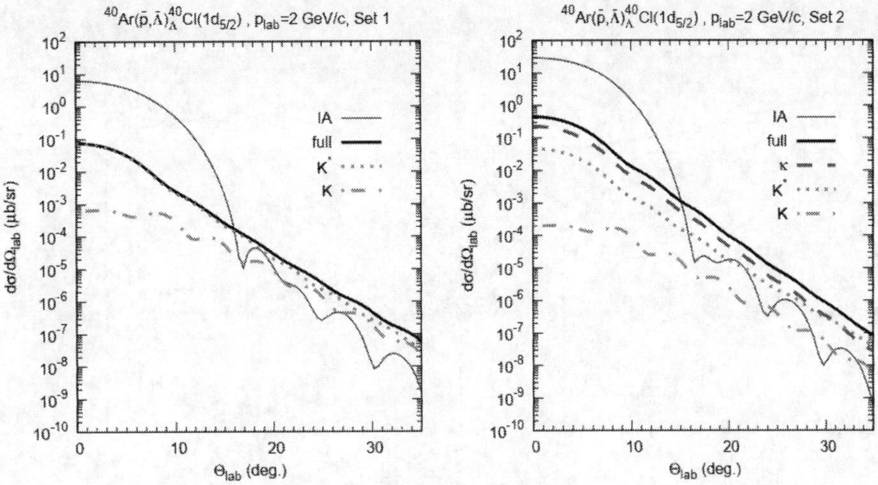

Figure 4. Angular differential cross section of the reaction ^{40}Ar$(\bar{p}, \bar{\Lambda})^{40}_{\Lambda}$Cl at p_{lab} = 2 GeV/c with $1d_{5/2}$ Λ state. As indicated, the IA calculation, the full calculation (with absorption), and the separate meson contributions to the full calculation are shown by different lines. The left and right panels display the results without (Set 1) and with (Set 2) κ meson, respectively.

The largest cross section is obtained for the $^{40}_{\Lambda}$Cl hypernucleus with Λ in the $1d_{5/2}$ state. The differential angular distribution for this case is analyzed in more detail in Fig. 4. From the comparison of the full and IA calculations we observe that the absorption of $\bar{p}$ and $\bar{\Lambda}$ has a quite significant effect: it reduces the cross section drastically, amounting at forward angles to about two orders of magnitude, and washes out the diffractive structures. Similar effects of the absorption are present also for the other Λ states (not shown).

A deeper insight into the production mechanism is obtained by decomposing the total reaction amplitude into the different meson exchange parts. From the partial meson exchange contributions, shown in Fig. 4, it is remarkable that for Set 1 the kaon contribution is small and the spectrum is dominated by K^*, even at large angles, while on first sight one would expect the opposite. For example, for $\bar{p}A$ collisions at $p_{\text{lab}} = 2$ GeV/c the $\bar{\Lambda}$ produced at $\Theta_{\text{lab}} = 30°$ carries away the momentum transfer of ~ 1 GeV/c. This corresponds approximately to $\Theta = 90°$ in center-of-mass frame if translated into the $\bar{p}p \to \bar{\Lambda}\Lambda$ reaction in free space. Thus, we should expect that the kaon exchange should be a factor of five larger than the K^* exchange. This surprising result can be understood by the fact that the momentum transfer to the $\bar{\Lambda}$ is provided by the nucleus as a whole while the hyperon is almost at rest. The exchange of a pseudoscalar meson is suppressed in this case since it proceeds through the lower of the proton and Λ Dirac spinors which are suppressed by factors of the order of the baryon masses. In contrast, in the case of the free space $\bar{p}p \to \bar{\Lambda}\Lambda$ process at $\Theta = 90°$ the Λ is produced with finite momentum and, thus, the upper and lower components of its Dirac spinor are of comparable magnitude which favors the pseudoscalar meson exchange. The situation is very different in the case of Set 2. Here, κ plays the dominant role both for the free scattering $\bar{p}p \to \bar{\Lambda}\Lambda$ and for the hypernucleus production since scalar exchange is not suppressed in recoilless kinematics.

The robust signal of the κ exchange is visible in the momentum dependence of the hypernucleus production cross section, as seen in Fig. 5. In calculations without κ the cross section is dominated

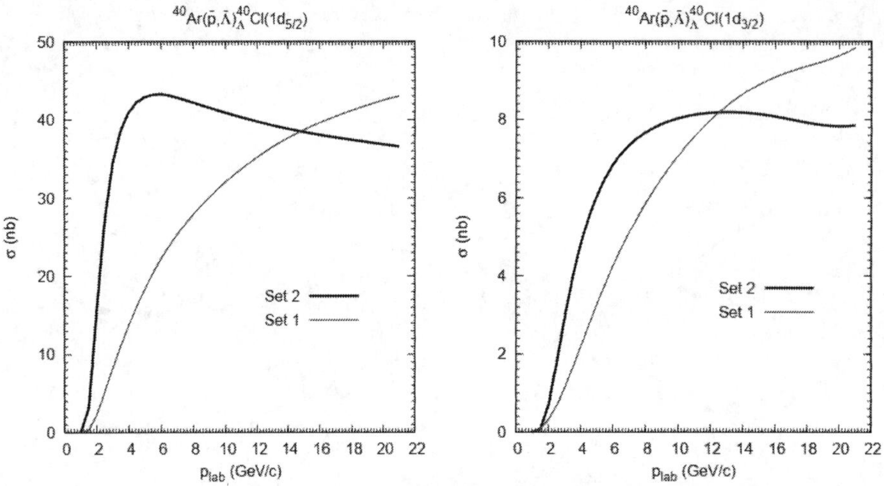

Figure 5. Beam momentum dependence of the $^{40}_{\Lambda}$Cl d-state hypernucleus production cross section in $\bar{p}\,^{40}$Ar collisions. The thick and thin solid lines show, respectively, the results with (Set 2) and without (Set 1) κ exchange.

by K^* exchange which leads to a growing cross section with increasing beam energy. Using set 2, the κ meson dominates at moderate beam momenta ~ 1.5...3 GeV/c. Its contributions are seen as a characteristic shoulder in $p_{\rm lab}$-dependence of the hypernuclear production cross section and even as the appearance of the maximum for $1d_{5/2}$ Λ state.

As we see from Fig. 6, the cross section of coherent hypernucleus production in different states is much larger when the κ exchange is included. This is a pure quantum coherence effect since the angle-integrated $\bar{p}p \to \bar{\Lambda}\Lambda$ cross sections differ by ~ 15% only at $p_{\rm lab}$ = 2 GeV/c (Fig. 2) while the hypernuclear production cross sections differ by almost one order of magnitude for Set 1 and Set 2.

4 Summary

In summary, the coherent hypernucleus production in $\bar{p}A$ collisions was investigated. The production dynamics of the elementary and the in-medium annihilation amplitudes were described in a covariant meson exchange approach. Initial and final state interactions of the scattered baryons have been taken into account by eikonal theory. Baryon bound states were obtained by a RMF-model. The approach was applied to hypernuclear production in coherent reactions on the medium-heavy ^{40}Ar target nucleus in the momentum range $p_{\rm lab}$ ~ 1.5...20 GeV/c. The total hypernucleus production cross sections populating a fixed quantum state generally grow with increasing beam momentum from several nb to a few of 10 nb with a certain sensitivity on the Λ bound state. The dynamics of the ^{40}Ar($\bar{p}$, $\bar{\Lambda}$)$^{40}_{\Lambda}$Cl reaction on the $1d_{3/2}$ valence shell proton favors the production of Λ states with $j = \ell + 1/2$ with the cross sections increasing with orbital angular momentum ℓ.

The pseudoscalar (K) exchange is strongly suppressed for the reactions replacing a bound proton by a bound Λ hyperon. The production mechanism is governed, in fact, by the exchange of natural parity vector and scalar strange mesons. Including only K^* exchange produces smooth and structureless cross sections increasing steadily with beam momentum for all possible bound Λ states. However, if the exchange of the scalar πK channel – described by the κ meson – is taken into account we find

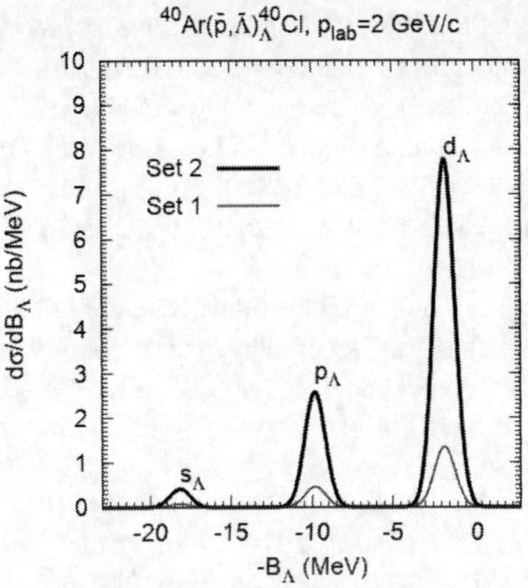

Figure 6. The Λ binding energy spectrum of the $^{40}_{\Lambda}$Cl hypernuclei coherently produced in $\bar{p}\,^{40}$Ar collisions at p_{lab} = 2 GeV/c. The smooth curves are obtained by multiplying the angle-integrated cross sections for the hypernucleus production in $1s_{1/2}$, $1p_{3/2}$ and $1d_{5/2}$ states by the Gaussians of a width FWHM=1.5 MeV which is a typical experimental energy resolution. Thick and thin solid lines correspond to the calculations with (Set 2) and without (Set 1) κ exchange.

that at the beam momenta in the range of p_{lab} = 4...6 GeV/c a rather sudden transition from increase to saturation occurs or, as in the case of the $^{40}Cl(1d_{5/2})$ Λ state, a maximum is emerging. These results strongly suggest that the coherent hypernuclear production in $\bar{p}A$ annihilation reactions could be a suitable tool to test in quite detail the dynamics of the production process, down to the possibility to identifying contributions from scalar πK correlation as described by the κ meson.

Acknowledgments

This work was supported by Deutsche Forschungsgemeinschaft (DFG) under Grant No. Le439/9-2.

References

[1] H. Bando, T. Motoba, J. Zofka, Int. J. Mod. Phys. **A5**, 4021 (1990)
[2] J. Cugnon, P. Deneye, J. Vandermeulen, Nucl. Phys. **A513**, 636 (1990)
[3] T. Gaitanos, A.B. Larionov, H. Lenske, U. Mosel, Nucl. Phys. **A881**, 240 (2012), 1111.5748
[4] T. Gaitanos, H. Lenske, Phys. Lett. **B737**, 256 (2014), 1408.6051
[5] T. Gaitanos, C. Moustakidis, G.A. Lalazissis, H. Lenske, Nucl. Phys. **A954**, 308 (2016), 1602.08905
[6] H. Lenske, M. Dhar, T. Gaitanos, X. Cao, Progress in Particle and Nuclear Physics **98C**, 119 (2018)
[7] A. Gal, E.V. Hungerford, D.J. Millener, Rev. Mod. Phys. **88**, 035004 (2016), 1605.00557

[8] R. Shyam, H. Lenske, U. Mosel, Nucl. Phys. **A764**, 313 (2006), nucl-th/0505043
[9] S. Bender, R. Shyam, H. Lenske, Nucl. Phys. **A839**, 51 (2010), 0910.4868
[10] R. Shyam, H. Lenske, U. Mosel, Phys. Rev. **C77**, 052201 (2008), 0710.4888
[11] K.A. Olive et al. (Particle Data Group), Chin. Phys. **C38**, 090001 (2014)
[12] R.G.E. Timmermans, T.A. Rijken, J.J. Deswart, Nucl. Phys. **A479**, 383C (1988)
[13] R.G.E. Timmermans, T.A. Rijken, J.J. de Swart, Phys. Rev. **D45**, 2288 (1992)
[14] J. Haidenbauer, U.G. Meissner, Phys. Rev. **C72**, 044005 (2005), nucl-th/0506019
[15] A.B. Larionov, H. Lenske, Nucl. Phys. **A957**, 450 (2017), 1609.03343
[16] D. Garreta et al., Phys. Lett. **149B**, 64 (1984)
[17] M.K. Cheoun, B.S. Han, I.T. Cheon, B.G. Yu, Phys. Rev. **C54**, 1811 (1996)
[18] K. Tsushima, A. Sibirtsev, A.W. Thomas, G.Q. Li, Phys. Rev. **C59**, 369 (1999), [Erratum: Phys. Rev.C61,029903(2000)], nucl-th/9801063
[19] B.S. Han, M.K. Cheoun, K.S. Kim, I.T. Cheon, Nucl. Phys. **A691**, 713 (2001), nucl-th/9912011
[20] A.B. Larionov, H. Lenske, Phys. Lett. **B773**, 470 (2017), 1709.04370
[21] R. Shyam, H. Lenske, Phys. Rev. **D90**, 014017 (2014), 1406.7071
[22] R. Shyam, H. Lenske, Phys. Rev. **D93**, 034016 (2016), 1509.04786
[23] J.J. de Swart, Rev. Mod. Phys. **35**, 916 (1963), [Erratum: Rev. Mod. Phys.37,326(1965)]
[24] O. Dumbrajs, R. Koch, H. Pilkuhn, G.c. Oades, H. Behrens, J.j. De Swart, P. Kroll, Nucl. Phys. **B216**, 277 (1983)
[25] R. Machleidt, K. Holinde, C. Elster, Phys. Rept. **149**, 1 (1987)
[26] G.A. Lalazissis, J. Konig, P. Ring, Phys. Rev. **C55**, 540 (1997), nucl-th/9607039
[27] C. Patrignani et al. (Particle Data Group), Chin. Phys. **C40**, 100001 (2016)
[28] R. Buettgen, K. Holinde, A. Mueller-Groeling, J. Speth, P. Wyborny, Nucl. Phys. **A506**, 586 (1990)
[29] A. Baldini et al., *Landolt-Börnstein, V. 12* (Springer, Berlin, 1987)
[30] B. Jayet, M. Gailloud, P. Rosselet, V. Vuillemin, S. Vallet, M. Bogdanski, E. Jeannet, C.J. Campbell, J. Dawber, D.N. Edwards, Nuovo Cim. **A45**, 371 (1978)
[31] C.M. Keil, F. Hofmann, H. Lenske, Phys. Rev. **C61**, 064309 (2000), nucl-th/9911014
[32] N.K. Glendenning, D. Von-Eiff, M. Haft, H. Lenske, M.K. Weigel, Phys. Rev. **C48**, 889 (1993), nucl-th/9211012

EPJ Web of Conferences **181**, 01027 (2018) https://doi.org/10.1051/epjconf/201818101027
EXA2017

Heavy-hadron interactions from Lattice QCD

Daniel Mohler[1,2,*]

[1]Helmholtz-Institut Mainz, 55099 Mainz, Germany
[2]Johannes Gutenberg Universität Mainz, 55099 Mainz, Germany

Abstract. I review recent progress in heavy hadron spectroscopy and interactions from ab-initio Lattice QCD calculations. After motivating lattice spectroscopy calculations for heavy-hadrons by contrasting recent LHCb results for charmed and doubly-charmed baryons with lattice predictions, selected results from scattering calculations for heavy-light mesons and for charmonia are presented. I close with a discussion of recent Lattice QCD predictions of explicitly exotic doubly-heavy states.

1 Introduction

In Lattice Quantum Chromodynamics, the action of QCD is regularized by a 4-dimensional Euclidean space-time lattice. It is an *ab-initio*, non-perturbative approach which can be improved systematically and thereby allows for a full control of systematic uncertainties. While calculations on fine lattices with dynamical quarks require large computational capabilities, progress in both hardware and algorithms have allowed for modern calculations to determine quantities of phenomenological relevance from QCD. An example is given by the flavor physics results collected in the FLAG review [1]. Most Lattice QCD calculations work in the isospin limit, and the light up and down as well as the heavier strange (referred to as 2+1 flavor simulations) or even light, strange and charm quarks (referred to as 2+1+1 flavor simulations) are included in the creation of gauge configurations (dynamical simulation). While many lattice simulations still employ heavier than physical light-quark masses corresponding to heavier than physical pions, calculations at physical pion masses are now feasible and many results in the FLAG review already result from ensembles with close-to-physical pion masses.

More recently, the isospin splittings from both non-degenerate up- and down-quark masses, as well as from their different charge under QED have been taken into account. An impressive example is provided in Figure 1 from a simulation by the Budapest-Marseille-Wuppertal (BMW) collaboration [2], which shows Isospin splittings for various hadrons determined from a dynamical QCD+QED simulation.

In this review we will focus on recent progress in the determination of the heavy-hadron spectrum from Lattice QCD, with particular emphasis on extracting resonances and bound states from heavy-hadron scattering in a finite volume. The latter is based on a finite-volume technique first worked out by Lüscher [3, 4], that relates the energy shifts in a hypercubic finite volume to infinite-volume scattering amplitudes. In its original form, the phase shift for elastic scattering is inferred from rest-frame calculations in multiple spatial volumes L^3.

*e-mail: damohler@uni-mainz.de

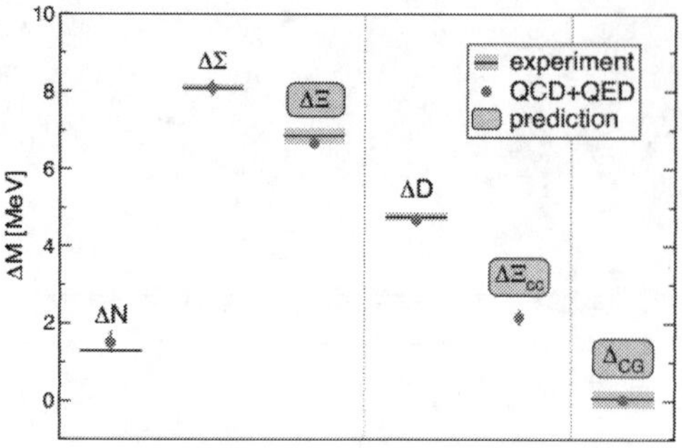

Figure 1. Isospin splitting from QCD+QED by the BMW collaboration. The error bars denote the full uncertainty estimate while the black bars and shaded regions show the experiment values. The splitting $\Delta\Xi_{cc}$ is of importance for the discussion in Section 2. Figure taken from [2]

This original method has been generalized in many ways. One kind of improvement focuses on obtaining more information from available lattices, by making use of calculations in both the rest frame and in a number of moving frames (for a first generalization see [5]). More important, the method has also been generalized to allow for coupled-channel scattering with an arbitrary number of two-hadron channels, for the treatment of $2 \leftrightarrow 1$ and $2 \leftrightarrow 2$ transitions (such as $\pi\pi \to \pi\gamma^*$), and to allow for 3-particle scattering. For an account of these generalizations please refer to a recent review [6] and to the references therein.

One requirement for successfully applying these finite volume methods to QCD is the availability of an efficient way to calculate all (valence) quark-line diagrams. In particular, the channels with attractive interaction, which are most relevant phenomenologically, typically involve diagrams where quarks propagate from a given timeslice back to the same timeslice. (The simplest of these would be the "disconnected diagrams" arising in the case of an isoscalar meson.) Recent techniques such as the *Distillation* [7] and *Stochastic Distillation* (or LapH) [8] methods allow the determination of these diagrams, while at the same time offering a large freedom for the construction of the lattice operators used for the creation and annihilation of the hadrons. For heavy quarks, another important ingredient is the development of improved heavy-quark actions (see for example [9, 10]), which allow for reasonably small discretization effects on currently available lattices with a lattice spacing a of roughly 0.05 to 0.12 fm.

In section 2, I will provide a motivation for modern lattice spectroscopy calculations, by taking a look at existing lattice calculations for heavy baryons containing charm quarks, and contrasting these calculations with recent results from the LHCb experiment. In section 3, I will provide an overview of lattice calculations of heavy-light mesons, where a number of results for bound-states and resonances in meson-meson scattering exist [11–16]. In section 4, exploratory results for charmonium and charmonium-like states are presented. Before concluding, I present an account of Lattice QCD predictions for explicitly exotic doubly-heavy hadrons predicted by several groups in Section 5. It should be noted that many of the results covered in this review are exploratory in nature, i.e. not all systematic uncertainties have been investigated. In particular some results discussed are obtained at unphysical light-quark masses, and many of the studies discussed did not perform an extrapolation to zero lattice spacing. As such most results are qualitative in nature. In addition, these results provide an outlook on future Lattice QCD results.

2 Charmed heavy baryons

Recently, the LHCb collaboration observed a doubly-charmed Cascade baryon Ξ_{cc}^{++} [17]. They observe a clear signal in data from LHC runs at both 13 TeV and 8 TeV, and determine the mass of the hadron to be $3621.40 \pm 0.72 \pm 0.27 \pm 0.14$ MeV. A number of years ago, an isospin partner to this state has been claimed by the SELEX collaboration [18, 19] [1]. The Ξ_{cc} state from SELEX has mass of approximately 3519 MeV [18, 19], and was not seen in other subsequent experiments [21–25], casting some doubt on this result.

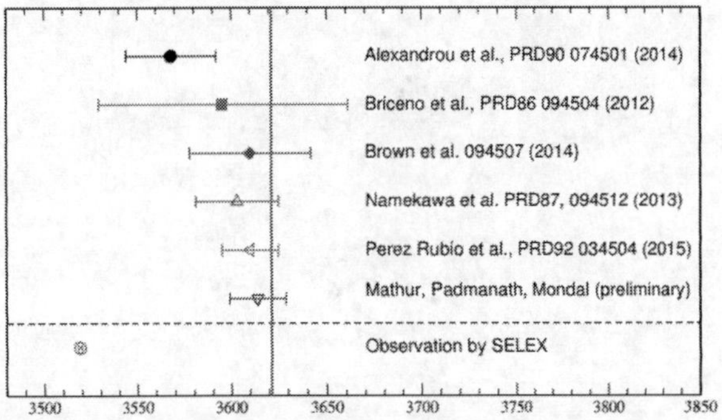

Figure 2. Overview of selected Lattice QCD predictions for the Ξ_{cc} in the isospin limit. Full symbols are chosen for calculations with good control of all relevant systematic uncertainties, while open symbols indicate that the results are results at a single (but reasonably fine) lattice spacing rather than continuum results. The results are compared to the state observed by SELEX [18, 19] (green circle) and to the recent observation by LHCb [17] (magenta line).

Before the observation by the LHCb collaboration, a number of lattice calculations yielded predictions for the mass of Ξ_{cc} in the isospin limit. The most advanced of these determinations [26–31] are collected in Figure 2, together with the state claimed by SELEX (green circle) and the observation by LHCb (magenta vertical band). While having a sizeable uncertainty, which largely stems from the limited set of gauge configurations, the various lattice determinations agree within uncertainties and the more precise determinations show considerable tension with the observation by SELEX, but agree quite well with the observed mass from LHCb. Beyond these more comprehensive calculations, a host of older predictions with a somewhat less well determined systematic uncertainty exist (see for example [32] for an impressive early prediction). Note that in addition to this doubly charmed baryon, many other charmed baryons are predicted in the same literature, some of which can likely be observed by LHCb in the future.

In another recent observation by LHCb, 5 narrow charmed Ω_c baryons have been discovered [33]. Figure 3 shows the discovery plot from LHCb with clearly observed and well separated peaks. While the spin-parity of these states has not yet been determined from experiment, the fact that the observed baryons seem to be quite narrow allows for a qualitative comparison to finite volume energies determined from 3-quark Euclidean space correlation functions calculated on the lattice.[2]

[1]Note that this Isospin splitting has been determined from Lattice QCD in [2] and has been found to be quite small, as one would in general expect. For a discussion of doubly-heavy baryon isospin splittings also see [20]. From a theoretical point of view it is therefore essentially impossible that this state could be an isospin partner.

[2]This approach essentially treats these hadrons as stable under the strong interaction, and will therefore only be applicable to very narrow resonances, as the naive uncertainty estimate of such a prediction is of the order of the width of the resonance.

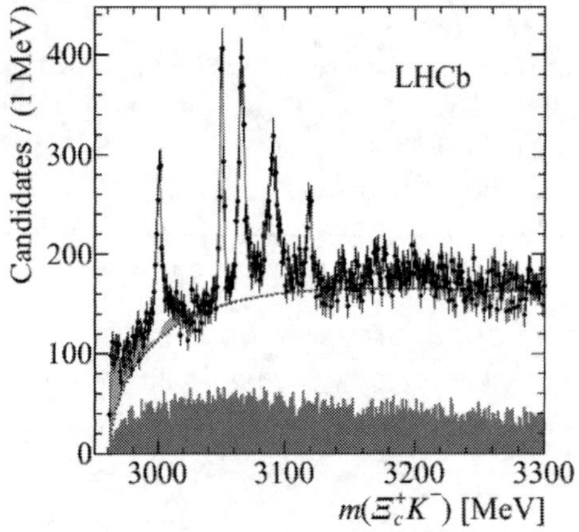

Figure 3. Discovery plot for the five narrow Ω_C baryons observed by LHCb. Figure taken from [33].

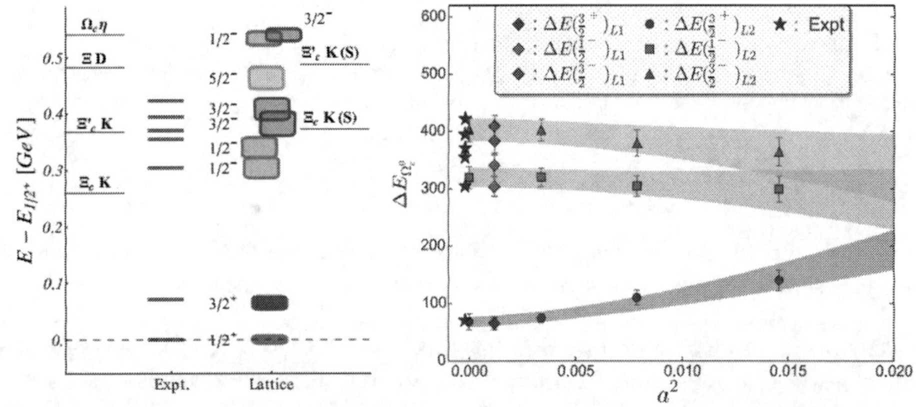

Figure 4. Left: Comparisons of energy splittings in the Ω_c baryon spectrum seen in experiment and the lowest energy levels determined on the lattice. For the lattice data the J^P quantum numbers are denoted and the lowest S-wave scattering thresholds are plotted with black lines. Right: Continuum extrapolation for three of the energy levels using a different lattice action. Both Figures taken from [34].

This is done in a recent letter [34], where the pattern of excitations from two different lattice calculations is compared to the excitations observed by LHCb. In a first calculation, the authors use a large basis of interpolating fields for charmed baryons and determine the continuum spin with the method first described in [35]. The results are collected in the left pane of Figure 4, where the mass-splittings to the $J^P = \frac{1}{2}^+$ ground state are compared to experiment. Their result suggests that the lowest two states observed by LHCb might have $J^P = \frac{1}{2}^-$, while the next two might have $J^P = \frac{3}{2}^-$ and the last one might have $J^P = \frac{5}{2}^-$.

To check that this picture is not spoiled by large discretization effects, the authors perform another calculation using a smaller interpolator basis and multiple lattice spacings. While their basis is not large enough to resolve the closely spaced energy levels with the same spin-parity and the spin $\frac{5}{2}$ state, the results shown in the right pane of Figure 4 qualitatively confirms the pattern observed previously.

3 Heavy-light hadrons

In this section I will briefly review existing studies determining bound states and resonances of mesons with a heavy charm or bottom quark and a light quark. One of the first candidates for an exotic heavy-light hadron was the $D^*_{s0}(2317)$ with $J^P = 1^+$ [36], whose mass is not only much lighter than expected from older potential model calculations such as [37], but also almost the same than the mass of the corresponding $J^P = 0^+$ D-meson. Short after its discovery, it has been noted that the close-by D-meson-kaon threshold is of special importance for this state [38].

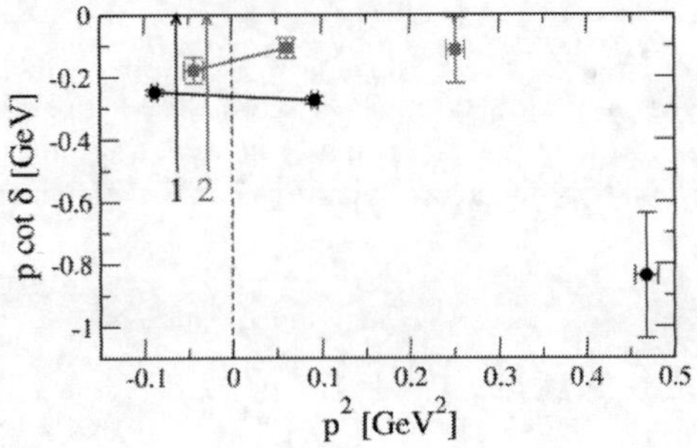

Figure 5. Determination of the scattering length and effective range used to extract the $D^*_{s0}(2317)$ pole position. Ensemble (1) black circles, Ensemble (2) red squares; the vertical arrows indicate the square of the binding momenta of the bound state for ensembles (1) and (2), and the dashed line indicates the threshold. Figure from [13].

In lattice QCD the $D^*_{s0}(2317)$ and the $D_{s1}(2460)$ belonging the to same $j = \frac{1}{2}$ doublet in the heavy-quark limit have been investigated in [12, 13], extracting the finite volume energy spectrum with a combined basis of quark-antiquark and D^*-K interpolating fields. Using Lüscher's formula and an effective range approximation

$$p \cot \delta_0(p) = \frac{2}{\sqrt{\pi}L} Z_{00}\left(1; \left(\frac{L}{2\pi}p\right)^2\right)$$
$$\approx \frac{1}{a_0} + \frac{1}{2} r_0 p^2 ,$$

the two data points close-to threshold were used to determine the scattering length and effective range for two ensembles of gauge configurations with different lattice spacing (labeled as ensemble (1) and ensemble (2) below), pion mass, and volume. The resulting data is shown in Figure 5.

It should be stressed that this is a somewhat rough determination, as discretization effects, which are sizeable for charm-light states with the lattice action used remain uncontrolled, and the pion mass in the simulations is not quite physical. The left part of Figure 6 shows how the resulting energy splittings with respect to the spin-averaged ground states compare to experiment for the two ensembles used. Unlike results from previous lattice calculations, the statistical uncertainty is well under control and the pattern agrees qualitatively with the pattern from experiment. Indeed simple order of magnitude estimates of the discretization uncertainties expected within the Fermilab prescription for heavy quarks [10] suggest that discretization effects alone could account for the differences observed.

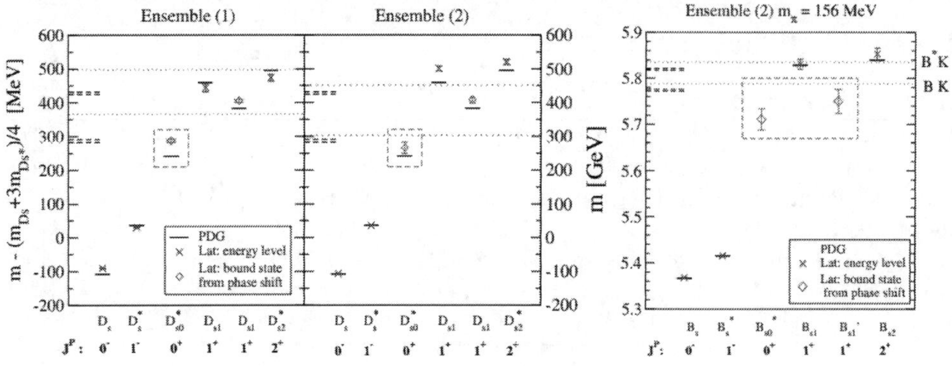

Figure 6. Left: Mass splittings in the low-lying D_s spectrum. The lattice results for bound-state poles on Ensembles (1) and (2) are the magenta data points. Mass splittings in experiment are indicated by the black lines. Right: Mass splittings in the low-lying B_s spectrum. The magenta data points are lattice predictions.

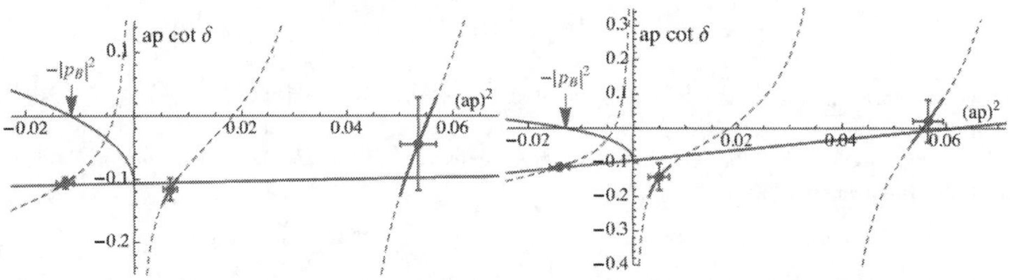

Figure 7. Effective range fits used to extract the pole positions for the predicted positive parity B_s mesons. The left pane shows the data for S-wave scattering in the A_1 irreducible representation, while the right pane shows the S-wave in T_1. Figure from [14]

Unlike for charm-strange mesons, the expected discretization effects for B_s states are quite small. In [14] the positive parity B_s cousins of the aforementioned D_s states, which are not yet seen in experiment, have been predicted. Figure 7 shows the resulting lattice data together with an effective range approximation to the points resulting from the three lowest energy levels. The predicted spectrum is shown in the right part of Figure 6, where the uncertainties for the points in magenta now contains a full uncertainty estimate detailed in [14].

More recently, the positive parity $D^*_{s0}(2317)$ and $D_{s1}(2460)$ were revisited by different authors in [16], using lattices with multiple volumes and two different pion masses. The use of multiple lattice volumes leads to a larger number of points close-to threshold, and allows for a test of the effective range approximation, improving upon the previous results. Their results are shown in Figure 8.

The r.h.s. part of Figure 8 shows the finite volume behavior and contrasts the data from [16] with the data from one of the ensembles used in [13], and with experiment, for both the $D^*_{s0}(2317)$ and the $D_{s1}(2460)$. The extracted binding energies are roughly compatible with the

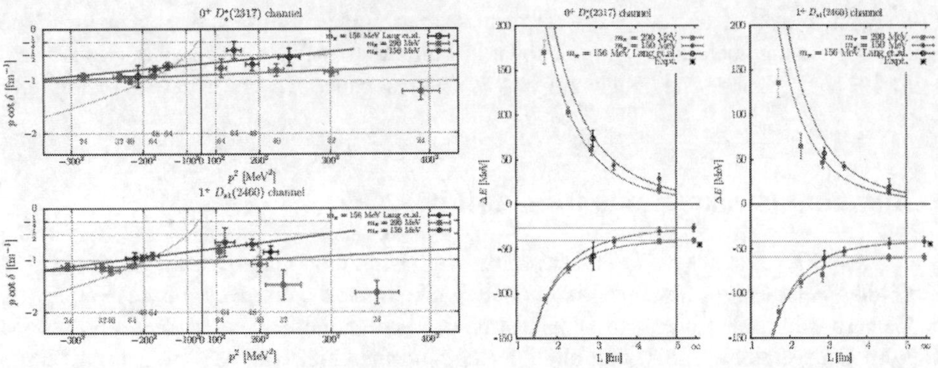

Figure 8. Left: Effective range fit used for the determination of the $D^*_{s0}(2317)$ and the $D_{s1}(2460)$ pole positions. The binding momentum of the bound state pole can be read-off from the intersection with the grey dashed line. The spatial extents L of the lattices used are indicated in the Figure. Right: Finite volume behavior of the data. Figure taken from [16].

binding energies observed in experiment. It should be stressed that there has been no attempt yet to fully to control the uncertainty from the lattice discretization. Note however that the two existing lattice studies employ somewhat different methods and agree qualitatively.

Unlike for the D_s spectrum, the D mesons belonging to a heavy-quark $j = \frac{1}{2}$ doublet, the $D^*_0(2400)$ and $D_1(2430)$ are broad resonances in nature [1]. These have previously been studied in [11], assuming elastic $D\pi$-scattering. More recently, the Hadron Spectrum collaboration studied coupled channel $D\pi$, $D\eta$, D_sK scattering in [15]. They employed lattice gauge configurations with multiple lattice volumes at a pion mass of 391 MeV. The use of anisotropic Wilson-clover lattices [39] allowed the determination in a number of moving frames in addition to the rest frame, which together with the multiple lattice volumes results in the the dense coverage of different center of momentum energies needed for the coupled-channel treatment.

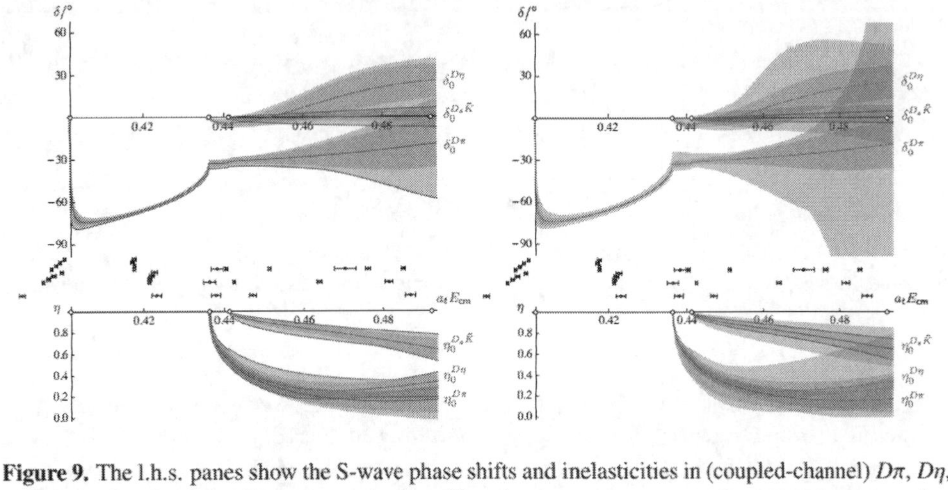

Figure 9. The l.h.s. panes show the S-wave phase shifts and inelasticities in (coupled-channel) $D\pi$, $D\eta$, D_sK scattering for a specific parameterization. The r.h.s. panes incorporate the uncertainty stemming from the choice of parameterization. Figure taken from [15].

They parameterize their results in terms of scattering phase shifts and inelasticities, The results of these parameterizations are shown in Figure 9. At this pion mass they observe a shallow $J^P = 0^+$ bound state, and a narrow $J^P = 2^+$ resonance. For more details please refer to [15].

4 Charmonium-like states from Lattice QCD

Candidates for exotic mesons are most prolific in the charmonium spectrum previously believed to be well described by potential models, which made the discovery of the exotic X, Y, and Z charmonium-like states particularly striking. Among these, charged Z_c states, like the $Z_c(3900)$ discovered by BESIII [40] and quickly confirmed by Belle [41] and using CLEO-c data [42] are most obviously of an exotic nature, as they contain a heavy charm-anticharm pair and are charged, requiring them to be made of at least 4 quarks. Nevertheless the structure of these states is heavily debated.

In [43] the finite volume spectrum with quantum numbers $I^G J^{PC} = 1^+1^{+-}$ was determined using a large basis of meson-meson and four quark interpolating fields (this included tetraquark interpolators of a $3_c \times \bar{3}_c$ structure), with the aim of extracting all meson-meson states below $\approx$ 4.3 GeV. Unlike the studies of heavy-light mesons in the previous section, no rigorous extraction of scattering amplitudes was attempted, as the large number of open channels and the resulting high density of energy levels makes this very challenging. Furthermore, 3-particle channels were neglected completely. The strategy was instead to count the low-lying energy levels, and to identify them according to their overlaps, with the hope to see an extra-level that could be related to the presence of a Z_c meson resonance pole.

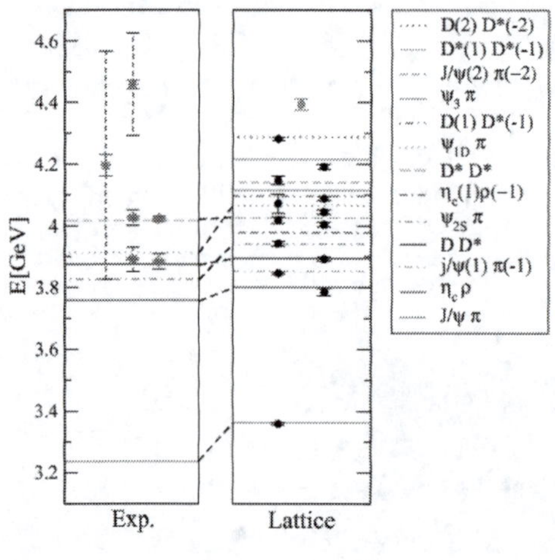

Figure 10. Left: collection of Z_c states from experiment. The horizontal lines illustrate the various thresholds. Right: lattice spectrum compared to the non-interacting meson-meson energies (horizontal lines). Figure from [43].

The right pane of Figure 10 shows the resulting lattice spectrum compared to the non-interacting two-meson levels. The resulting spectrum is compared to observed meson states taken from [44] in the left pane. Note that no additional level is seen below 4.3 GeV, while 13 energy levels close to non-interacting meson meson energies are seen. While an additional energy level would have been a smoking-gun signature for a Z_c state at roughly the same energy, not much can be inferred from the absence of such a level.

A different approach is pursued in [45, 46], where a coupled channel extension of the HALQCD method (for the original method see [47]) is used to investigate coupled-channel scattering $J/\Psi\pi$, $\eta_c\rho$, $\bar{D}D^*$ with quantum numbers $I^G(J^{PC}) = 1^+1^{+-}$ (same as above). In the HALQCD method potentials $V(r)$, functions of the distance r, are calculated as an intermediate quantity. The Schrödinger Equation is then solved for the given potentials and the scattering observables are extracted. For their study of coupled channel scattering in the $Z_c(3900)$ channel, 2+1 flavor gauge configurations with lattice spacing $a = 0.0907(13)$ and three different pion masses $m_\pi = 410, 570, 700$ MeV were used.

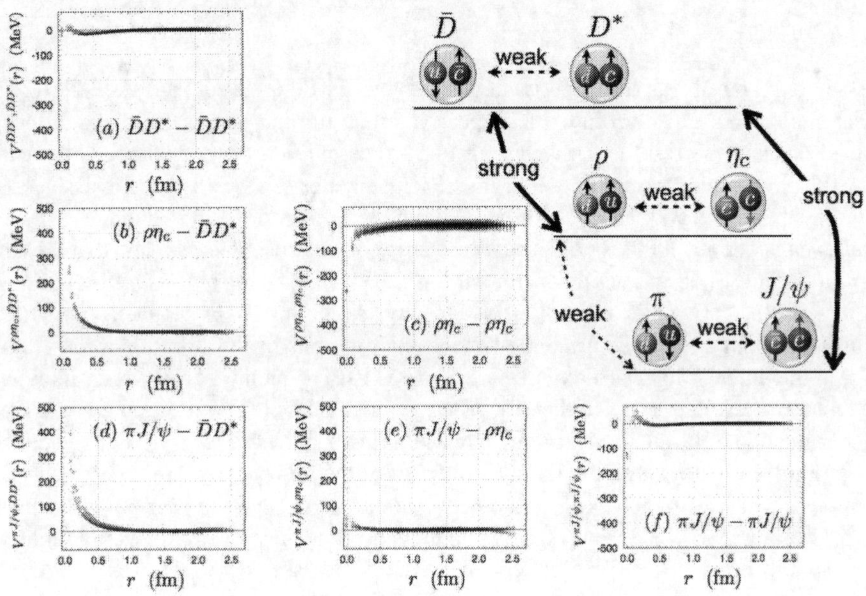

Figure 11. Coupled-channel S-wave potentials from the HALQCD method. Figure taken from [45].

Figure 11 shows the resulting diagonal and off-diagonal potentials and illustrates their pattern: While not much is happening for the diagonal, some of the off-diagonal potentials are quite strong.

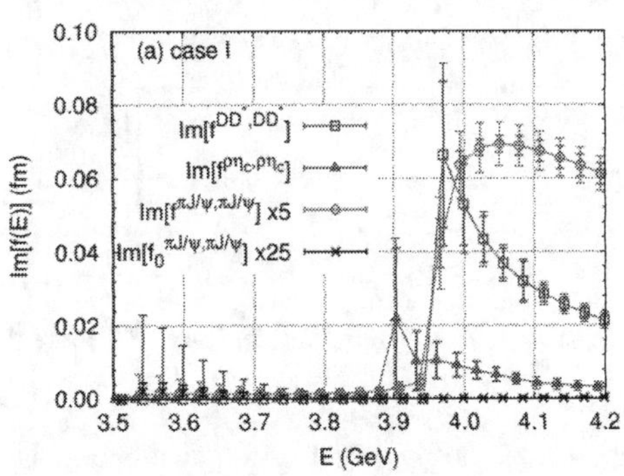

Figure 12. Two-body invariant mass spectra in the respective channels. Figure taken from [45].

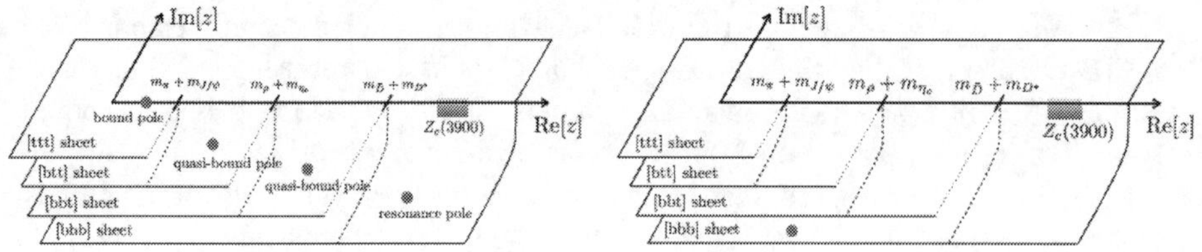

Figure 13. Left: Illustration of example pole-structures in the sheet-notation used in [46]. Right: Pole structure determined from lattice data at unphysical pion mass. Figure taken from [46].

The two-body invariant mass spectra are shown in Figure 12. The results look qualitatively similar to the pattern in experiment. Note that turning off the off-diagonal potentials instead leads to the results depicted with black crosses. The authors proceed to search for poles in the complex plane on the various Riemann sheets. For all quark masses, a pole is found far below DD^* threshold. The pole structure is illustrated in Figure 13. The authors conclude that the $Z_c(3900)$ is not a usual resonance but a threshold cusp, and that the structure seen in the invariant mass comes from strong $\pi J/\Psi - \bar{D}D^*$ and $\rho\eta_c - \bar{D}D^*$ couplings. The authors also state that an analysis at close-to-physical pion mass is planned. Note that these results are not in obvious contradiction to the energy levels from [43].

Another interesting charmonium state is the $X(3915)$ which was previously interpreted as the first radial excitation χ'_{c0} of the $J^{PC} = 0^{++}$ χ_{c0} by the PDG. This assignment was based on a determination of its quantum numbers by BaBar [48], and was heavily criticized on theoretical grounds [49, 50]. In particular, no evidence for the *fall-apart mode* $X(3915) \to \bar{D}D$ has been seen, which would be expected to be the dominant decay mode for a χ'_{c0}. Furthermore, the spin splitting $m_{\chi_{c2}(2P)} - m_{\chi_{c0}(2P)}$ between the meson in question an the well established χ'_{c2} is too small. Instead the $X(3915)$ is seen in the OZI suppressed decay $X(3915) \to \omega J/\psi$. Also its width should be significantly larger than $\Gamma_{\chi_{c2}(2P)}$. More recently, Zhou *et al.* [51] argued that what is dubbed $X(3915)$ is the $\chi_{c2}(2P)$ state already known, and just like [49] suggest that a broader state is hiding in the experiment data. More recently Belle [52] have observed a state with mass $M = 3862^{+26+40}_{-32-13}$ MeV and with a width of $\Gamma = 201^{+154+88}_{-067-82}$ MeV.

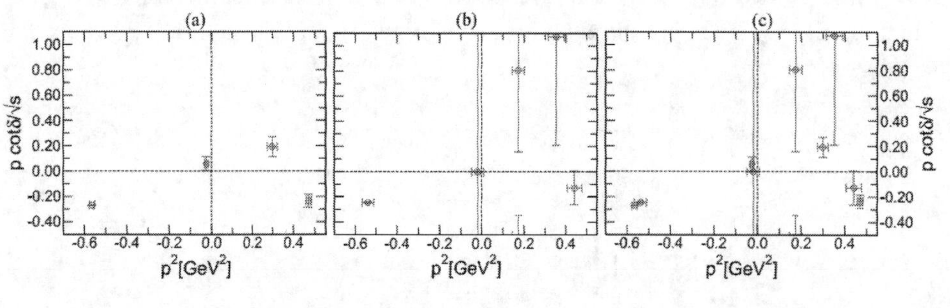

Figure 14. $p \cot\delta/\sqrt{s}$ versus p^2 scattering in S-wave. p is the D-meson momentum. Left: Ensemble (1); Middle: Ensemble (2); Right: Both datasets in one plot. For more details please refer to [53]

Motivated by the idea of a single broad χ'_{c0} resonance decaying predominantly into $\bar{D}D$, the authors of [53] studied elastic $\bar{D}D$ scattering in the $J^{PC} = 0^{++}$ channel. Figure 14 shows the resulting data for the two ensembles used. Various possibilities are discussed in detail in [53], and the conclusion is that the lattice data suggests a fairly narrow resonance with 3.9 GeV $\leq M \leq$ 4.0 GeV and $\Gamma <$ 100 MeV. Notice that only the statistical uncertainty is quantified, and that non-negligible uncertainties stem also from the lattice discretization used and from pion mass dependence of the results. Overall neither the experiment data nor the lattice QCD results are conclusive and further results from experiment, as well as further lattice QCD results will be needed to clarify the situation.

5 Exotic doubly-heavy hadrons

In recent years a number of groups started to study explicitly exotic doubly-heavy hadrons using lattice QCD. In [54, 55] the authors studied doubly-charmed tetraquarks, while doubly-bottom four-quark states are studied in [56–60]. In this chapter I will highlight some of the newer results.

The authors of [59, 60] study B meson - B meson interactions with static bottom quarks. They determine the potentials of two static antiquarks in the presence of two light quarks on the lattice and search for bound states (rather than resonances). The potentials are calculated on lattice gauge configurations with a lattice spacing of $a = 0.079$ fm and a pion mass $m_\pi \approx 650, 480, 340$ MeV. The lattice potentials are fit with a 3-parameter form (keeping p fixed)

$$V(r) = -\frac{\alpha}{r}\exp\left(-\left(\frac{r}{d}\right)^p\right) + V_0\,. \tag{1}$$

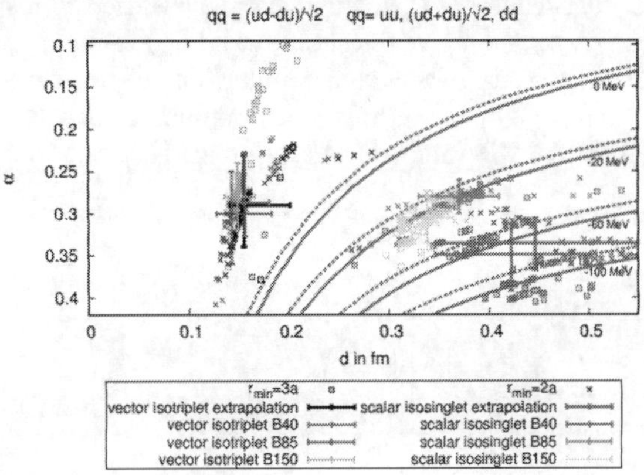

Figure 15. Binding energy isolines for the scalar isosinglet and vector isotriplet channels. Figure taken from [59].

In a first publication [59] the authors study the scalar isosinglet and vector isotriplet channels neglecting effects of the heavy $\bar{b}$ spin. Solving a non-relativistic Schrödinger equation, they find the vector isotriplet channel to be unbound, both at all three of the pion masses separately and in the limit of physical pion masses. In the scalar isosinglet case they instead observe binding for all pion masses and obtain the binding energy 90^{+43}_{-36} MeV after a chiral extrapolation. The results from this publication are shown in Figure 15. More recently a similar procedure (solving the coupled channel Schrödinger equations) was used to also include effects of the heavy $\bar{b}$ spin [60]. A lower and somewhat less significant binding energy $E_B = 59^{+30}_{-38}$ MeV is obtained in this case.

In [61] the authors study doubly bottom $J^P = 1^+$ tetraquarks with NRQCD b-quarks. They us lattice gauge configurations with three different pion masses and 164 MeV $\leq M_\pi \leq$ 415 MeV at a single lattice spacing. They obtain energy levels that suggest rather deeply

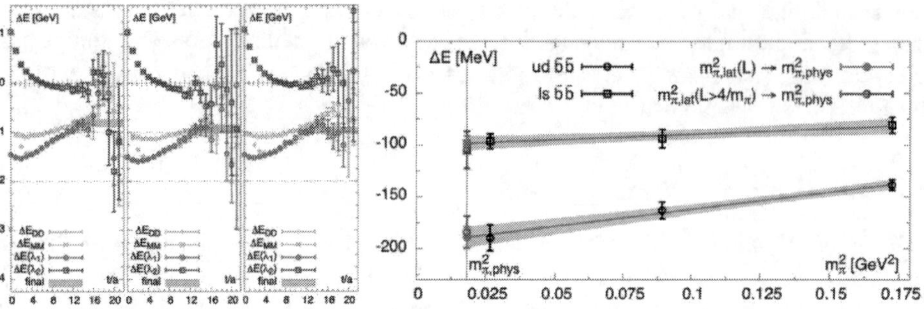

Figure 16. Left: $ls\bar{b}\bar{b}$ effective binding energies extracted from the ground state (red), and behavior for the excited state (blue). The final fit results are indicated by the red band. Right: Chiral extrapolation of the resulting binding energies. Figure taken from [61].

bound states for four-quark systems with both $ud\bar{b}\bar{b}$ and $ls\bar{b}\bar{b}$ flavor structure. The left part of Figure 16 shows the lattice energy levels they obtain and the chiral extrapolation of their results is shown in the right part of the same Figure. It should be noted that the binding energies are extracted from ratios of correlation functions. In this procedure care has to be taken [62]. A somewhat peculiar feature of the data is that the energies corresponding to the first excited state are naively expected above threshold for a bound state, however no significant shift is visible in the data. It is also expected that finite volume effects may significantly alter the observed binding energies. Given these caveats the authors quote 189(10) and 98(7) MeV as the binding energies for the $ud\bar{b}\bar{b}$ and $ls\bar{b}\bar{b}$ states respectively.

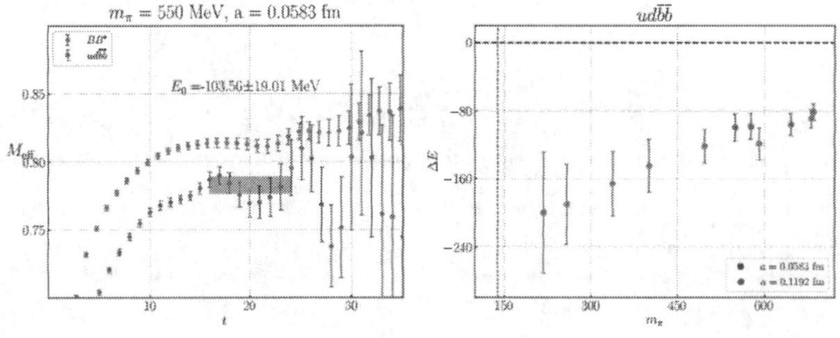

Figure 17. Effective masses of the $ud\bar{b}\bar{b}$ and (non-interacting) BB^* system illustrating clearly that the interaction is attractive. Right: Preliminary results for the resulting binding energies. Figure taken from [63].

In [63] the authors present preliminary results for the same system using overlap valence quarks (for light, strange and charm) on staggered gauge configurations and lattice NRQCD for the bottom-quarks. For their current results they use two HISQ gauge ensemble with pion masses 305 and 319 MeV and lattice spacings $a \approx 0.12$ fm and $a \approx 0.06$ fm respectively. On each of those ensembles they work with valence pions at various pion masses (partially quenched setup). Example results for their effective masses ares shown in the left pane of Figure 17.

Their preliminary results at rather low statistics confirm the existence of a bound state state at heavier than physical pion mass. These results are shown in Figure 17. The binding energies at comparable quark mass agree roughly with the values presented in [61], both of which are much larger than the value obtained in the static limit in [59].

6 Summary

I reviewed recent progress in lattice QCD calculations of heavy hadrons. One of the highlights from recent years has been the prediction of the mass of the Ξ_{cc} (and of a number of other charmed baryons) by various groups [26–32], which has been presented in Section 2.

With regard to heavy-light mesons (see Section 3), recent simulations employing Lüscher's finite volume method for the study of charmed-light mesons show a qualitative agreement with experiment [11–13, 15, 16]. To achieve this agreement, it is vital to use a diverse basis of interpolating field operators of both quark-antiquark and meson-meson structure. For bottom-strange mesons a quantitative prediction for the not yet observed positive parity $j = \frac{1}{2}$ doublet has been presented [14]. These states appear as QCD bound states in BK and B^*K scattering, and are the bottom analogues of the $D^*_{s0}(2317)$ and $D_{s1}(2460)$.

In Sections 4 and 5 exploratory results for exotic charmonia and doubly-heavy hadrons have been presented. Among these, evidence for the existence of doubly-heavy bound states with flavor content $ud\bar{b}\bar{b}$ and $ls\bar{b}\bar{b}$ [59, 61, 63] is especially intriguing.

Acknowledgements

I would like to thank Parikshit Junnarkar and Nilmani Mathur for sending me (at the time of the talk) unpublished material. I would also like to thank C. B. Lang and M. Padmanath for their careful reading of the manuscript.

References

[1] S. Aoki et al., Eur. Phys. J. **C77**, 112 (2017), 1607.00299
[2] S. Borsanyi et al., Science **347**, 1452 (2015), 1406.4088
[3] M. Lüscher, Nucl. Phys. **B354**, 531 (1991)
[4] M. Lüscher, Nucl. Phys. **B364**, 237 (1991)
[5] K. Rummukainen, S.A. Gottlieb, Nucl. Phys. **B450**, 397 (1995), hep-lat/9503028
[6] R.A. Briceno, J.J. Dudek, R.D. Young (2017), 1706.06223
[7] M. Peardon, J. Bulava, J. Foley, C. Morningstar, J. Dudek, R.G. Edwards, B. Joo, H.W. Lin, D.G. Richards, K.J. Juge (Hadron Spectrum), Phys. Rev. **D80**, 054506 (2009), 0905.2160
[8] C. Morningstar, J. Bulava, J. Foley, K.J. Juge, D. Lenkner, M. Peardon, C.H. Wong, Phys. Rev. **D83**, 114505 (2011), 1104.3870
[9] A.X. El-Khadra, A.S. Kronfeld, P.B. Mackenzie, Phys. Rev. **D55**, 3933 (1997), hep-lat/9604004
[10] M.B. Oktay, A.S. Kronfeld, Phys. Rev. **D78**, 014504 (2008), 0803.0523
[11] D. Mohler, S. Prelovsek, R.M. Woloshyn, Phys. Rev. **D87**, 034501 (2013), 1208.4059
[12] D. Mohler, C.B. Lang, L. Leskovec, S. Prelovsek, R.M. Woloshyn, Phys. Rev. Lett. **111**, 222001 (2013), 1308.3175
[13] C.B. Lang, L. Leskovec, D. Mohler, S. Prelovsek, R.M. Woloshyn, Phys. Rev. **D90**, 034510 (2014), 1403.8103

[14] C.B. Lang, D. Mohler, S. Prelovsek, R.M. Woloshyn, Phys. Lett. **B750**, 17 (2015), 1501.01646

[15] G. Moir, M. Peardon, S.M. Ryan, C.E. Thomas, D.J. Wilson, JHEP **10**, 011 (2016), 1607.07093

[16] G.S. Bali, S. Collins, A. Cox, A. Schäfer, Phys. Rev. **D96**, 074501 (2017), 1706.01247

[17] R. Aaij et al. (LHCb), Phys. Rev. Lett. **119**, 112001 (2017), 1707.01621

[18] M. Mattson et al. (SELEX), Phys. Rev. Lett. **89**, 112001 (2002), hep-ex/0208014

[19] A. Ocherashvili et al. (SELEX), Phys. Lett. **B628**, 18 (2005), hep-ex/0406033

[20] S.J. Brodsky, F.K. Guo, C. Hanhart, U.G. Meissner, Phys. Lett. **B698**, 251 (2011), 1101.1983

[21] S.P. Ratti, Nucl. Phys. Proc. Suppl. **115**, 33 (2003), [,33(2003)]

[22] B. Aubert et al. (BaBar), Phys. Rev. **D74**, 011103 (2006), hep-ex/0605075

[23] R. Chistov et al. (Belle), Phys. Rev. Lett. **97**, 162001 (2006), hep-ex/0606051

[24] Y. Kato et al. (Belle), Phys. Rev. **D89**, 052003 (2014), 1312.1026

[25] R. Aaij et al. (LHCb), JHEP **12**, 090 (2013), 1310.2538

[26] C. Alexandrou, V. Drach, K. Jansen, C. Kallidonis, G. Koutsou, Phys. Rev. **D90**, 074501 (2014), 1406.4310

[27] R.A. Briceno, H.W. Lin, D.R. Bolton, Phys. Rev. **D86**, 094504 (2012), 1207.3536

[28] Z.S. Brown, W. Detmold, S. Meinel, K. Orginos, Phys. Rev. **D90**, 094507 (2014), 1409.0497

[29] Y. Namekawa et al. (PACS-CS), Phys. Rev. **D87**, 094512 (2013), 1301.4743

[30] P. Pérez-Rubio, S. Collins, G.S. Bali, Phys. Rev. **D92**, 034504 (2015), 1503.08440

[31] S. Mondal, M. Padmanath, N. Mathur, *Spectroscopy of Charmed and Bottom Hadrons using Lattice QCD*, in *35th International Symposium on Lattice Field Theory (Lattice 2017) Granada, Spain, June 18-24, 2017* (2017), 1712.08446, http://inspirehep.net/record/1644787/files/arXiv:1712.08446.pdf

[32] R. Lewis, N. Mathur, R.M. Woloshyn, Phys. Rev. **D64**, 094509 (2001), hep-ph/0107037

[33] R. Aaij et al. (LHCb), Phys. Rev. Lett. **118**, 182001 (2017), 1703.04639

[34] M. Padmanath, N. Mathur, Phys. Rev. Lett. **119**, 042001 (2017), 1704.00259

[35] J.J. Dudek, R.G. Edwards, M.J. Peardon, D.G. Richards, C.E. Thomas, Phys. Rev. Lett. **103**, 262001 (2009), 0909.0200

[36] B. Aubert et al. (BaBar), Phys. Rev. Lett. **90**, 242001 (2003), hep-ex/0304021

[37] S. Godfrey, N. Isgur, Phys. Rev. **D32**, 189 (1985)

[38] E. van Beveren, G. Rupp, Phys. Rev. Lett. **91**, 012003 (2003), hep-ph/0305035

[39] H.W. Lin et al. (Hadron Spectrum), Phys. Rev. **D79**, 034502 (2009), 0810.3588

[40] M. Ablikim et al. (BESIII), Phys. Rev. Lett. **110**, 252001 (2013), 1303.5949

[41] Z.Q. Liu et al. (Belle), Phys. Rev. Lett. **110**, 252002 (2013), 1304.0121

[42] T. Xiao, S. Dobbs, A. Tomaradze, K.K. Seth, Phys. Lett. **B727**, 366 (2013), 1304.3036

[43] S. Prelovsek, C.B. Lang, L. Leskovec, D. Mohler, Phys. Rev. **D91**, 014504 (2015), 1405.7623

[44] N. Brambilla et al., Eur. Phys. J. **C74**, 2981 (2014), 1404.3723

[45] Y. Ikeda, S. Aoki, T. Doi, S. Gongyo, T. Hatsuda, T. Inoue, T. Iritani, N. Ishii, K. Murano, K. Sasaki (HAL QCD), Phys. Rev. Lett. **117**, 242001 (2016), 1602.03465

[46] Y. Ikeda (HAL QCD) (2017), 1706.07300

[47] N. Ishii, S. Aoki, T. Hatsuda, Phys. Rev. Lett. **99**, 022001 (2007), nucl-th/0611096

[48] J.P. Lees et al. (BaBar), Phys. Rev. **D86**, 072002 (2012), 1207.2651

[49] F.K. Guo, U.G. Meissner, Phys. Rev. **D86**, 091501 (2012), 1208.1134
[50] S.L. Olsen, Phys. Rev. **D91**, 057501 (2015), 1410.6534
[51] Z.Y. Zhou, Z. Xiao, H.Q. Zhou, Phys. Rev. Lett. **115**, 022001 (2015), 1501.00879
[52] K. Chilikin et al. (Belle), Phys. Rev. **D95**, 112003 (2017), 1704.01872
[53] C.B. Lang, L. Leskovec, D. Mohler, S. Prelovsek, JHEP **09**, 089 (2015), 1503.05363
[54] Y. Ikeda, B. Charron, S. Aoki, T. Doi, T. Hatsuda, T. Inoue, N. Ishii, K. Murano, H. Nemura, K. Sasaki, Phys. Lett. **B729**, 85 (2014), 1311.6214
[55] A.L. Guerrieri, M. Papinutto, A. Pilloni, A.D. Polosa, N. Tantalo, PoS **LATTICE2014**, 106 (2015), 1411.2247
[56] Z.S. Brown, K. Orginos, Phys. Rev. **D86**, 114506 (2012), 1210.1953
[57] P. Bicudo, M. Wagner (European Twisted Mass), Phys. Rev. **D87**, 114511 (2013), 1209.6274
[58] P. Bicudo, K. Cichy, A. Peters, B. Wagenbach, M. Wagner, Phys. Rev. **D92**, 014507 (2015), 1505.00613
[59] P. Bicudo, K. Cichy, A. Peters, M. Wagner, Phys. Rev. **D93**, 034501 (2016), 1510.03441
[60] P. Bicudo, J. Scheunert, M. Wagner, Phys. Rev. **D95**, 034502 (2017), 1612.02758
[61] A. Francis, R.J. Hudspith, R. Lewis, K. Maltman, Phys. Rev. Lett. **118**, 142001 (2017), 1607.05214
[62] T. Iritani et al., JHEP **10**, 101 (2016), 1607.06371
[63] P. Junnarkar, M. Padmanath, N. Mathur, *Heavy light tetraquarks from Lattice QCD*, in *35th International Symposium on Lattice Field Theory (Lattice 2017) Granada, Spain, June 18-24, 2017* (2017), 1712.08400, http://inspirehep.net/record/1644795/files/arXiv:1712.08400.pdf

EPJ Web of Conferences **181**, 01028 (2018) https://doi.org/10.1051/epjconf/201818101028
EXA2017

Antinucleon-nucleon interaction in chiral effective field theory

Johann Haidenbauer[1,⋆]

[1] *Institute for Advanced Simulation, Institut für Kernphysik, and Jülich Center for Hadron Physics, Forschungszentrum Jülich GmbH, D-52425 Jülich, Germany*

Abstract. A study of the antinucleon-nucleon interaction within chiral effective field theory is presented. This novel approach suggested by Weinberg for investigating the nucleon-nucleon interaction can be adapted straightforwardly to the antinucleon-nucleon system. The antinucleon-nucleon potential is derived up to next-to-next-to-next-to-leading order in the chiral expansion. The low-energy constants associated with the arising contact interactions are fixed by a fit to phase shifts and inelasticities provided by a recently published phase-shift analysis of antiproton-proton scattering data. The achieved description of the antinucleon-nucleon amplitudes is excellent and of a quality comparable to the one found in case of the nucleon-nucleon interaction at the same order.

1 Introduction

Over the last decade there has been a renewed interest in antinucleon-nucleon ($\bar{N}N$) scattering phenomena. It was prompted not least by measurements of the $\bar{p}p$ invariant mass in the decays of heavy mesons such as J/ψ, ψ', and B, and of the reaction cross section for $e^+e^- \to \bar{p}p$. In several of those reactions a sizeable near-threshold enhancement in the mass spectrum was found [1–4]. Those enhancements nourished speculations about $\bar{p}p$ bound states, new resonances, or even more exotic objects in some parts of the physics community. Others noted that such data could provide a unique opportunity to test the $\bar{p}p$ interaction at very low energies [5–11]. Another boost comes, of course, from activities related to the $\bar{N}N$ interaction planned for the Facility for Antiproton and Ion Research (FAIR) in Darmstadt whose construction is finally on its way. Specifically, the PANDA project aims at a study of the interactions between antiprotons and fixed target protons and nuclei in the momentum range of 1.5−15 GeV/c using the high energy storage ring HESR [12]. In this context also the proposal by the PAX Collaboration [13] should be mentioned, though it is not in the core program of FAIR. It intends to exploit the method of spin-filtering in storage rings to achieve a polarized antiproton beam which would then allow one to measure $\bar{N}N$ double-spin observables.

The renewed interest in $\bar{N}N$ phenomena led also to the construction of new $\bar{N}N$ potentials, phenomenological ones [14, 15] but also of interactions [16] that are based on modern and systematic approaches like chiral effective field theory (EFT) [17]. Indeed, one of the consequences of a growing interest in $\bar{N}N$ physics was perhaps that an update of the Nijmegen partial-wave analysis (PWA) of antiproton-proton scattering data has been presented [18]. Since for the new PWA the resulting phase shifts and inelasticities are explicitly given, a straightforward application of chiral EFT to studies of the $\bar{N}N$ force became feasible [19, 20].

⋆e-mail: j.haidenbauer@fz-juelich.de

Chiral EFT has been suggested by Weinberg in the early 1990s [21] as a systematic approach for investigating nucleon-nucleon (*NN*) scattering. It facilitates a rather accurate and consistent description of the *NN* interaction and nuclear few-body systems [22, 23], see also the overviews in Refs. [17, 24]. Its most salient features are that it exploits the symmetries and symmetry-breaking pattern of QCD and that there is an underlying power counting which allows one to improve calculations systematically by going to higher orders in a perturbative expansion. The emerging chiral potential for the *NN* force consists of pion exchanges and a series of contact interactions with an increasing number of derivatives. The latter represent the short-range part of the *NN* force and are parameterized by low-energy constants (LECs), that need to be fixed by a fit to data. The reaction amplitude is obtained from solving a regularized Lippmann-Schwinger equation for the derived interaction potential.

The $\bar{N}N$ interaction is closely connected to that in the *NN* system via *G*-parity. Specifically, the *G*-parity transformation (a combination of charge conjugation and a rotation in the isospin space) relates that part of the $\bar{N}N$ potential which is due to pion exchanges to the one in the *NN* case in an unambiguous way. Thus, like in the *NN* case, the long-range part of the $\bar{N}N$ potential is completely fixed by the underlying chiral symmetry of pion-nucleon dynamics.

In this contribution I review a $\bar{N}N$ potential derived in a chiral EFT approach up to next-to-next-to-next-to leading order (N^3LO). Its evaluation has been done in complete analogy to that of the *NN* interaction published in Ref. [23] and is likewise based on the Weinberg power counting. In Ref. [19] our group had already studied the $\bar{N}N$ force within chiral EFT up to next-to-next-to leading order (N^2LO). It had been found that the approach works very well. Indeed, the overall quality of the description of the $\bar{N}N$ amplitudes achieved in Ref. [19] is comparable to the one found in case of the *NN* interaction at the same order. By going to a higher order one expects to be able to describe the $\bar{N}N$ interaction over a larger energy range. Furthermore, in Ref. [23] a simple approach for estimating the uncertainty due to the truncation of the chiral expansion is proposed. This approach can be adopted for performing an analogous analysis for our $\bar{N}N$ results.

2 The $\bar{N}N$ interaction in chiral EFT

The derivation of the chiral $\bar{N}N$ potential using the Weinberg power counting is outlined in detail in Ref. [20], see also Ref. [23] for the corresponding application to the *NN* system. Therefore, I will be very brief here. As already mentioned above, the chiral potential contains pion exchanges and a series of contact interactions with an increasing number of derivatives. Up to N^3LO there are contributions from one-, two- and three-pion exchanges. Those are identical to the ones that appear in the *NN* potential [23]. However, there is a sign change in case of an odd number of exchanged pions due to its negative *G* parity, i.e. $V^{\bar{N}N} = -V_\pi + V_{2\pi} - V_{3\pi}$, with V_π etc. being the corresponding contributions to the *NN* force. On the other hand, the contact interaction, $V^{\bar{N}N}_{cont}$, cannot be taken over simply from the *NN* case. As said above, the contact terms represent effectively the short-range part of the interaction and, therefore, the *G* parity of the indivdual contributions remains unresolved. Thus, the strength parameters associated with the arising contact terms, the LECs, need to be determined in a fit to $\bar{N}N$ data. We fix them by fitting to the phase shifts and inelasticity parameters of the PWA of Zhou and Timmermans [18]. How this is done is described in detail in Ref. [20]. Note that there are more independent LECs in the $\bar{N}N$ case than in *NN* because in the former system there is no restriction from the Pauli principle. In addition, in our approach $\bar{N}N$ annihilation is likewise parameterized by contact terms, see Refs. [19, 20] for explicit expressions.

Once the potential is established, the reaction amplitude is obtained from the solution of a relativistic Lippmann-Schwinger (LS) equation. It reads in partial-wave projected form

$$\begin{aligned} T_{L''L'}(p'',p';E_k) &= V_{L''L'}(p'',p') \\ &+ \sum_L \int_0^\infty \frac{dpp^2}{(2\pi)^3} V_{L''L}(p'',p) \frac{1}{2E_k - 2E_p + i0^+} T_{LL'}(p,p';E_k) . \end{aligned} \tag{1}$$

Here p'' and p' are the (moduli of the) center-of-mass $\bar{N}N$ momenta in the final and initial states, respectively, and $E_k = \sqrt{M_p^2 + k^2}$, where k is the on-shell momentum and We adopt a relativistic scattering equation so that our amplitudes fulfill the relativistic unitarity condition at any order, as done also in the NN sector [22, 24]. On the other hand, relativistic corrections to the potential are calculated order by order. They appear first at N^{3}LO in the Weinberg scheme, see Appendix A in Ref. [20]. Note that the $\bar{N}N$ system can be in uncoupled spin-singlet or triplet waves (where $L'' = L' = L = J$) or in coupled partial waves (where $L'', L', L = J-1, J+1$).

Since the integral in the LS equation (1) is divergent for the chiral potentials [22, 24] a regularization needs to be introduced. We utilize the regularization scheme of Ref. [23], where a local regulator is used for the pion-exchange contributions and a nonlocal regulator for the contact terms:

$$\begin{aligned} &V_{n\pi}(q) \to V_{n\pi}(r) \times f_R(r) \to V_{n\pi}^{reg}(q); \qquad (\vec{q} = \vec{p}\,' - \vec{p}) \\ &V_{cont} = V(p',p) \to V(p',p) \times f_\Lambda(p',p) = V_{cont}^{reg} . \end{aligned} \tag{2}$$

The explict form of the regulation functions is $f_R(r) = \left[1 - exp(-r^2/R^2)\right]^6$ and $f_\Lambda(p,p') = exp(-(p'^2 + p^2)/\Lambda^2)$. The cutoff radius R is varied in the range R = 0.7-1.2 fm [23] where $\Lambda = 2/R$ is used for relating the momentum-space cutoff parameter with the cutoff radius.

Since the fit is done to phase shifts and inelasticity parameters of Ref. [18], which are given for definite isospins, at this stage the LS equation is solved in the isospin basis. For the calculation of observables we solve the LS equation in the particle basis, in order to incorporate the correct physical thresholds of the $\bar{p}p$ and $\bar{n}n$ channels. Note that the $\bar{n}n$ channel opens at the antiproton laboratory momentum of p_{lab} = 98.7 MeV/c (T_{lab} = 5.18 MeV). Furthermore, the Coulomb interaction in the $\bar{p}p$ channel is taken into account appropriately via the Vincent-Phatak method.

3 Results

A complete overview of our investigation of $\bar{N}N$ scattering up to N^{3}LO in chiral EFT can be found in Ref. [20]. Here we focus on specific aspects and present only some exemplary results. One of those aspects is the influence of the regularization scheme. While ideally the results should be completely independent of the employed regularization, this cannot be achieved within the scheme [23] adopted in our study [20]. Nonetheless, one can monitor the effects of keeping the cutoff finite, and the artifacts introduced by that, by analyzing the residual cutoff dependence of the observables [23]. Pertinent results are presented in Fig. 1, where phase shifts are displayed for partial waves up to an orbital angular momentum of $L = 2$. We use the spectral notation $^{(2S+1)}L_J$ and indicate the isospin I separately. Note that instead of the inelasticity parameters the imaginary part of the phases and mixing angles are shown and subscripts R and I are used for δ in order to distinguish between the real and imaginary parts. The employed cutoffs are R = 0.7, 0.8, 0.9, 1.0, 1.1, and 1.2 fm and the results are based on the chiral potential up to N^{3}LO.

One can see that for most partial waves the cutoff dependence is fairly weak for $T_{lab} \lesssim 300$ MeV ($p_{lab} \lesssim 800$ MeV/c). Indeed, the small residual cutoff dependence that we observe here is comparable

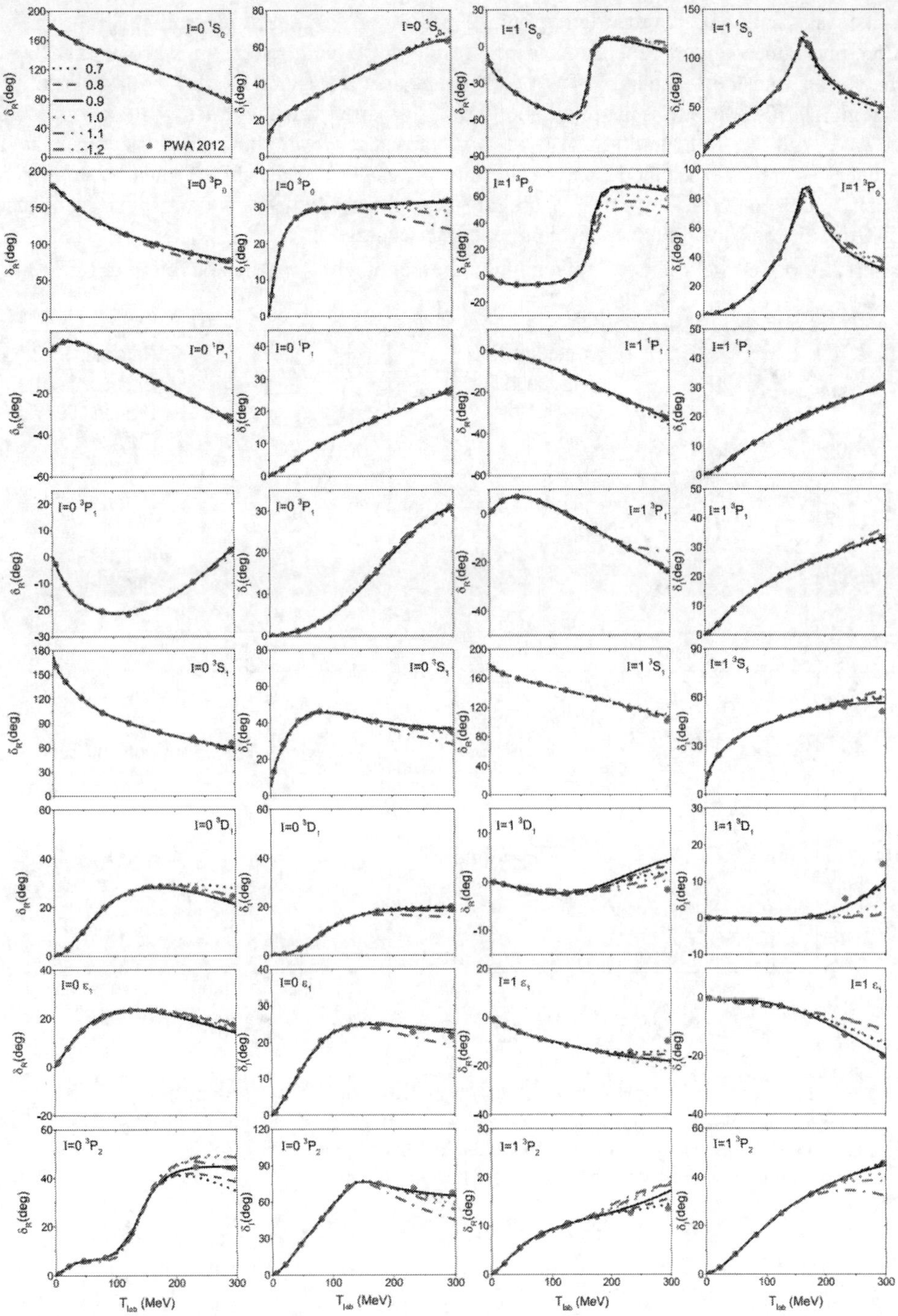

Figure 1. Real and imaginary parts of various $\bar{N}N$ phase shifts at N^{3}LO for cutoffs $R = 0.7 - 1.2$ fm. The filled circles represent the solution of the $\bar{p}p$ PWA [18].

to the likewise small variation reported in Ref. [23] for the NN interaction. Only in a few cases there is a more pronounced cutoff dependence of the results for laboratory energies above 150-200 MeV. This has to do with the fact that the PWA [18] suggests a resonance-like behavior of some phases in this region. This concerns most prominently the $^1S_0$ partial wave with isospin $I = 1$ and the $^3P_0$ partial wave with $I = 1$. In addition, also a few other partial waves show a conspicuous behavior at higher energies in the sense that the energy dependence changes noticeably. Typical examples are the inelasticity parameters for the $I = 0$ $^3P_0$ and $^3P_2$ partial wave, where the corresponding δ_I increase rapidly from the threshold, but then level out at higher energies.

Results for partial waves up to a total angular momentum of $J = 4$ can be found in Ref. [20].

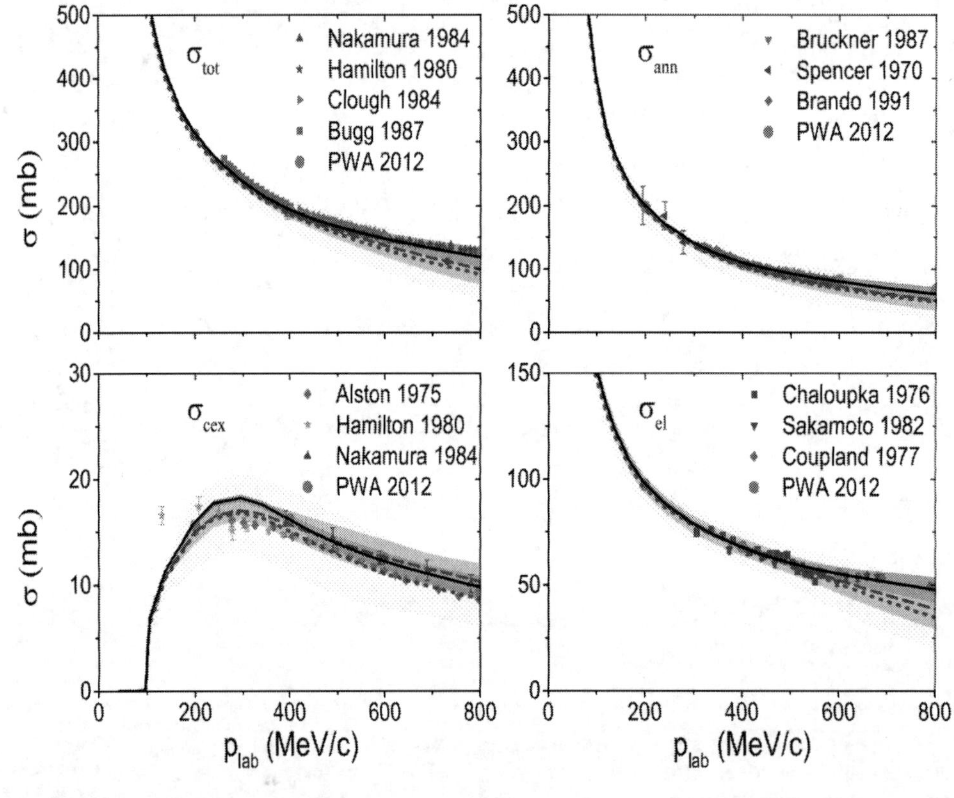

Figure 2. Total (σ_{tot}) and integrated elastic (σ_{el}), charge-exchange (σ_{cex}), and annihilation (σ_{ann}) cross sections for $\bar{p}p$ scattering. Results at N^{3}LO (black solid line), N^{2}LO (blue dashed line), and NLO (red dotted line) are shown. The estimated uncertainty is indicated by bands: N^{3}LO (magenta), N^{2}LO (cyan), and NLO (yellow). The filled circles represent the solution of the $\bar{p}p$ PWA [18].

Results for the integrated cross sections are presented in Fig. 2 for the chiral interaction with cutoff $R = 0.9$ fm up to NLO (red dotted line), N^{2}LO (blue dashed line), and N^{3}LO (solid black line). Obviously, at N^{3}LO our results agree rather well with the ones of the PWA (filled circles), even up to $p_{lab} = 800$ MeV/c. Indeed, also the charge-exchange cross section is nicely reproduced, though it is much smaller than the other ones. The amplitude for this process is given by the difference of the isospin $I = 0$ and $I = 1$ amplitudes and its description requires a delicate balance between the interactions in the corresponding isospin channels.

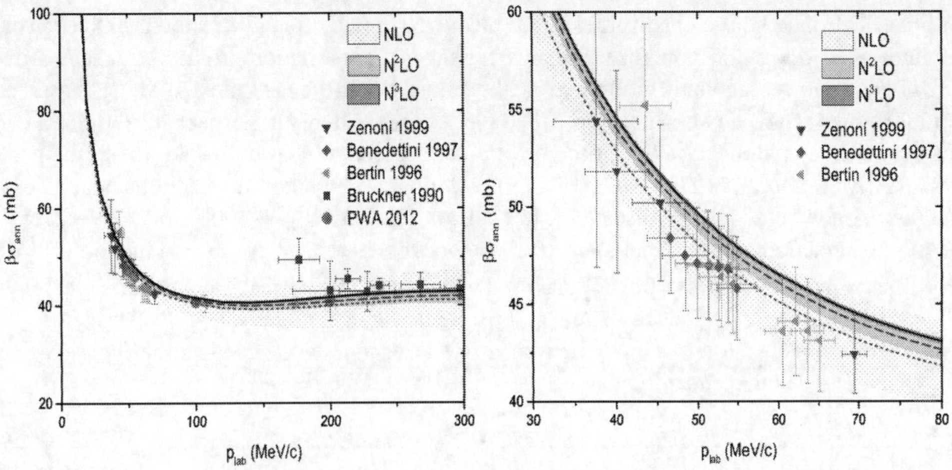

Figure 3. $\bar{p}p$ annihilation cross section multiplied by the velocity β of the incoming $\bar{p}$. Same description of curves as in Fig. 2. The results of the PWA [18] are indicated by circles.

The bands represent the estimated uncertainty of our results. Here we follow the suggestion of Epelbaum et al. [23] and use as measure the expected size of the higher-order corrections together with the actual size of the higher-order corrections. Then the uncertainty of an observable $X(p)$ is given by

$$\begin{aligned}
\Delta X^{LO} &= Q^2|X^{LO}| \\
\Delta X^{NLO} &= \max\left(Q^3|X^{LO}|, Q^1|\delta X^{NLO}|\right); \quad \delta X^{NLO} = X^{NLO} - X^{LO} \\
\Delta X^{N^2LO} &= \max\left(Q^4|X^{LO}|, Q^2|\delta X^{NLO}|, Q^1|\delta X^{N^2LO}|\right); \quad \delta X^{N^2LO} = X^{N^2LO} - X^{NLO} \\
\Delta X^{N^3LO} &= \max\left(Q^5|X^{LO}|, Q^3|\delta X^{NLO}|, Q^2|\delta X^{N^2LO}|, Q^1|\delta X^{N^3LO}|\right); \quad \delta X^{N^3LO} = X^{N^3LO} - X^{N^2LO},
\end{aligned}$$

where X^{LO}, etc., are the results for that observable at the corresponding order. The expansion parameter Q is defined by $Q = \max\left(\frac{k}{\Lambda_b}, \frac{m_\pi}{\Lambda_b}\right)$ with k being the $\bar{N}N$ on-shell momentum. The quantity Λ_b is the breakdown scale, i.e. the momentum where the chiral EFT expansion is expected to break down. For it we take over the value $\Lambda_b = 600$ MeV from Ref. [23], deduced from the analysis of corresponding NN results. Considering the bands for the interactions with increasing order – yellow (NLO), blue (N^{2}LO), magenta (N^{3}LO) – one can see that there is a clear convergence of our results for all cross sections when going to higher orders.

Results for differential observables, specifically for differential cross sections, analyzing powers, and depolarization parameters, can be found in Ref. [20], where they are compared directly with experimental information.

4 Predictions

As can be seen in Table III of Ref. [18] no data below 100 MeV/c (corresponding to $T_{lab} = 5.32$ MeV) have been included in the partial-wave analysis. Indeed, there is only very limited experimental information at lower energies. It consists of measurements of the $\bar{p}p$ annihilation cross section at very low energy and of data on the level shifts and widths of antiproton-proton atoms.

We have calculated those observables based on our $\bar{N}N$ interaction [20], which can be considered as predictions of our chiral potential. Results for the $\bar{p}p$ annihilation cross section are shown in Fig. 3, where the cross section is multiplied by the velocity β of the incoming $\bar{p}$. Values based on the amplitudes of the PWA are also included (filled circles). An interesting aspects of those data is that one can see the anomalous behavior of the reaction cross section near threshold due to the presence of the attractive Coulomb force. Usually the cross sections for exothermic reactions behave like $1/\beta$ so that $\beta\sigma_{ann}$ is then practically constant, cf. left side of Fig. 3 for $p_{lab} \approx 100-300$ MeV/c. However, the Coulomb attraction modifies that to a $1/\beta^2$ behavior for energies very close to the threshold.

For illustration we show also predictions for $\bar{n}p$ scattering, see Fig. 4. Data for this reaction were not included in the PWA [18] because they are less precise than the ones for $\bar{p}p$. Nevertheless, the $\bar{n}p$ system is interesting because it is a pure isospin $I = 1$ state so that one can test the $I = 1$ component of the $\bar{N}N$ amplitude independently.

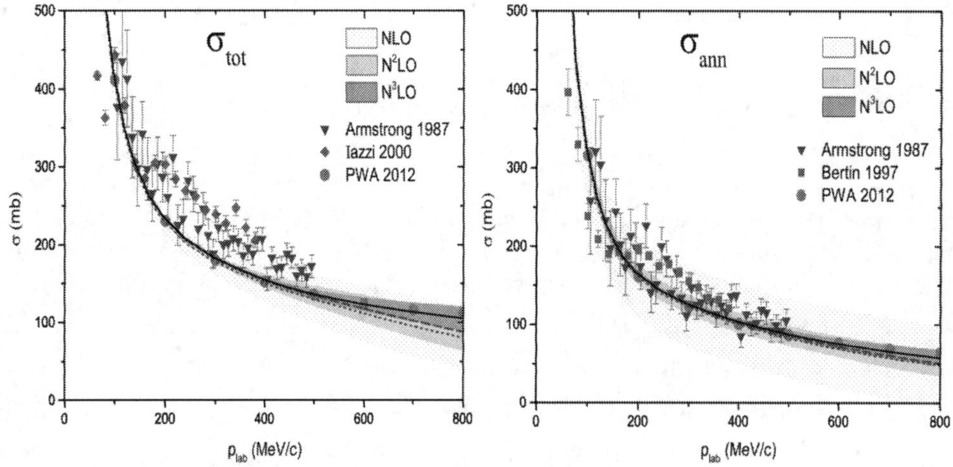

Figure 4. Total (σ_{tot}) and integrated annihilation (σ_{ann}) cross sections for $\bar{n}p$ scattering. Same description of curves as in Fig. 2.

Our results for the $\bar{p}p$ atomic level shifts and widths are summarized in Table 1 and compared with measurements [25–28]. In the pertinent calculation the Trueman formula [29] is utilized and it is applied to the theory results with the second-order term taken into account for the S-waves. It has been found in Ref. [30] that values obtained in this way agree rather well with direct calculations. In this context let us emphasize that the results in Table 1, including those for the N^{2}LO interaction from Ref. [19], are deduced, of course, from a calculation in particle basis. In particular, the Coulomb force in $\bar{p}p$ is taken into account and likewise the p-n mass difference that leads to separated thresholds for the $\bar{p}p$ and $\bar{n}n$ channels. The corresponding results given in our earlier study of the $\bar{N}N$ interaction within chiral EFT [19] are from a calculation in the isospin basis.

Given that experimental evidence on level shifts and widths in antiprotonic hydrogen was not taken into account in the PWA [18], it is remarkable that there is a rather good agreement between our predictions and experiments, especially with the more recently published measurement by Augsburger et al. [26]. In this context, let us mention that additional assumptions have to be made in order to derive the splitting of the $^1S_0$ and $^3S_1$ level shifts from the experiment [28, 31]. This caveat has to be kept in mind when comparing the theory results with experiments.

Table 1. Hadronic shifts and broadenings in hyperfine states of $\bar{p}$H for the chiral potentials with R = 0.9 fm. For comparison N^{2}LO predictions of our previous chiral potential are included, based on the cutoff combination $(\Lambda, \tilde{\Lambda})$ = (450,500) MeV [19]. The experimental information is taken from Refs. [25–28].

	NLO	N^{2}LO	N^{3}LO	N^{2}LO [19]	Experiment
$E_{^1S_0}$ (eV)	−448	−446	−443	−436	−440(75) [26]
					−740(150) [25]
$\Gamma_{^1S_0}$ (eV)	1155	1183	1171	1174	1200(250) [26]
					1600(400) [25]
$E_{^3S_1}$ (eV)	−742	−766	−770	−756	−785(35) [26]
					−850(42) [27]
$\Gamma_{^3S_1}$ (eV)	1106	1136	1161	1120	940(80) [26]
					770(150) [27]
$E_{^3P_0}$ (meV)	17	12	8	16	139(28) [28]
$\Gamma_{^3P_0}$ (meV)	194	195	188	169	120(25) [28]
E_{1S} (eV)	−670	−688	−690	−676	−721(14) [26]
Γ_{1S} (eV)	1118	1148	1164	1134	1097(42) [26]
E_{2P} (meV)	1.3	2.8	4.7	2.3	15(20) [28]
Γ_{2P} (meV)	36.2	37.4	37.9	27	38.0(2.8) [28]

Table 2. Bound states and resonances (in MeV) predicted by recently published $\bar{N}N$ potentials. In Ref. [11] only the $^1S_0$ partial wave is considered.

	N^{2}LO [19, 34]	El-Bennich [14]	Entem [15]	Milstein [11]
$^{11}S_0$	-	-4.8-i 26	-	22-i 33
$^{31}S_0$	-37-i 47	-	-	-
$^{13}S_1$	$+(4.8\cdots21.3)-\mathrm{i}\,(60.6\cdots74.9)$	-	-	
$^{11}P_1$	-	1877±i 13	-	
$^{13}P_0$	$-(3.7\cdots0.2)-\mathrm{i}\,(22.0\cdots26.4)$	1876±i 5	1895±i 17	
$^{33}P_0$	-	1871±i 11	-	
$^{13}P_1$	-	1872±i 10	-	
$^{33}P_1$	-	-4.5-i 9	-	

One of the topics that has escorted studies of the $\bar{N}N$ interaction over many decades is the possible existence of $\bar{N}N$ bound states, also called baryonium states, see, e.g., Ref. [32] for some historical remarks. So far there is no clear and generally accepted evidence for their existence. Nonetheless, let me summarize some recent predictions for $\bar{N}N$ bound states and resonances in Table 2 for curiosity reasons. Besides our results from the N^{2}LO potential in [19], those of a meson-exchange potential [14] and an $\bar{N}N$ interaction based on the constituent quark model [15] are included.

In the table the spectroscopic notation ${}^{(2I+1)(2S+1)}L_J$ is used for identifying the partial waves. In case of bound states the energy E_B with respect to the $\bar{p}p$ threshold is given, $M_p + M_{\bar{p}} = 1876.574$ MeV, while for resonances the total energy is provided. Notice that there are cases like that of the ${}^{13}S_1$–${}^{13}D_1$ state predicted by the N^{2}LO interaction from Ref. [19], or the ${}^{11}S_0$ state given in Ref. [11], where there is a pole which corresponds to a positive value of the binding energy! The positive sign of the real part of E_B indicates that the poles are actually located above the $\bar{N}N$ threshold. But they move below the threshold when the imaginary part of the potential is switched off and that is the reason why they are referred to as bound states. To be precise these are unstable bound states in the terminology of Ref. [33]. Since those poles lie on the physical sheet they do not correspond to resonances.

Evidently, and not unexpectedly, conclusions are difficult to draw given the large spread of the results. Indeed, the only "unambiguous" candidate could be the state in the ${}^{13}P_0$ partial wave where all potentials predict either a bound state or a resonance. It should be said that the bound states in the ${}^{1}S_0$ partial waves in Refs. [11, 34] are primarily a result of attempting to reproduce the $\bar{p}p$ invariant mass spectrum measured in the reaction $J/\psi \to \gamma\bar{p}p$ by the BESIII Collaboration [4], which exhibits a strong near-threshold enhancement. However, the obtained results for the binding energy differ significantly, and even appear in different isospin states. Moreover, neither of them seems to correspond to the $X(1835)$ (also denoted as $X(\bar{p}p)$), a state that the BESIII Collaboration itself sees as being possibly responsible for the enhancement of the $\bar{p}p$ mass spectrum [4, 35]. Finally, note that the width of the S-wave states, $\Gamma = -2\,\mathrm{Im}\,E_B$, are in general rather large.

5 Summary

In this contribution I have presented results of a study of the antinucleon-nucleon interaction within chiral effective field theory [19, 20]. This novel approach suggested by Weinberg has been applied rather successfully to the nucleon-nucleon interaction and can be adapted straightforwardly to the antinucleon-nucleon system. So far the antinucleon-nucleon potential has been derived up to next-to-next-to-next-to-leading order in the chiral expansion. The low-energy constants associated with the arising contact interactions are fixed by a fit to phase shifts and inelasticities provided by a recently published phase-shift analysis of antiproton-proton scattering data. The achieved description of the antinucleon-nucleon amplitudes is excellent and of a quality comparable to the one found in case of the nucleon-nucleon interaction at the same order. As a special application of the antinucleon-nucleon potential predictions for the level shifts and width of antiproton-proton atoms have been presented. Further applications of our chiral $\bar{N}N$ potentials, notably to the reactions $J/\psi \to \gamma\bar{p}p$ [34], $e^+e^- \to \bar{p}p$ [36], and $e^+e^- \to 5\pi$, 6π [37], can be found in the literature.

Acknowledgements

I would like to thank Ling-Yun Dai, Xian-Wei Kang, and U.-G. Meißner for collaboration on the topics covered by my talk. This work is supported in part by the DFG and the NSFC through funds provided to the Sino-German CRC 110 "Symmetries and the Emergence of Structure in QCD" and by the BMBF (contract No. 05P2015 -NUSTAR R&D).

References

[1] J. Z. Bai et al. [BES Collaboration], Phys. Rev. Lett. **91**, 022001 (2003).

[2] B. Aubert et al. [BaBar Collaboration], Phys. Rev. D **72**, 051101 (2005).

[3] B. Aubert et al. [BaBar Collaboration], Phys. Rev. D **73**, 012005 (2006).
[4] M. Ablikim *et al.* [BESIII Collaboration], Phys. Rev. Lett. **108**, 112003 (2012).
[5] A. Sibirtsev, J. Haidenbauer, S. Krewald, U.-G. Meißner and A. W. Thomas, Phys. Rev. D **71**, 054010 (2005).
[6] B. Loiseau and S. Wycech, Phys. Rev. C **72**, 011001 (2005).
[7] J. Haidenbauer, U.-G. Meißner and A. Sibirtsev, Phys. Rev. D **74**, 017501 (2006).
[8] D. R. Entem and F. Fernández, Phys. Rev. D **75**, 014004 (2007).
[9] J.-P. Dedonder, B. Loiseau, B. El-Bennich, and S. Wycech, Phys. Rev. C **80**, 045207 (2009).
[10] J. Haidenbauer and U.-G. Meißner, Phys. Rev. D **86**, 077503 (2012).
[11] A. I. Milstein and S. G. Salnikov, Nucl. Phys. A **966**, 54 (2017).
[12] W. Erni et al. [Panda Collaboration], arXiv:0903.3905 [hep-ex].
[13] V. Barone et al. [PAX Collaboration], hep-ex/0505054.
[14] B. El-Bennich, M. Lacombe, B. Loiseau and S. Wycech, Phys. Rev. C **79**, 054001 (2009).
[15] D.R. Entem and F. Fernández, Phys. Rev. C **73**, 045214 (2006).
[16] G. Y. Chen and J. P. Ma, Phys. Rev. D **83**, 094029 (2011).
[17] E. Epelbaum, H.-W. Hammer and U.-G. Meißner, Rev. Mod. Phys. **81**, 1773 (2009).
[18] D. Zhou and R. G. E. Timmermans, Phys. Rev. C **86**, 044003 (2012).
[19] X. W. Kang, J. Haidenbauer and U.-G. Meißner, JHEP **1402**, 113 (2014).
[20] L. Y. Dai, J. Haidenbauer and U.-G. Meißner, JHEP **1707**, 078 (2017).
[21] S. Weinberg, Nucl. Phys. **B 363**, 3 (1991).
[22] E. Epelbaum, W. Glöckle, U.-G. Meißner, Nucl. Phys. A **747**, 362 (2005).
[23] E. Epelbaum, H. Krebs and U.-G. Meißner, Eur. Phys. J. A **51**, 53 (2015).
[24] R. Machleidt and D. R. Entem, Phys. Rept. **503**, 1 (2011).
[25] M. Ziegler et al. [ASTERIX Collaboration], Phys. Lett. B **206**, 151 (1988).
[26] M. Augsburger et al., Nucl. Phys. A **658**, 149 (1999).
[27] K. Heitlinger et al., Z. Phys. A **342**, 359 (1992).
[28] D. Gotta et al., Nucl. Phys. A **660**, 283 (1999).
[29] T. L. Trueman, Nucl. Phys. **26**, 57 (1961).
[30] J. Carbonell, J.-M. Richard and S. Wycech, Z. Phys. A **343**, 325 (1992).
[31] D. Gotta, Prog. Part. Nucl. Phys. **52**, 133 (2004).
[32] E. Klempt, F. Bradamante, A. Martin and J.-M. Richard, Phys. Rept. **368**, 119 (2002).
[33] A. M. Badalian, L. P. Kok, M. I. Polikarpov, Y. A. Simonov, Phys. Rep. **82**, 31 (1982).
[34] X. W. Kang, J. Haidenbauer and U.-G. Meißner, Phys. Rev. D **91**, 074003 (2015).
[35] M. Ablikim *et al.* [BESIII Collaboration], Phys. Rev. Lett. **117**, 042002 (2016).
[36] J. Haidenbauer, X.-W. Kang and U.-G. Meißner, Nucl. Phys. A **929**, 102 (2014).
[37] J. Haidenbauer, C. Hanhart, X. W. Kang and U.-G. Meißner, Phys. Rev. D **92**, 054032 (2015).

Two Recent Developments in QCD Spectroscopy

Kamal K. Seth[1,*]

[1] *Northwestern University, Evanston IL 60208, USA*

Abstract. The presentation has two parts, the first, and main part, is devoted to the presentation of results from the first measurements of the production of Λ, Σ, Ξ and Ω hyperons at large timelike momentum transfers of 13.6, 14.2 and 17.4 GeV/c^2 made using e^+e^- annihilation data taken at the CESER election-positron collider at Cornell using the CLEOc detector. The measurements reveal interesting features of hyperon production systematics and timelike form factors, and provide evidence for diquark correlations in hyperon structure.
The second part of the presentation is devoted to measurements of photoproduction of J/ψ near threshold energies, $E_G = 8 - 12$ GeV, at the new photon physics facility, GlueX, at the Jefferson Laboratory. The first preliminary results are presented. These are expected to shed light on the gluon structure of protons, gluons which mediate the production of J/ψ in the reaction, $\gamma + p \to J/\psi + p, J/\psi \to e^+e^-$.

1 Introduction

The quark-gluon model of hadrons was introduced by Gell-Mann in 1956 to explain the structure of the quark-antiquark mesons and the three quark baryons which were the only hadrons known at that time. However, Gell-Mann suggested that other color-neutral hadrons containing larger number of quarks and antiquarks, as well as hybrids containing valence gluons, and glueballs containing only gluons should exist. These predictions launched numerous searches for these unconventional hadrons. Among the first of these was the search for six-quark "dibaryons". Many were proposed, and numerous claims and counter-claims were published. Unfortunately, none survived, and I myself claim responsibility for the demise of many of them [1]. Similarly, despite many dedicated searches during the last twenty years most claims for glueballs have not survived either [2].

These failures did not stop the searches by stubborn physicists that we are. But, so far no dibaryons or glueballs have been convincingly identified, and I will not talk about them.

However, a silver lining has recently emerged. Several convincing citings of the unconventional hadrons have been reported. An excellent summary of these exist in the 2011 compilation of QWG [3]. Subsequent updates have been presented in this conference by Belle and LHCb. With that said, I want to talk about only two developments in hadron spectroscopy in which I have been personally involved.

2 Hyperons

The major part of my talk is devoted to hyperons.

*e-mail: kseth@northwestern.edu

The universe is built of baryons. Before 1947 only two baryons, the proton and neutron, made of up and down quarks, were known. The 1947 discovery of the first strange particles, the kaon and the Lambda, and consequently, of the strange quark enriched the field of baryons immensely. By 1960, when the theoretically predicted Ω^- was discovered, all seven light baryons, containing strange quarks, Λ^0, Σ^0, Σ^+, Σ^-, Ξ^0, Ξ^- and Ω^-, known as hyperons, were known. However, even more than 50 years after their discovery very little more than the static properties of their ground states as listed in Table 1 are known [4].

Most of our extensive knowledge of nucleon structure comes from lepton scattering by nucleon and nuclear targets [5]. Unfortunately, hyperons are not available as targets, and this is responsible in large part for the lack of our understanding of the structure of hyperons.

In 1960 Cabibo and Gatto [6] pointed out that electron-positron colliders were being planned at various laboratories, and they offered opportunity of overcoming the lack of target disadvantage of hyperons; one could measure timelike form factors of hyperons in $e^+e^- \to B\bar{B}$ ($B \equiv$ hyperon) measurements. To put this opportunity in perspective, we note that four momentum transfers is defined as

$$Q(4\text{mom.})^2 = q(3\text{mom.})^2_{\text{space}} - (\text{energy})^2_{\text{time}} \quad (1)$$

It can be positive and spacelike, as in scattering measurements, or negative and timelike, as in production measurements, as illustrated in Figure 1.

Form factors are analytic functions of momentum transfer $|Q|^2$, and $B\bar{B}$ pair production experiments can be analyzed in the same formalism as the scattering experiments, i.e., in terms of the Dirac form factor, $F_1(|Q|^2)$, and the Pauli form factor, $F_2(|Q|^2)$, or equivalently, in terms of

Table 1. The hyperon.

Hyperon	Quarks	Mass, M(MeV)	Mag.mom. (μ_N)	Main Decay
Proton, p	uud	938.272(<0.001)	2.793(<0.001)	stable
Λ^0	uds	1115.683(6)	-0.613(4)	$p\pi^-$(64%)
Σ^0	uds	1192.642(24)	1.61(8)	$\Lambda^0\gamma$(100%)
Σ^+	uus	1189.37(7)	2.458(10)	$p\pi^0$(52%)
Σ^-	dds	1197.449(30)	-1.160(25)	$n\pi^-$(99.8%)
Ξ^0	uss	1314.86(20)	-1.250(14)	$\Lambda^0\pi^0$(99.5%)
Ξ^-	dss	1321.71(7)	-0.6507(25)	$\Lambda^0\pi^-$(99.9%)
Ω^-	sss	1672.45(29)	-2.02(5)	$\Lambda^0 K^-$(69%)

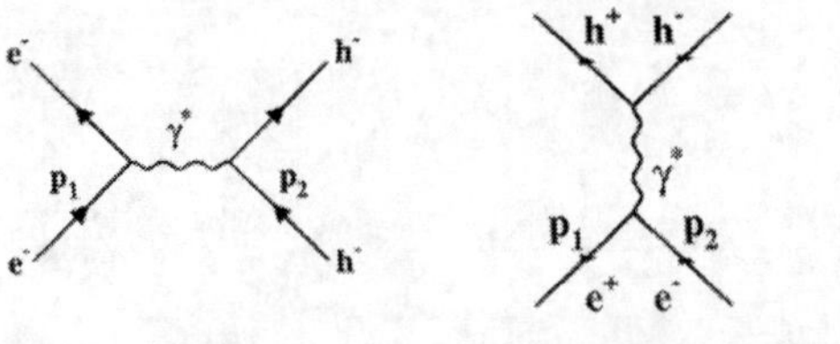

Figure 1. Spacelike and timelike four momentum transfers.

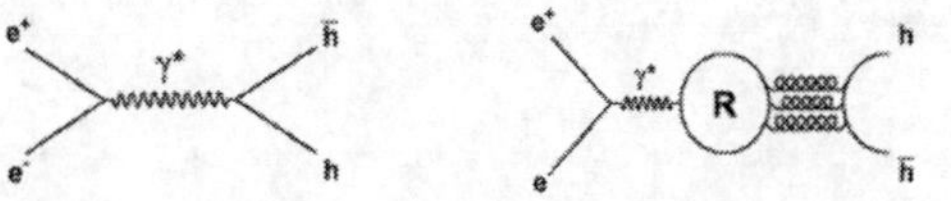

Figure 2. Electromagnetic decay and Hadronic decay via gluons.

Table 2. The estimation events from $\psi(3770)$ and $\psi(4170)$.

	$\Lambda\bar{\Lambda}$	$\Sigma^+\bar{\Sigma}^+$	$\Sigma^0\bar{\Sigma}^0$	$\Xi^-\bar{\Xi}^-$	$\Xi^0\bar{\Xi}^0$	$\Omega^-\bar{\Omega}^-$
$\psi(3770)$	3.0	1.4	1.2	1.2	0.6	0.3
$\psi(4170)$	2.0	1.0	0.9	0.9	0.4	0.2

the electric and magnetic form factors

$$G_E(|Q|^2) = F_1(|Q|^2) + (s/m^2)F_2(|Q|^2), \quad (2)$$

and

$$G_M(|Q|^2) = F_1(|Q|^2) + F_2(|Q|^2). \quad (3)$$

It took 30 years for the first successful measurement of $\Lambda^0\bar{\Lambda}^0$ hyperon production by the DM2 Collaboration at Orsay [7], and seventeen more years by the BaBar Collaboration at SLAC [8] to report measurements of $G_M(|Q|^2)$ of Λ^0, Σ^0, and the Λ^0 Σ^0 transition form factors. Because both these measurements were made near threshold energies $\sqrt{s} \approx 2.4$ GeV, and only a few events $4\Lambda^0\bar{\Lambda}^0$ by DM2, ≤ 22 by SLAC were observed, they were not suitable for QCD based analyses. No further progress in hyperon production studies was made until at CLEO in 2005 we made measurements of pair production of hyperons at the $\psi(2S)$ resonance, $\sqrt{s} = 3.69$ GeV, $|Q|^2 = 13.59$ GeV^2 [9].

We must remember, however, that unlike for spacelike form factors, G_E and G_M for timelike form factors do not relate to spatial distributions of charge and magnetic moment. Instead they relate to the helicity correlations between the particle antiparticle pair produced. $F_2(|Q|^2)$ denotes photon coupling to parallel spins and F_1 to antiparallel spins of the pair.

Hadronic decays at resonances proceed via gluons and have large yields. To measure electromagnetic form factors we require the decays to be electromagnetic, which have much smaller yields. To measure form factors we must measure e^+e^- annihilation at non-resonance energies, or at those resonances where it can be demonstrated that resonance yields are negligibly small, as $\psi(3770)$ and $\psi(4170)$ which mainly decay to $D\bar{D}$. Using the experimentally confirmed pQCD relation, schematically, illustrated below,

$$\frac{B(\psi(n')) \to \text{hadrons}}{B(\psi(n)) \to \text{hadrons}} = \frac{B(\psi(n')) \to \text{leptons}}{B(\psi(n)) \to \text{leptons}}, \quad (4)$$

we estimate the expected number of resonance events as in Table 2, i.e., at $\psi(3770)$ and $\psi(4170)$ the resonance contribution is indeed negligible for all hyperons, and the observed hyperon yield is entirely electromagnetic.

We have now made the world's first measurements of the pair production of Λ^0, Σ^0, Σ^+, Ξ^0, Ξ^- and Ω^- hyperons at large momentum transfers of 13.6, 14.2 and 17.4 GeV^2, and with good statistics. These measurements provide insight into the systematics of pair production of hyperons, the dependence of their cross section on their s-quark content, evidence for diquark correlations, and their timelike form factors. We use e^+e^- annihilation data taken at the CESR collider using the CLEO-c detector. The near-4pi acceptance CLEO-c detector, illustrated in Figure 3, has cylindrical geometry and consists of a CsI electromagnetic calorimeter, drift chambers, and a RICH detector, all in a 1 Tesla solenoidal magnetic field. The data used in the measurements are listed in Table 3.

We identify the hyperons by detecting their major decay products, $\Lambda^0 \to p\pi^-$(64%), $\Sigma^+ \to p\pi^0$(52%), $\Sigma^0 \to \Lambda\gamma$(100%), $\Xi^- \to \Lambda\pi^-$(100%), $\Xi^0 \to \Lambda\pi^0$(100%), $\Omega^- \to \Lambda K^-$(68%). The $\psi(2S)$ resonance decay into hyperons has a prolific yield, and although it is not the subject

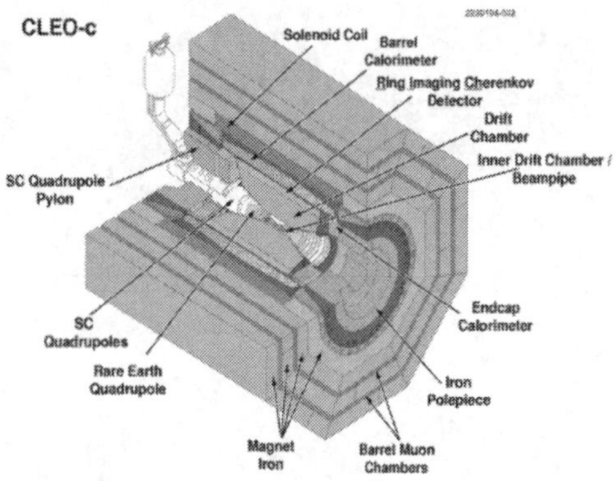

Figure 3. CLEO-c detector.

Table 3. The data used in the measurements.

	$\sqrt{s}$ (GeV)	$\|Q\|^2$ (GeV^2)	Luminosity (pb^{-1})
$\psi(2S)$	3.69	13.6	48
$\psi(3770)$	3.77	14.2	805
$\psi(4160)$	4.17	17.4	586

of my talk, Figures 4 and 5 illustrate the steps in hyperon identification very effectively.

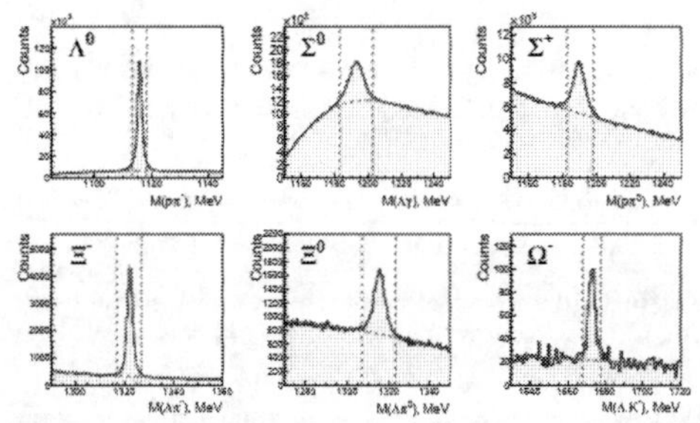

Figure 4. The six panels show the raw invariant mass distribution for $\psi(2S)$ data.

The six panels in Figure 6 show the invariant mass distribution of $B\bar{B}$ pair production events of Figures 4. The six panels in Figure 7 show event distribution for $\psi(3770)$ data for electromagnetic production of hyperons with backgrounds due to inclusive production of single hyperons.

In Table 4 we present the summary of pair production cross sections. We note that the cross sections for electromagnetic production at $\psi(3770)$ are smaller by factors ≥ 200 than for the resonance production at $\psi(2S)$. The VDM theoretical predictions of Körner and Kuroda [10] for $\psi(3770)$ are smaller by orders of magnitude than our measured values. The top panel in Figure 8 illustrates the variation of cross sections for hyperon pair production at

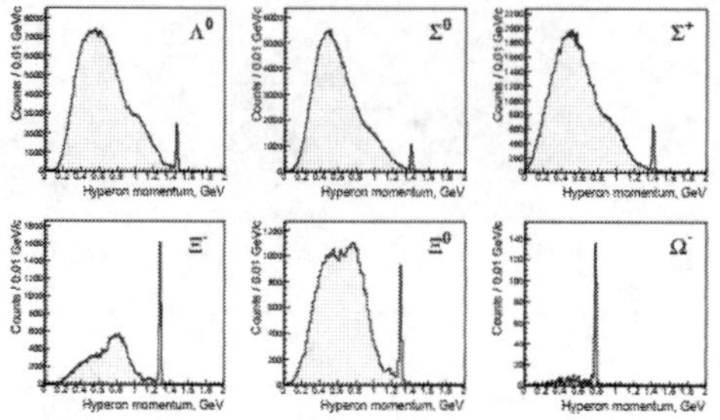

Figure 5. The six panels show the momentum distribution for $\psi(2S)$ data. The large yields on left correspond to inclusive production and the sharp peaks on right to exclusive $B\bar{B}$ pair production.

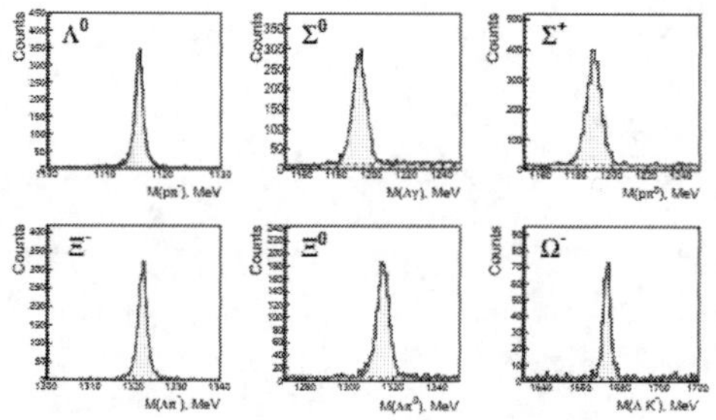

Figure 6. $\psi(2S)$ pair production: the numbers of observed events are $N(\Lambda, \Sigma^0, \Sigma^+) = 6531, 2645, 1874$, $N(\Xi^-, \Xi^0, \Omega) = 3580, 1242, 326$.

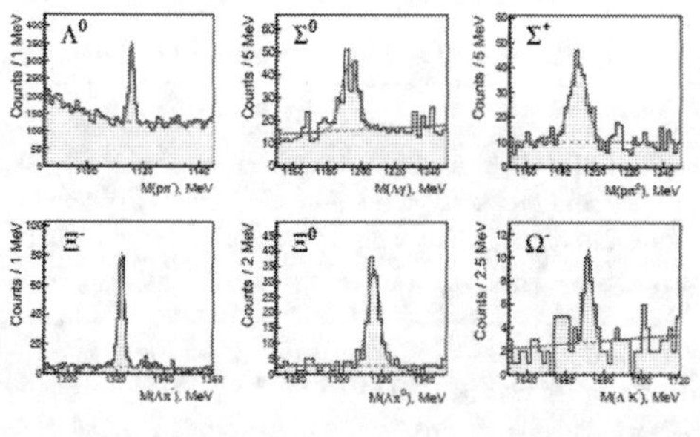

Figure 7. $\psi(3770)$ pair production: the numbers of observed events are $N(\Lambda, \Sigma^0, \Sigma^+) = 498, 142, 200$, $N(\Xi^-, \Xi^0, \Omega) = 240, 111, 20$. Notice an order or more decrease than the resonance yield at $\psi(2S)$.

$\psi(3770)$. It shows that $\sigma(\Sigma^0)$ is much smaller than the general trend of the data for $J = 1/2$ hyperons, and nearly factor 2.3 ± 0.4 smaller than $\sigma(\Lambda^0)$ which has the same uds quark construct. The bottom panel in Figure 8 illustrates an unexpected feature, the ratio $\sigma(3770)/\sigma(4170)$ does not vary as $1/s^5$ as pQCD predicts, but shows steps with increasing number of s-quarks in the hyperonss. More about this very important observation later.

Table 4. Summary of pair production cross sections (picobarns), and $G_M(3770)$.

	p	Λ^0	Σ^0	Σ^+	Ξ^-	Ξ^0	Ω^-	$\Lambda^0\Sigma^0$
$\psi(2S)$	196(12)	244.7(106)	145.6(77)	151.4(74)	199.9(100)	131.6(82)	33.7(28)	8.1(16)
$\psi(3770)$	0.46(4)	1.13(10)	0.46(8)	0.97(10)	0.78(7)	0.68(9)	0.11(3)	0.43(9)
VDM Theory $\psi(3770)$ [10]	0.069	0.010	–	0.081	0.064	0.014	0.006	0.042
$G_M(3770)$	0.88(4)	1.40(6)	0.91(7)	1.31(7)	1.20(5)	1.12(6)	0.53(8)	0.77(8)

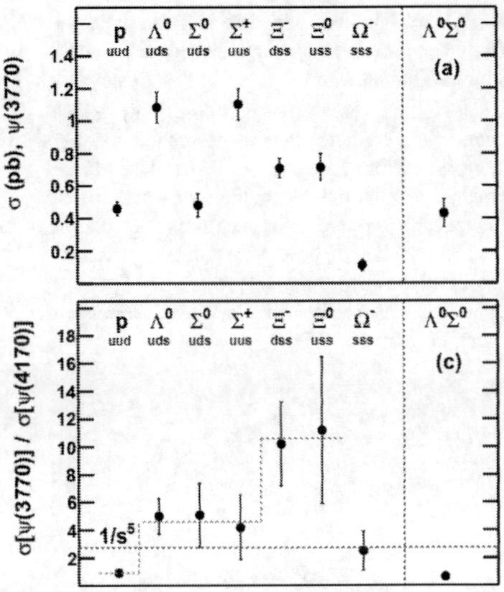

Figure 8. The cross sections from $\psi(3770)$ and $\psi(4170)$ data. Note steps at number of strange quark, $S = 0, 1, 2$.

As for nucleons, the timelike form factors are related to cross sections in terms of form factors G_E and G_M, which now refer to correlations between the helicities of the baryon and antibaryon

$$d\sigma/d\Omega = \frac{\alpha^2\beta_B}{4s}[|G_M^B(s)|^2(1+\cos^2\theta) + (4m_B^2/s)|G_E^B(s)|^2\sin^2\theta], \quad (5)$$

$$\sigma_{B\bar{B}} = \frac{4\pi\alpha^2\beta_B}{3s}\left(|G_M^B(s)|^2 + (2m_B^2/s)|G_E^B(s)|^2\right). \quad (6)$$

Because of small yield of hyperons from electromagnetic decays, it is generally not possible to determine G_E and G_M, or G_E/G_M separately, and most experimental data are analyzed by assuming $G_E/G_M = 0$ or 1.

BaBar has recently analyzed the angular distributions for their ISR based production of $\Lambda\bar{\Lambda}$ pairs in two vs bins. They obtained $|G_E/G_M| = 1.73^{+0.99}_{-0.57}$ for the $\sqrt{s} = 2.23 - 2.40$ GeV/c^2 bin with 115 events, and $|G_E/G_M| = 0.71^{+0.66}_{-0.71}$ for the $\sqrt{s} = 2.40 - 2.80$ GeV/c^2 bin with 61 events. They considered both of them as consistent with $G_E/G_M = 1$, and analyzed their data with that assumption.

We note that because of their large errors BaBar's results are equally consistent with $G_E/G_M = 0$. We have analyzed our data for Λ^0, Ξ^0, Ξ^- production, and obtained $G_E/G_M = 0$ in all three cases, with 90% confidence limits: $\Lambda^0 < 0.17, \Xi^0 < 0.32, \Xi^- < 0.29$. Unexpected as this result is, it is consistent with the recent Jlab observation, $G_E = 0$ at $|Q|^2 \approx 8$ GeV^2 for spacelike form factor of proton. We have analyzed our data for $|Q|^2$=14.2 and 17.4 GeV^2 assuming $G_E = 0$. The results for $\psi(3770)$ are presented in Table 4. Unfortunately, unlike for the proton there are no measurements of spacelike form factors hyperons to compare with.

We note that these are the first measurements ever of these timelike form factors of hyperons at such large momentum transfer. There are no VDM or QCD based predictions for them. We note that G_M for all $J = 1/2$ hyperons are the same within a factor ~ 1.5. We note that $G_M(\Lambda)/G_M(\Sigma^0) = 1.5 \pm 0.1$.

Figure 9 illustrates variation of $G_M(3770)$, and the steps in the ratio $G_M(3770)/G_M(4170)$ with increasing number of strange quarks. Figure 10 shows the rapid decrease of inclusive hyperon production from Λ^0, to $\Sigma^{0,+}$, to $\Xi^{0,+}$, and finally to $J = 3/2\ \Omega^-$.

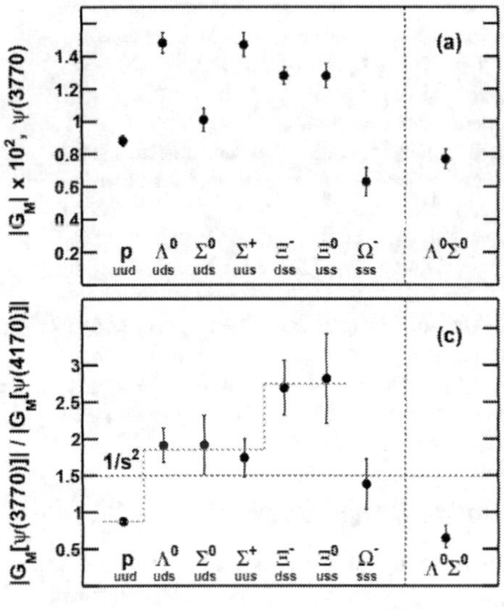

Figure 9. The G_M results from $\psi(3770)$ and $\psi(4170)$ data. Note steps at $S = 0, 1, 2$.

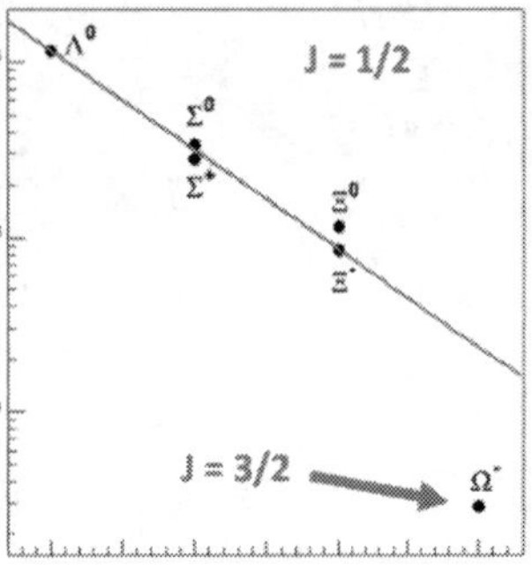

Figure 10. An interesting trend for inclusive events produced with $\mathcal{L}$=48 pb^{-1} at $\psi(2S)$.

3 Diquark in hyperons

Our most important result concerns the evidence for diquark correlations in hyperon pair production. The importance of certain configurations of flavor, spin, and isospin of two quarks in the structure of hadrons has been recognized for a long time [12]. One dramatic example of the role of diquarks was provided by the Fermilab observation that the timelike form factor of the proton was twice as large as the spacelike form factor at the same large momentum transfer $|Q|^2$ [13], and its successful explanation by Kroll et al. [14] in terms of the diquark-quark structure of the proton.

Recently Wilczek and colleagues [15] have emphasized the role of diquarks in QCD in terms of isoscalar "good", and isovector "bad" diquarks. They predicted that the "good" diquark in Λ^0 with isospin 0 compared to the "bad" diquark in Σ^0 with isospin 1, would lead to enhancement of Λ^0 over Σ^0 in production experiments. They cited the observation of $\Lambda^0/\Sigma^0 = 3.5 \pm 1.7$ in the LEP experiment in support of this prediction.

Our measurements provide strong independent support for the role of diquarks in Λ^0/Σ^0 hyperon production. We observe $\sigma(\Lambda^0)/\sigma(\Lambda^0) = 2.46 \pm 0.46$ at $|Q|^2 = 14.2$ GeV^2 (in exclusive pair production), $\sigma(\Lambda^0)/\sigma(\Lambda^0) = 2.56 \pm 1.40$ at $|Q|^2 = 17.4$ GeV^2 (in exclusive pair production), $\sigma(\Lambda^0)/\sigma(\Lambda^0) = 4.1 \pm 0.6$ at $|Q|^2 = 13.6$ GeV^2 (in inclusive pair production). Our data provide the opportunity to consider diquark pairs other than the up/down diquarks, and we expect that they will lead to a deeper understanding of diquark correlations. Lattice calculations are also expected to address diquark correlations soon [16].

4 Threshold photoproduction of J/ψ

The second part of my talk is not about a new subject like hyperons, but an old subject: Threshold photoproduction of J/ψ. The discovery of J/ψ in $e^+e^- \to \gamma^* \to J/\psi$ launched the modern era in QCD spectroscopy. Despite the fact that all particle physics experiments cut their teeth on the detection of J/ψ, it remains true that J/ψ production mechanisms are not well understood. There are theoretical models to be sure, color singlet model, color evaporation model, factorization models, etc., but serious problems in quantitative understanding of J/ψ production remain. Of particular interest is understanding photoprouction of J/ψ at energies near threshold, $E_\gamma \sim 8.5$ GeV, because at small momentum transfers coherent electro production of vector mesons like ϕ, J/ψ, etc. provides valuable insight into the gluon structure function of the target [17].

Gluon distribution functions at small x have been of interest in relation to studies of deconfinement in QGP, the phenomena of color transparency and others, and good precision data on J/ψ photo production near threshold energies has long been needed to distinguish between models of gluon structure functions.

Brodsky et al. have made a more detailed study of J/ψ photoproduction and predicted that near threshold the J/ψ production cross sections have very different dependence on the momentum of photons depending on whether two or three gluons carry the targets momentum to the charm quarks [18] (Figure 11). The existing data consists of just two small statistics measurements by Cornell [19] and SLAC [20], and they are too sparse to distinguish between the two models (Figure 12).

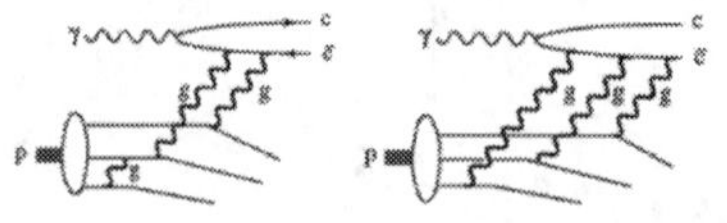

Figure 11. J/ψ production in two/three gluons processes.

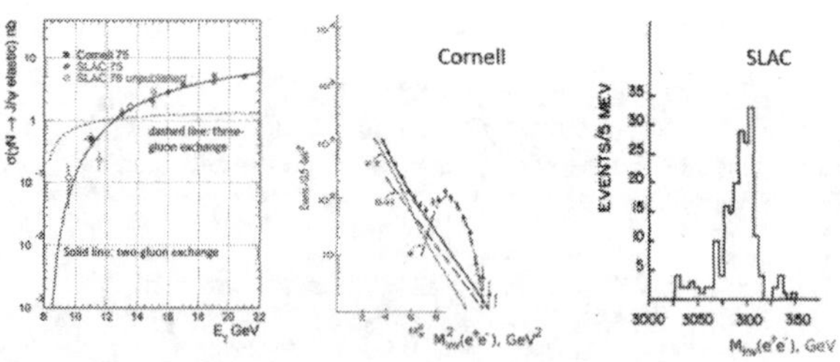

Figure 12. Left: Theoretical predictions for 2/3 gluon production of J/ψ. Right: Existing measurements by Cornell [19] and SLAC [20].

At the Jefferson lab we now have polarized and unpolarized electron beams of energies up to 12 GeV available, and a facility called GlueX has been constructed, dedicated to photo production experiments (Figure 13). This has made it possible to fill the gap in threshold measurements of J/ψ photo production. We have made the first such measurements of $\gamma + p \to p + J/\psi, J/\psi \to e^+e^-$, and I want to present the first results of these measurements which we believe shed valuable light on the gluon content of protons and their role on J/ψ photo production. Figure 13 shows the photon beam production and the schematic of the GlueX detector.

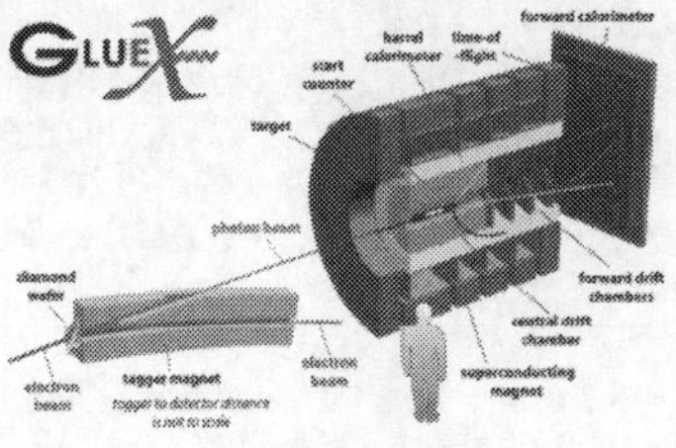

Figure 13. The GlueX detector.

We have made the first successful measurements of $\gamma + p \to p + J/\psi, J/\psi \to e^+e^-$ with tagged photons of energies between 8 and 12 GeV at the GlueX facility at Jlab. I will not bore you with the details of event selection for these, but here are a few details: (1) At least 3 charged tracks are required in the event. (2) The yield of the e^+e^- decays are overwhelmed by more than three orders of magnitude larger production of the Bethe-Heitler production of $\pi^+\pi^-$ pairs. To reject the $\pi^+\pi^-$ BH background we use the quantity E/p, with E from EM calorimeter and p from drift chambers. E/p$\sim$ 1 for electrons, and is much smaller for pions. We require E/p> 0.8 for the selected events, which selects e^+e^- events very effectively, and provides very good rejection of the pion background.

Brodsky et al [18] pointed out that the negative charge conjugation J/ψ can be populated in photoproduction mainly by transition of two gluons from the proton target $\gamma + 2g$. If convincing evidence is found for three gluon transition $\gamma + 3g$, the gluons have to carry orbital angular momentum in addition to spin. This possibility is of great fundamental importance to QCD in many contexts, e.g., in context of the famous problem of proton spin, and it has been long studied by means of deep inelastic scattering. If our measurements of J/ψ photoproduction in the threshold region provide evidence for 3-gluon transition, we will have made a major contribution to QCD itself.

We have made two measurements of J/ψ photoproduction in the threshold region, in Feb. 2016 (E_γ = 8 − 12 GeV), and in Feb. 2017 (E_γ = 8 − 11.5 GeV), whose invariant mass spectra are shown in the Figure 14. Population of the vector resonances J/ψ and ϕ are clearly observed. The detailed plot of our observed J/ψ spectrum is shown in Figures along with the results of the earlier referred measurements of SLAC and Cornell, for comparison. I have to disappoint you by telling you that we are not yet able to quote production cross sections or their all important energy dependence. We expect to be able to do so in about three months. Keep tuned.

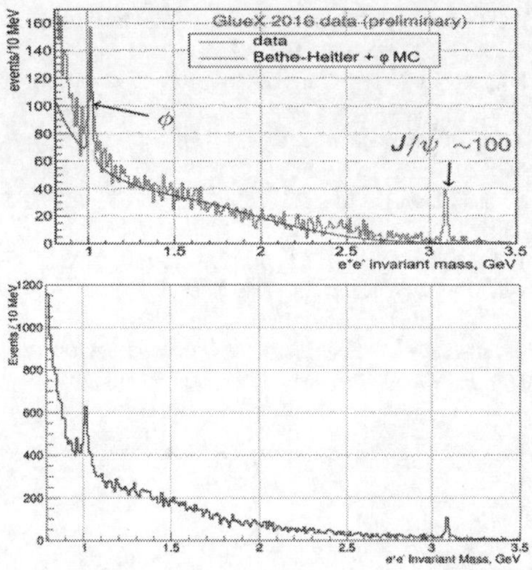

Figure 14. The J/ψ from GlueX in 2016 and 2017.

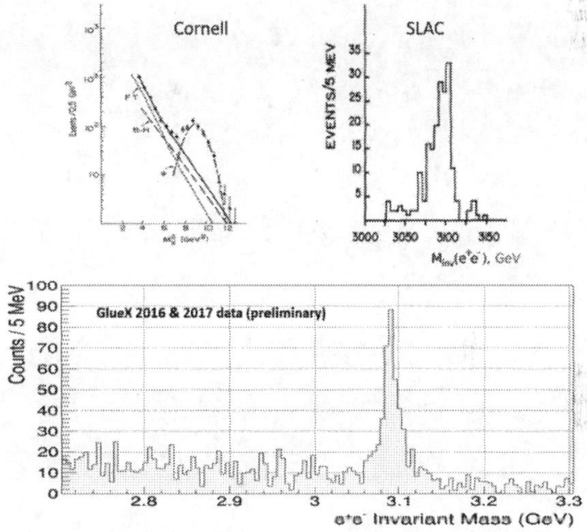

Figure 15. The preliminary result from GlueX.

References

[1] Kamal K. Seth, Proc. Few Body XII Conference, TRIUMF (1989) TRI 89-2, p. C35
[2] S. Dobbs et al., Phys. Rev. D **91**, 052006 (2015)
[3] N. Brambilla et al., Eur. Phys. J. C **71**, 1534 (2011)
[4] C. Patrignani et al. Particle Data Group, Chinese Phys. C, 40 (2016)
[5] For reviews of spacelike data, see: J. Arrington et al., J. Phys G **34**, 523 (2007); C. F. Perdristat et al., Prog. Particle and Nucl. Phys. **59**, 694 (2007)
[6] N. Cabibbo and R. Gatto, Phys. Rev. Lett. **313**, 4 (1960); Phys. Rev. **124**, 1577 (1961)
[7] D. Bisello et al. [DM2 Collaboration], Z. Phys. C **48**, 23 (1990)
[8] B. Aubert et al. [BaBar Collaboration], Phys. Rev. D **76**, 092006 (2007)

[9] T. K. Pedlar et al. [CLEO Collaboration], Phys. Rev. D **72**, 051108 (2005)

[10] J. G. Körner and M. Kuroda, Phys. Rev. D **16**, 2165 (1977)

[11] A. Z. Dubnickova et al, Czech. J. Phys. **43**, 1177 (1993)

[12] M. Anselmino et al., Rev. Mod. Phys. **65**, 1199 (1993)

[13] T. A. Armstrong et al. PRL **70**, 1212 (1993); M. Ambrogiani et al., Phys. Rev. D **60**, 032002 (1999)

[14] P. Kroll et al., Phys. Lett. B **316**, 546 (1993)

[15] R. Jaffe and F. Wilczek, Phys. Rev. Lett. **91**, 232003 (2003); F. Wilczek, Diquarks as inspiration and as objects, in: M. Shifman, et al. (Eds.), From Fields to Strings, vol. 1. World Scientific, Singapore, 2005, pp. 77-93, arXiv: hep-ph/0409168; A. Selem et al., arXiv: hep-ph/0602128;

[16] For lattice calculations of hyperons, see R. Babich et al., Phys. Rev. D **76**, 074021 (2007); J. Green et al., Proc. Sci. (LATTICE 2010), 140 [arXiv:1012.2353]

[17] Brodsky, et al. PRD **50**, 3134 (1994); D. Kharzeev, et al., Nucl. Phys. **A661**, 568 (1999); L. Frounkfuzt and M. Strikman, PRD **66**, 031502 (R) (2002)

[18] S. J. Brodsky et al., Phys. Lett. B **498**, 23 (2001)

[19] Gittleman et al., PRL **35**, 1616 (1975)

[20] Camerini et. al., PRL **35**, 483 (1975)

EPJ Web of Conferences **181**, 01030 (2018) https://doi.org/10.1051/epjconf/201818101030
EXA2017

Searches for magnetic monopoles and beyond with MoEDAL at the LHC

Vasiliki A. Mitsou[1,*]
On behalf of the MoEDAL Collaboration

[1]Instituto de Física Corpuscular (IFIC), CSIC – Universitat de València,
Parc Científic de la U.V., C/ Catedrático José Beltrán 2,
E-46980 Paterna (Valencia), Spain

Abstract. The MoEDAL experiment at the LHC is optimised to detect highly-ionising particles such as magnetic monopoles, dyons and (multiply) electrically-charged stable massive particles predicted in a number of theoretical scenarios. MoEDAL, deployed in the LHCb cavern, combines passive nuclear track detectors with magnetic monopole trapping volumes, while cavern backgrounds are being monitored with an array of MediPix detectors. The detector concept and its physics reach is presented with emphasis given to recent results on monopole searches providing the best limits on high magnetic charges in colliders. The potential to search for heavy, long-lived supersymmetric electrically-charged particles and multi-charged states is also discussed.

1 Introduction

MoEDAL (Monopole and Exotics Detector at the LHC) [1, 2], the 7th experiment at the Large Hadron Collider (LHC) [3], is designed to search for manifestations of new physics through highly-ionising particles in a manner complementary to ATLAS, CMS and ALICE [4]. The most important motivation for the MoEDAL experiment is to pursue the quest for magnetic monopoles and dyons at LHC energies. In addition, the experiment is designed to search for any massive, stable or long-lived, slow-moving particle [5] with single or multiple electric charges that arise in various scenarios of physics beyond the Standard Model (SM). For an extended and detailed account of the MoEDAL discovery potential, the reader is referred to the *MoEDAL Physics Review* [6]. Emphasis is given here on recent MoEDAL results, based on the exposure of magnetic monopole trapping volumes to 8 TeV proton-proton collisions during Run 1 and 13 TeV collisions during Run 2.

The structure of this paper is as follows. In Section 2, magnetic monopoles and monopolia are briefly discussed. Section 3 provides a brief description of the MoEDAL detector, whilst Section 4 presents the MoEDAL results on monopole searches. Section 5 is dedicated to supersymmetric models predicting massive (meta)stable states. The paper concludes with a summary and an outlook in Section 6.

*e-mail: vasiliki.mitsou@ific.uv.es

2 Magnetic monopoles

The MoEDAL detector is designed to fully exploit the energy-loss mechanisms of magnetically charged particles [7–10] in order to optimise its potential to discover these messengers of new physics. There are various theoretical scenarios in which magnetic charge would be produced at the LHC [6]: (light) 't Hooft-Polyakov monopoles [9, 11], electroweak monopoles [12–14], global monopoles [15–18] and monopolium [8, 19–21]. Magnetic monopoles that carry a non-zero magnetic charge and dyons possessing both magnetic and electric charge are among the most fascinating hypothetical particles. Even though there is no generally acknowledged empirical evidence for their existence, there are strong theoretical reasons to believe that they do exist, and they are predicted by many theories including grand unified theories and superstring theory [22, 23].

The theoretical motivation behind the introduction of magnetic monopoles is the symmetrisation of the Maxwell equations and the explanation of the charge quantisation [7]. Dirac showed that the mere existence of a monopole in the universe could offer an explanation of the discrete nature of the electric charge, leading to the Dirac Quantisation Condition (DQC),

$$\alpha\, g = \frac{N}{2} e, \quad N = 1, 2, ..., \tag{1}$$

where e is the electron charge, $\alpha = \frac{e^2}{4\pi\hbar c\varepsilon_0} = \frac{1}{137}$ is the fine structure constant (at zero energy, as appropriate to the fact that the DQC pertains to long (infrared) distances from the centre of the monopole), ε_0 is the vacuum permittivity, and g is the monopole magnetic charge. In Dirac's formulation, magnetic monopoles are assumed to exist as point-like particles and quantum mechanical consistency conditions lead to Eq. (1), establishing the value of their magnetic charge. Although monopoles symmetrise Maxwell equations in form, there is a numerical asymmetry arising from the DQC, namely that the basic magnetic charge is much larger than the smallest electric charge. A magnetic monopole with a single Dirac charge (g_D) has an equivalent electric charge of $\beta(137e/2)$. Thus for a relativistic monopole the energy loss is around 4,700 times (68.5^2) that of a minimum-ionising electrically-charged particle. The monopole mass remains a free parameter of the theory.

A possible explanation for the lack of experimental confirmation of monopoles is Dirac's proposal [7, 8, 19] that monopoles are not seen freely because they form a bound state called *monopolium* [20, 21, 24] being confined by strong magnetic forces. Monopolium is a neutral state, hence it is difficult to detect directly at a collider detector, although its decay into two photons would give a rather clear signal for the ATLAS and CMS detectors [25, 26], which however would not be visible in the MoEDAL detector. Nevertheless according to a novel proposal [27], the LHC radiation detector systems can be used to turn the LHC itself into a new physics search machine by detecting final-state protons $pp \to pXp$ (X are centrally produced particles), exiting the LHC beam vacuum chamber at locations determined by their fractional momentum losses. At high central masses $M_X > 150$ GeV, the photon-photon process dominates making this technique appealing for detecting monopolia, which are expected to be produced in $\gamma\gamma$ fusion [24, 25].

3 The MoEDAL detector

The MoEDAL detector [2] is deployed around the intersection region at Point 8 (IP8) of the LHC (IP8) in the LHCb experiment Vertex Locator (VELO) [28] cavern. A three-dimensional depiction of the MoEDAL experiment is presented in Fig. 1. It is a unique and largely passive

LHC detector comprised of four sub-detector systems. It is worth emphasising that the detector components are renewed and installed/re-installed into upgraded configurations during technical stops throughout Run 1 (2010–2012) and Run 2 (2015–2018).

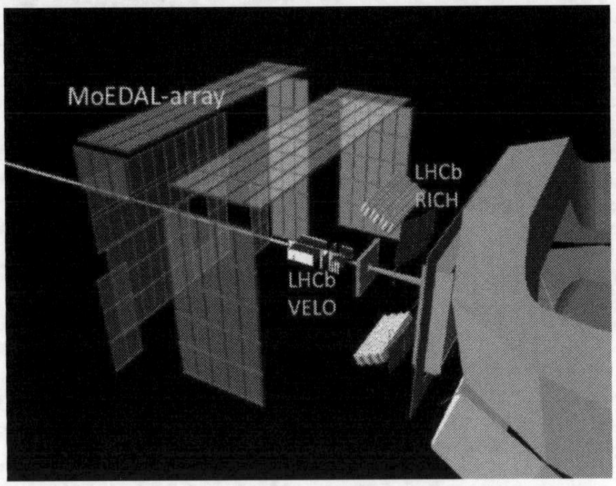

Figure 1. A three-dimensional schematic view of the MoEDAL detector around the LHCb VELO region at Point 8 of the LHC.

3.1 Low-threshold nuclear track detectors

The main sub-detector system is made of a large array of CR39®, Makrofol® and Lexan® plastic stacks surrounding the intersection area. The passage of a highly-ionising particle through these nuclear track detectors (NTDs) is marked by an invisible damage zone along the trajectory. The damage zone is revealed as a cone-shaped etch-pit when the plastic detector is chemically etched. Then the sheets of plastics are scanned looking for aligned etch pits in multiple sheets. The MoEDAL NTDs have a threshold of $Z/\beta \sim 5$, where Z is the charge and $\beta = v/c$ the velocity of the incident particle. During proton-proton collision running, the only source of known particles that are highly ionising enough to leave a track in MoEDAL NTDs are spallation products with range that is typically much less than the thickness of one sheet of the NTD stack. In that case the ionising signature will be that of a very low-energy electrically-charged *stopped* particle. This signature is distinct to that of a *penetrating* electrically or magnetically charged particle that will usually traverse every sheet in a MoEDAL NTD stack, accurately demarcating a track that points back to the collision point with a resolution of $\sim$ 1 cm. The part of the Run 2 NTD deployment which rests on top of the LHCb VELO is visible in Fig. 2. This is the closest possible location to the interaction point and represents a novelty of this run with respect to earlier installations during Run 1.

The basic detector unit of the MoEDAL experiment is a stack of ten sheets of plastic NTDs, consisting of four sheets of CR39®, each $\sim$ 0.5 mm thick, three sheets of Makrofol® (each $\sim$ 0.5 mm thick) with Lexan® (each $\sim$ 0.2 mm thick) forming the first, middle and end sheets of the stack.

3.2 Very high-charge catcher

Another new feature of the Run 2 deployment is the installation of a high-threshold NTD array ($Z/\beta \sim 50$): the Very High Charge Catcher (VHCC). The VHCC sub-detector, consisting of two flexible low-mass stacks of Makrofol® in an aluminium foil envelope, is deployed in the forward acceptance of the LHCb experiment between the LHCb RICH1 detector and the

Figure 2. Part of the Run 2 NTD deployment on top of the LHCb VELO.

Figure 3. The VHCC between RICH1 and TT installed for Run 2.

Trigger Tracker (TT), as shown in Fig. 3. It is the only NTD (partly) covering the forward region, adding only $\sim 0.5\%$ to the LHCb material budget while enhancing considerably the overall geometrical coverage of MoEDAL NTDs.

3.3 Magnetic trappers

A unique feature of the MoEDAL detector is the use of paramagnetic magnetic monopole trappers (MMTs) to capture electrically- and magnetically-charged highly-ionising particles. Such volumes installed in IP8 for the 2015 proton-proton collisions is shown in Fig. 4. The aluminium absorbers of the MMTs are subject to an analysis looking for magnetically-charged particles at a remote SQUID magnetometer facility [29, 30]. The search for the decays of long-lived electrically charged particles that are stopped in the trapping detectors will subsequently be carried out at a remote underground facility.

A trapping detector prototype was exposed to 8 TeV proton-proton collisions for an integrated luminosity of 0.75 fb^{-1} in 2012. For the 2015 run at 13 TeV, the MMT was upgraded to an array consisting of 672 square aluminium rods with dimension $19 \times 2.5 \times 2.5$ cm^3 for a total mass of 222 kg in 14 stacked boxes that were placed 1.62 m from the IP8 LHC interaction point under the beam pipe on the side opposite to the LHCb detector. The results for the Run 2 configuration are presented in Section 4.

3.4 TimePix radiation monitors

The only non-passive MoEDAL sub-detector system comprises an array of TimePix pixel device arrays (256×256 square pixels with a pitch of 55 μm) distributed throughout the MoEDAL cavern at IP8, forming a real-time radiation monitoring system of highly-ionising beam-related backgrounds. A photo of its readout setup for the 2015 installations is shown in Fig. 5. Each pixel of the TimePix chip comprises a preamplifier, a discriminator with threshold adjustment, synchronisation logic and a 14-bit counter. The operation of TimePix in time-over-threshold mode allows a 3D mapping of the charge spreading effect in the whole volume of the silicon sensor, thus differentiating between different types of particles species from mixed radiation fields and measuring their energy deposition [31].

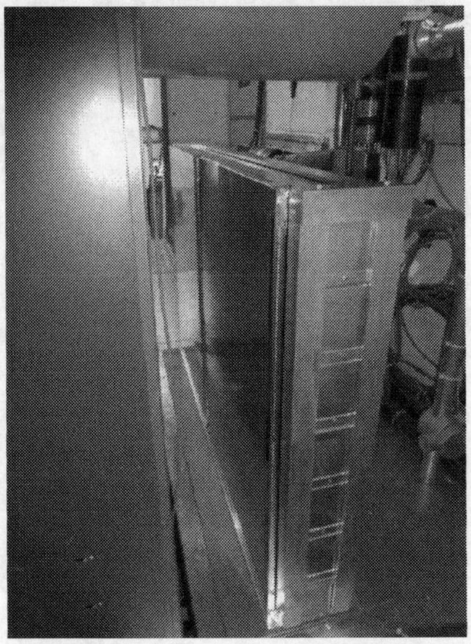

Figure 4. Deployment of the MMT for the LHC Run 2.

Figure 5. Run 2 deployment of TimePix chips in MoEDAL.

4 Searches for monopoles in MoEDAL

The high ionisation of slow-moving magnetic monopoles and dyons, implies quite characteristic trajectories when such particles interact with the MoEDAL NTDs, which can be revealed during the etching process [2, 6]. In addition, the high magnetic charge of a monopole (which is expected to be at least one Dirac charge $g_D = 68.5e$ (*cf.* Eq. (1)) implies a strong magnetic dipole moment, which in turn may result in a strong binding of the monopole with the $^{27}_{13}$Al nuclei of the aluminium MoEDAL MMTs. In such a case, the presence of a monopole trapped in an aluminium bar of an MMT would be detected through the existence of a persistent current, defined as the difference between the currents in the SQUID of a magnetometer before and after the passage of the bar through the sensing coil.

In the context of the MMT exposure during Run 2, no magnetic charge exceeding $0.5g_{\rm D}$ was detected in any of the exposed samples when passed through the ETH Zurich SQUID facility, allowing limits to be placed on monopole production. Model-independent cross-section limits have been obtained in fiducial regions of monopole energy and direction for $1g_{\rm D} \leq |g| \leq 6g_{\rm D}$ with the 8 TeV analysis [32]. Model-dependent cross-section limits are obtained for Drell-Yan (DY) pair production of spin-½ and spin-0 monopoles for $1g_{\rm D} \leq |g| \leq 5g_{\rm D}$ at 13 TeV [33], as shown in Fig. 6. Caution, however, should be exerted here in the sense that the non-perturbative nature of the large magnetic Dirac charge of the monopole invalidate any perturbative treatment based on Drell-Yan calculations of the pertinent cross sections and hence any result based on the latter is only indicative, due to the lack of any other concrete theoretical treatment. This situation may be resolved if thermal production in heavy-ion collisions —that does not rely on perturbation theory— is considered [34].

The weaker limits for $|g| = g_{\rm D}$ displayed in Fig. 6 when compared to higher charges are mostly due to loss of acceptance from monopoles punching through the trapping volume. For higher charges, monopoles ranging out before reaching the trapping volume decrease the acceptance for DY monopoles with increasing charge and reaches below 0.1% for a charge of $6g_{\rm D}$. The spin dependence is solely due to the different event kinematics: more central and more energetic monopoles for spin 0.

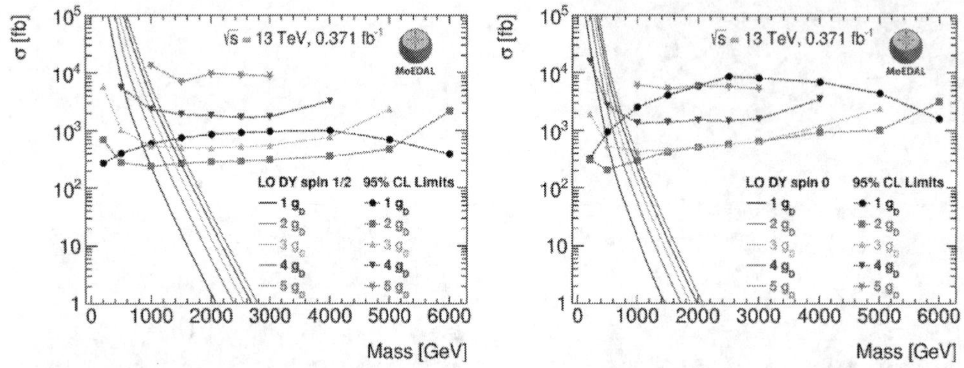

Figure 6. Cross-section upper limits at 95% confidence level (dotted lines) for DY monopole production as a function of mass for spin-½ (left) and for spin-0 monopoles (right). The various marker styles and colours correspond to different monopole charges. The solid lines represent DY cross-section calculations at leading order. From Ref. [33].

Under the assumption of Drell-Yan cross sections, mass limits are derived for $1g_D \leq |g| \leq 4g_D$ at the LHC, complementing previous results from ATLAS Collaboration [35, 36], which placed limits for monopoles with magnetic charge $|g| \leq 1.5g_D$ (*c.f.* Fig. 7). The ATLAS bounds are more stringent that those of MoEDAL for $|g| = 1g_D$ due to the higher luminosity delivered in ATLAS and the loss of acceptance in MoEDAL for small magnetic charges. On the other hand, higher charges are difficult to be probed in ATLAS due to the limitations of the electromagnetic-calorimeter-based level-1 trigger deployed for such searches. A comparison of the limits on monopole production cross sections set by other colliders with those set by MoEDAL is presented in Ref. [23], while general limits including searches in cosmic radiation are reviewed in Ref. [37].

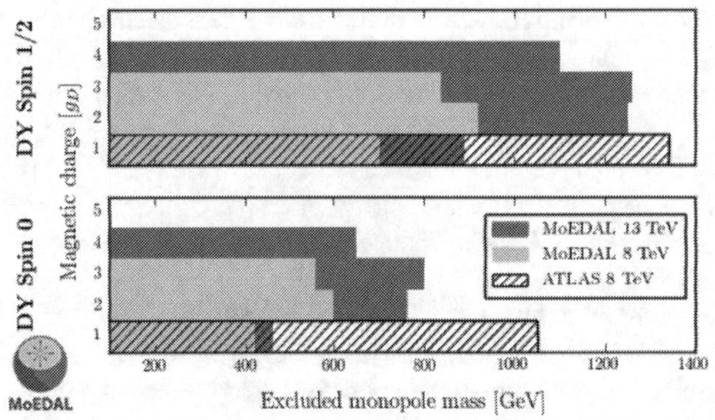

Figure 7. Excluded monopole masses for DY production for spin-½ (top) and spin-0 (bottom) monopoles. The MoEDAL results obtained at 8 TeV (yellow, light grey) [32] and 13 TeV (red, dark grey) [33] are superimposed on the ATLAS 8 TeV limits (hatched area) [36].

5 Electrically-charged long-lived particles in supersymmetry

Supersymmetry (SUSY) is an extension of the Standard Model which assigns to each SM field a superpartner field with a spin differing by a half unit. SUSY provides elegant solutions to several open issues in the SM, such as the hierarchy problem, the identity of dark matter, and the grand unification. SUSY scenarios propose a number of massive slowly moving electrically charged particles. If they are sufficiently long-lived to travel a distance of at least $O(1\text{m})$ before decaying and their $Z/\beta \gtrsim 5$, they may be detected in the MoEDAL NTDs. No highly-charged particles are expected in such a theory, but there are several scenarios in which supersymmetry may yield massive, long-lived particles that could have electric charges ± 1, potentially detectable in MoEDAL if they are produced with low velocities ($\beta \lesssim 0.2$) .

The lightest supersymmetric particle (LSP) is stable in models where R parity is conserved, where R is a multiplicative quantum number defined as: $R = (-1)^{3B+L+2s}$, with s the spin, B the baryon number, and L the lepton number. The LSP should have no strong or electromagnetic interactions, otherwise it would bind to conventional matter and be detectable in anomalous heavy nuclei [38]. Possible weakly-interacting neutral candidates in the Minimal Supersymmetric Standard Model (MSSM) include the sneutrino, which has been excluded by LEP and direct searches, the lightest neutralino $\tilde{\chi}_1^0$ (a mixture of spartners of the Z, H and γ) and the gravitino $\tilde{G}$.

5.1 Supersymmetric scenarios with R-parity violation

Several scenarios featuring metastable charged sparticles might be detectable in MoEDAL. One such scenario is that R parity *may not be exact*, since there is no exact local symmetry associated with either L or B, and hence no fundamental reason why they should be conserved. One could consider various ways in which L and/or B could be violated in such a way that R is violated, as represented by the following superpotential terms:

$$W_{RV} = \lambda''_{ijk}\bar{U}_i\bar{D}_j\bar{D}_k + \lambda'_{ijk}L_iQ_j\bar{D}_k + \lambda_{ijk}L_iL_j\bar{E}_k + \mu_iL_iH, \quad (2)$$

where $Q_i, \bar{U}_i, \bar{D}_i, L_i$ and $\bar{E}_i$ denote chiral superfields corresponding to quark doublets, antiquarks, lepton doublets and antileptons, respectively, with i, j, k generation indices. The simultaneous presence of terms of the first and third type in Eq. (2), namely λ and λ'', is severely restricted by lower limits on the proton lifetime, but other combinations are less restricted. The trilinear couplings in Eq. (2) generate sparticle decays such as $\tilde{q} \to \bar{q}\bar{q}$ or $q\ell$, or $\tilde{\ell} \to \ell\ell$, whereas the bilinear couplings in Eq. (2) generate Higgs-slepton mixing and thereby also $\tilde{q} \to q\ell$ and $\tilde{\ell} \to \ell\ell$ decays [39]. For a nominal sparticle mass ~ 1 TeV, the lifetime for such decays would exceed a few nanoseconds for $\lambda, \lambda', \lambda'' < 10^{-8}$.

If R parity is broken, the LSP would be unstable, and might be charged and/or coloured. In the former case, it might be detectable directly at the LHC as a massive slowly-moving charged particle. In the latter case, the LSP would bind with light quarks and/or gluons to make colour-singlet states, the so-called *R-hadrons*, and any charged state could again be detectable as a massive slowly-moving charged particle. If $\lambda \neq 0$, the prospective experimental signature would be similar to a stau next-to-lightest sparticle (NLSP) case to be discussed later. On the other hand, if λ' or $\lambda'' \neq 0$, the prospective experimental signature would be similar to a stop NLSP case, yielding the possibility of charge-changing interactions while passing through matter. This could yield a metastable charged particle, created whilst passing through the material surrounding the intersection point, that would be detected by MoEDAL.

5.2 Metastable lepton NLSP in the CMSSM with a neutralino LSP

However, even if R parity *is* exact, the NLSP may be long lived. This would occur, for example, if the LSP is the gravitino, or if the mass difference between the next-to-lightest supersymmetric particle (NLSP) and the neutralino LSP is small, offering more scenarios for long-lived charged sparticles. In *neutralino dark matter* scenarios based on the constrained minimal supersymmetric standard model (CMSSM), for instance, the most natural candidate for the NLSP is the lighter stau slepton $\tilde{\tau}_1$ [40], which could be long lived if $m_{\tilde{\tau}_1} - m_{\tilde{\chi}^0_1}$ is small. There are several regions of the CMSSM parameter space that are compatible with the constraints imposed by unsuccessful searches for sparticles at the LHC, as well as the discovered Higgs boson mass and the LSP relic density allowed by cosmological observations. It was found in a global analysis that these two possibilities are favoured [41]: (a) a region where the relic density is controlled by rapid annihilation through direct-channel heavy Higgs resonances due to the relatively large higgsino component of the LSP, and (b) a strip where the relic LSP density is reduced by coannihilations with near-degenerate staus and other sleptons.

In the coannihilation region of the CMSSM, the lighter $\tilde{\tau}_1$ is expected to be the lightest slepton [40], and the $\tilde{\tau}_1 - \tilde{\chi}^0_1$ mass difference may well be smaller than m_τ: indeed, this is required at large LSP masses. In this case, the dominant stau decays for $m_{\tilde{\tau}_1} - m_{\tilde{\chi}^0_1} >$ 160 MeV are expected to be into three particles: $\tilde{\chi}^0_1 \nu \pi$ or $\tilde{\chi}^0_1 \nu \rho$. If $m_{\tilde{\tau}_1} - m_{\tilde{\chi}^0_1} < 1.2$ GeV, the $\tilde{\tau}_1$ lifetime is calculated to be so long, in excess of $\sim$ 100 ns, that it is likely to escape the detector before decaying, and hence would be detectable as a massive, slowly-moving charged particle [42, 43].

5.3 Metastable sleptons in gravitino LSP scenarios

On the other hand, in *gravitino dark matter* scenarios with more general options for the pattern of supersymmetry breaking, other options appear quite naturally, including the lighter selectron or smuon, or a sneutrino [44], or the lighter stop squark $\tilde{t}_1$ [45]. If the gravitino $\tilde{G}$ is the LSP, the decay rate of a slepton NLSP is given by

$$\Gamma(\tilde{\ell} \to \tilde{G}\ell) = \frac{1}{48\pi M_*^2} \frac{m_{\tilde{\ell}}^5}{M_{\tilde{G}}^2} \left[1 - \frac{M_{\tilde{G}}^2}{m_{\tilde{\ell}}^2}\right]^4, \tag{3}$$

where M_* is the Planck scale. Since M_* is much larger than the electroweak scale, the NLSP lifetime is naturally very long.

Gravitino (or axino) LSP with a long-lived charged stau may arise in gauge mediation and minimal supergravity models [46]. Large part of the parameter space potentially attractive for long-lived slepton searches with MoEDAL are compatible with cosmological constraints on the dark-matter abundance in superweakly interacting massive particle scenarios [47].

5.4 Long-lived gluinos in split supersymmetry

The above discussion has been in the context of the CMSSM and similar scenarios where all the supersymmetric partners of Standard Model particles have masses in the TeV range. Another scenario is "split supersymmetry", in which the supersymmetric partners of quarks and leptons are very heavy, of a scale m_s, whilst the supersymmetric partners of SM bosons are relatively light [48]. In such a case, the gluino could have a mass in the TeV range and hence be accessible to the LHC, but would have a very long lifetime:

$$\tau \approx 8 \left(\frac{m_s}{10^9 \text{ GeV}}\right)^4 \left(\frac{1 \text{ TeV}}{m_{\tilde{g}}}\right)^5 \text{ s}. \tag{4}$$

Long-lived gluinos would form long-lived gluino R-hadrons including gluino-gluon *(gluinoball)* combinations, gluino-$q\bar{q}$ *(mesino)* combinations and gluino-qqq *(baryino)* combinations. The heavier gluino hadrons would be expected to decay into the lightest species, which would be metastable, with a lifetime given by Eq. (4), and it is possible that this metastable gluino hadron could be charged.

In the same way as stop hadrons, gluino hadrons may flip charge through conventional strong interactions as they pass through matter, and it is possible that one may pass through most of a conventional LHC tracking detector undetected in a neutral state before converting into a metastable charged state that could be detected by MoEDAL.

5.5 MoEDAL complementarity to ATLAS, CMS and ALICE

There are several considerations supporting the complementary aspects of MoEDAL w.r.t. ATLAS and CMS when discussing the observability of (meta-)stable massive electrically-charged particles. Most of them stem from MoEDAL being "time-agnostic" due to the passive nature of its detectors. Therefore a signal from very slowly moving particles will not be lost due to arriving in several consecutive bunch crossings. In addition, ATLAS and CMS perform trigger-based analyses relying either on triggering on accompanying "objects", e.g. missing transverse momentum, or by developing and deploying specialised triggers. The offline selection may require imposing additional cuts to reject cosmic-ray muons and/or enhance the reconstruction of highly-ionising particles.

Therefore, the efficiency may be lower and the probed parameter space may be reduced in ATLAS and CMS. MoEDAL, on the other hand, is mainly limited by the lower luminosity delivered at Point 8, by the geometrical acceptance of the detectors, especially the MMTs, and by the requirement of passing the Z/β threshold of NTDs. In general, ATLAS and CMS have demonstrated to cover high-velocities $\beta \gtrsim 0.2$, while MoEDAL is sensitive to ones with $\beta \lesssim 0.2$. Hence, by dropping selection criteria imposed by ATLAS and CMS when searching for long-lived SUSY particles, MoEDAL may extend the SUSY parameter space explored by LHC [49].

When discussing the detection of particles stopped *(trapped)* in material that they may decay later, different possibilities are explored. CMS and ATLAS look in empty bunch crossings for decays of trapped particles into jets. MoEDAL MMTs may be monitored in a underground/basement laboratory for tracks arising from such decays. The background in the latter case, coming from cosmic rays, should be easier to control and assess. The probed lifetimes should be larger due to the unlimited monitoring time.

6 Summary and outlook

MoEDAL is going to extend considerably the LHC reach in the search for (meta)stable highly ionising particles. The latter are predicted in a variety of theoretical models and include: magnetic monopoles, SUSY long-lived spartners, D-matter [50], quirks, strangelets, Q-balls, etc [6, 51]. Such particles may be light enough to be producible at the LHC energies. In this paper we have described searches for monopoles and have discussed the MoEDAL relevance for long-lived partners in some SUSY models.

MoEDAL is optimised to probe precisely all such long lived states, unlike the other LHC experiments [4]. Furthermore it combines different detector technologies: plastic nuclear track detectors (NTDs), trapping volumes and pixel sensors [2]. The first physics results, pertaining to magnetic monopole trapping detectors, obtained with LHC Run 1 data [32], and the corresponding analysis at 13 TeV have already been published [33]. The MoEDAL

Collaboration is preparing new analyses with higher statistical precision from LHC Run 2, with other detectors (NTDs) and with a large variety of interpretations involving not only magnetic but also electric charges.

Acknowledgments

The author is grateful to the EXA2017 organisers for the kind invitation to present this talk. She acknowledges support by the Spanish MINEICO under the project FPA2015-65652-C4-1-R, by the Generalitat Valenciana through the MoEDAL-supporting agreement CON.21.2017-09.02.03 and the Excellence Project PROMETEO/2017/033, by the Severo Ochoa Excellence Centre Project SEV-2014-0398, and by a 2017 Leonardo Grant for Researchers and Cultural Creators, BBVA Foundation.

References

[1] The MoEDAL experiment web page: http://moedal.web.cern.ch/

[2] J. Pinfold *et al.* [MoEDAL Collaboration], "Technical Design Report of the MoEDAL Experiment," CERN-LHCC-2009-006, MoEDAL-TDR-001, and references therein.

[3] L. Evans and P. Bryant, JINST **3**, S08001 (2008).

[4] A. De Roeck, A. Katre, P. Mermod, D. Milstead and T. Sloan, Eur. Phys. J. C **72**, 1985 (2012).

[5] M. Fairbairn, A. C. Kraan, D. A. Milstead, T. Sjostrand, P. Z. Skands and T. Sloan, Phys. Rept. **438**, 1 (2007); S. Burdin, M. Fairbairn, P. Mermod, D. Milstead, J. Pinfold, T. Sloan and W. Taylor, Phys. Rept. **582**, 1 (2015).

[6] B. Acharya *et al.* [MoEDAL Collaboration], Int. J. Mod. Phys. A **29**, 1430050 (2014).

[7] P. A. M. Dirac, Proc. Roy. Soc. Lond. A **133**, 60 (1931).

[8] P. A. M. Dirac, Phys. Rev. **74**, 817 (1948).

[9] G. 't Hooft, Nucl. Phys. B **79**, 276 (1974); A. M. Polyakov, JETP Lett. **20**, 194 (1974) [Pisma Zh. Eksp. Teor. Fiz. **20**, 430 (1974)].

[10] B. Julia and A. Zee, Phys. Rev. D **11**, 2227 (1975); Y. Nambu, Nucl. Phys. B **130**, 505 (1977); E. Witten, Phys. Lett. **86B**, 283 (1979); G. Lazarides, M. Magg and Q. Shafi, Phys. Lett. **97B**, 87 (1980); R. d. Sorkin, Phys. Rev. Lett. **51**, 87 (1983); D. J. Gross and M. J. Perry, Nucl. Phys. B **226**, 29 (1983); J. S. Schwinger, Science **165**, 757 (1969); J. Preskill, Ann. Rev. Nucl. Part. Sci. **34**, 461 (1984); A. Achucarro and T. Vachaspati, Phys. Rept. **327**, 347 (2000) [Phys. Rept. **327**, 427 (2000)]; T. W. Kephart, C. A. Lee and Q. Shafi, JHEP **0701**, 088 (2007); D. G. Pak, P. M. Zhang and L. P. Zou, Int. J. Mod. Phys. A **30**, no. 27, 1550164 (2015); A. Rajantie, JHEP **0601**, 088 (2006).

[11] V. Vento and V. S. Mantovani, "On the magnetic monopole mass," arXiv:1306.4213 [hep-ph] (2013).

[12] Y. M. Cho and D. Maison, Phys. Lett. B **391**, 360 (1997); W. S. Bae and Y. M. Cho, J. Korean Phys. Soc. **46**, 791 (2005).

[13] Y. M. Cho, K. Kimm and J. H. Yoon, Mod. Phys. Lett. A **31**, no. 09, 1650053 (2016); Phys. Lett. B **761**, 203 (2016).

[14] J. Ellis, N. E. Mavromatos and T. You, Phys. Lett. B **756**, 29 (2016); J. Ellis, V. Sanz and T. You, JHEP **1407**, 036 (2014).

[15] M. Barriola and A. Vilenkin, Phys. Rev. Lett. **63**, 341 (1989).

[16] A. K. Drukier and S. Nussinov, Phys. Rev. Lett. **49**, 102 (1982).

[17] P. O. Mazur and J. Papavassiliou, Phys. Rev. D **44**, 1317 (1991); N. E. Mavromatos and J. Papavassiliou, Eur. Phys. J. C **78**, no. 1, 68 (2018).

[18] N. E. Mavromatos and S. Sarkar, Phys. Rev. D **95**, no. 10, 104025 (2017).
[19] Y. B. Zeldovich and M. Y. Khlopov, Phys. Lett. **79B**, 239 (1978).
[20] C. T. Hill, Nucl. Phys. B **224**, 469 (1983).
[21] V. K. Dubrovich, Grav. Cosmol. Suppl. **8N1**, 122 (2002).
[22] A. Rajantie, Contemp. Phys. **53**, 195 (2012).
[23] A. Rajantie, Phys. Today **69**, no. 10, 40 (2016).
[24] L. N. Epele, H. Fanchiotti, C. A. Garcia Canal and V. Vento, Eur. Phys. J. C **56**, 87 (2008); Eur. Phys. J. C **62**, 587 (2009).
[25] L. N. Epele, H. Fanchiotti, C. A. G. Canal, V. A. Mitsou and V. Vento, Eur. Phys. J. Plus **127**, 60 (2012).
[26] L. N. Epele, H. Fanchiotti, C. A. G. Canal, V. A. Mitsou and V. Vento, "Can the 750 GeV enhancement be a signal of light magnetic monopoles?," arXiv:1607.05592 [hep-ph] (2016).
[27] M. Kalliokoski, J. W. Lämsä, M. Mieskolainen and R. Orava, "Turning the LHC Ring into a New Physics Search Machine," arXiv:1604.05778 [hep-ex] (2016).
[28] A. A. Alves, Jr. *et al.* [LHCb Collaboration], JINST **3**, S08005 (2008).
[29] M. D. Joergensen, A. De Roeck, H.-P. Hachler, A. Hirt, A. Katre, P. Mermod, D. Milstead and T. Sloan, "Searching for magnetic monopoles trapped in accelerator material at the Large Hadron Collider," arXiv:1206.6793 [physics.ins-det] (2012).
[30] A. De Roeck, H. P. Hächler, A. M. Hirt, M.-D. Joergensen, A. Katre, P. Mermod, D. Milstead and T. Sloan, Eur. Phys. J. C **72**, 2212 (2012).
[31] A. Sopczak *et al.*, IEEE Trans. Nucl. Sci. **62**, 3225 (2015).
[32] B. Acharya *et al.* [MoEDAL Collaboration], JHEP **1608**, 067 (2016).
[33] B. Acharya *et al.* [MoEDAL Collaboration], Phys. Rev. Lett. **118**, 061801 (2017).
[34] O. Gould and A. Rajantie, arXiv:1705.07052 [hep-ph] (2017).
[35] G. Aad *et al.* [ATLAS Collaboration], Phys. Rev. Lett. **109**, 261803 (2012).
[36] G. Aad *et al.* [ATLAS Collaboration], Phys. Rev. D **93**, 052009 (2016).
[37] L. Patrizii and M. Spurio, Ann. Rev. Nucl. Part. Sci. **65**, 279 (2015).
[38] H. Goldberg, Phys. Rev. Lett. **50**, 1419 (1983) Erratum: [Phys. Rev. Lett. **103**, 099905 (2009)]; J. R. Ellis, J. S. Hagelin, D. V. Nanopoulos, K. A. Olive and M. Srednicki, Nucl. Phys. B **238**, 453 (1984).
[39] V. A. Mitsou, PoS **PLANCK 2015**, 085 (2015).
[40] J. R. Ellis, K. A. Olive, Y. Santoso and V. C. Spanos, Phys. Lett. B **565**, 176 (2003).
[41] O. Buchmueller *et al.*, Eur. Phys. J. C **72**, 2243 (2012); Eur. Phys. J. C **74**, 2922 (2014).
[42] T. Jittoh, J. Sato, T. Shimomura and M. Yamanaka, Phys. Rev. D **73**, 055009 (2006) Erratum: [Phys. Rev. D **87**, 019901 (2013)].
[43] S. Kaneko, J. Sato, T. Shimomura, O. Vives and M. Yamanaka, Phys. Rev. D **78**, 116013 (2008) Erratum: [Phys. Rev. D **87**, 039904 (2013)].
[44] J. R. Ellis, K. A. Olive and Y. Santoso, JHEP **0810**, 005 (2008).
[45] J. L. Diaz-Cruz, J. R. Ellis, K. A. Olive and Y. Santoso, JHEP **0705**, 003 (2007).
[46] K. Hamaguchi, M. M. Nojiri and A. de Roeck, JHEP **0703**, 046 (2007).
[47] J. L. Feng, S. Iwamoto, Y. Shadmi and S. Tarem, JHEP **1512**, 166 (2015).
[48] N. Arkani-Hamed and S. Dimopoulos, JHEP **0506**, 073 (2005); G. F. Giudice and A. Romanino, Nucl. Phys. B **699**, 65 (2004) Erratum: [Nucl. Phys. B **706**, 487 (2005)].
[49] J. Mamuzic, N. E .Mavromatos, V. A. Mitsou, J. L. Pinfold, R. Ruiz de Austri, K. Sakurai, O. Vives, "Long-lived charged supersymmetric particles and MoEDAL," *in preparation.*

[50] G. Shiu and L. T. Wang, Phys. Rev. D **69**, 126007 (2004); J. R. Ellis, N. E. Mavromatos and D. V. Nanopoulos, Gen. Rel. Grav. **32**, 943 (2000); Phys. Lett. B **665**, 412 (2008); J. R. Ellis, N. E. Mavromatos and M. Westmuckett, Phys. Rev. D **70**, 044036 (2004); *ibid.* **71**, 106006 (2005); N. E. Mavromatos, S. Sarkar and A. Vergou, Phys. Lett. B **696**, 300 (2011); N. E. Mavromatos, V. A. Mitsou, S. Sarkar and A. Vergou, Eur. Phys. J. C **72**, 1956 (2012).

[51] V. A. Mitsou [MoEDAL Collaboration], EPJ Web Conf. **95**, 04042 (2015); PoS **EPS-HEP2015**, 109 (2015); PoS CORFU **2016**, 028 (2017); N. E. Mavromatos and V. A. Mitsou [MoEDAL Collaboration], EPJ Web Conf. **164**, 04001 (2017).

EDM Experiments at Storage Rings

Hans Ströher[1*]

[1]Institut für Kernphysik, Forschungszentrum Jülich, 52425 Jülich, Germany

Abstract. According to our present understanding, the early Universe contained the same amount of matter and anti-matter and, if the Universe had behaved symmetrically as it developed, every particle would have been annihilated by one of its antiparticles. One of the great mysteries in the natural sciences is therefore, why matter dominates over antimatter in the visible Universe. The breaking of the combined charge-conjugation and parity symmetries (CP-violation, CPV) in the Standard Model of particle physics (SM) is insufficient to explain this and further sources of CPV must be sought. These could manifest themselves in electric dipole moments (EDMs) of elementary particles, which occur when the centroids of positive and negative charges are mutually and permanently displaced. An EDM observation would also be an indication for physics beyond the SM. No EDM has been observed so far; after its discovery, investigations on different systems will be required to pin down CPV sources. In this respect the idea to search for EDMs of charged hadrons (i.p. proton and deuteron) in a new class of precision storage rings has recently been put forward. In addition, such searches bear the potential to reach sensitivities of 10^{-29} e·cm. The EDM measurement principle, the time development of the polarization vector subject to a perpendicular electric field, is simple, but the smallness of the effect makes this an enormously challenging project. A stepwise approach, from R&D for key-technologies towards the holy grail of a double-beam precision storage ring with counter-rotating beams, is needed. The ongoing research of the JEDI Collaboration at the cooler synchrotron COSY of Forschungszentrum Jülich (Germany) provides invaluable information: recent experimental milestones are presented and discussed.

1 Introduction

A new class of precision storage rings is required to search for electric dipole moments (EDMs) of charged particles with unprecedented sensitivity. In recent years, first decisive steps towards the design and construction of such a facility have been taken by establishing the required key technologies. One future milestone is to deliver the first directly measured EDMs for both the proton and the deuteron using the existing cooler storage ring COSY at Forschungszentrum Jülich (FZJ) in Jülich (Germany), thus providing the proof-of-principle for this method.

2 Background

Permanent EDMs of particles violate both time reversal (T) and parity (P) invariance and, on the basis of the CPT theorem, they also violate the combined symmetry CP (CPV). Such a symmetry breaking is thought to be responsible for the different behaviour of particles and antiparticles, leading, e.g., to the apparent matter-antimatter asymmetry in the Universe. CPV is found in the electroweak part of the Standard Model of particle physics (SM) but, since SM-CPV is much too weak to explain the matter-antimatter asymmetry, other sources must be sought. An obvious candidate to investigate is an EDM of a fundamental particle: no EDM has yet been observed, but they must exist. Finding a finite EDM would very probably also indicate new physics, i.e., interactions not contained in the SM. Even after a possible EDM observation, different systems would have to be investigated in order to identify the CPV-source.

EDMs are very small – the best current limit for the neutron is 3 x 10^{-26} e·cm [1]; in addition to free particles, there are searches in atoms and molecules that have reached even smaller upper limits – and the aim for charged particles is at least 10^{-29} e·cm. Although the measurement principle – the time development of the polarization vector subject to a perpendicular electric field – is simple, this represents an enormously challenging project due to the smallness of the expected effect.

It should be noted that, whereas charged particles potentially offer the highest sensitivity with discovery potential, up to now no EDM measurements for protons and deuterons in a storage ring have been performed. However, an upper limit of 10^{-19} e·cm was obtained for muons as a side activity of the measurement of the anomalous magnetic moment ($(g\text{-}2)_\mu$) in a storage ring experiment.

Given the history of neutron EDM searches over the past 50 years or so, it is evident that a corresponding level of

* Corresponding author: h.stroeher@fz-juelich.de

sensitivity cannot be obtained in one giant leap – a stepwise approach, in which the next generation measurement is based on the expertise and technological know-how gained in the preceding one, is compulsory. The steps towards the holy grail of a charged particle EDM search, using a double-beam precision storage ring with counter-rotating beams, comprise:

1. Research and development of all the key techniques at an existing conventional single-beam storage ring. COSY, the cooler synchrotron and storage ring at the Forschungszentrum Jülich, is the ideal place for these investigations, which involve spin-coherence time optimization, precision polarimetry development, beam tracking measurements, etc.;
2. Precision spin-tracking simulations: an essential requirement for an assessment of the capabilities of the final precision ring is to provide realistic simulations, e.g., for benchmarking EDM test installations in COSY;
3. A proof-of-principle experiment: this measurement will use COSY-Jülich without major transformations except for improvements of the beam position monitor (BPM) system and the assembly of a radiofrequency (RF) Wien-filter which (in an ideal storage ring) would induce spin rotation due to an EDM;
4. A first direct EDM measurement (for the proton and the deuteron), again exploiting COSY-Jülich, but here upgraded with a short electrostatic deflector inside the ring. The beam chicane could be inserted in one of the straight sections of COSY. This would test one of the key techniques of the final ring and should lead to an EDM upper limit in the order of 10^{-24} e·cm. It would also provide a reality check on further key items, e.g., the spin coherence time and polarimetry;
5. Conceptual design report (CDR) and technical design report (TDR) for the final dedicated storage ring, including cost estimates for building and operating the facility;
6. Construction and commissioning of the new facility for EDM searches, once funding has been secured.

In the following, recent achievements for step 1., obtained by the JEDI (Jülich Electric Dipole moment Investigations) collaboration [2] at COSY-Jülich, are presented. It should be mentioned that this research is supported by an Advanced Grant of the European Research Council (ERC AdG "srEDM") [3], which the reader is referred to for further information.

3 Experimental achievements

3.1 Ultra-high precision spin tune measurement

A new method to determine the spin tune was established and tested [4]. In an ideal planar magnetic storage ring, the spin tune – defined as the number of spin precessions per turn – is given by $\nu_s = \gamma G$ (γ is the Lorentz factor, G the gyromagnetic anomaly). At 970 MeV/c, the deuteron spins coherently precess at a frequency of about 120 kHz in COSY. The spin tune was deduced from the up-down asymmetry of deuteron-carbon scattering. In a time interval of 2.6 s, the spin tune was determined with a precision of the order 10^{-8}, and to 10^{-10} for a continuous 100 s accelerator cycle. This renders the new method a precision tool for accelerator physics; observing and controlling the spin motion of particles to high precision is mandatory, in particular, for the measurement of electric dipole moments of charged particles in a storage ring.

3.2 Record in-plane polarization lifetime

A deuteron beam polarization lifetime near 1000 seconds in the horizontal plane of the magnetic storage ring COSY has been observed [5]. This long spin coherence time was maintained through a combination of beam bunching, electron cooling, sextupole field corrections, and the suppression of collective effects through beam current limits. This record lifetime is required for a storage ring search for an intrinsic electric dipole moment on the deuteron at a statistical sensitivity level approaching 10^{-29} e·cm.

3.3 Feedback system for an active control of the spin tune

The successful use of feedback from a spin polarization measurement to the revolution frequency of a 0.97 GeV/c bunched and polarized deuteron beam in the Cooler Synchrotron (COSY) storage ring has been realized in order to control both the precession rate (≈ 120 kHz) and the phase of the horizontal polarization component [6]. Real time synchronization with a radio frequency (RF) solenoid made possible the rotation of the polarization out of the horizontal plane, yielding a demonstration of the feedback method to manipulate the polarization. In particular, the rotation rate shows a sinusoidal function of the horizontal polarization phase (relative to the RF solenoid), controlled to within a one standard deviation range of $\sigma = 0.21$ rad. The minimum possible adjustment was 3.7 mHz out of a revolution frequency of 753 kHz, which changes the precession rate by 26 mrad/s. Such capability meets the requirement for the use of storage rings to look for an intrinsic electric dipole moment of charged particles.

3.4 Spin tune mapping as a novel tool to probe spin dynamics in storage rings

An extended paper entitled "Spin tune mapping as a novel tool to probe the spin dynamics in storage rings" has recently been published [7]. It was motivated by the fact that precision experiments, such as the search for electric dipole moments of charged particles using storage rings, demand understanding of the spin dynamics with unprecedented accuracy. A new method artificially applied longitudinal magnetic fields, called "spin tune mapping", has been developed. The technique was experimentally tested in 2014 at COSY and, for the first time, the angular orientation of the stable spin axis at two different locations in the ring has been determined to an unprecedented accuracy of better than 2.8 μrad.

4 Data base experiment

For the deuteron EDM search, a polarimeter (reaction) is required that maximizes efficiency and sensitivity to the deuteron vector polarized beam in the storage ring. Such a polarimeter is likely to be based on a thick carbon target, because the spin-orbit effects in elastic scattering are large. The data required to optimize the polarimeter design have been taken with the refurbished forward part of the WASA-at-COSY detector [8] for a variety of deuteron energies and carbon as well as CH_2 ribbon targets. Analysis of the data has started – once completed its results will be used in simulations of various polarimeter designs.

5 Technical developments

5.1 RF Wien filter

A novel waveguide RF Wien filter [9] is required for the proof-of-principle EDM measurement (the so called precursor experiment; see below) in a conventional storage ring like COSY. By design the orthogonality of the electric and magnetic field is accomplished. In order to maintain an optimal Lorentz-force compensation, the beam width must be kept as small as possible. Inside COSY this is achieved at the so called PAX low-ß section [10].

5.2 Electrostatic deflector

For the final EDM ring, high-field electrostatic deflectors have to be developed. The deflector development for a final deuteron EDM ring, which requires combined *ExB* deflectors, is focused on high-field devices of roughly 17 MV/m. Such high electrostatic field gradients could already be achieved on a test bench at RWTH Aachen University [11]. In addition, a first technical design for an electromagnetic deflector was completed and a prototype test device will be produced to further investigate and optimize the field gradient of high electrostatic fields in superimposed magnetic fields.

5.3 EDM polarimeter

A new polarimeter, based on LYSO scintillator modules with SiPM readout, is under development at our institute [12]. Test with extracted deuteron beams from COSY setup. The complete system, currently comprising 52 modules, is planned to supersede the currently used COSY polarimeters (EDDA, WASA) in the near future.

6 Planned precursor experiment

For 2018, the JEDI collaboration plans a first measurement of the deuteron EDM at COSY. In a pure magnetic storage ring such as COSY, an EDM will generate an oscillation of the vertical polarization component. For a 970 MeV/c deuteron beam with the spin precession frequency of 120 kHz, a tiny amplitude is expected, e.g., $3 \cdot 10^{-10}$ for an EDM of $d = 10^{-24}$ e·cm. To allow for a build-up of the vertical polarization proportional to the EDM, a so called radio-frequency (RF) Wien-filter has to be operated [8]. A prototype was successfully installed and operated in COSY in 2014. A new device with a stronger magnetic field (0.05 T·mm) was developed and constructed together with the *Institut für Hochfrequenztechnik* (IHF) at RWTH Aachen University and ZEA-1 in Jülich. This new RF Wien-filter was installed in COSY in May 2017. A first commissioning run was successfully conducted in June 2017.

Acknowledgements

The author wishes to acknowledge the support by the European Research Council via the ERC AdG "srEDM" (#694340). He also thanks all involved members of the JEDI collaboration and of the Institut für Kernphysik of Forschungszentrum Jülich for their dedication and persistence towards this long-term project.

References

1. J. M. Pendlebury et al. Phys. Rev. D 92, 092003 (2015)
2. JEDI collaboration at COSY-Jülich;
   see: http://collaborations.fz-juelich.de/ikp/jedi/
3. H. Ströher (PI): ERC AdG srEDM (#694340);
   see: http://sredm-ercgrant.de/
4. D. Eversmann et al. (JEDI Collaboration) Phys. Rev. Lett. 115, 094801 (2015)
5. G. Guidoboni et al. (JEDI Collaboration) Phys. Rev. Lett. 117, 054801 (2016)
6. N. Hempelmann et al. (JEDI Collaboration) Phys. Rev. Lett. 119, 014801 (2017)
7. A. Saleev et al. (JEDI Collaboration) Phys. Rev. Accel. Beams 20, 072801 (2017)
8. WASA-at-COSY;
   see: http://collaborations.fz-juelich.de/ikp/wasa/
9. J. Slim et al., Nucl. Instr. Meth. A 828, 116 (2016);
   J. Slim et al., Nucl. Instr. Meth. A 859, 52 (2017)

10. C. Weidemann et al., Phys. Rev. ST-AB 18, 020101 (2015)
11. Institut für Kernphysik (IKP) Annual Report 2015 and 2016, Berichte des Forschungszentrums Jülich: Jül-4393, Jül-4398

    see: http://www.fz-juelich.de/ikp/EN/Service/
12. JEDI contributions to CALOR2016 (accessible via Ref. [2] above)

Low-energy electron-positron collider to search and study ($\mu^+\mu^-$) bound state

*Anton Bogomyagkov**, *Vladimir Druzhinin*, *Eugene Levichev*, *Alexander Milstein*, and *Sergej Sinyatkin*

Budker Institute for Nuclear Physics, Novosibirsk 630090, Russia

Abstract. We discuss a low energy e^+e^- collider for production of the not yet observed ($\mu^+\mu^-$) bound system (dimuonium). Collider with large crossing angle for e^+e^- beams intersection produces dimuonium with non-zero momentum; therefore, its decay point is shifted from the beam collision area providing effective suppression of the elastic e^+e^- scattering background. The experimental constraints define subsequent collider specifications. We show preliminary layout of the accelerator and obtained main parameters. High luminosity in chosen beam energy range allows to study $\pi^\pm$ and η-mesons.

1 Introduction

Currently some authors consider a lepton atom ($\mu^+\mu^-$)[a] as a strong candidate for casting light on disagreement between the anomalous magnetic moment measurement and Standard Model prediction, the muonic hydrogen proton's charge radius discrepancy, etc. [2, 3]. Even if this expectation will not come true, a discovery of a new lepton bound state, which has not been observed yet contrary to positronium (e^+e^-) and muonium (μ^+e^-), and its study is by far a challenging and interesting scientific enterprise. Like positronium, dimuonium is a Bohr atom and its study (including transition spectroscopy, lifetime precise measurement, etc.) allows verifying QED and quantum mechanics computations with great accuracy. In such a research, ($\mu^+\mu^-$) has higher new-physics reach potential in comparison with other exotic atoms. It has larger reduced mass $\mu = m_\mu/2$ than (e^+e^-) and (μ^+e^-), for which $\mu \approx m_e$; therefore, it is considerably more compact (the dimuonium Bohr radius is 200 times smaller than the positronium one). In contrast to (pe^-) and ($p\mu^-$) systems, ($\mu^+\mu^-$) experiences no strong interaction obscuring exploration of tiny QED effects. Study of heavier and more compact taonium ($\tau^+\tau^-$) would also be interesting but its production is extremely difficult even if compared with ($\mu^+\mu^-$).

In the past, many of proposed ($\mu^+\mu^-$) production channels have been discussed [1, 4-13] including those from electron and positron collision [1, 5, 10]. The $e^+e^-\rightarrow$ ($\mu^+\mu^-$) production mechanism is particularly interesting because it contains no hadrons which, otherwise, should be untangled in the reconstruction process. However, a e^+e^- collider for the dimuonium experimental study has never been discussed in details.

We propose a preliminary design of the electron-positron collider for ($\mu^+\mu^-$) search and study. Low beam energy $E_b \approx 400$ MeV results in a compact machine, inexpensive in both manufacture and maintenance. Large crossing angle of the intersecting beams induces a non-zero dimuonium momentum, which is a distinct feature of our design (originally proposed in [10]). In this case, the ($\mu^+\mu^-$) production and decay points are separated along the beams directions bisector (for equal beam energies). The distance between the atom formation and its annihilation into e^+e^- pair effectively suppresses elastic scattering background from the primary beams and allows unambiguous detection of dimuonium decay.

The collider design is rather challenging, because large collision is a source of various effects in e^+e^- beams interaction, absent in the head-on collisions. Low beam energy and high bunch intensity result in the strong intra-beam scattering (IBS) and reduced Touschek lifetime. We discuss and estimate all these issues below.

Besides the ($\mu^+\mu^-$) research, the beam energy and collider configuration allow studying such processes as $e^+e^-\rightarrow$ *pions* and $e^+e^-\rightarrow$ η,η' substantially expanding experimental program of the facility.

2 Dimuonium

We briefly present the major properties of ($\mu^+\mu^-$) atom relevant to the collider design. More detailed information can be found elsewhere (see for instance [14] and references therein).

Dimuonium is one of the simplest hydrogen-like atoms and its spectrum in the first approximation is similar to the positronium spectrum with respectful mass substitution. Unlike positronium, dimuonium annihilates into e^+e^- pair, which makes its formation and spectrum more sensitive to vacuum polarization effects caused by

[a] Dubbed "dimuonium", "bimuonium" or "true muonium".

* Corresponding author: author@e-mail.org

virtual electron-positron pairs production. Fig.1 shows dimuonium energy levels.

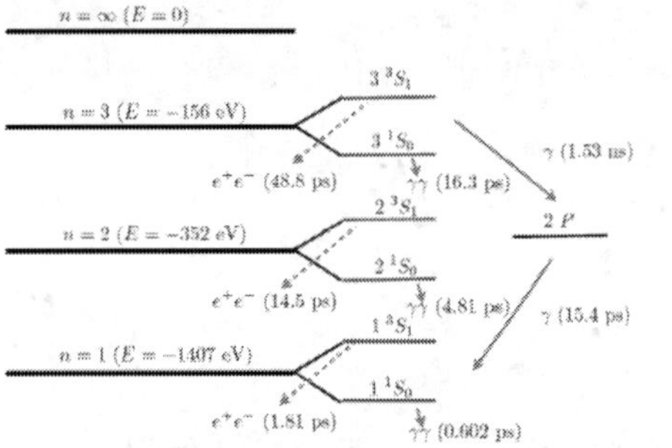

Fig. 1. The dimuonium energy states diagram [10] (not in scale).

Dimuonium Bohr radius is $R_{\mu\mu} = 512$ fm (for reference the positronium Bohr radius is $R_{ee} = 0.106$ nm). The ($\mu^+\mu^-$) atom exists in two spin states: paradimuonium, which is a singlet with total muon-antimuon spin S = 0, and orthodimuonium (triplet S = 1). For paradimuonium ground state with the dominant decay mode $1^1S_0 \to \gamma\gamma$ the lifetime is 0.602 ps while for orthodimuonium it is of 1.81 ps (dominant decay is $1^3S_1 \to e^+e^-$). Muon weak decay is very slow with the lifetime of 2.2 μs, and thus, dimuonium is a unique metastable system available for precision QED tests.

Cross section of ($\mu^+\mu^-$) production in the electron-positron collision for exact resonance condition (the beam energy $E_b = m_\mu - \Delta_{\mu\mu}/2$, m_μ is the muon mass and $\Delta_{\mu\mu} \approx 1.41$ keV is the binding energy) is [1]

$$\sigma_{\mu\mu n} = 3.3 \cdot 10^{-25} n^{-3}\ \text{cm}^2 \tag{1}$$

The cross section (1) rapidly decreases with the principal quantum number n. Radiative corrections reduce the cross section, the ground state reduction is [1]

$$\sigma_{\mu\mu 1R} = 0.27\sigma_{\mu\mu 1}. \tag{2}$$

However, detailed simulation of the ($\mu^+\mu^-$) production rate with account for the beam energy and spatial distribution [16] recovers the cross section reduction due to the radiative corrections (within 10% accuracy), thus for convenience to estimate the dimuonium production rate below we use the cross section (1).

3 Large angle beam crossing

To study the beams collision with an arbitrary angle we apply results from [15]. Two leptons with four-momenta $(E_1, \boldsymbol{p}_1)$ and $(E_2, \boldsymbol{p}_2)$ collide as shown in Fig.2.

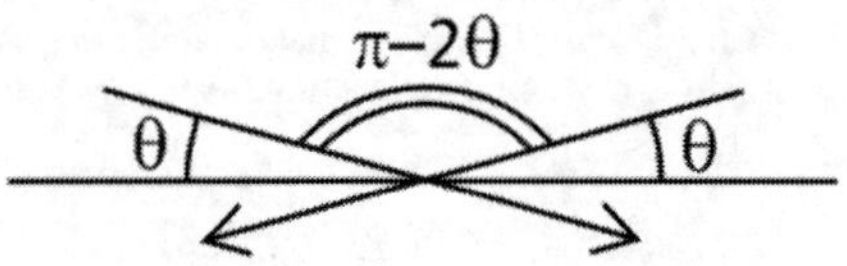

Fig. 2. Electron positron ccollision at arbitrary angle θ.

Head-on collision corresponds to $\theta = 0$ and parallel momenta at $\theta = \pi/2$. Assuming Gaussian beam energy spread with r.m.s. $\sigma_\delta = \sigma_E/E_0$ and where E_0 is beam average energy, Gaussian angular spreads in horizontal and vertical planes with r.m.s. $\sigma_{x'}$ and $\sigma_{y'}$ we obtain invariant mass mean and r.m.s. values

$$\begin{aligned}\langle M\rangle_{x',y',\delta} &= 2E_0\cos\theta \\ &\quad - \frac{E_0}{2}\Big(\sigma_\delta^2\cos\theta \\ &\quad - \sigma_{x'}^2\cos\theta - \sigma_{y'}^2\frac{\cos 2\theta}{\cos\theta}\Big),\end{aligned} \tag{3}$$

$$\sigma_M^2 = 2E_0^2(\sigma_\delta^2\cos^2\theta + \sigma_{x'}^2\sin^2\theta). \tag{4}$$

For short bunches (a *hour-glass* effect is neglected) with equal population N, colliding with the frequency f_0 at the arbitrary angle θ the peak luminosity is

$$L_0 = \frac{N^2}{4\pi\sigma_x\sigma_y\sqrt{1+\varphi^2}}f_0, \tag{5}$$

where $\sigma_{x,y}$ is transverse r.m.s. beam size at the interaction point (IP), $\varphi = \sigma_z\tan\theta/\sigma_x$ is Piwinski angle, σ_z is bunch length. For large Piwinski angle $\varphi \gg 1$ and the peak luminosity

$$L_0 \approx \frac{N^2}{4\pi\sigma_y\sigma_z\tan\theta}f_0 \tag{6}$$

is less than head-on luminosity and does not depend on the horizontal beam size at IP.

Using approximate expression for ($\mu^+\mu^-$) production cross section

$$\begin{aligned}\sigma(x'_{1,2}, y'_{1,2}, \delta_{1,2}) = \Gamma_{\mu\mu}\sigma_{\mu\mu}\delta\big(M(x'_{1,2}, y'_{1,2}, \delta_{1,2}) \\ - m_{\mu\mu}\big),\end{aligned} \tag{7}$$

we calculated dimuonium production rate as convolution of cross section with differential luminosity ([15])

$$\dot{N}_{\mu\mu} = \frac{L_0\Gamma_{\mu\mu}\sigma_{\mu\mu}}{2\sqrt{\pi}\sigma_M}. \tag{8}$$

Non-zero e^+e^- crossing angle boosts the atom to move along the bisector of the beam lines (for equal e^+e^- energies); therefore, with given dimuonium lifetime at rest $\tau_{0\mu\mu} = \hbar/\Gamma_{\mu\mu}$, the annihilation point is separated from the formation point by

$$l = c\,\tau_{0\mu\mu}\beta_{\mu\mu}\gamma_{\mu\mu} = c\,\tau_{0\mu\mu}\tan\theta = \frac{c\hbar}{\Gamma_{\mu\mu}}\tan\theta, \tag{9}$$

where c is speed of light, $\beta_{\mu\mu}$ and $\gamma_{\mu\mu}$ are dimuonium normalized velocity and Lorentz factor.

4 Beam-beam effects

Contrary to the head-on collision, the opposite bunch forces in the large crossing angle collision are not purely transverse, and depend on longitudinal coordinates of the suffering particle. Authors of [17, 18, 19] considered beam-beam effects for arbitrary crossing angle and showed that transverse influence is suppressed, but longitudinal is emphasized for large angles. The beam-beam tune shifts for flat beams ($\sigma_y \ll \sigma_x$) clearly illustrate this fact:

$$\xi_y = \frac{Nr_e}{2\pi\gamma}\frac{\beta_y}{\sigma_y\sqrt{\sigma_x^2 + \sigma_z^2\tan^2\theta}}, \tag{10}$$

$$\xi_x = \frac{Nr_e}{2\pi\gamma}\frac{\beta_x}{\sigma_x^2 + \sigma_z^2\tan^2\theta}, \tag{11}$$

where $\beta_{x,y}$ are betatron functions at IP, r_e is classical electron radius, and

$$\xi_z = -\frac{N r_e}{2\pi\gamma}\frac{\sigma_{z0}\tan^2\theta}{\sigma_{\delta 0}(\sigma_x^2+\sigma_{z0}^2\tan^2\theta)}\frac{\alpha}{|\alpha|}, \quad (12)$$

where $\sigma_{\delta 0}$ and σ_{z0} are the energy spread and bunch length of the "strong" beam affecting the test particle (weak-strong approximation). Bunch population N, Lorentz factor γ, and IP beam sizes σ_x and σ_z also belong to the "strong" bunch.

For some (not necessary large) intersection angle θ we have $\sigma_z \tan\theta \gg \sigma_z$, and the conventional transverse beam-beam effects (10) (11) are suppressed. Longitudinal beam-beam effects (12) increase, and the synchrotron tune shift

$$\Delta\nu_s = \xi_z \approx -\frac{N r_e}{2\pi\gamma}\frac{1}{\sigma_{\delta 0}\sigma_{z0}}\frac{\alpha}{|\alpha|} \quad (13)$$

might be as large as the unperturbed synchrotron tune ν_s, causing the phase instability of the test bunch. Even below this limit, when the particle motion is still stable, the "strong" bunch electric field distorts RF bucket, modifies longitudinal bunch distribution, and reduces luminosity.

In canonical coordinates (z, p_z), where z is particle position with respect to the bunch center and p_z is conjugate momentum, the linear Hamiltonian of the test particle longitudinal motion is

$$H_z \approx -\frac{\alpha\, p_z^2}{2} - \frac{\nu_s^2}{\alpha R^2}\frac{z^2}{2} - \frac{2\xi_z\nu_s}{\alpha R^2}\frac{z^2}{2}. \quad (14)$$

The second term relates to accelerating RF system, the third one corresponds to the colliding bunch electrical field, α is momentum compaction factor, and R is orbit average radius. To keep the bunch field effect negligible one needs $\xi_z \ll \nu_s/2$, which means either large synchrotron tune ν_s or small longitudinal beam-beam parameter (13) and finally limits the "strong" bunch population according to

$$N < \frac{2\pi R\,\gamma}{r_e}\frac{\alpha\,\sigma_\delta^2}{2}. \quad (15)$$

This limit within the factor of two corresponds to the condition for coherent instability of two bunches intersecting at large angle found in [18].

From the Hamiltonian (14) one can see another possibility to suppress the longitudinal beam-beam effects. For positive momentum compaction $\alpha > 0$ the beam-beam synchrotron tune shift $\xi_z < 0$, and the "strong" bunch field flattens RF potential well, increases "weak" bunch length, and destroys stability. Alternatively, for magnetic lattice with $\alpha < 0$ the third term in (14) has the same sign as a second one, giving deeper RF potential well and smaller "weak" bunch length.

High order expansion of the synchrotron motion Hamiltonian shows that for $\alpha > 0$ the "strong" bunch field modulates RF potential well and splits the "weak" bunch into sub-bunches. Simulation with the help of the beam-beam computer code LIFETRAC [20] confirmed this effect as well as above estimation of the bunch intensity limit.

5 Background and choice of collider parameters

The main source of the background is elastic (Bhabha) $e^+e^- \to e^+e^-$ scattering with cross section $\sigma_{Bhabha} = 22000$ nb for crossing angles from 45° to 135°. For different collision schemes production cross section of 1^3S_1 dimuonium state is $\sigma_{\mu\mu}(m_{\mu\mu}) = 0.23 \div 11$ nb. Background to signal ratio is than $(3 \div 130) \times 10^3$. Crossing angle collisions is the only way to suppress background, it separates dimuonium decay point from the origin point. For estimation of the signal to background ratio $\dot N_{\mu\mu}/\dot N_{ee}$, where $\dot N_{ee}$ is elastic scattering rate we use collision scheme shown on Fig. 3.

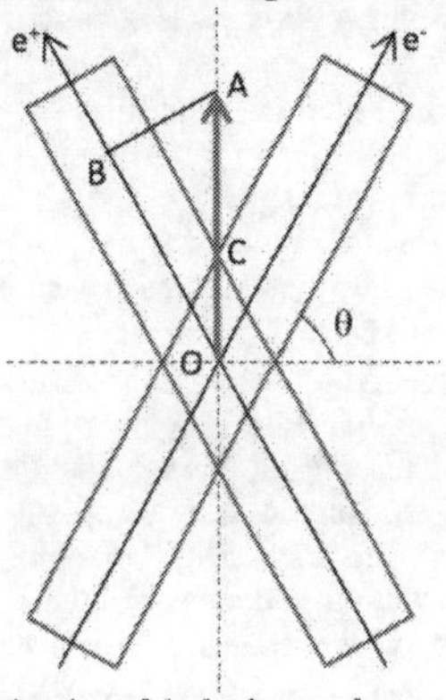

Fig. 3. On estimation of the background rate.

Points O and A are those of dimuonium production and decay; the distance $OA = l$ is defined in (9). At the point A, where e⁺e⁻ scattering mimics dimuonium decay, the density of the primary particles (assuming normal distribution) drops according to $N_1 = N_0 \exp(-n_x^2/2)$, where $n_x = AB/\sigma_x$, $AB = l\cos\theta = c\,\tau_{0\mu\mu}\sin\theta$, and background rate $\dot N_{ee} \propto N_1^2$. Using σ_M from (4) and substituting beam size and divergence with $\sigma_{x'}^2 = \varepsilon_x/\beta_x$ and $\sigma_x^2 = \varepsilon_x\beta_x$, ε_x is horizontal emittance, we obtain signal to background ration estimation

$$\frac{\dot N_{\mu\mu}}{\dot N_e} \propto \frac{\exp\left(\frac{(c\,\tau_{0\mu\mu})^2\sin^2\theta}{\sigma_x^2}\right)}{\sqrt{\sigma_\delta^2\cos^2\theta + (\sigma_x^2/\beta_x^2)\sin^2\theta}}. \quad (16)$$

To increase numerator in (16) one needs the large crossing angle $\sin\theta \to 1$ $(\theta \to 90°)$ and small horizontal beam size $\sigma_x^2 = \varepsilon_x\beta_x$ at the IP. To reduce denominator (better resolution) one needs small beam divergence $\sigma_{x'}^2 = \varepsilon_x/\beta_x$ at the IP. Large intersection angle increases required beam energy (3), enlarges the energy resolution (4), and strongly degrades the peak luminosity (5). The choice of collider parameters requires careful balance of pros and cons. Another important issue is existing injection facility [21] with capacity $(0.5 \div 1) \times 10^{10}$ e^+/s in the energy range of $E_{inj} \approx 350 \div 450$ MeV. Top-up injection (at the energy of experiment) allows reaching average luminosity close to the peak one. Despite the luminosity drop with the crossing angle increase, reduction of the beam interaction length permits decrease of the vertical beta β_y

at the IP (like for Crab-Waist collision [22]), this partially recovers the luminosity loss. In addition, the beam energy enlargement suppresses intra-beam scattering (IBS) and results in the collider performance enhancement.
The major specifications of the electron-positron collider for dimuonium production and exploration are:

1. The crossing angle of 75° allows ($\mu^+\mu^-$) production for the beam energy of E_b = 408 MeV (m_μ = 105.7 MeV). At 70° $\pi^+\pi^-$ ($m_{\pi\pm}$ = 139.6 MэB) can be generated at the same beam energy. The angle modification is available either by mechanical rearrangement of the interaction area or using corrector magnets.
2. The crossing angle of 75° provides 1^3S_1 ($\mu^+\mu^-$) decay path l = 2 mm, it is enough to detect the atom decay to e^+e^-. For the next dimuonium terms, the lifetime (and the escape length) increases as n^3 (see Fig.1), however production rate falls down insofar.
3. To increase the signal-background ratio (16) the horizontal beam size at the IP should be $\sigma_x \ll c\,\tau_{0\,\mu\mu} = 0.54$ mm, so we take $\sigma_x < 0.15$ mm.
4. For large crossing angle, horizontal angular spread at the IP is the main source of the invariant mass resolution degradation. To reduce the angular spread with the small beam size collider has to provide low horizontal emittance.
5. Lattice with negative momentum compaction $\alpha < 0$ have been studied at several colliders (for instance, at KEKB [23] or at DAΦNE [24]). This operation mode has never been used as regular one. The explanation is that in such a mode microwave instability current threshold reduces, giving larger beam energy spread and transverse size (especially in vertical direction) and smaller luminosity [24]. Therefore, in our design we adopt a conservative but robust solution with positive α. Another argument in favor of $\alpha > 0$ is that the negative compaction lattice requires extra space for matching sections and the orbit length increases.
6. For high luminosity one needs highly populated e^+e^- bunches but the longitudinal beam-beam effects limit the bunch intensity according to (15). In our specifications we take the bunch population $N_0 \leq 3.5\times10^{10}\ e^+/e^-$. To arrange a multi-bunch operation two separate storage rings are necessary.
7. Low vertical beta-function at the IP benefits to the high luminosity. However, extremely low beta causes strong chromatic effect in the final focus quadrupoles, which needs to be compensated by specially arranged chromatic correction sections. To reduce collider size we choose β_y = 2 mm and avoid final focus local chromaticity correction.
8. Reverse of the beam circulation direction in one of the rings changes the crossing angle to 15°, and makes it possible to study with high luminosity energy range from η-meson (m_η = 547.8 MeV, $E_b \approx$ 284 MeV at θ = 15°) to η′-meson ($m_{\eta'}$ = 957.7 MeV, $E_b \approx$ 496 MeV).

6 Collider

A simple and symmetric configuration of the large crossing angle collider consists of two identical bulb-shape storage rings with two intersection points. Fig. 4 shows e^+e^- orbits for such configuration. Each ring includes two straight interaction sections and two closing arcs: inward and outward. The collider footprint is 12×6 м², and the ring circumference is about 23 m.

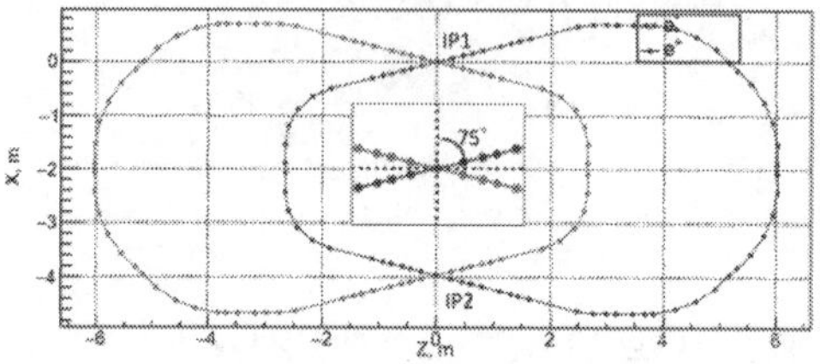

Fig. 4. Orbits of e^+e^- collider rings.

Fig. 5 presents the magnet sequence and the lattice functions along the ring. The inward arc lattice (IP1-IP2) is a Double Bend Achromat (DBA) with the IP betatron functions β_y = 2 mm and β_x = 20 cm. The maximum vertical beta in the first defocusing final focus quadrupole is not large ($\beta_y \approx$ 50 m) and the local chromatic section is unnecessary. The outward Multipole Bend Achromat (MBA) arc (IP2-IP1) provides low radiative emittance [25]. Table 1 gives the main parameters of the ring at 408 MeV.

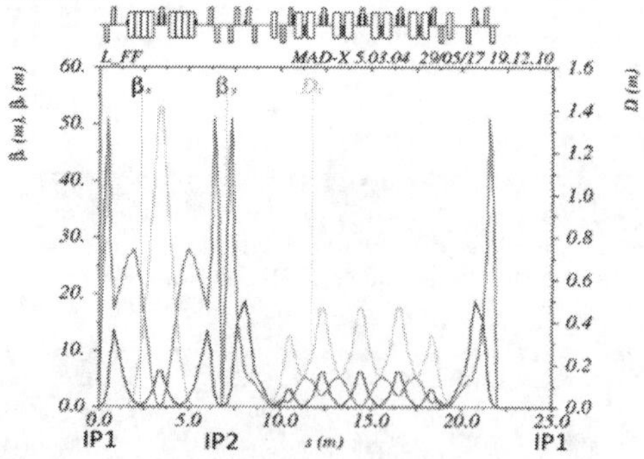

Fig. 5. Lattice functions of the storage ring.

The momentum compaction factor is quite large α = 0.064, so condition $\xi_z \ll \nu_s/2$ is well satisfied. The longitudinal beam-beam effects define the maximum bunch intensity (15) as $N_{max} = 1.5\times10^{11}$. We have chosen $N = 3.5\times10^{10}$ and it is rather adequate for reliable operation. The transverse beam-beam effects are negligible; therefore, other reasons define the choice of the betatron tune point (for instance, size of dynamic aperture or momentum acceptance). The lattice provides radiative emittance of ε_x = 26 nm, which IBS increases by factor of four. IBS elongates the bunch length from natural 5.4 mm to 11.6 mm. Touschek beam lifetime for the pa-rameters given in Table 1 is around 1500 s.

The primary e^+e^- collisions beyond the point A in Fig.3 determine elastic scattering background rate.

Table 1. Collider main parameters at 408 MeV.

Beam energy (MeV)	408
Circumference (m)	23
Bunch intensity/current (mA)	3.5×10^{10}/73
Revolution frequency/period (MHz)/(ns)	13.04/76.7
RF harmonic number/frequency (MHz)	26/338.98
Energy loss per turn (keV)	2.3
RF voltage (kV)	450
RF acceptance	2%
Synchrotron tune	1.71×10^{-2}
Momentum compaction α	6.4×10^{-2}
Damping time hor/ver/long (ms)	17.3/27.3/22.1
Damping partition hor/long	1.6/1.4
Horizontal emittance (without/with IBS) (nm)	26/90
Energy spread (without/with IBS), $\times10^4$	4/8.4
Bunch length (without/with IBS) (mm)	5.4/11.6
Betatron coupling	0.3%
IP horizontal angular spread $\sigma_{x'}$, $\times 10^4$	6.7
Invariant mass resolution (keV)	390
Hor/vert betatron function at IP (mm)	200/2
Hor/vert betatron size at IP (μm)	130/0.7
Hor/vert beam-beam parameter (ξ_x/ξ_y)	$2\times10^{-6}/1.2\times10^{-3}$
Longitudinal beam-beam parameter ξ_z	-2×10^{-3}
Peak luminosity for 1 bunch ($cm^{-2}s^{-1}$)	4×10^{30}
Peak luminosity for 20 bunches ($cm^{-2}s^{-1}$)	8×10^{31}

Assuming Gaussian distribution of the beam particles along the dimuonium escape line with standard deviation $OC = \sigma_x/\cos\theta$, the number of the electrons in the beam tail beyond the A divided by the total amount of beam electrons is

$$\frac{N_{>A}}{N_{\Sigma}} \propto 1 - \mathrm{erf}\left(\frac{OA}{\sqrt{2}\,OC}\right) = 1 - \mathrm{erf}\left(\frac{l\,\cos\theta}{\sqrt{2}\,\sigma_x}\right). \quad (17)$$

Our parameters provide $OA > 2$ mm and suppress the particle density in the beam tail (and the elastic scattering background rate) by 4.5 standard deviations for the ground state 1^3S_1.

There are eight bending magnets in each ring. Two magnets compose the DBA cell of the inward arc. Six magnets compose the regular MBA cell in the outward arc and two additional magnets cancel dispersion function in the interaction region. All bending magnets have the same magnetic field (1.27 T at 408 MeV) and low negative gradient to reduce the number of defocusing quadrupoles, to increase horizontal partition number and to reach smaller emittance. The regular magnets in the MBA cell are split into two halves with defocusing sextupoles installed in between, at the azimuth with the best ratio of the vertical-to-horizontal betas. Twenty quadrupoles combined in 10 families determine the ring optical functions. The maximum gradient in the arc quadrupoles is rather tolerable (≈ 15 T/m) but the first (defocusing) final focus quadrupole QD0, providing the vertical beta mm at the IP, is challenging. It should combine the maximum gradient of - 32 T/m with extreme compactness to accommodate detector equipment.

Fig. 6 is an artist sketch of the interaction area. It consists of the experimental vacuum vessel with the beams intersecting inside, final focus quadrupoles, vacuum tubes and equipment (pumps, valves, bellows, etc.), beam diagnostics equipment (beam position monitors). The experimental chamber is a flat box with 0.5 mm thick beryllium windows on the box top and bottom to pass $e^\pm$ produced by the dimuonium atoms decay (Fig.6).

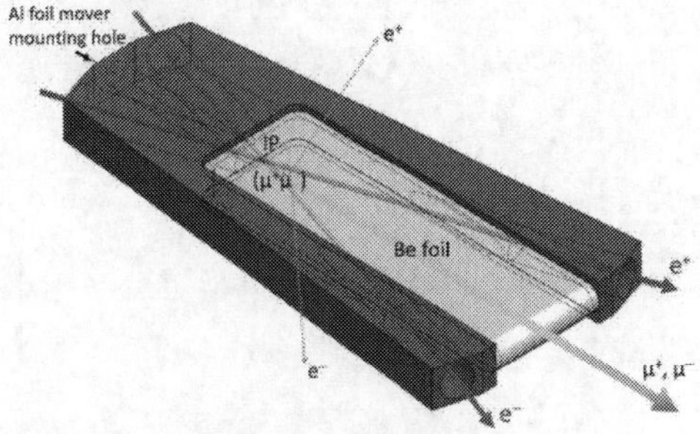

Fig. 6. Experimental chamber.

Detector set up for ($\mu^+\mu^-$) study looks simple and compact; it consists of the tracking systems above and below the median plane, magnetic spectrometer, etc.

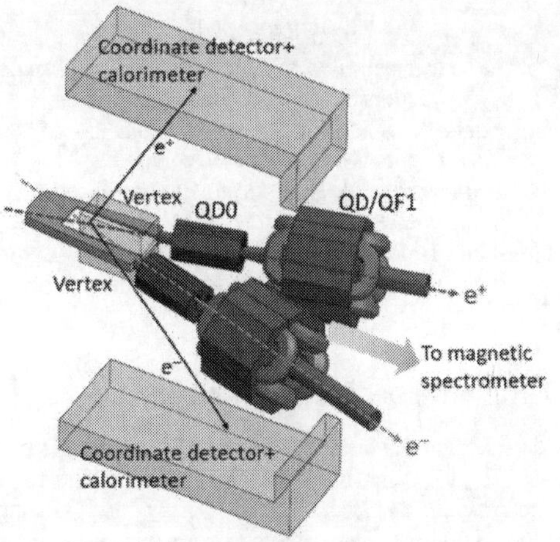

Fig. 7. Detector and interaction region equipment.

Four thin-walled stainless steel vacuum tubes with QD0 defocusing quadrupoles over them join the experimental chamber. The quadrupoles are built of permanent magnets and provide the maximum gradient up to - 35 T/m in the Ø30 mm aperture while being very compact and allowing enough room for passing $\mu^+\mu^-$ (or $\pi^+\pi^-$) from the IP toward the magnetic spectrometer. The next quadrupole (QD/QF1 on Fig. 7) is conventional electromagnet and serves either as a focusing (at the low energy) or as an additional defocusing (at the high energy) quadrupole.

Table 2 clearly demonstrates that production rate of dimuonium allows observation of the atom in the reasonable time and investigation of its properties, including spectrum. Fig. 8 illustrates ($\mu^+\mu^-$) states distribution along the escape line.

Table 2. Dimuonium production rate estimation for collider parameters from Table 1.

$(\mu^+\mu^-)$ rate	1 hour	4 months
Totally (1S/2S/3S)	65/8.1/2.4	187k/23k/6.9k
For l >2.3 mm	21/7/2.3	59k/20k/6.6k

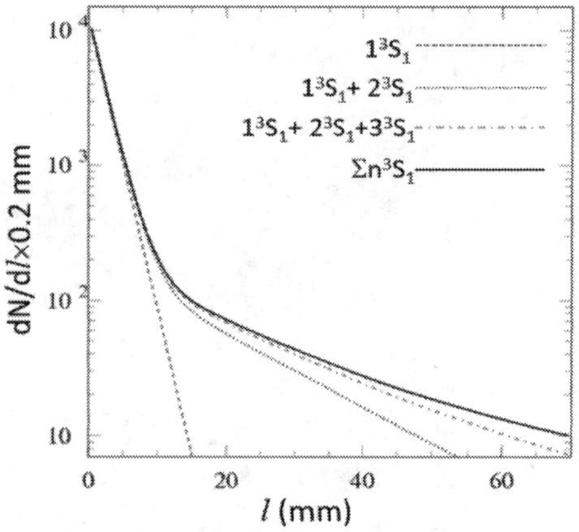

Fig. 8. $(\mu^+\mu^-)$ distribution along the escape path.

The beam energy 408 MeV under the crossing angle of 70° allows formation of $\pi^+\pi^-$ pairs. The collider parameters are the same as for 75°. To study pions two scenarios are possible: (a) both collision points are modified mechanically from 75° to 70° after the end of the dimuonium research program, (b) simultaneous operation when the beams collide at 75° in the first point to study $(\mu^+\mu^-)$ and at 70° in the second point to study $\pi^+\pi^-$.

7 Reverse beam configuration

Reverse of one of the beams changes collision angle from 75° to 15° and opens additional possibilities for experiments. In the injection facility energy range (300÷500 MeV) collider with 15° beam crossing allows experiments in the range of $E_{CM} \approx 500 \div 1000$ MeV which includes ρ, ω, η and η′-mesons, etc. Such experiments require a general-purpose detector with solenoid field, which is very much different from the $(\mu^+\mu^-)$ detector set up. Thus, the interaction region needs reconfiguration including vacuum chamber, final focus magnets, etc. toward the conventional e^+e^- colliders arrangement. Also, injection polarity in one of the rings should be reversed. As there are no crucial difficulties for that, we do not consider the new configuration in detail and for further estimation apply the lattice from the previous section. We assume adjustment of the new lattice with 15° crossing angle to the optics in Fig.5 and estimate collider parameters by scaling with the energy (considering IBS). Table 3 shows new collider specifications (excluding those repeating the Table 1: circumference, beam current, RF frequency, etc.) for the minimum (η-meson, $m_\eta = 547.8$ MeV, $E_b \approx 284$ MeV) and maximum (η′-meson, $m_{\eta'} = 957.7$ MeV, $E_b \approx$ 496 MeV) beam energies.

According to (6) the acute intersection angle significantly increases luminosity. Transverse beam-beam effects although but remain small compared to the head-on collision. The longitudinal beam-beam effects do not change much with respect to 75° collision.

Table 3. Collider main parameters for 15° collision angle

Beam energy (MeV)	283.59 (η)	495.78 (η′)
Invariant mass (MeV)	547.86	957.76
Energy loss per turn (keV)	0.535	4.997
RF voltage (kV)	300	550
Synchrotron tune	1.67×10^{-2}	1.71×10^{-2}
Horizontal emittance (without/with IBS) (nm)	11.4/105	34.8/75
Energy spread (without/with IBS), $\times10^4$	2.8/10.6	4.8/8.4
Bunch length (without/with IBS) (mm)	3.7/14.2	6.3/11
IP horizontal angular spread $\sigma_{x'}\times10^4$	8.3	7.1
Invariant mass resolution (keV)	420	580
Hor/vert beam-beam parameter (ξ_x/ξ_y)	3×10^{-4} / 1.4×10^{-2}	3×10^{-4} / 1.3×10^{-2}
Longitudinal beam-beam parameter ξ_z	-1.8×10^{-3}	-1.7×10^{-3}
Peak luminosity for 1 bunch ($cm^{-2}s^{-1}$)	3.3×10^{31}	5.2×10^{31}
Peak luminosity for 20 bunches ($cm^{-2}s^{-1}$)	6.6×10^{32}	1×10^{33}

8 Conclusion

We have considered preliminary design of the electron-positron collider for production and study of $(\mu^+\mu^-)$ bound state and capable to perform other experiments with high luminosity in the center-of-mass energy range $E_{CM} \approx 500 \div 1000$ MeV. Two rings collider with large crossing angle is quite compact, the ring orbit length ~20÷30 m, and not expensive in development and maintenance. Despite many challenging issues still needing investigation (dynamic aperture, collective effects, magnet and vacuum systems design, etc.); the project seems interesting for both fundamental research and collider technologies development.

Collider performance and dimuonium production rate allow not only detection of new leptonic atom for the first time but also study $(\mu^+\mu^-)$ properties (including precise measurement of the lifetime, spectroscopic experiments, atoms passing through a thin foil, etc.).

9 Acknowledgements

We are grateful to

- Yu.A. Pupkov for his important advice to consider extremely low energy e^+e^- collider,
- N.A. Vinokurov, S.I. Serednikov, E.P. Solodov for many fruitful discussions and valuable comments,

- E.A. Perevedentsev for introduction to and D.N. Shatilov for simulation of the longitudinal beam-beam effects,
- V.A. Kiselev for estimation of the injection schemes and efficiency,
- P.D. Vobly for preliminary design of the quadrupole with permanent magnets.

References

1. V.N.Baier and V.S.Synakh, Bimuonium production in electron-positron collisions, SOVIET PHYSICS JETP, 14, N5, 1962, pp.1122-1125.
2. H. Lamm, True muonium: the atom that has it all, arXiv 1509.09306v1, 30 Sep 2016.
3. H. Lamm and R.F. Lebed, True Muonium ($\mu^+\mu^-$) on the Light Front, arXiv hep-ph/1311.3245v3, 12 Nov 2014.
4. L.L. Nemenov. Atomic decays of elementary particles. Yad. Fiz., 15:1047-1050, 1972.
5. J.W. Moffat. Does a Heavy Positronium Atom Exist? Phys. Rev. Lett., 35:1605, 1975.
6. E. Holvik and H.A. Olsen. Creation of Relativistic Fermionium in Collisions of Electrons with Atoms. Phys. Rev., D35:2124, 1987.
7. G.A. Kozlov. On the problem of production of relativistic lepton bound states in the decays of light mesons. Sov. J. Nucl. Phys., 48:167-171, 1988.
8. I.F. Ginzburg, U.D. Jentschura, S.G. Karshenboim et al. Production of bound $\mu^+\mu^-$ systems in relativistic heavy ion collisions. Phys. Rev., C58:3565-3573, 1998.
9. N. Arteaga-Romero, C. Carimalo, and V.G. Serbo. Production of bound triplet $\mu^+\mu^-$ system in collisions of electrons with atoms. Phys. Rev., A62:032501, 2000.
10. S.J. Brodsky and R.F. Lebed. Production of the Smallest QED Atom: True Muonium ($\mu^+\mu^-$). Phys. Rev. Lett., 102:213401, 2009.
11. Y.Chen and P.Zhuang. Dimuonium ($\mu^+\mu^-$) Production in a Quark-Gluon Plasma. 2012.
12. A.Banburski and P.Schuster. The Production and Discovery of True Muonium in Fixed-Target Experiments. Phys. Rev., D86:093007, 2012.
13. S.C. Ellis and J. Bland-Hawthorn. On the possibility of observable signatures of leptonic onium atoms from astrophysical sources. 2015.
14. H. Lamm, True muonium on the light front, PhD Dissertation, Arizona State University, May 2016, 115 pages.
15. A. Bogomyagkov and E. Levichev. Collision monochromatization in e^+e^- colliders, Phys. Rev. Accel. Beams 20, 051001, May 2017.
16. V.P. Druzhinin, private communication, March 2017.
17. J.R.Rees, Longitudinal effects of colliding beam space charge force in electron-positron storage rings with crossing angles. SLAC-PUB-781, July 1970.
18. V.V. Danilov et al. Longitudinal effects in beam-beam interaction for an ultra-high luminosity regime. Workshop on beam radiation interaction, UCLA, May 13-16, 1991, World Scientific, pp.1-10.
19. A. Drago et al. Synchrotron oscillation damping by beam-beam collision in DAΦNE, PRST AB 14, 092803 (2011).
20. D. Shatilov, Beam-beam simulation at large amplitudes and lifetime determination. Part. Accel. 52, p.65, 1996.
21. D.E. Berkaev et al., "VEPP-5 Injection Complex: Two Colliders Operation Experience", in Proc. 8th Int. Part.Acc.Conf. (IPAC'17), Copenhagen, Denmark, May 2017, pp. 2982-2984.
22. P. Raimondi et al. Beam-Beam Issues for Colliding Schemes with Large Piwinski Angle and Crabbed Waist, LNF-07/003 (IR) 2007.
23. M. Hosaka et al., Nucl.Instr.Meth. A407 (1998) 234-240.
24. M. Zobov et al., DAΦNE experience with negative momentum compaction, Proc.of EPAC 2006, Edinburgh, Scotland, 989-991.
25. D. Einfeld et al. Design of a Diffraction Limited Light Source (DIFL), in Proc of PAC'95, Dallas, USA (1996)

Toward the measurement of the hyperfine structure of muonic hydrogen in the FAMU experiment.

Multi-pass cavity optimization for experiments with pulsed sources.

Dimitar Bakalov[1,*] and *Mihail* Stoilov[1]

[1]Institute for Nuclear Research and Nuclear Energy, Bulgarian Academy of Sciences, Tsarigradsko ch. 72, Sofia 1784, Bulgaria

Abstract. We consider a simplified model of the optical multi-pass cavity that is being currently developed by the FAMU collaboration for the measurement of the hyperfine splitting in the ground state of muonic hydrogen and of the Zemach radius of the proton. The model is focused on the time distribution of the events of laser-stimulated hyperfine transitions in the muonuc atom and may be helpful in the preliminary design of the FAMU experimental set-up and, more generally, in the optimization of multi-pass optical cavities for experiments with pulsed lasers.

1 Introduction

The laser excitation of the ortho $S = 1$ hyperfine sub-level of the ground state of the muonic hydrogen atom from the para $S = 0$ sub-level (where $\mathbf{S} = \mathbf{S}_p + \mathbf{S}_\mu$ is the total spin of the muonic hydrogen atom $\mu^- p$) is an ambitious project of the FAMU collaboration [1]: the measurement of the resonance transition frequency will provide top-accuracy data on the Zemach radius of the proton[2]. This is a very weak M1 magnetic dipole transition with probability of only $2 \times 10^{-5}(E/\mathrm{J})(S/\mathrm{m}^2)^{-1}(\mathrm{T}/°K)^{-1/2}$ for laser pulse energy E, laser beam cross section S and target temperature T [3], for which an optical multi-pass cavity needs to be used to enhance the probability of laser-stimulated transitions to a reasonable level.

The measurement of the hyperfine splitting in the ground state of muonic hydrogen consists in the detailed study of the chain of processes occurring when muonic hydrogen atoms in a mixture of hydrogen and oxygen interact with laser radiation tuned at a frequency around the hyperfine transition resonance frequency [3]. The observable quantity used as signature is the time distribution of the events of muon transfer from hydrogen to oxygen [4] recognized by the emission of characteristic X-rays : the maximal deviation of the latter from the background time distribution in absence of laser indicates that the laser source is tuned at the resonance frequency.

The FAMU experiment will use the pulsed muon source of the RIKEN-RAL muon facility [5], a tunable pulsed mid-infrared laser in the 6.8 μm range [6], and a multi-pass cavity with mirrors of very high reflectance. The work on the optimization of the experimental set-up showed that in experiments with pulsed sources the time distribution of the laser energy in the multi-pass cavity is of primary importance. We present here a simplified model of the

*e-mail: dbakalov@inrne.bas.bg

phenomena in this case that allows to qualitatively determine the optimal parameters of the cavity in dependence of the laser pulse parameters. Particular attention is paid to the optimal choice of the measurement time gate, i.e. the time interval in which the time distributions of the characteristic X-rays with and without laser are to be observed.

2 Modelling the time profile of the electromagnetic fields in a multi-pass cavity

2.1 Preliminary remarks

The probability dp for the excitation of the ortho hyperfine ground state of $\mu^- p$ with a oscillating magnetic field of resonance frequency ν_0 in the time interval dt such that $\nu_0\, dt \gg 1$, under the condition that the width of the laser line is smaller that the Doppler broadening of the hyperfine transition line, can be put in the form:

$$dp = K|\mathbf{B}(\mathbf{r})|^2\, dt \tag{1}$$

where $\mathbf{B}(r)$ is the amplitude of the magnetic field carried by the laser plane wave at the position $\mathbf{r}$ of the $\mu^- p$ atom, and the value of the dimension coefficient K is expressed in terms of the masses and magnetic moments of the proton and the muon, the temperature and ν_0 [3]

$$K = \sqrt{\frac{(m_p + m_\mu)c^2}{32\pi k T \nu_0^2}} \left(\frac{\mu_B}{\hbar}\left(\frac{m_e}{m_p}\mu_p + \frac{m_e}{m_\mu}\mu_\mu\right)\right)^2 \tag{2}$$

Accordingly, the probability dP that a para-to-ortho transition will occur in the elemental volume $d^3\mathbf{r}$ (i.e. the expected number of laser-stimulated spin-flip events) is

$$dP = C(\mathbf{r})\, K|\mathbf{B}(\mathbf{r})|^2\, dt\, d^3\mathbf{r}, \tag{3}$$

where $C(\mathbf{r})$ is the number density of $\mu^- p$ atoms. At distances of the order of the wave length $\lambda_0 = c/\nu_0$, the amplitude $\mathbf{B}(\mathbf{r})$ itself varies much faster than the atomic density $C(\mathbf{r})$, so we consider the probability $d\bar{P}$, obtained from dP by averaging over a volume ΔV such that $\Delta V \gg \lambda_0^3$:

$$d\bar{P}/dt = \frac{1}{\Delta V}\int\limits_{\Delta V} \frac{dP}{dt}\, d^3\mathbf{r} = \frac{K}{\Delta V}\int\limits_{\Delta V} d^3\mathbf{r}\, C(\mathbf{r})\, |\mathbf{B}(\mathbf{r})|^2 \approx K\, C(\mathbf{r}) \frac{1}{\Delta V}\int\limits_{\Delta V} d^3\mathbf{r}\, |\mathbf{B}(\mathbf{r})|^2 = K\, C(\mathbf{r})\, \overline{|\mathbf{B}(\mathbf{r})|^2}, \tag{4}$$

where $\overline{|\mathbf{B}(\mathbf{r})|^2}$ is the amplitude of the magnetic field averaged at the wavelength scale. Denote by $N(t_1, t_2; V)$ the number of laser-stimulated spin-flip events that occur in the volume V during the measurement time interval $[t_1, t_2]$:

$$N(t_1, t_2; V) = \int\limits_{t_1}^{t_2} dt \int\limits_{V} \frac{d\bar{P}}{dt}\, d^3\mathbf{r} = K \int\limits_{t_1}^{t_2} dt \int\limits_{V} C(\mathbf{r}) \overline{|\mathbf{B}(\mathbf{r})|^2}\, d^3\mathbf{r}, \tag{5}$$

and by $N(t_1, t_2)$ – the number of events in the whole cavity (of which, in the general case, only part of which is irradiated by the laser source). For uniform muonic atom density $C(\mathbf{r}) = C_0$ this yields

$$N(t_1, t_2; V) = K\, C_0 \int\limits_{t_1}^{t_2} dt \int\limits_{V} \overline{|\mathbf{B}(\mathbf{r})|^2}\, d^3\mathbf{r}. \tag{6}$$

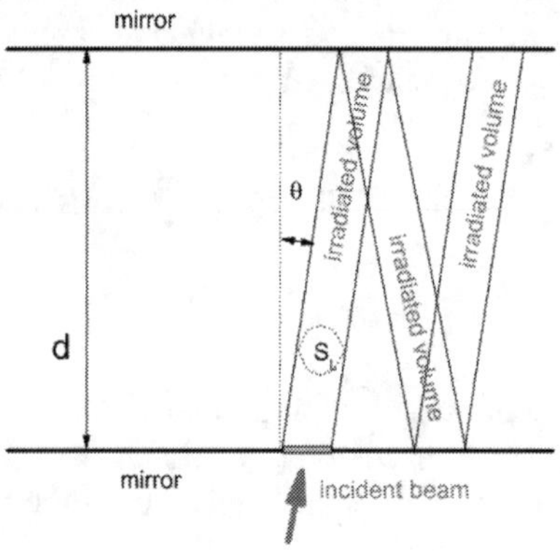

Figure 1. A schematic representation of the laser beam propagation in the optical multi-pass cavity. The distance between the two flat parallel mirrors is d; the beam incidence angle is ϑ, and its cross section is s_L.

In a multi-pass cavity the oscillating magnetic field $\mathbf{B}(\mathbf{r},t)$ at $\mathbf{r}$ is the *incoherent sum* of the magnetic fields carried by the multiple reflected laser beams that irradiate the point $\mathbf{r}$:

$$\mathbf{B}(\mathbf{r},t) = \sum_n \mathbf{B}_n \exp i(\mathbf{k}_n.\mathbf{r} - \omega_0 t + \delta_n) + c.c., \tag{7}$$

where $\mathbf{B}_n$, δ_n and $\mathbf{k}_n$ are the amplitude, phase and wave vector of the n-th plane wave, all of them with the same $\omega_0 = 2\pi\nu_0$ and $|\mathbf{k}_n| = k$. It can be shown for two summands that $\overline{|\mathbf{B}(\mathbf{r})|^2} = |\mathbf{B}_1|^2 + |\mathbf{B}_2|^2 + O((\theta k \Delta V)^{-1/3})$ (where θ is the angle between $\mathbf{k}_1$ and $\mathbf{k}_2$, $\mathbf{k}_1.\mathbf{k}_2 = k^2\cos\theta$, and ΔV is the smoothing volume of Eq.(4)); the latter is directly extended to any number of plane waves crossing at non-zero angles:

$$\overline{|\mathbf{B}(\mathbf{r})|^2} \approx \sum_n |\mathbf{B}_n|^2 \tag{8}$$

2.2 A simple mathematical model of the multi-pass cavity

We shall consider a simplified model of the multi-pass cavity aimed at qualitatively reproducing only the *time* dependence of $N(t_1,t_2;V)$. The evaluation of the *spatial* distribution of the spin-flip events requires specific simulation codes and the detailed knowledge of the $\mu^- p$-atom number density spatial distribution that are out of the scope of this talk; accordingly, we adopt the approximation of uniform number density of the muonic atoms, assume V to be the *whole* cavity volume $\bar{V}$, and consider only the time distribution of $N(t_1,t_2) \equiv N(t_1,t_2,\bar{V})$.

We assume that the cavity consists of two parallel mirrors at distance d. The laser beam with cross section s_L is injected in the cavity at incidence angle α (see Fig. 1) and is multiply reflected by the mirrors of reflectance R. The details of the curved mirror edges that force the beam not to leave the cavity volume are not considered at all. The irradiated volume $V^* \in V$ may be thought to consists of cylindric segments with base surface s_L and height $d/\cos\alpha \approx d$ for incidence angles $\alpha \ll \pi/2$; the time light travels through a segment is approximately $\tau_d = d/c$. We assume that the target gas is perfectly transparent for the laser light, so that $\overline{|\mathbf{B}(\mathbf{r})|^2}$ does not vary along a segment (except for the zone of overlap of two segments – an effect that can be accounted for at a later stage). At each reflection the value of $\overline{|\mathbf{B}(\mathbf{r})|^2}$ is suppressed by the factor R.

Let the laser pulse of duration τ_L enter the cavity at time $t_0 = 0$, and the the value of the averaged magnetic field at entrance be $\overline{|\mathbf{B}(\mathbf{r})|^2} = B_0^2$. Between $t = t_0$ and $t = \tau_L$ the irradiated volume V, as well as the number of events $N(t_0, t; V^*)$ increase monotonously: at each τ_d a new segment is irradiated. The value of $N(t_0, t; V)$ in this time interval is approximately

$$N(t_0, t) = \alpha \left(\frac{c}{1-R} t - d \frac{R(1 - R^{tc/d})}{(1-R)^2} \right), \quad 0 \le t \le \tau_L, \tag{9}$$

where c is the speed of light and, for simplicity, we introduced the notation $\alpha = K\, C_0\, B_0^2\, s_L$. After the end of the laser pulse, for $t \ge \tau_L$, the irradiated spot propagates through the cavity but its volume V remains unchanged, and after each reflection, i.e. at every τ_d, the magnetic field in the segments is suppressed by the factor R, that leads to

$$N(\tau_L, t) = \alpha \frac{d}{(1-R)^2} \left(1 - R^{c\tau_L/d}\right)\left(1 - R^{c(t-\tau_L)/d}\right), \quad t \ge \tau_L. \tag{10}$$

From here $N(0, t) = N(0, \tau_L) + N(\tau_L, t)$. For large measurement times $t \gg \tau_d$

$$\lim_{t\to\infty} N(0, t) = \frac{\alpha}{1-R} c\tau_L + \frac{\alpha}{1-R} d \left(1 - R^{\tau_L/\tau_d}\right) \gtrsim \frac{\alpha}{1-R} c\tau_L. \tag{11}$$

Without a multi-pass cavity, the number of laser-stimulated spin-flip events at the same physical conditions would be $N_0 = K\, C_0\, B_0^2\, s_l\, c', \tau_L = \alpha\, c\tau_L$. The "amplification effect" of the multi-pass cavity is therefore described with the factor

$$A(t) = N(0, t)/N_0 = (1-R)^{-1} \times \left(1 + \frac{\tau_d}{\tau_L}\left(1 - R^{\tau_L/\tau_d}\right)\left(1 - (1-R)^{-1} R^{(t-\tau_L)/\tau_d}\right)\right), \quad t \ge \tau_L, \tag{12}$$

the whole dependence on t being hidden in the rightmost factor. Fig. 2 is a plot of $A(t)$ for the typical values of the parameters $\tau_L = 20$ ns, $R = 0.998$ and $R = 0.999$, and the set of values $d = 5(5)20$ cm.

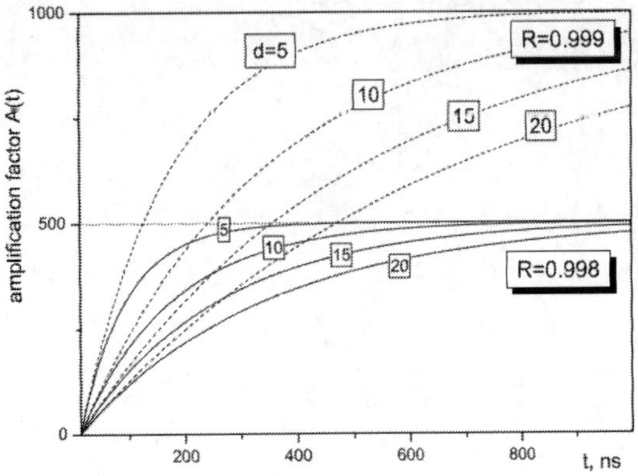

Figure 2. Amplification factor $A(t)$ of the number of laser-stimulated spin-flip events in a multi-pass cavity as function of the measurement time t, for laser pulse length $\tau_L = 20$ ns, mirror reflectance $R = 0.998$ and $R = 0.999$, and distance between the mirrors $d = 5(5)20$ cm.

3 Optimization of the measurement time gate

Eq. (11) gives the upper limit of the amplification factor: $A(t) \simeq (1-R)^{-1}$; Fig. 2 shows that the time needed to reach the maximal amplification grows fast with the distance d between mirrors and with the mirror reflectivity R. Knowing the explicit dependence of the number of spin-flip events on the measurement time t may therefore help select the optimal measurement time gate $[0, t]$ for which the number of laser-stimulated spin-flip events is maximal. In such an optimization problem one should, of course, take into account the finite muon decay rate λ_0 as well. As shown of Fig. 3, the muon decay partly suppresses the gain of laser-stimulated spin-flip events due to the multi-pass cavity.

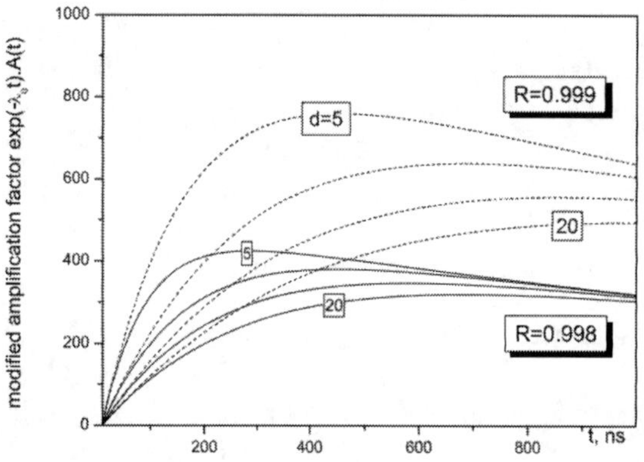

Figure 3. Modified amplification factor $\exp(-\lambda_0 t)A(t)$ of the number of laser-stimulated spin-flip events in a multi-pass cavity as function of the measurement time t, for laser pulse length $\tau_L = 20$ ns, mirror reflectance $R = 0.998$ and $R = 0.999$, and distance between the mirrors $d = 5(5)20$ cm.; $\lambda_0 = 0.45\mu s^{-1}$ is the muon decay rate.

To optimize the efficiency of the FAMU experiment, however, one should maximize the *signal-to-noise ratio* Δ/σ rather than just the number of laser-stimulated spin-flip events. In the experimental method of FAMU, the "noise" σ (also referred to as "background") mainly consists of characteristic X-rays emitted during the relaxation of the $\mu^- O$ atoms formed through muon transfer from *thermalized* muonic hydrogen atoms (and therefore independent of the laser light), while the "signal" Δ is the variation of the the background distribution due to X-rays from $\mu^- O$ atoms formed by muon transfer from *epithermal* $\mu^- H$ [3, 7]. Obviously, the latter is proportional to the number of laser-stimulated events, while the "noise" depends only on the number of muonic atoms in target:

$$\Delta \sim \exp(-\lambda_0 t)A(t), \ \sigma^2 \sim \int_0^t \exp(-\lambda_0 t')\, dt' = (1 - \exp(-\lambda_0 t))/\lambda_0, \tag{13}$$

so that the time dependence $S(t)$ of the signal-to-noise ratio is:

$$S(t) = \frac{\Delta}{\sigma} \sim \frac{\exp(-\lambda_0 t)}{\sqrt{1 - \exp(-\lambda_0 t))}} A(t). \tag{14}$$

The noise σ grows monotonously with the duration t of the measurement time gate while the signal Δ reaches its maximum at some finite moment (see Fig. 3). We can therefore expect $S(t)$ to have pronounced maxima for any value of d and R, that will determine the optimal time gate for measurement of the hyperfine splitting of muonic hydrogen in the FAMU approach.

Fig. 4 confirms these qualitative predictions. We also note that:

- The maximal achievable signal-to-noise ratio is increased as the distance d between the mirrors decreases;
- The optimal time gate length t grows with inter-mirror distances d;.

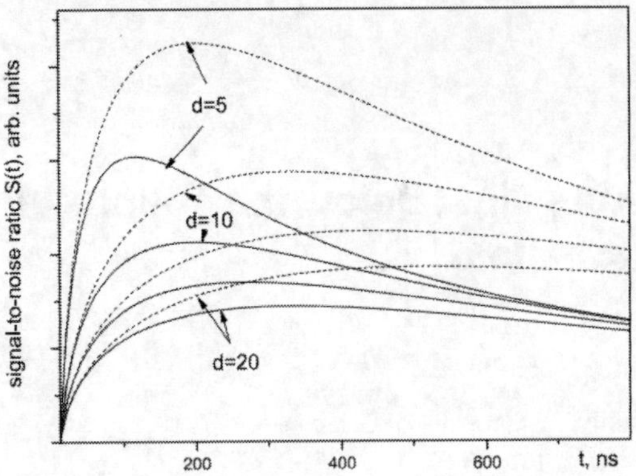

Figure 4. Dependence of the signal-to-noise ratio $S(t) = \Delta/\sigma$ (in arbitrary units) on the measurement time gate t, for a set of inter-mirror distances $d = 5(5)20$ cm, mirror reflectivity $R = 0.998$ (solid lines) or $R = 0.999$ (dashed lines), and laser pulse length τ_L=20 ns.

4 Conclusions

The simplified model of the optical multi-pass cavity considered here completely neglects the effects of the spatial distribution of the muonic hydrogen atoms and of the laser radiation throughout the cavity, as well as some other details of the undergoing physical processes, and cannot be directly applied in simulations of the FAMU experiment. Still, it reveals important peculiarities related to the use of multi-pass cavities in experiments with pulsed laser and muon sources, and can be helpful in the preliminary design of the experimental set-up and the approximate choice of the optimal measurement time gate and inter-mirror distance.

5 Acknowledgments

The authors acknowledge the support of Grant 08-17 of the Bulgarian Science Fund.

References

[1] `https://webint.ts.infn.it/ricerca/exp/famu/`

[2] A. Adamczak, G. Baccolo, D. Bakalov, *et al.*, JINST 11, P05007 (2016).

[3] A. Adamczak, *et al.*, Nucl. Instrum. Meth. **B 281**, 72 (2012).

[4] E. Mocchiutti, V. Bonvicini, R. Carbone, *et al.*, to appear in JINST (2017).

[5] T. Matsuzaki, *et al.*, Nucl. Instr. Meth. A 465 (2001) 365.

[6] L.Stoychev, M. Danailov, A. Demidovich, *et al.*, Proc. SPIE 9135, Laser Sources and Applications II, 91350J (May 1, 2014); doi:10.1117/12.2052110

[7] A. Adamczak, D. Bakalov, K. Bakalova, E. Polacco, C. Rizzo, Hyperfine Interactions **136**, 1 (2001)

Three-body resonance states just below the antiproton and hydrogen dissociation threshold

Takuma Yamashita[1,*] and *Yasushi* Kino[1,**]

[1]Department of Chemistry, Tohoku University, Sendai 980-8578, Japan

Abstract. We analyze two shallow resonance states below the antiproton hydrogen dissociation threshold with a non-adiabatic three-body calculation. Rearrangement correlation between initial channel and protonium formation channel is explicitly included in the total wavefunction. The lower resonance state is in good agreement with the resonance position and width calculated with the R-matrix theory. The higher resonance state which is newly found is closer to the threshold and much narrower than the former resonance. A polarization effect of the hydrogen atom is found to be indispensable to support the resonance state. The accuracy of the present calculation is evaluated by the extended virial theorem. The resonance states calculated in the present work gives shallower relative energy below the dissociation threshold than the Born-Oppenheimer calculation, suggesting that the electron motion which is ignored in latter calculation would give positive energy because the electron is unbound inside the distance.

1 Introduction

An antiproton ($\overline{\rm p}$) scattering off a hydrogen atom has attracted attentions as a fundamental case of the matter-antimatter reaction [1–5] supported by the antiproton beam experiments [6–8]. Recent experimental progress in the strict test for the CPT symmetry using antiprotons and antihydrogen atoms ($\overline{\rm H}$) is remarkable [9–14]. One of the significant basements for these experiments is the synthesis, confinement and manipulation of cold antiprotons and antihydrogen atoms [15–19]. The cold sources of the antiproton and antihydrogen atom can open a vast new scientific field stimulated by interactions between matter and antimatter. Understanding of the fundamental cold antiproton/antihydrogen reaction with atoms is, therefore, of importance in particle, nuclear and atomic physic and quantum chemistry.

There have been a basic question about formation of particle-antiparticle complex, such as $\overline{\rm H}$H and $\overline{\rm p}$H, in particle-antiparticle collision. Theoretical studies of the four-body compound $\overline{\rm H}$H based on the Born-Oppenheimer (BO) approximation [20–22] are pioneering works for this question; however, the BO approximation involves a fundamental difficulty near the critical distance. When the motion of the antiproton and proton in $\overline{\rm H}$H is separated from that of the positron and electron, a dipole moment of the antiproton-proton system decreases as a decrease of the size of the system. The critical distance ρ_0 where the antiproton-proton dipole can bind the electron is reported to be $\rho_0 = 0.639$ a.u. [23, 24] and for the two leptons,

*e-mail: t.ymst@dc.tohoku.ac.jp
**e-mail: y.k@m.tohoku.ac.jp

electron and positron, $\rho_0 = 0.7427$ a.u. [25, 26]. The BO potential vanishes inside the critical distance; namely, the electron motion is ignored when the distance between the two nuclei is smaller than the critical distance. Since the BO states do not include continuum states of the electron or Ps, they are considered to be bound states in the BO approximation. The interaction between the Pn and e^-/Ps, however, cannot be neglected; the complexes $\bar{\text{p}}$H and $\bar{\text{H}}$H should be treated as resonance states whose decay mode is electron/positronium detachment with Pn formation, Pn + e^-/Ps. The Pn is in the highly excited state having principal quantum number of $n > 30$ because a ground state of the Pn is located more than 1 keV lower than the resonance state.

The identification of the near-threshold resonance states of $\bar{\text{p}}$H (and its charge-conjugated system p$\bar{\text{H}}$) and $\bar{\text{H}}$H close to the thresholds $\bar{\text{p}}$+H (p+$\bar{\text{H}}$) and $\bar{\text{H}}$+H is also a serious concern in the realistic cold scattering [27–31]. A recent theoretical investigation for $\bar{\text{p}}$H was shown in a R-matrix scattering calculation [32] of Pn + e^- which reported several resonance states below the $\bar{\text{p}}$ + H threshold. A non-adiabatic calculation of $\bar{\text{H}}$H was performed with complex/real scaling method that includes the explicit description of Pn + Ps states [33].

In this work, we perform the non-adiabatic three-body calculation for S-wave resonance states of $\bar{\text{p}}$H located just below the $\bar{\text{p}}$+H threshold by $\sim$ 1 meV. Since the dominant interaction between the antiproton and hydrogen atom is characterized by the induced dipole moment caused by the electric field of the antiproton that is in proportion to r^{-4} where r is the relative distance between the antiproton and hydrogen atom, the three-body system may have bound (resonance) states below the $\bar{\text{p}}$ + H dissociation threshold. In order to reveal these resonance states, the complex scaling method (CSM) is used to determine the resonance positions and widths.

Atomic units (a.u.; $m_e = \hbar = e = 1$) are used throughout this paper, except where mentioned otherwise.

2 Theory

The total Hamiltonian for $\bar{\text{p}}$H is written as

$$H = \hat{K}_{\bar{\text{p}}} + \hat{K}_{\text{p}} + \hat{K}_{\text{e}^-} - \hat{K}_{\text{CM}} - \frac{1}{r_{\text{pe}^-}} - \frac{1}{r_{\bar{\text{p}}\text{p}}} + \frac{1}{r_{\bar{\text{p}}\text{e}^-}}, \tag{1}$$

where $\hat{K}_x$ (x = {$\bar{\text{p}}$, p, e^-}) is a kinetic energy operator associated with point vector of particle x, $\hat{K}_{\text{CM}}$ is the kinetic energy operator for center of mass of the system and r_{xy} is relative distance between particle x and y. The proton(antiproton) masses are set to be $m_{\text{p}}(m_{\bar{\text{p}}})$ = 1836.152 673 89.

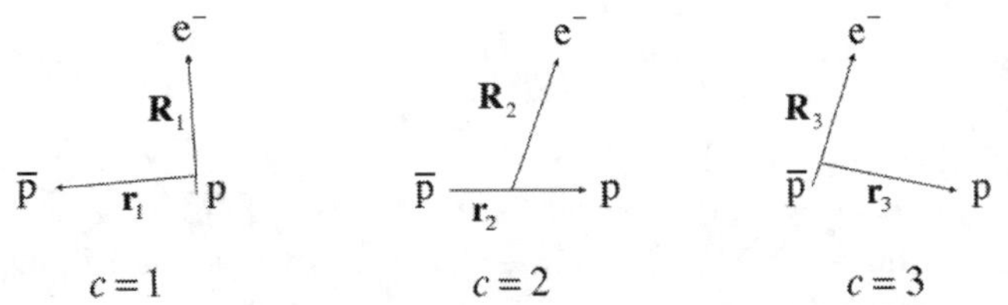

Figure 1. Three sets of Jacobian coordinates for $\bar{\text{p}}$H.

In order to describe the all inter-particle correlation of the near-threshold resonance states precisely, we adopt a Gaussian expansion method (GEM) [34]. Recently the GEM approach successfully revealed the near-threshold resonances in positronic atom systems [35–37]. Besides, this approach can be applied to the four-body system $\overline{\mathrm{H}}\mathrm{H}$ in the same theoretical framework [33, 38]. The three-body wavefunction for vibrational quantum number υ and total angular momentum J, $\Psi_{\upsilon J}$, consists of a summation of channel functions,

$$\Psi_{\upsilon J} = \sum_{cl} \psi_{cl}(\mathbf{r}_c, \mathbf{R}_c). \tag{2}$$

where $(\mathbf{r}_c, \mathbf{R}_c)$ are sets of Jacobian coordinates shown in Fig. 1. The first channel $c = 1$ is suitable to describe the configuration $\overline{\mathrm{p}} + \mathrm{H}$ in which the electron and proton interact with each other, and the $\overline{\mathrm{p}}$ interacts with the center of mass of the electron and proton. The second one $c = 2$ is suitable to describe the structure in which the electron interacts with the highly excited protonium, $\mathrm{Pn} + \mathrm{e}^-$. The third channel $c = 3$ explicitly introduces the repulsive correlation between the antiproton and electron. The S-wave ($J = 0$) channel wavefunction having an angular momentum of l on the c-th Jacobian coordinate is expanded in terms of Gaussian basis function with sine and cosine functions,

$$\psi_{cl}(\mathbf{r}_c, \mathbf{R}_c) = \sum_{j=1}^{N_{cl}} R_c^l \exp\left(-\frac{R_c^2}{\beta_{cl,j}^2}\right) \sum_{i=1}^{n_{cl}} \left[A_{ij} r_c^l \exp\left(-\frac{r_c^2}{\alpha_{cl,i}^2}\right) \cos\left(\gamma \frac{r_c^2}{\alpha_{cl,i}^2}\right) \right.$$
$$\left. + B_{ij} r_c^l \exp\left(-\frac{r_c^2}{\alpha_{cl,i}^2}\right) \sin\left(\gamma \frac{r_c^2}{\alpha_{cl,i}^2}\right) \right] \sum_m (-1)^m Y_{lm}(\hat{\mathbf{r}}_c) Y_{lm}(\hat{\mathbf{R}}_c). \tag{3}$$

The non-linear parameters $\alpha_{cl,i}$ and $\beta_{cl,j}$ are given according to the geometrical progression to describe both short-range correlation and long-range tail behavior. The linear coefficients A_{ij} and B_{ij} are determined by the variational principle for eigenenergies,

$$\langle \Psi_{\upsilon 0} \mid H \mid \Psi_{\upsilon' 0} \rangle = E_{\upsilon 0} \delta_{\upsilon \upsilon'}. \tag{4}$$

Since the resonance states concerned here are accosted with highly excited Pn whose number of radial nodes are more than 30, radial wave functions for r_c should have many nodes. Thus, a fragment of the channel function $c = 2$,

$$\phi_{nl}(\mathbf{r}_2) = \sum_{i=1}^{n_{2l}} \left[a_i r_2^l \exp\left(-\frac{r_2^2}{\alpha_{2l,i}^2}\right) \cos\left(\gamma \frac{r_2^2}{\alpha_{2l,i}^2}\right) + b_i r_2^l \exp\left(-\frac{r_2^2}{\alpha_{2l,i}^2}\right) \sin\left(\gamma \frac{r_2^2}{\alpha_{2l,i}^2}\right) \right] Y_{lm}(\hat{\mathbf{r}}_2), \tag{5}$$

should have an ability to reproduce highly excited eigenstates of Pn. The Schrödinger equation for Pn is written as

$$\left(\hat{K}_{\mathbf{r}_2} - \frac{1}{r_2}\right) \phi_{nl} = \epsilon_n \phi_{nl}, \tag{6}$$

where $\hat{K}_{\mathbf{r}_2}$ is a kinetic energy operator associated with $\mathbf{r}_2$ and n is the principal quantum number of Pn. We show the ϵ_n convergences for Pn (ns) states in Fig. 2 against the number of total Gaussian basis functions $2n_{2l}$. One can see that the sufficient number of basis functions $2n_{2l} \geq 160$ provides proper eigenenergies of the highly excited Pn ($n \geq 30$).

Resonance energies and widths are determined by the help of complex scaling method [39]. A complex dilation operator is defined by

$$U(\eta, \theta) f(\mathbf{r}) = \eta^{\frac{3}{2}} e^{\frac{3}{2} i\theta} f(\eta e^{i\theta} \mathbf{r}), \tag{7}$$

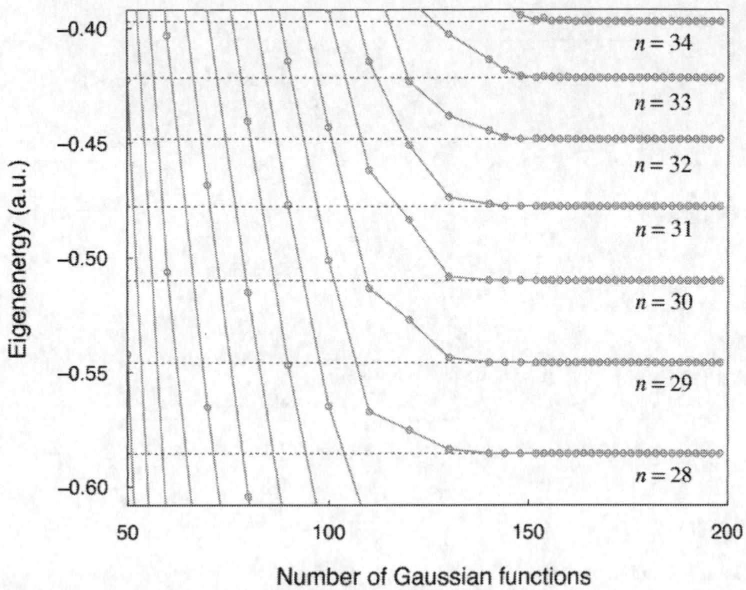

Figure 2. Eigenenergy convergences for Pn (*ns*) in Eq. (6) are shown against the number of Gaussian functions $2n_{2l}$ (solid line with points). The dashed lines are the exact eigenenergies.

where $f(\mathbf{r})$ is an arbitrary function described by a coordinate $\mathbf{r}$, and η and θ are real factors for scaling. A complex scaled Schrödinger equation is given by

$$U(\eta, \theta)HU(\eta, \theta)^{-1}\Psi_0(\eta, \theta) = E(\eta, \theta)\Psi_0(\eta, \theta), \tag{8}$$

where $E(\eta, \theta)$ is a complex eigenenergy and $\Psi_0(\eta, \theta) = U(\eta, \theta)\Psi_0$ is a complex-scaled total wavefunction. The resonance energy and width can be determined by the complex eigenvalue $E(\eta, \theta') = E_r - i\Gamma/2$ when

$$\left.\frac{\mathrm{d}E(\eta, \theta)}{\mathrm{d}\theta}\right|_{\theta=\theta'} \approx 0 \tag{9}$$

is satisfied for any η.

3 Results and discussion

We find two S-wave resonance states just below the $\bar{\mathrm{p}}$ + H(1s) dissociation threshold. Hereafter, we distinguish the resonance states as S_{-1} and S_{-2} with energies $E_{-1} = -0.499\,728\,85(4) - i0.000\,000\,77(2)$ and $E_{-2} = -0.499\,775(1) - i\,0.000\,022(2)$, respectively, where numbers in the parentheses indicate the numerical error of the last digit.

With a trial function (noted as $\Psi_0^{(a)}$) consisting of three channel functions,

$$\Psi_0^{(a)} = \psi_{10} + \psi_{11} + \psi_{20}, \tag{10}$$

we calculate complex eigenenergies $E(\eta, \theta)$ for $0 \leq \theta \leq 0.2$ by interval $\Delta\theta = 0.01$. We find a resonance state corresponding to S_{-1} state just below the $\bar{\mathrm{p}}$ + H(1s) threshold, $E_{\mathrm{th}} = -0.499\,727\,84$. Series of complex eigenenergies for resonance state S_{-1} calculated with a

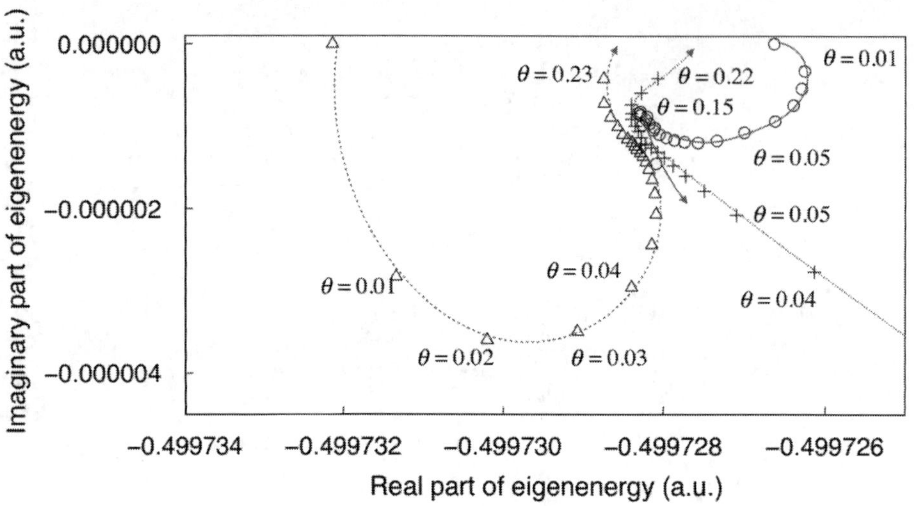

Figure 3. Complex eigenenergies ($E(\eta, \theta)$) are shown (circle: $\eta = 1.0$, cross: $\eta = 1.02$, triangle: $\eta = 1.04$). The regular interval of complex scaling parameter θ is $\Delta\theta = 0.01$. Complex energy for resonance state S_{-1} is found to be (−0.499 728 36(3), 0.000 001 7(4)).

Table 1. Resonance energy and width for S_1 state calculated with trial functions consisting of different set of channel functions are shown together. The total number of basis functions is given by the sum of channel functions $N_{\rm tot} = \sum_{cl} 2N_{cl}n_{cl}$.

$N_{\rm tot}$	Ψ	E_r	Γ
7200	$\psi_{10} + \psi_{20}$	–	–
10800	$\psi_{10} + \psi_{20} + \psi_{30}$	–	–
10800	$\psi_{10} + \psi_{11} + \psi_{20}$	−0.499 728 36(3)	0.000 001 7(4)
14400	$\psi_{10} + \psi_{11} + \psi_{20} + \psi_{30}$	−0.499 728 85(4)	0.000 001 55(4)

trial function $\Psi_0^{(a)}$ are shown in Fig. 3. Three series with different η's pass around a stationary point which corresponds to a complex energy of the resonance, as increasing θ.

The resonance state S_{-1} is examined by several different trial functions shown in Table 1. When we construct the trial function $\Psi_0^{(b)}$ without the channel function ψ_{11}, namely,

$$\Psi_0^{(b)} = \psi_{10} + \psi_{20}, \tag{11}$$

the resonance state is not found in the same energy region. The channel function ψ_{10} gives the configuration in which the relative motion between the antiproton and the center of mass of the hydrogen atom has angular momentum $l = 0$, and the ψ_{20} gives the configuration in which the relative motion between the electron and the center of mass of the protonium has angular momentum $l = 0$. Since the channel function ψ_{11} plays a role to explicitly describe the excitation of the hydrogen atom by the electric field of the antiproton, this resonance state is expected to be supported by the induced polarization attraction. If we replace ψ_{11} in the trial function $\Psi_0^{(a)}$ by ψ_{30} which can introduce an explicit description between repelling

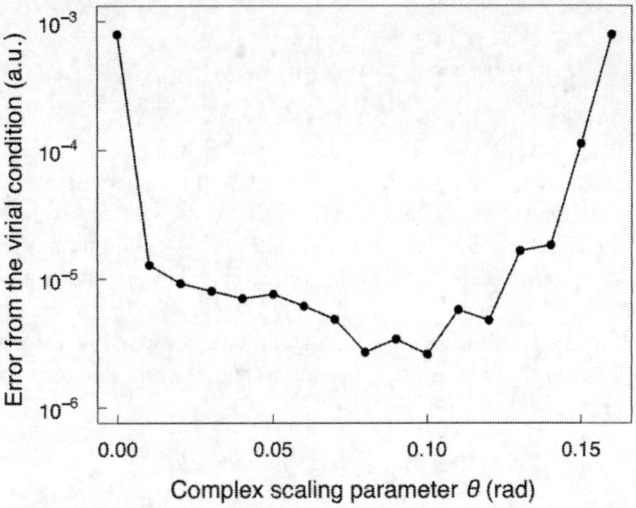

Figure 4. The deviation from the virial condition | $2\langle K\rangle^{(\eta,\theta)} + \langle V\rangle^{(\eta,\theta)}$ | (circles with a solid line) is shown against the complex scaling parameter θ for the resonance state S_{-1} with the trial function in Eq. (13).

particles, antiproton and electron,

$$\Psi_0^{(c)} = \psi_{10} + \psi_{20} + \psi_{30}, \tag{12}$$

gives no resonance state corresponding to the S_{-1} resonance state. Therefore, the role of polarization of hydrogen atom introduced by ψ_{11} is indispensable to support the resonance state. The most precise result is obtained by the four-channel trial function $\Psi_0^{(d)}$ given as

$$\Psi_0^{(d)} = \psi_{10} + \psi_{11} + \psi_{20} + \psi_{30}. \tag{13}$$

The resonance energy and width of the resonance state S_{-1} is determined to be $E_r = -0.499\,728\,85(4)$ and $\Gamma = 0.000\,001\,55(4)$. The energy relative to the $\bar{p} + H(1s)$ threshold is $E_{th} - E_r = 0.000\,001\,01(4)$. Since the resonance state S_{-1} is located below the $\bar{p} + H(1s)$ threshold, the resonance width is associated with the electron detachment decay with Pn formation Pn + e^-. The lifetime τ is given as $\tau = \hbar/\Gamma = 16$ ps.

The accuracy of the resonance state S_{-1} is examined by an extended virial theorem. Since the all inter-particle interactions in the Hamiltonian Eq. (1) are Coulombic interaction, the virial condition $2\langle K\rangle + \langle V\rangle = 0$ is satisfied, where K and V are kinetic and potential energy operators, respectively. For resonance states in the complex scaling treatment, the extended virial theorem gives a similar virial condition around the resonance state [40], namely,

$$2\langle K\rangle^{(\eta,\theta)} + \langle V\rangle^{(\eta,\theta)} = 0, \tag{14}$$

where the symbols $\langle K\rangle^{(\eta,\theta)}$ and $\langle V\rangle^{(\eta,\theta)}$ are defined as

$$\langle K\rangle^{(\eta,\theta)} = \int d\tau \Psi(\eta,\theta) U(\eta,\theta) K \Psi(\eta,\theta), \tag{15}$$

and

$$\langle V\rangle^{(\eta,\theta)} = \int d\tau \Psi(\eta,\theta) U(\eta,\theta) V \Psi(\eta,\theta), \tag{16}$$

where $d\tau = d\mathbf{r}_1 d\mathbf{R}_1 = d\mathbf{r}_2 d\mathbf{R}_2 = d\mathbf{r}_3 d\mathbf{R}_3$. Deviation of the $2\langle K\rangle^{(\eta,\theta)} + \langle V\rangle^{(\eta,\theta)}$ from zero can be used to estimate an accuracy of the calculated complex eigenenergy $\langle K\rangle^{(\eta,\theta)} + \langle V\rangle^{(\eta,\theta)}$. In Fig. 4, the expectation values calculated with $\Psi_0^{(d)}$ are shown against the scaling angle θ. One can see that the expectation value $\mid 2\langle K\rangle^{(\eta,\theta)} + \langle V\rangle^{(\eta,\theta)} \mid$ has minimum value at $\theta = 0.10$ indicating the most accurate point in the series of the calculation. The relative accuracy can be estimated by $\mid 2\langle K\rangle^{(\eta,\theta)} + \langle V\rangle^{(\eta,\theta)} \mid / \mid E_{-1} \mid \sim 5.9 \times 10^{-6}$.

For the further analysis of the resonance state S_{-1}, we compare the resonance states with the BO vibrational states. In the framework of the BO approximation, the total wavefunction of $\bar{p}$H molecule is expressed as

$$\Psi_0^{BO} = \phi_{lep}(r_2; \mathbf{r}_1)\phi_{nuc}(r_2)Y_{00}(\hat{\mathbf{r}}_1), \tag{17}$$

where $\mathbf{r}_1$ and $\mathbf{r}_2$ are defined by the channel $c = 1$ and $c = 2$. The leptonic wavefunction $\phi_{lep}(r_2; \mathbf{r}_1)$ provides the BO energy $\epsilon_{BO}(r_2)$ [23]. An effective potential for the nuclear wavefunction $\phi_{nuc}(r_2)$ is given by using $\epsilon_{BO}(r_2)$,

$$V_{BO}(r_2) = \begin{cases} -\frac{1}{r_2} & (r_2 < R_{FT}) \\ \epsilon_{BO}(r_2) - \frac{1}{r_2} & (R_{FT} \le r_2) \end{cases} \tag{18}$$

where $R_{FT} \sim 0.639$ is the critical distance where $\epsilon_{BO}(r_2) = 0$. The asymptotic form of $V_{BO}(r_2)$ satisfies a boundary condition $V_{BO}(r_2 \to \infty) = -\alpha_d/2r_2^4 + \epsilon_H$ where $\alpha_d = 9/2$ is a polarizability of the hydrogen atom with infinite nuclear mass. The energies of BO states E_{BO} are given by the following equation

$$(K_{\mathbf{r}_2} + V_{BO}(r_2))\,\phi_v^{BO} = E_{BO}\phi_v^{BO}, \tag{19}$$

where v is the vibrational quantum number.

The potential $V_{BO}(r_2)$ provides a series of vibrationally excited states ($v = 1 - 39$, BO states) [32]. We compare the resonance energy and width with the BO and recent R-matrix calculations for Pn + e$^-$ scattering in Table 2. The resonance energy and the BO energy are given as, relative to the $\bar{p}$ + H threshold, $E_{th} = -0.499\,727\,8$ and $E_{th}^{BO} = -0.5$, respectively. The highest BO state $v = 39$ is located just below the dissociation threshold by 0.000 048 eV. The resonance state S_{-1} gives a similar but shallower energy 0.000 027 eV relative to the $\bar{p}$ + H threshold while the corresponding resonance state was not reported in the R-matrix calculation [32]. For the resonance state S_{-2} located below the S_{-1} state, complex scaling trajectories are shown in Fig. 5. The energy of the resonance state S_{-2} relative to the $\bar{p}$ + H threshold is 0.001 274 eV that is well consistent with the R-matrix calculation 0.001 276 eV and is shallower than the BO state $v = 38$.

Since the width of the resonance state S_{-1} is two orders of magnitude narrower than that of S_{-2}, the resonance state S_{-1} is much more stable than the S_{-2}. Generally, for the narrow resonance states, the CSM based on the bound state approximation may be more suitable than scattering calculations like R-matrix calculation. The half of the resonance width $\Gamma/2 = 0.000\,022$ eV for the S_{-1} resonance state is almost comparable to the resonance energy relative to the threshold, 0.000 027 eV. Besides, the resonance width of the S_{-2} resonance state corresponding to $v = 38$ is almost the half of the relative resonance energy by the $\bar{p}$ + H threshold. Therefore, the resonance state in scattering calculation near the threshold could

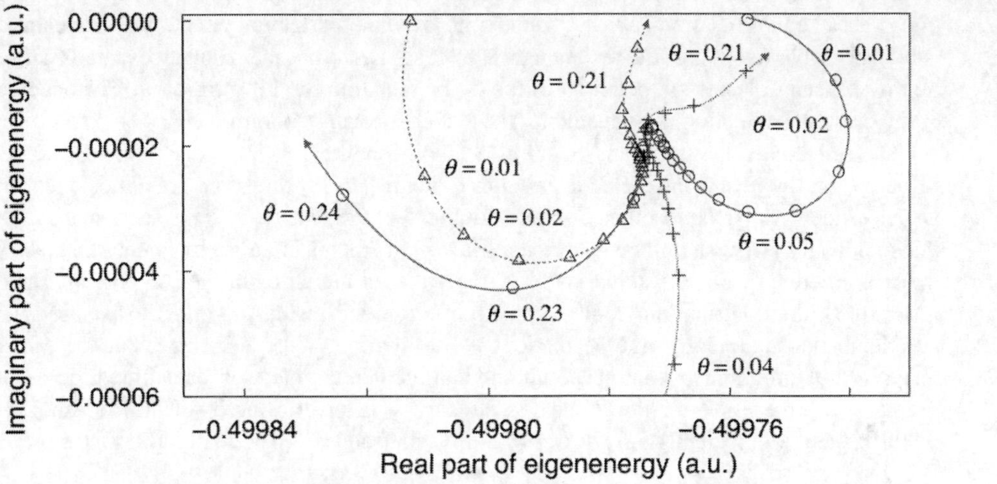

Figure 5. Complex eigenenergies ($E(\eta, \theta)$) for the S_{-2} resonance state are shown (circle: $\eta = 1.0$, cross: $\eta = 1.02$, triangle: $\eta = 1.04$). The regular interval of complex scaling parameter θ is $\Delta\theta = 0.01$. The complex energy of the S_{-2} resonance state is found to be (−0.499 775(1), −0.000 022(2)).

Table 2. Resonance positions and widths calculated with complex scaling method (CSM), Born-oppenheimer (BO) and R-matrix calculations are shown in electron-volt (eV). The positions are relative to the $\bar{p}$ + H threshold.

	This work	Literature [32]		
	CSM		BO	R-Matrix
E_{-1}	−0.000 027 (0.000 043)	$\upsilon = 39$	−0.000 048	–
E_{-2}	−0.001 274 (0.001 175)	$\upsilon = 38$	−0.001 616	−0.001 276 (0.001 05)

be influenced by the existence of the open dissociation channel $\bar{p}$ + H, and also by the S_{-2} resonance tail.

For the resonance states S_{-1} and S_{-2}, one can see that the non-adiabatic calculation gives shallower energy than the BO calculation, which implies that the adiabatic picture cannot be adoptable quantitatively to the interpretation of the compound $\bar{p}$H.

4 Conclusion

We non-adiabatically analyze the resonance states just below the $\bar{p}$ + H dissociation threshold with GEM and CSM. The total wavefunction is constructed by several channel functions defined in rearrangement Jacobian coordinates that describes both major particle correlations $\bar{p}$ + H and Pn + e^- explicitly. The highly excited states of Pn in the dissociation states are reproduced by a number of oscillation Gaussian functions.

We find two resonance states S_{-1} and S_{-2} below the $\bar{p}$ + H(1s) threshold. The energy and width of the lower resonance state S_{-2} are in good agreement with the recent R-matrix calculation [32], and the other resonance state S_{-1} is newly found. The resonance state S_{-1}

located 0.000 027 eV below the $\overline{\mathrm{p}}$ + H(1s) threshold is examined by several trial wavefunctions. The channel function having a role to introduce a polarization effect of hydrogen atom is indispensable to support the resonance state. It suggests that the resonance state is caused by the induced polarization potential of the hydrogen atom by the antiproton. The accuracy of the present calculation is evaluated by the extended virial theorem.

The resonance states S_{-1} and S_{-2} are located near the $\upsilon = 39$ and $\upsilon = 38$ BO states, respectively. However, since relative energies to the $\overline{\mathrm{p}}$ + H(1s) threshold are almost the same values as the widths, the resonance states should affect the scattering cross section above the threshold. This effect should be significant for experiments using ultra low energy $\overline{\mathrm{p}}$ because the cross section of $\overline{\mathrm{p}}$ and residual hydrogen atoms may increase. In the BO potential given in Eq. (18), the contribution from the electron is omitted inside the critical distance. The non-adiabatic calculation, which can take the contribution of the electron inside the critical distance to the total energy into account and can include the electron detachment decay associated with the Pn formation, improves the energy relative to the $\overline{\mathrm{p}}$ + H(1s) threshold by 0.000 021 a.u. (equivalently 6.6 K) for $\upsilon = 39$ BO state and 0.000 342 a.u. (108 K) for $\upsilon = 38$ BO state.

Acknowledgment

This work was financially supported by a Grant-in-Aid for Scientific Research from the Ministry of Education, Culture, Sports, Science and Technology (MEXT), Japan, a Kakenhi Grant-in-Aid from the Japan Society for the Promotion of Science (JSPS), and by the Division for Interdisciplinary Advanced Research and Education (DIARE), Tohoku University. The computation was conducted on the supercomputers at Kyushu University.

References

[1] A. Ermolaev, Phys. Lett. A **149**, 151 (1990)
[2] N. Toshima, Phys. Lett. A **175**, 133 (1993)
[3] J. Azuma, N. Toshima, K. Hino, A. Igarashi, Phys. Rev. A **64**, 062704 (2001)
[4] A.B. Voitkiv, J. Ullrich, Phys. Rev. A **67**, 062703 (2003)
[5] K. Sakimoto, Physica Scripta **2004**, 271 (2004)
[6] T. Walcher, Ann. Rev. Nucl. Part. Sci. **38**, 67 (1988)
[7] C. Amsler, F. Myhrer, Ann. Rev. Nucl. Part. Sci. **41**, 219 (1991)
[8] H. Knudsen, U. Mikkelsen, K. Paludan, K. Kirsebom, S.P. Møller, E. Uggerhøj, J. Slevin, M. Charlton, E. Morenzoni, Phys. Rev. Lett. **74**, 4627 (1995)
[9] S. Ulmer, C. Smorra, A. Mooser, K. Franke, H. Nagahama, G. Schneider, T. Higuchi, S.V. Gorp, K. Blaum, Y. Matsuda et al., Nature **524**, 196 (2005)
[10] M. Ahmadi, B.X.R. Alves, C.J. Baker, W. Bertsche, E. Butler, A. Capra, C. Carruth, C.L. Cesar, M. Charlton, S. Cohen et al., Nature **541**, 506 (2016)
[11] M. Ahmadi, B.X.R. Alves, C.J. Baker, W. Bertsche, E. Butler, A. Capra, C. Carruth, C.L. Cesar, M. Charlton, S. Cohen et al., Nature **548**, 66 (2017)
[12] M. Diermaier, C.B. Jepsen, B. Kolbinger, C. Malbrunot, O. Massiczek, C. Sauerzopf, M.C. Simon, J. Zmeskal, E. Widmann, Nat. Commun. **8**, 15749 (2017)
[13] H. Nagahama, C. Smorra, S. Sellner, J. Harrington, T. Higuchi, M. Borchert, T. Tanaka, M. Besirli, A. Mooser, G. Schneider et al., Nat. Commun. **8**, 14084 (2017)
[14] C. Smorra, A. Mooser, M. Besirli, M. Bohman, M. Borchert, J. Harrington, T. Higuchi, H. Nagahama, G. Schneider, S. Sellner et al., Phys. Lett. B **769**, 1 (2017)

[15] M. Amoretti, C. Amsler, G. Bonomi, A. Bouchta, P. Bowe, C. Carraro, C.L. Cesar, M. Charlton, M.J.T. Collier, M. Doser et al., Nature **419**, 456 (2002)
[16] T.A. Collaboration, Nat. Phys. **7**, 558 (2011)
[17] C. Smorra, K. Blaum, K. Franke, Y. Matsuda, A. Mooser, H. Nagahama, C. Ospelkaus, W. Quint, G. Schneider, S. Van Gorp et al., Hyperfine Interact. **228**, 31 (2014)
[18] C. Smorra, A. Mooser, K. Franke, H. Nagahama, G. Schneider, T. Higuchi, S. Gorp, K. Blaum, Y. Matsuda, W. Quint et al., Int. J. Mass Spectrom. **389**, 10 (2015)
[19] C. Smorra, K. Blaum, L. Bojtar, M. Borchert, K. Franke, T. Higuchi, N. Leefer, H. Nagahama, Y. Matsuda, A. Mooser et al., Eur. Phys. J. Special Topics **224**, 3055 (2015)
[20] B. Zygelman, A. Saenz, P. Froelich, S. Jonsell, A. Dalgarno, Phys. Rev. A **63**, 052722 (2001)
[21] P. Berggren, H. Stegeby, A. Voronin, P. Froelich, J. Phys. B: At. Mol. Opt. Phys. **41**, 155202 (2008)
[22] H. Stegeby, K. Piszczatowski, H. Karlsson, R. Lindh, P. Froelich, Cent. Eur. J. Phys. **10**, 1038 (2012)
[23] J.E. Turner, V.E. Anderson, K. Fox, Phys. Rev. **174**, 81 (1968)
[24] J.E. Turner, Am. J. Phys. **45**, 758 (1977)
[25] E.A.G.Armour, J.M.Carr, V.Zeman, J. Phys. B: At. Mol. Opt. Phys. **31**, L679 (1998)
[26] K. Strasburger, J. Phys. B: At. Mol. Opt. Phys. **37**, 4483 (2004)
[27] A. Voronin, J. Carbonell, Hyperfine Interact. **115**, 143 (1998)
[28] P. Froelich, S. Jonsell, A. Saenz, B. Zygelman, A. Dalgarno, Phys. Rev. Lett. **84**, 4577 (2000)
[29] B. Zygelman, A. Saenz, P. Froelich, S. Jonsell, Phys. Rev. A **69**, 042715 (2004)
[30] A.Y. Voronin, P. Froelich, Phys. Rev. A **77**, 022505 (2008)
[31] E. Garrido, D. Fedorov, A. Jensen, Phys. Lett. B **684**, 132 (2010)
[32] K. Sakimoto, Phys. Rev. A **90**, 032514 (2014)
[33] H. Stegeby, K. Piszczatowski, J. Phys. B: At. Mol. Opt. Phys. **49**, 014002 (2016)
[34] E. Hiyama, Y. Kino, M. Kamimura, Prog. Part. Nucl. Phys. **51**, 223 (2003)
[35] T. Yamashita, Y. Kino, Eur. Phys. J. D **70**, 1 (2016)
[36] M. Umair, S. Jonsell, J. Phys. B: At. Mol. Opt. Phys. **50**, 044001 (2017)
[37] T. Yamashita, M. Umair, Y. Kino, J. Phys. B: At. Mol. Opt. Phys. **50**, 205002 (2017)
[38] K. Piszczatowski, A. Voronin, P. Froelich, Phys. Rev. A **89**, 062703 (2014)
[39] Y.K. Ho, Phys. Rep. **99**, 1 (1983)
[40] E. Brändas, P. Froelich, Phys. Rev. A **16**, 2207 (1977)

EPJ Web of Conferences **181**, 01035 (2018) https://doi.org/10.1051/epjconf/201818101035
EXA2017

A New Silicon Drift Detector For Kaonic Deuterium Measurements

On behalf of the E57 collaboration at J-PARC

Carina Trippl[1,*] and *Johann* Zmeskal[1,**]

[1]Stefan Meyer Institute, Vienna, Austria

Abstract. The interaction of antikaons with nucleons and nuclei in the low-energy regime is a very active research field in hadron physics. Specially, the antikaon-nucleon interaction close to threshold provides crucial information on the interplay between spontaneous and explicit chiral symmetry breaking in low-energy QCD and touching one of the fundamental problems in hadron physic today - the still unsolved question of how hadron masses are generated. To study this problem, we have developed a new X−ray detector system and constructed an experimental apparatus to measure the kaonic deuterium 1s ground state shift and width with an accuracy of 60 eV and 140 eV, respectively.

1 Introduction

For kaonic hydrogen atoms a detectable energy shift of the 1s ground state has been found induced due to the strong interaction, as well as an observable energy broadened ground state level, caused by nuclear absorption. Measuring these observables kaonic hydrogen atoms offer the unique possibility to determine s−wave antikaon−nucleon scattering lengths at almost zero energy. Although the importance of a kaonic deuterium X−ray measurement has been well recognised, no experimental results have yet been obtained due to the difficulty of such a measurement. A determination of the isospin dependent (I=0 and I=1) antikaon−nucleon scattering lengths will be only possible by combining the result of the kaonic hydrogen and deuterium measurements. This combined result will provide the most stringent constraints on the antikaon−nucleon interaction at low energy, promising a breakthrough for this field.

1.1 State-of-the-art

The K^-p interaction is well understood from the recent results of kaonic hydrogen obtained from KpX [1] at KEK, DEAR [2] and finally from SIDDHARTA [3] at DAΦNE and the theoretical calculations based on these results [4], [5]. A milestone was the KpX experiment solving the long standing kaonic hydrogen puzzle (the old kaonic atom experiments were in contradiction with scattering data and theory). Although the importance of kaonic deuterium X−ray spectroscopy has been well recognised for more than 30 years (R.H. Dalitz [6]), no experiments have been performed so far due to the difficulty of the X−ray measurement.

*e-mail: Carina.Trippl@oeaw.ac.at
**e-mail: Johann.Zmeskal@oeaw.ac.at

"The necessity to perform measurements of the kaonic deuterium ground state observables is justified by the fact that, unlike the case of pionic atoms, the measurement of only the kaonic hydrogen spectrum does not allow – even in principle – to extract independently both s–wave antikaon–nucleon scattering lengths a_0 and a_1.", quoting from U.–G. Meisser [7]. We have now designed and constructed an apparatus which will allow us to measure the kaonic deuterium X–ray spectrum at J-PARC [8].

2 K⁻d Apparatus for J-PARC

The K⁻d setup consists of three main parts: The **Cryogenic Target Cell**, which will be filled with gaseous deuterium, is surrounded by 48 newly developed **Silicon Drift Detectors** (SDDs) connected to 24 **Amplifier Boards**, which are stationed outside of the vacuum chamber. In Figure 1 the setup is shown schematically. The distance between SDDs and amplifier boards is due to the dimensions of the E15 setup inside a magnet, in which the K⁻d apparatus will be placed.
For detailed information on Silicon Drift Detectors see chapter 3.

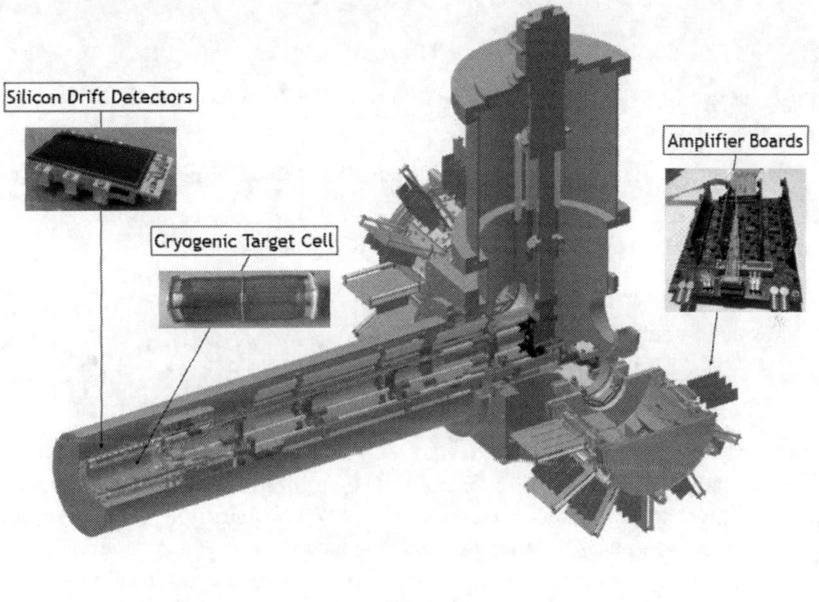

Figure 1. Schematic setup of the E57 Experiment at J-PARC

Kaons, which come from the K1.8BR beam at J-PARC, are stopped inside the cryogenic target cell and form kaonic deuterium atoms. The target cell, with dimensions of 65 mm in diameter and 200 mm in length, is cooled down to 29 K and is able to withstand a pressure of 0.35 MPa. The walls of the target cell are made of two layers of a 25 μm thick Kapton foil and the cell has been tested successfully to withstand a pressure up to 0.7 MPa. To achieve a good stopping rate of kaons, but still keep the X-ray yield (the probability that the kaon reaches the 1s ground state, which is inverse proportional to the gas density) as high as possible an optimal gas density has been found by performing a dedicated MC simulation to be in the order of 5% of the liquid deuterium density. Therefore, the wall of the lightweight

cryogenic target cell have been constructed as thick as necessary to safely hold a target pressure of 5% gas density, but still thin enough to provide sufficiently high transition for K_α photons, in the order of 70-80% at 6 keV.
The amplifier boards are arranged on two sides outside of the vacuum chamber. The amplification is performed in two steps: Firstly, a single amplification is done, and secondly, in addition to the amplification the shaping of the signal is carried out. Typically, the shaping time of the signal can be set between 1 to 5 μs. Furthermore, the adjustment of trigger and reset thresholds are necessary as differences of individual SDD channels have to be compensated.
Due to the large distance between the SDDs and amplifier boards in the final setup at the J-PARC accelerator, a line-driver has been developed to connect the SDD arrays inside the vacuum chamber with the amplifier boards stationed outside of it. The line-driver is a 1 : 1 amplifier with low power consumption, placed closely to the SDDs, and is connected via a one metre cable to the amplifier boards outside the vacuum chamber. For all tests performed at the Stefan Meyer Institute the configuration for the final measurement has been used.

3 Newly developed Silicon Drift Detectors

The main requirements for newly developed Silicon Drift Detectors are:

- Optimised active to total area ratio in the order of 80% (compared with 20% of the previously used SDDs)
- Drift time less than 300 ns at low temperatures (compared with 800 ns of the previously used SDDs)
- Robust against high charge load (better letch up performance)
- Peak stability better than ± 5eV
- Energy resolution better than 180 eV
- High rate capability 20 kHz

A dedicated Monte Carlo simulation (using Geant 4) has shown that with this assumption a K^-d experiment at J-PARC will be feasible. The requirements for peak stability, energy resolution and high rate capability are similar as for the former used SDDs. An optimised active to total area will allow an efficient packing density around the target cell increasing the detection efficiency, while the reduced drift time is essential for additional background reduction. The more simple design without FET of the newly developed SDDs, but with a small CMOS preamplifier chip (CUBE) mounted in close vicinity to the anode, will reduce the letch up probability drastically. In addition, a lower working temperature of the new SDDs will reduce the drift time to 300 ns at about 100 K. For the final setup the working temperature of the SDD should be as low as 50 K.

3.1 SDD Layout

The detection area around the target cell will be covered by 48 Silicon Drift Detector modules. One SDD array consists of eight square SDD units, each providing an active detection area of 8×8 mm^2. Thus, the complete active area of the detectors adds up to 246 cm^2. Due to the special design of the layout, the dead area of one SDD array is minimised, with only a margin of 1 mm on all four sides of the array. Another advantageous feature of the new Silicon Drift

Detectors is the special layout of the support structure. With the small bulges on two opposite sides of the SDD the arrangement of the SDD arrays around the target cell is simplified and additionally reduces the dead area of the detector.
As usual, three different voltages have to be applied to provide the electric field which forces the electrons to the collection anode and the holes to the back.
The newly developed Silicon Drift Detectors are directly connected to a CUBE preamplifier on the support structure. Compared to the previously used JFET structure of the SDD chip, the now used CMOS is not as dependent on the applied voltages and shows less changes due to temperature fluctuations. This means that both peak position as well as energy resolution (width) should show fluctuations of only a few eV. This is necessary to allow the summation of almost 400 SDD channels for the determination of the K^-d peak position in the order of 50 eV and the peak width in the order of 100 eV.
A high intrinsic efficiency for X-rays in the energy range from 5 – 15 keV is provided due to a small thickness of the SDD arrays of 450 μm, which makes them also less sensitive to MIPs compared to e.g. HP Ge detectors. It is necessary for the SDDs to cope with high count rates up to 100 kHz in combination with readout electronics, and to have a high yield of active to total area.

3.2 SDD tests at the Stefan Meyer Institute

Several test measurements have been performed at the Stefan Meyer Institute, including tests on stability as well as dependence on the applied voltages. The test setup is shown below (Figure 2). It is important to emphasise that the setup uses the same distances as in the final setup at J-PARC for all tests, including the line-driver element and the one metre long cable.

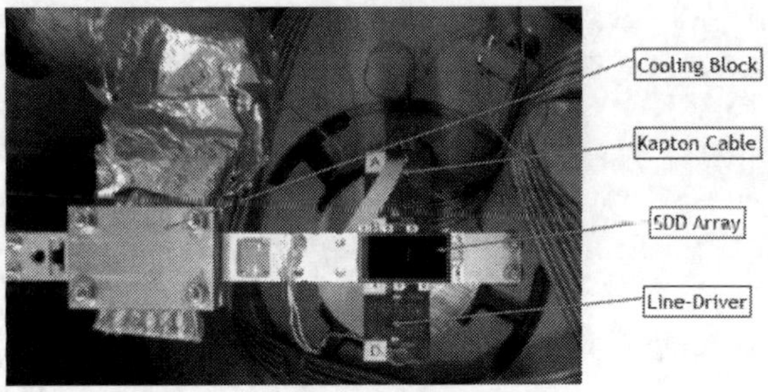

Figure 2. Test setup of the E57 experiment at the Stefan Meyer Institute including the cooling block, one single SDD array and the line-driver

The first tests have been dedicated to improve the energy resolution of the SDDs. For the energy calibration a Fe-55 source with an additional titanium foil has been used to produce characteristic X-Rays. In Figure 3 the energy calibration of one SDD unit (channel 2) can be seen. The first two peaks of the spectrum correspond to the Ti K_α and the Ti K_β transitions;

the third and fourth peak are the Mn K_α and the Mn K_β lines.
At a temperature T = (165 ± 2) K an energy resolution FWHM = (152.59 ± 0.45) eV has been achieved.

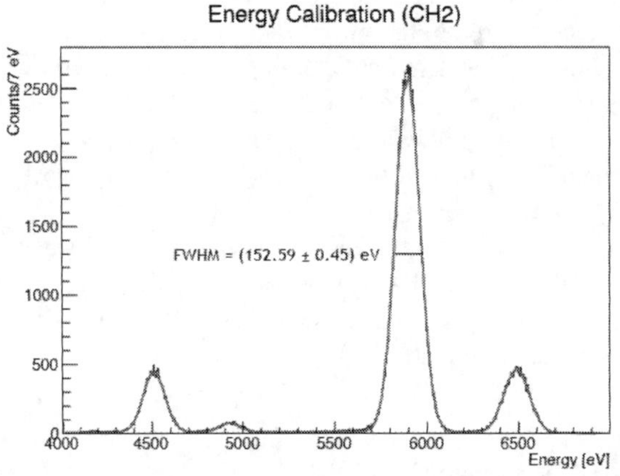

Figure 3. Energy calibration of one SDD channel (channel 2) obtained with a Fe-55 source and an additional Ti foil.

The second test measurement was performed to determine the voltage configuration applied to the SDDs to be able to achieve the best energy resolution. For this the voltage applied to the innermost ring, R_1, was set to be constant on $R_1 = -16.0$ V; the voltage applied to the outermost ring, R_N, and the back voltage, R_B, have been varied in steps of 5 V, always keeping a difference of 15 V and 20 V between R_N and R_B. The result can be seen in Figure 4 . There is very little change between the different configurations. However, the best value for the width for all channels was achieved for a voltage supply of $R_N = -90.0$ V and $R_B = -70.0$ V.

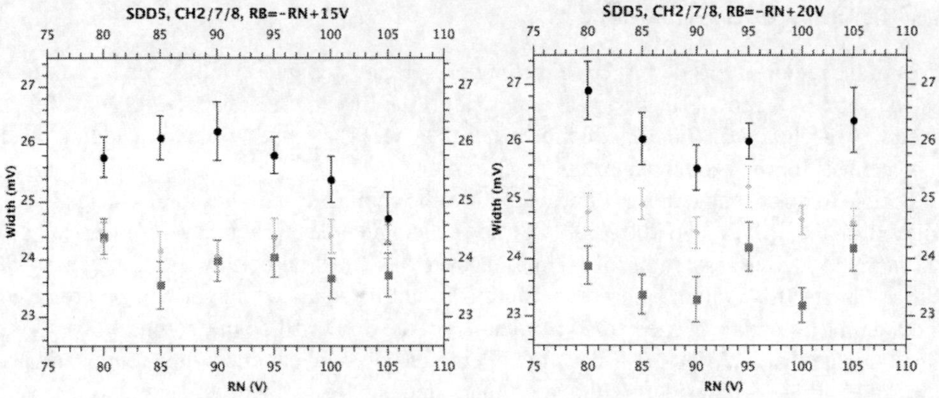

(a) (b)

Figure 4. Voltage configuration of three SDD channels (black = CH2, red = CH7, green = CH8) with different R_B to R_N interval.

Finally, to determine the stability of the Silicon Drift Detectors when operating, the Mn K_α drift and width has been analysed over a period of almost seven days. The data has been divided into packages of eight hours, showing an insignificant variation of the Mn K_α width in the order of ±0.2 ADC channel (Figure 5). The same is true for the Mn K_α peak, which varied only in the order of ±0.25 ADC channel (= ±1.0 eV). Both results indicate very good stability of the SDD over a long period of time and therefore ensure the excellent performance of the SDDs during the experiment to determine the K-d peak position and peak width.

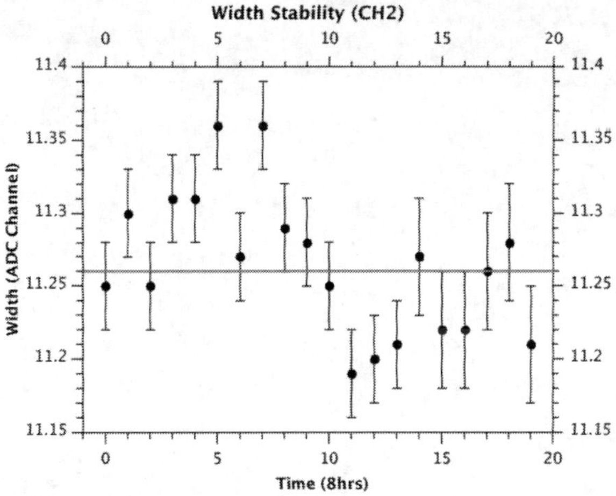

Figure 5. Width Stability of the Mn K_α width of one SDD channel (channel 2). The red line indicates the average width of the data sample corresponding to 148.0 eV with fluctuations of ±1.0 eV.

4 Summary and Outlook

Most of the requirements described in the first paragraph of chapter 3 have been successfully tested. The SDDs provide a good peak (and width) stability in the order of ± 1 eV as well as an energy resolution of 160 eV, which is better than necessary, but will help in adding up all 384 channels for the kaonic deuterium measurement.
To be able to reach temperatures of around 50 K, the cooling system has to be changed. This will reduce the drift time to 200 - 300 ns, thus reducing the background events drastically.
Furthermore, a time resolution (drift time) measurement will be performed. Using a Sr−90 source (an electron emitter) placed in front of a scintillator a start signal will be produced, and the time difference between the scintillator and the SDD will be measured.
Finally, more than 40 SDDs still have to undergo testing to check their functioning and stability. Nevertheless, considering the good measurements performed until now, we are very optimistic to reproduce these positive results.

References

[1] M. Iwasaki et al., Phys. Rev. Lett. **78** (1997) 3067
[2] DEAR Collaboration, Phys. Rev. Lett. **94** (2005)
[3] SIDDHARTA Collaboration, Phys. Lett. **B 704** (2011) 113
[4] Y. Ikeda, T. Hyodo, and W. Weise, Phys. Lett. **B 706** (2011) 63
[5] M. Mai and U.−G. Meißner, Nucl. Phys. **A 900** (2013) 51
[6] R.H. Dalitz et al., "*Proceedings of the International Conference on Hypernuclear and Kaon Physics*", ed. by B. Povh (M.P.I., Heidelberg 1982), p. 201
[7] U.−G. Meißner, U. Raha and A. Rusetsky, Eur. Phys. J. **C 47** (2006) 473
[8] J-PARC P57 proposal (2014)

EPJ Web of Conferences **181**, 01036 (2018) https://doi.org/10.1051/epjconf/201818101036
EXA2017

Observation of $\pi^- K^+$ and $\pi^+ K^-$ atoms

Valeriy Yazkov[1,*] on behalf of the DIRAC collaboration

[1]D. V. Skobeltsyn Institute of Nuclear Physics, M. V. Lomonosov Moscow State University, Leninskie Gory 1, 119991 Moscow, Russia

Abstract. Experiment DIRAC at CERN PS detects 349 ± 62 pairs from $\pi^- K^+$ and $\pi^+ K^-$ atoms and makes observation of exotic atoms consist of pion and kaon. It allows to measure a difference of S-wave pion-kaon scattering length with isospin 1/2 and 3/2: $|a_0^{1/2} - a_0^{3/2}|$. Values of pion-kaon scattering lengths are predicted in a frame of ChPT and LQCD. Therefore investigation of $\pi^- K^+$ and $\pi^+ K^-$ atoms gives possibility to check these predictions for simplest hadron-hadron system with *s*-quark.

1 Introduction

πK-atom ($A_{\pi K}$) is a hydrogen-like atom consisting of K^+ (K^-) and π^- (π^+) mesons with a Bohr radius of a_B = 249 fm, Bohr momentum of $p_B \simeq 0.8$ MeV/c and a ground state Coulomb binding energy of $E_B = 2.9$ keV .

The πK-atom lifetime (in ground state 1S), $\tau = \frac{1}{\Gamma}$ is dominated by the annihilation process into $\pi^0 K^0$ ($\pi^0 \bar{K}^0$). There is a relation between the width of $A_{\pi K}$ decay and S-wave πK scattering lengths for isospin 1/2 ($a_0^{1/2}$) and 3/2 ($a_0^{3/2}$) [1]:

$$\Gamma_{1S,\pi^0 K^0} = \frac{1}{\tau_{1S}} = 8\alpha^3 \mu^2 p^* (a_0^-)^2 (1 + \delta_K). \tag{1}$$

Here S-wave isospin-odd πK scattering length $a_0^- = \frac{1}{3}(a_0^{1/2} - a_0^{3/2})$ is defined in pure QCD for the quark masses $m_u = m_d$, α is the fine structure constant, μ is the reduced mass of the $\pi^\pm K^\mp$ system, p^* is the outgoing π^0 momentum in the πK atom system, and δ_K accounts for corrections, due to isospin breaking, at order α and quark mass difference ($m_u - m_d$).

Chiral Perturbation Theory (ChPT) describes QCD processes at low energies. ChPT in 1-loop approximation predicts S-wave pion-kaon scattering lengths [2, 3]:

$$a_0^{1/2} = 0.19 \pm 0.2,\ a_0^{3/2} = -0.05 \pm 0.02,\ a_0^{1/2} - a_0^{3/2} = 0.23 \pm 0.01 \tag{2}$$

in units of inverse pion mass.

ChPT with $L^{(2)}, L^{(4)}, L^{(6)}$ in 2-loop approximation predicts S-wave scattering length difference [4]:

$$a_0^{1/2} - a_0^{3/2} = 0.267\,. \tag{3}$$

Scattering length difference also has been predicted with dispersion analysis, using Roy-Steiner equations and experimental data in GeV range [5]:

$$a_0^{1/2} - a_0^{3/2} = 0.269 \pm 0.015\,. \tag{4}$$

*e-mail: valeri.yazkov@cern.ch

In the framework of lattice QCD, predictions for πK scattering length and their combination a_0^- have been obtained: $a_0^{1/2} = 0.1725^{+0.0026}_{-0.0157}$, $a_0^{3/2} = -0.0574^{+0.0029}_{-0.0060}$ [6], $a_0^{1/2} = 0.183 \pm 0.039$, $a_0^{3/2} = -0.0602 \pm 0.0040$ [7], $a_0^- = \frac{1}{3}(a_0^{1/2} - a_0^{3/2}) = 0.0811 \pm 0.0143$ [8], $a_0^- = \frac{1}{3}(a_0^{1/2} - a_0^{3/2}) = 0.0745 \pm 0.0020$ [9].

Prediction of scattering length difference in (4) together with $\delta_K = 0.040 \pm 0.022$ [1] provides (1) an estimation of lifetime of $A_{\pi K}$ in ground state:

$$\tau = (3.5 \pm 0.4) \times 10^{-15}\,. \tag{5}$$

There is differences in predictions of $a_0^{1/2} - a_0^{3/2}$ value, obtained with ChPT and Roy-Steiner equation [4, 5] from one side and LQCD [6, 9] from another side. It is needed to test predictions experimentally and accuracy of experimental measurement is to be at the level 5%.

The measurement of the S-wave πK scattering lengths would test our understanding of the chiral $SU(3)_L \times SU(3)_R$ symmetry breaking of QCD (u, d and s quarks), while the measurement of $\pi\pi$ scattering lengths checks only the $SU(2)_L \times SU(2)_R$ symmetry breaking (u, d quarks). This is the principal difference between $\pi\pi$ and πK scattering.

Experimental data on the πK low-energy phases are absent. The only experimental pion-kaon scattering length measurement has been done with estimation of πK atom lifetime, using data collected in 2008-2010 with Nickel (Ni) target [10]:

$$|a_0^-|M_\pi = 0.107^{+0.093}_{-0.035}\,. \tag{6}$$

Using all the data since 2007 and optimizing data handling and analysis, the observation of the πK atom could be achieved for the first time with a significance of more than 5 standard deviations [11]. On the basis of the same data sample, πK atom lifetime and πK S-wave isospin-odd scattering length have been measured with improved accuracy [12].

2 Method of πK atom observation and investigation

A method of investigation for $\pi^+\pi^-$, πK and other atoms, consisted of two oppositely charged mesons, has been proposed in [13]. Pairs of K^+ (K^-) and π^- (π^+) mesons are producing in proton-target interactions. Pairs, which are generated from fragmentation and strong decays ("short-lived" sources), are affected by Coulomb interaction in the final state. Some of them form Coulomb bound states — atoms, other are generated as free pairs ("Coulomb pairs"). Number of produced atoms (N_A) is proportional to a number of "Coulomb pairs" (N_C) with low relative momentum Q in a pair C.M. system: $N_A = K \cdot N_C$. The coefficient K is calculated with an accuracy better than 1% [14].

If at least one meson is generated from long-lived sources (electromagnetically or weakly decaying mesons or baryons: $\eta, \eta', K_s^0, \ldots$), then such pairs ("non-Coulomb pairs") are not affected by interaction in the final states.

After production, $A_{\pi K}$ travel through the target and could to annihilate into $\pi^0 K^0$, or to be ionised due to interaction with the target matter, producing specific "atomic pairs". These pairs have small relative momentum ($Q < 3$ MeV/c) and a number of such pairs n_A could be measured experimentally. Ratio of "atomic pair" number to a number of produced atoms is a breakup probability: $P_{\rm br}(\tau) = n_A/N_A = n_A/(K \cdot N_C)$ [15, 16]. In Fig 1 dependence of $A_{\pi K}$ breakup probability is shown for two Ni target are used in experiment DIRAC for pair laboratory momentum range $5.1 \div 8.5$ GeV/c. Value is averaged, using experimentally measured spectrum of atoms.

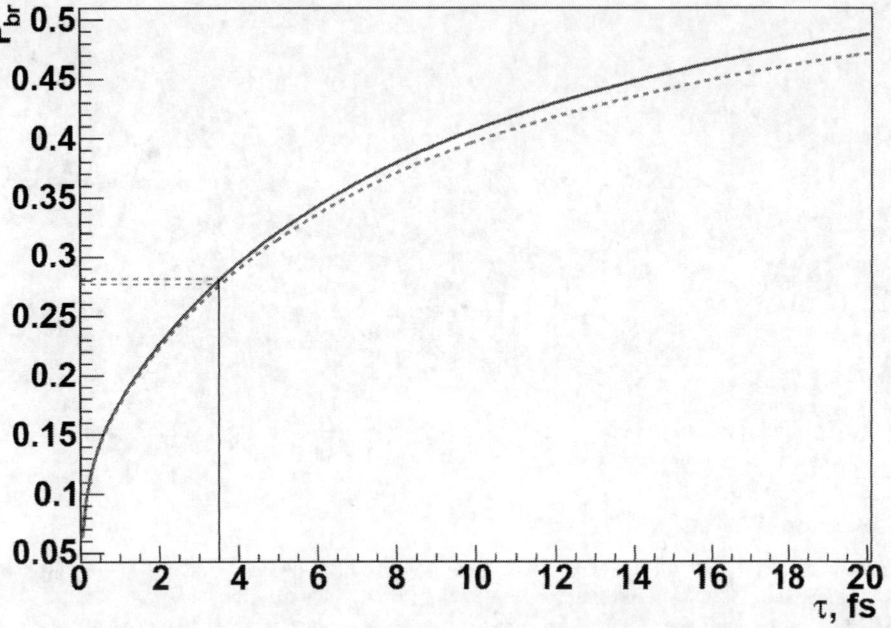

Figure 1. Dependence of the breakup probability P_{br} on $A_{\pi K}$ lifetime for 108μm (solid blue line) and 98μm (dashed red line) Ni targets, and an example how lifetime could be obtained from experimentally measured breakup probability

3 DIRAC setup

DIRAC setup was created to detect $\pi^+\pi^-$ with small relative momenta [17]. In 2004-2006 it has been modified in order to detect both $\pi^+\pi^-$ and πK pairs [18]. New detectors for particle identification have been added: Cherenkov detectors with heavy gas and aerogel for identification of K-mesons among background of pions and protons, correspondingly. Taking into account kinematic of πK "atomic pairs", new detectors cover only internal parts of each arm (see Fig. 2). Aerogel Cherenkov detector is mounted only in positive particle arm, because a flux of antiprotons is small relative to a flux of K^- mesons.

4 Selection πK events

Analysis procedure selects events which have signals of detectors expected for π^+K^- and π^-K^+ pairs. Fig. 3a presents the distribution of selected events over the difference of the particle production times for K^+ mesons in the range (4.4–4.5) GeV$/c$. The distribution is fitted by the simulated distribution of π^-K^+, $\pi^+\pi^-$, pπ^- and accidental pairs. Fig. 3b shows the fit for K^+ in the range (5.4–5.5) GeV$/c$. The contribution of misidentified pairs was estimated and accordingly subtracted [19]. Fraction of non-suppressed background pairs is calculated as function of momentum and used for estimation of systematic uncertainty (see below).

Fig. 4a illustrates the Q_L distribution of potential π^-K^+ pairs requiring a ChF signal and $Q_T < 4$ MeV$/c$. The dominant peak on the left side is due to pπ^- pairs from Λ decay. After requesting a ChA signal, the admixture of pπ^- pairs is decreased by a factor of 10 (Fig. 4b). By selecting proper TOFs between target and VH, background pπ^- and $\pi^+\pi^-$ pairs

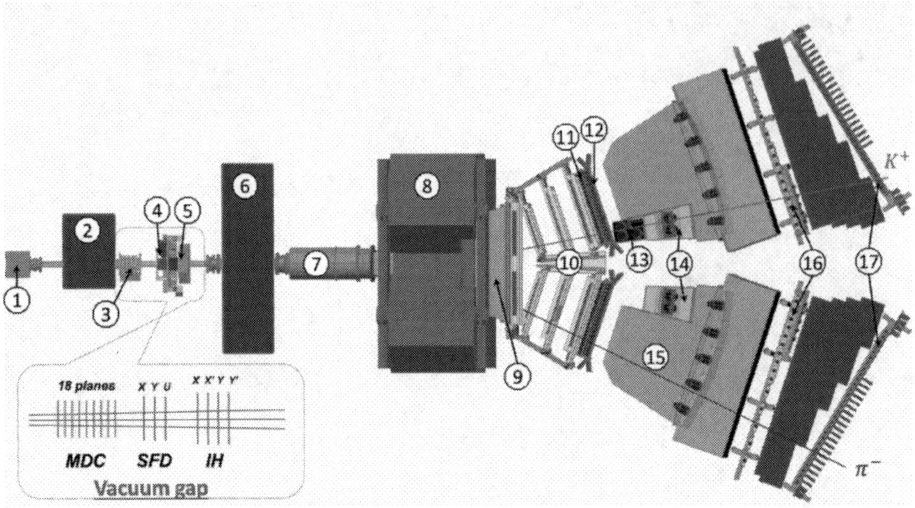

Figure 2. General view of upgraded DIRAC setup (1 – target station; 2 – first shielding; 3 – micro drift chambers (MDC); 4 – scintillating fiber detector (SFD); 5 – ionization hodoscope (IH); 6 – second shielding; 7 – vacuum tube; 8 – spectrometer magnet; 9 – vacuum chamber; 10 – drift chambers (DC); 11 – vertical hodoscope (VH); 12 – horizontal hodoscope (HH); 13 – aerogel Cherenkov detector (ChA); 14 – heavy gas Cherenkov detector (ChF); 15 – nitrogen Cherenkov detector (ChN); 16 – preshower detector (PSh); 17 – muon detector (Mu).

is substantially suppressed (Fig. 4c). In the final distribution, the well-defined $\pi^- K^+$ Coulomb peak at $Q_L = 0$ emerges beside the strongly reduced peak from Λ decays at $Q_L = -30$ MeV/c. The Q_L distribution of potential $\pi^+ K^-$ pairs shows a similar behavior. Applying the ChF and TOF criteria provides a sufficient background rejection. Fig. 5 presents the $\pi^+ K^-$ Coulomb peak at $Q_L = 0$ and a second peak from $\bar{\Lambda}$ decays at $Q_L = 30$ MeV/c.

For the final analysis, the DIRAC procedure selects events fulfilling the following criteria:

$$Q_T < 4\text{MeV}/c, |Q_L| < 20\text{MeV}/c\,.$$

5 Observation of $\pi^+ K^-$ and $\pi^- K^+$ atoms

Distributions of experimental data over relative momentum Q and its projections Q_L and Q_T have been fitted by a sum of simulated distributions of "atomic", "Coulomb" and "non-Coulomb" pairs [12]. Contributions of simulated distributions are free parameters of fit. In order to reproduce distribution of experimental pairs over relative momentum Q and its projections, simulation procedure takes into account resolution and efficiency of the setup detectors, multiplicity of background particles and noise signals, multiple scattering in Pt (run 2007) and Ni (2008-2010) targets, detector planes and partitions.

Fig. 6a presents the experimental and simulated Q distributions of $\pi^- K^+$ and $\pi^+ K^-$ pairs for the data obtained from the Pt target and Ni targets. One observes an excess of events above the sum of "Coulomb" and "non-Coulomb" pairs in the low Q region, where atomic pairs are expected: these excess spectrum is shown in Fig. 6b together with the simulated distribution of atomic pairs. Comparing the experimental with the simulated distributions, demonstrates good agreement.

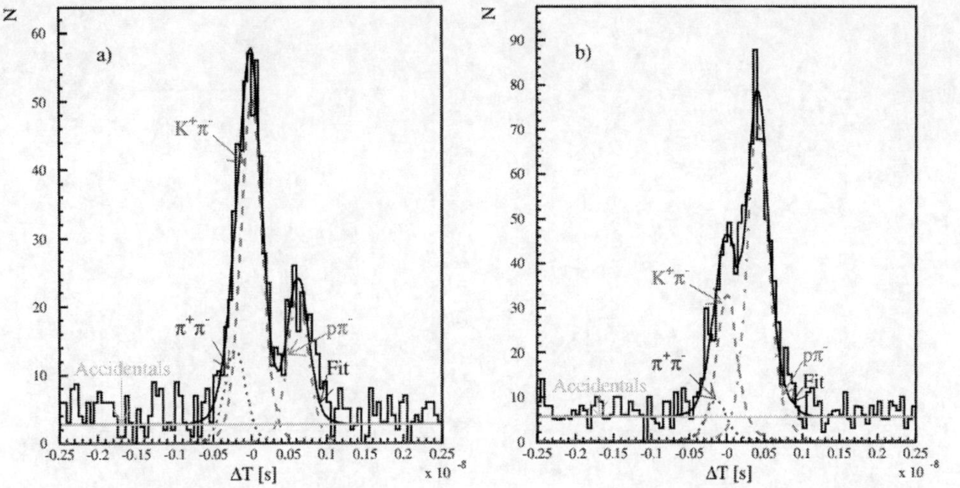

Figure 3. a) difference of particle generation times for events with positively charged particle momenta $(4.4 \div 4.5)$ GeV/c. Experimental data (histogram) are fitted by the event sum (black, solid): $\pi^- K^+$ (red, dashed), $\pi^+\pi^-$ (blue, dotted), $p\pi^-$ (magenta, dotted-dashed) and accidentals (green, constant). b) similar distributions for events with positively charged particle momenta $(5.4 \div 5.5)$ GeV/c.

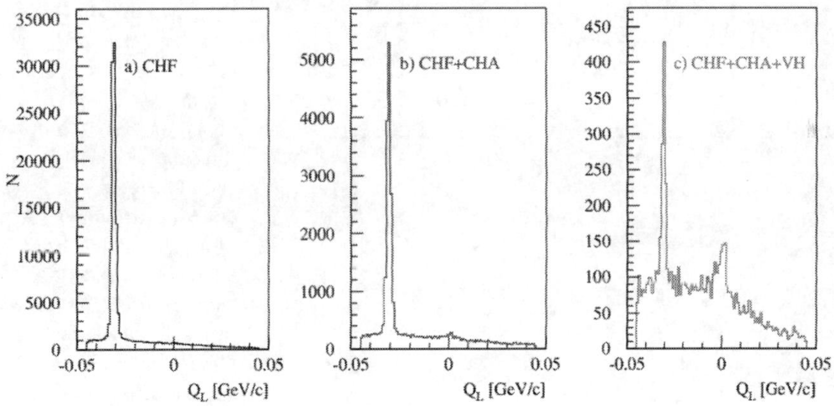

Figure 4. Q_L distribution of $\pi^- K^+$ pairs after applying different criteria: absence of signals in CHF and CHA detectors, proper TOFs between target and VH.

The numbers of atomic pairs, found in the $\pi^- K^+$ and $\pi^+ K^-$ data, are $n_A(\pi^- K^+) = 243 \pm 51$ ($\chi^2/n = 36/37$, n = number of degrees of freedom) and $n_A(\pi^+ K^-) = 106 \pm 32$ ($\chi^2/n = 42/37$). Total number of "atomic pairs" is $n_A(\pi K) = 349 \pm 61$ ($\chi^2/n = 41/37$). Quality of $\pi^- K^+$ and $\pi^+ K^-$ data fit, performed in assumption that "atomic pairs" are absent, is worse: $\chi^2/n = 73/38$.

Numbers of $\pi^+ K^-$ and $\pi^- K^+$ "atomic pairs" obtained with analysis of one-dimensional distributions over Q and $|Q_L|$ and two-dimensional ($|Q_L|$, Q_T) distribution are presented in

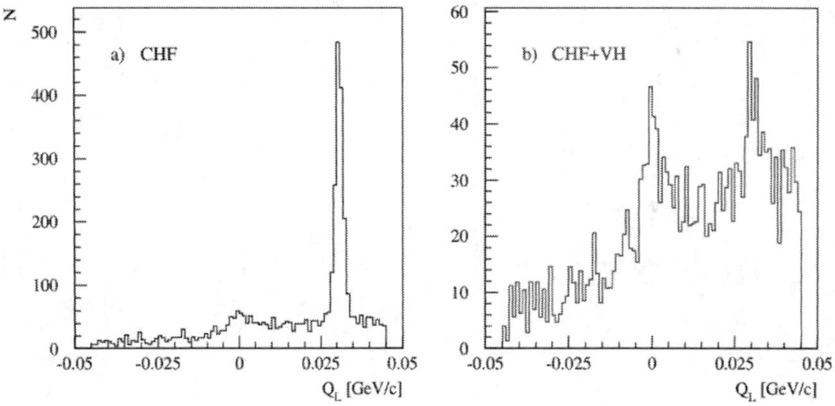

Figure 5. Q_L distribution of π^+K^- pairs after applying different criteria: absence of signals in CHF detector, proper TOFs between target and VH.

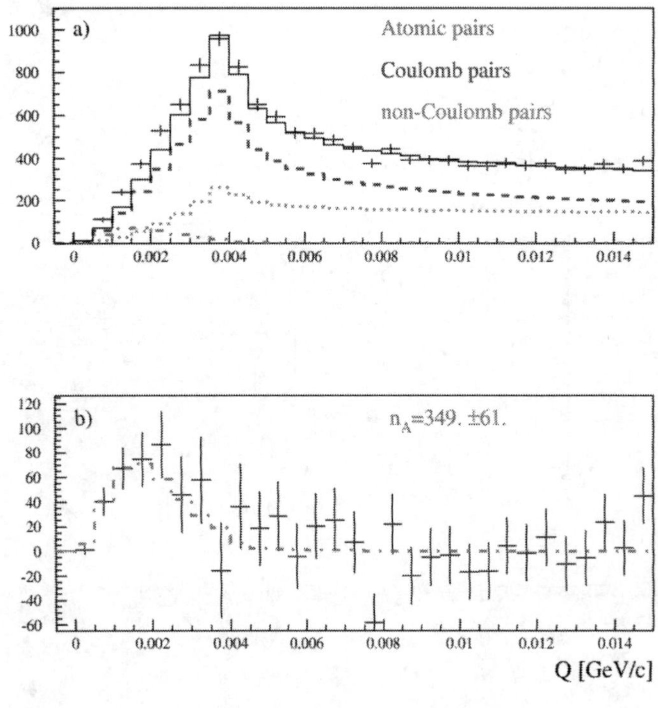

Figure 6. a) Distribution of π^+K^- and π^-K^+ pairs over Q, shown by points with error bars, is fitted by a sum of simulated distributions of "atomic" (red dotted-dashed), "Coulomb" (blue dashed) and "non-Coulomb" (magenta dotted) distributions. A sum of background distributions ("Coulomb" and "non-Coulomb") is shown by a solid black line. b) Difference of experimental and background distributions is shown together with simulated distributions of "atomic pairs".

Table 1 (Ni and Pt target together). The best statistical accuracy is achieved by analysis of Q and ($|Q_L|$, Q_T) distributions. Signal to error ratio is more than 5. The 1-dimensional $|Q_L|$ analysis for all πK data yields $n_A = 230\pm92$, which does not contradict the values, obtained in the other two statistically more precise analyses. Compared to the previous investigation [20], the Pt data was analysed including the upstream detectors. The consequence is a decrease of the statistics, but on the other hand an increase of the Q_T resolution. This better resolution

Table 1. "Atomic pair" numbers n_A obtained by analyzing the 1-dimensional Q and $|Q_L|$ distributions and the 2-dimensional ($|Q_L|$, Q_T) distribution. Only statistical errors are given.

Analysis	$\pi^- K^+$	$\pi^+ K^-$	$\pi^- K^+$ and $\pi^+ K^-$
Q	243 ± 51 (4.7σ)	106 ± 32 (3.3σ)	349 ± 61 (5.7σ)
$\lvert Q_L \rvert$	164 ± 79 (2.1σ)	67 ± 47 (1.4σ)	230 ± 92 (2.5σ)
$\lvert Q_L \rvert$, Q_T	237 ± 50 (4.7σ)	78 ± 32 (2.5σ)	314 ± 59 (5.3σ)

Table 2. Estimations of systematic errors, which are induced by different sources, for analysis of data distribution over relative momentum Q, its longitudinal projection $|Q_L|$ and two dimensional distribution over ($|Q_L|$, Q_T).

Sources of systematic errors	σ_Q^{syst}	$\sigma_{Q_L}^{syst}$	$\sigma_{\lvert Q_L \rvert, Q_T}^{syst}$
Uncertainty in Λ width correction	0.8	3.0	2.0
Uncertainty of multiple scattering in Ni (Pt) target	4.4	0.7	2.7
Accuracy of SFD simulation	0.2	0.0	0.1
Correction of Coulomb correlation function due to finite size production region	0.0	0.2	0.1
Uncertainty in πK pair laboratory momentum spectrum	3.3	5.4	7.8
Uncertainty in laboratory momentum spectrum of background pairs	6.6	1.6	5.4
Total	8.6	6.4	10.1

improves the quality of data. Concerning the Ni target, the increase of n_A, compared to [10], is caused by optimizing the time-of-flight criteria, which decreases "atomic pair" losses for the same fraction of background in the final distributions.

The evaluation of the "atomic pair" number n_A is affected by several sources of systematic errors [19, 21]. These uncertainties lead to differences in the shapes of experimental and MC distributions for "atomic", "Coulomb" and to a much lesser extent for "non-Coulomb" pairs. The shape differences induce a bias in the value of the fit parameter n_A, corresponding to a systematic error of the "atomic pair" number. Sources of systematic error and estimation of error values are listed in Table 2.

Taking into account both statistical and systematic errors, the one-dimensional $\pi^\pm K^\mp$ analysis in Q yields $n_A = 349 \pm 61(stat) \pm 9(syst) = 349 \pm 62(tot)$ "atomic pairs" (5.6 σ) for both combinations of charge and two-dimensional analysis in ($|Q_L|$, Q_T) yields $n_A = 314 \pm 59(stat) \pm 10(syst) = 314 \pm 60(tot)$ "atomic pairs" (5.2 σ). This is the first statistically significant observation πK atom — the dimesonic Coulomb bound states involving strangeness.

6 Measurement of S-wave isospin-odd πK scattering length

Experimentally measured numbers of "atomic pairs" n_A and produced atoms N_A allow (see section 2) to obtain breakup probability $P_{\rm br}$.

Table 3 contains the $P_{\rm br}$ values obtained in the Q and ($|Q_L|$,Q_T) analyses with statistical uncertainties.

Table 3. Experimental $P_{\rm br}$ from Q and $(|Q_L|, Q_T)$ analyses. Only statistical uncertainties are cited.

Data	RUN	Target (μm)	P_{br}^{Q}	$P_{br}^{\|Q_L\|,Q_T}$
π^+K^-	2007	Pt (25.7)	1.2 ± 1.3	0.27 ± 0.56
π^+K^-	2008	Ni (98)	0.53 ± 0.39	0.42 ± 0.38
π^+K^-	2009	Ni (108)	0.29 ± 0.20	0.33 ± 0.24
π^+K^-	2010	Ni (108)	0.33 ± 0.22	0.21 ± 0.20
π^-K^+	2007	Pt (25.7)	1.09 ± 0.52	1.44 ± 0.59
π^-K^+	2008	Ni (98)	0.32 ± 0.20	0.44 ± 0.22
π^-K^+	2009	Ni (108)	0.23 ± 0.16	0.16 ± 0.15
π^-K^+	2010	Ni (108)	0.41 ± 0.17	0.34 ± 0.16
$\pi^+K^-\&\pi^-K^+$	2007	Pt, 25.7	1.11 ± 0.48	0.83 ± 0.41

Sources of systematic uncertainties of breakup probability are mainly the same like for number of "atomic pairs" (see Table 2). There only one new source - uncertainty in the $P_{\rm br}(\tau)$ relation.

Estimations of systematic errors, induced by different sources [12], are presented in Table 4. The total errors were calculated as the quadratic sum. The procedure of the πK atom lifetime estimation, described below, includes all systematic errors, although their contributions are insignificant compared to the statistical errors.

Table 4. Estimated systematic errors of $P_{\rm br}$ for data collected with Pt and Ni targets in Q and $(|Q_L|, Q_T)$ analyses.

Source	Q		$(\|Q_L\|, Q_T)$	
	Pt	Ni	Pt	Ni
Uncertainty in Λ width correction	0.011	0.0006	0.073	0.0006
Uncertainty of multiple scattering in the target	0.0087	0.0051	0.014	0.0036
Accuracy of SFD simulation	0.	0.0002	0.	0.0003
Correction of the Coulomb correlation function on finite size production region	0.0001	0.0001	0.0002	0.0000
Uncertainty in πK pair lab. momentum spectrum	0.089	0.0052	0.25	0.0050
Uncertainty in the laboratory momentum spectrum of background pairs	0.22	0.0011	0.21	0.0011
Uncertainty in the $P_{\rm br}(\tau)$ relation	0.01	0.0055	0.01	0.0055
Total	0.24	0.0092	0.34	0.0084

For estimating the lifetime of $A_{\pi K}$ in the ground state, the maximum likelihood method [22] is applied [23]:

$$L(\tau) = \exp\left(-U^T G^{-1} U/2\right), \tag{7}$$

where $U_i = \Pi_i - P_{{\rm br},i}(\tau)$ is a vector of differences between measured Π_i ($P_{\rm br}$ in Table 3) and corresponding theoretical breakup probability $P_{{\rm br},i}(\tau)$ for a data sample i. The error matrix of U, named G, includes statistical (σ_i) as well as systematic uncertainties. Only the term corresponding to the uncertainty in the $P_{\rm br}(\tau)$ relation is considered as correlated between the Ni and Pt data, which is a conservative approach and overestimates this error. The other systematic uncertainties do not exhibit a correlation between the data samples from the Ni and Pt targets. On the other hand, systematic uncertainties of the Ni data samples are correlated. The likelihood functions of the $(|Q_L|, Q_T)$ and Q analyses are shown in Fig. 7.

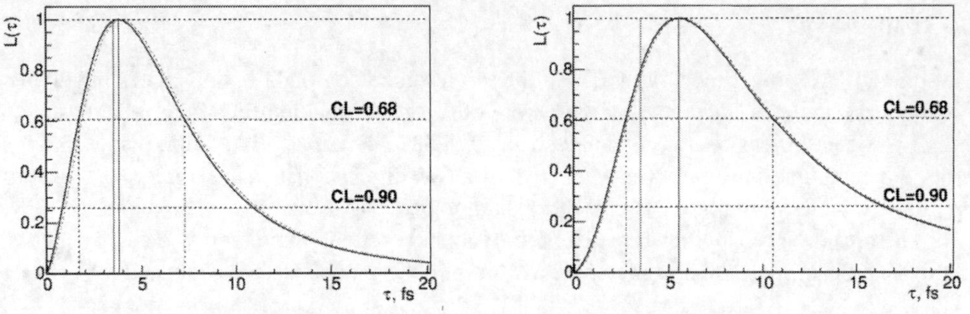

Figure 7. Likelihood functions $L(\tau)$ for $(|Q_L|, Q_T)$ (left) and Q (right) analyses with $Q_T < 4$ MeV/c. The likelihood functions on the basis of both statistical and systematic errors (dashed green line) and on the basis of only statistical error (solid blue line) are presented. The vertical blue lines indicate the best estimate for τ_{tot} and the corresponding confidence interval. The vertical red line is the theoretical prediction (5).

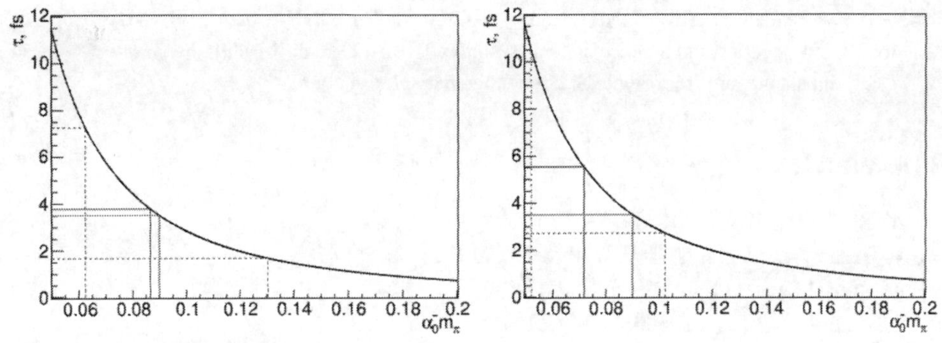

Figure 8. Ground state $A_{\pi K}$ lifetime τ_{1S} versus a_0^-. Experimental results (blue lines) are compared to the theoretical prediction (red lines). $(|Q_L|, Q_T)$ analysis (left) and Q analysis (right).

New estimation of $A_{\pi K}$ lifetime in ground state, based on two-dimensional analysis in $(|Q_L|, Q_T)$ [12]:

$$\tau = (3.8^{+3.3}_{-2.0}|_{stat}\ {}^{+1.0}_{-0.6}|_{syst})\ \text{fs} = (3.8^{+3.5}_{-2.1}|_{tot})\ \text{fs}\,, \tag{8}$$

which corresponds (see Fig. 8(left)) to isospin-odd πK scattering length estimation to be:

$$|a_0^-|M_\pi = 0.087^{+0.043}_{-0.024}|_{tot}\,. \tag{9}$$

Estimation, based on Q analysis [12]:

$$\tau = (5.5^{+4.9}_{-2.8}|_{stat}\ {}^{+0.9}_{-0.5}|_{syst})\ \text{fs} = (5.5^{+5.0}_{-2.8}|_{tot})\ \text{fs}\,, \tag{10}$$

provides (see Fig. 8(right)) the following value of the πK scattering length a_0^-:

$$|a_0^-|M_\pi = 0.072^{+0.031}_{-0.020}|_{tot}\,. \tag{11}$$

Measured values are compatible with theoretical predictions, taking into account the experimental precision.

7 Summary

In the DIRAC experiment at CERN, the dimesonic Coulomb bound states involving strangeness, the $\pi^- K^+$ and $\pi^+ K^-$ atoms, were observed for the first time with reliable statistics. The one-dimensional $\pi^\pm K^\mp$ analysis in Q yields 349 ± 62(tot) "atomic pairs" (5.6 σ) for both combinations of charge. Analogously, a two-dimensional analysis in ($|Q_L|, Q_T$) was performed with the result of 314 ± 60(tot) "atomic pairs" (5.2 σ).

The breakup probabilities for each atom type and each target are determined. By means of these probabilities, the lifetime of the πK atom in the ground state is evaluated to be:

$$\tau_{\rm tot} = (5.5^{+5.0}_{-2.8}\big|_{\rm tot}) \cdot 10^{-15}\text{s}\,,$$

and the S-wave isospin-odd πK scattering length deduced:

$$\left|a_0^-\right| = \frac{1}{3}\left|a_0^{1/2} - a_0^{3/2}\right| = (0.072^{+0.031}_{-0.020}\big|_{\rm tot})M_\pi^{-1}\,.$$

We are grateful to CERN for support and the PS team for the excellent performance of the accelerator. This work was funded by CERN, INFN (Italy), INCITE and MICINN (Spain), IFIN-HH (Romania), the Ministry of Education and Science and RFBR grant 01-02-17756-a (Russia), the Grant-in-Aid from JSPS and Sentanken-grant from Kyoto Sangyo University (Japan).

References

[1] J. Schweizer, Phys. Lett. B **587**, 33 (2004)
[2] V. Bernard *et al.*, Phys. Rev. D **43**, 2757 (1991)
[3] A. Roessl, Nucl. Phys. B **555**, 507 (1999)
[4] J. Bijnens *et al.*, JHEP **0405**, 036 (2004)
[5] P. Buttiker *et al.*, Eur. Phys. J. C **33**, 409 (2004)
[6] S.R. Beane *et al.*, Phys. Rev. D **77**, 094507 (2008)
[7] C.B. Lang *et al.*, Phys. Rev. D **86**, 054508 (2012)
[8] K. Sasaki *et al.*, Phys. Rev. D **89**, 054502 (2014)
[9] T. Janowski et al., PoS *LATTICE2014* 080 (2015)
[10] B. Adeva *et al.*, Phys. Lett. B **735**, 288 (2014)
[11] B. Adeva *et al.*, Phys. Rev. Lett. **117**, 112001 (2016)
[12] B. Adeva *et al.*, Phys. Rev. D **96**, 052002 (2017)
[13] L. Nemenov, Sov. J. Nucl. Phys. **41**, 629 (1985)
[14] L. Afanasyev and O. Voskresenskaya, Phys. Lett. B **453**, 302 (1999)
[15] L. Afanasyev and A. Tarasov, Phys. At. Nucl. **59**, 2130 (1996)
[16] M. Zhabitsky, Phys. At. Nucl. **71**, 1040 (2008)
[17] B. Adeva *et al.*, Nucl. Instrum. Methods A **515**, 467 (2003)
[18] B. Adeva *et al.*, Nucl. Instrum. Methods A **839**, 52 (2016)
[19] V. Yazkov and M. Zhabitsky, DN **2013-06** (DN = DIRAC Note), cds.cern.ch/record/1628544
[20] B. Adeva *et al.*, Phys. Lett. B **674**, 11 (2009)
[21] A. Benelli and V. Yazkov, DN **2016-01**, cds.cern.ch/record/2137645
[22] D. Drijard and M. Zhabitsky, DN **2008-07**, cds.cern.ch/record/1367888
[23] V. Yazkov and M. Zhabitsky, DN **2016-06**, cds.cern.ch/record/2252375

AUTHOR INDEX

AUTHOR INDEX